Structural Design & Drawing

Structural Design & Drawing

(Concrete Structure)

Volume 2

Dr. D. KRISHNAMURTHY
Ph.D. (Glas), ME (Cal) BE (Hons),
MIE (Inc). MACI, MIABSE (Eng)
Professor of Civil Engineering
M. A. College of Technology, Bhopal
(Presently Professor of Civil Engineering
Sulemaniah University
Sulemaniah)

CBS Publishers & Distributors Pvt. Ltd.

New Delhi • Bengaluru • Chennai • Kochi • Kolkata • Mumbai
Hyderabad • Uttarakhand • Nagpur • Patna • Pune • Jharkhand

ISBN: 81-239-0147-X

First Edition: 1985
Reprint: 1992, 1999, 2003, 2006, 2008, 2009, 2010,
2011, 2012, 2013, 2014, 2015, 2018

Published by **Satish Kumar Jain** and produced by **Varun Jain** for
CBS Publishers & Distributors Pvt. Ltd.,
4819/XI Prahlad Street, 24 Ansari Road, Daryaganj, New Delhi - 110002
delhi@cbspd.com, cbspubs@airtelmail.in • www.cbspd.com
Ph.: 23289259, 23266861, 23266867 • Fax: 011-23243014

Corporate Office: 204 FIE, Industrial Area, Patparganj, Delhi - 110 092
Ph: 49344934 • Fax: 011-49344935
E-mail: publishing@cbspd.com • publicity@cbspd.com

Branches:
- ***Bengaluru:*** 2975, 17th Cross, K.R. Road, Bansankari 2nd Stage, Bengaluru - 70 • Ph: +91-80-26771678/79 • Fax: +91-80-26771680 E-mail: cbsbng@gmail.com, bangalore@cbspd.com
- ***Chennai:*** No. 7, Subbaraya Street, Shenoy Nagar, Chennai - 600030 Ph: +91-44-26681266, 26680620 • Fax: +91-44-42032115 E-mail: chennai@cbspd.com
- ***Kochi:*** Ashana House, 39/1904, A.M. Thomas Road, Valanjambalam, Ernakulum, Kochi • Ph: +91-484-4059061-65 Fax: +91-484-4059065 • E-mail: cochin@cbspd.com
- ***Kolkata:*** 6-B, Ground Floor, Rameshwar Shaw Road, Kolkata - 700014 Ph: +91-33-22891126/7/8 • E-mail: kolkata@cbspd.com
- ***Mumbai:*** 83-C, Dr. E. Moses Road, Worli, Mumbai - 400018 Ph: +91-9833017933, 022-24902340/41 • E-mail: mumbai@cbspd.com

Representatives:

- Hyderabad: 0-9885175004
- Patna: 0-9334159340
- Jharkhand: 0-9811541605
- Nagpur: 0-9021734563
- Pune: 0-9623451994
- Uttarakhand: 0-9716462459

Printed at:
J.S. Offset Printers, Delhi (India)

Preface

Structural Engineering is a branch of Civil Engineering which has to deal with the design of buildings, bridges, water towers, retaining walls, silos, bunkers, chimneys and other types of Civil Engineering Structures.

Structural design of any structure may be divided into !—

(*i*) Assessment and distribution of loads which the structural component is required to support or subjected to.

(*ii*) Proportioning of trial section of the components or the appropriate arrangement of structural elements.

(*iii*) Computations of bending moments, shear forces and direct forces due to critical combinations of loads to which the structural member is subjected or is required to carry.

(*iv*) Selection of proper size and shape of the member to resist the forces safely keeping in view the economy in the design.

(*v*) Preparation of lay out of the structure and the finished working drawings with all dimensions for all members which will be required for executing the construction of the structure at site or for fabrication of the structure in the factory.

Students taking courses on structural design have usually no great difficulty in computations involved in the structural design, but they often find that the available text books afford very little guidance in the actual sequence in the design, appropriate choice of section on the basis of strength and economy and the detailed dimensioned drawings.

The author has endeavoured to provide these needs to the students, structural designers handling designs using concrete and steel as materials and also to provide a detailed guide for young engineers and architects engaged in consultancy services.

In the process of making a complete design special attention has been devoted to giving full details of each step. Each design is provided with fully dimensioned working drawings. Designs of all kinds of practical problems that the designer normally comes across are presented using a standard sequence along with the fully dimensioned drawings.

This volume deals with designs with detailed drawings fully dimensioned on practical problems on retaining walls, domed roofs, water tanks on ground, underground, overhead and reservoirs, silos, bunkers, slab and *T*-beam bridges. This book is intended to serve as a text book to civil engineering students of final year, degree course and to those appearing for Section 'B' of the professional examinations and as a reference book for practising engineers in public works, public health departments and structural designers in consulting firms.

AUTHOR

Contents

Notations for Drawings

(*i*) G_1—MR—10—185
G_1– Grade—1—Steel
MR – Mild steel round bars
10—10 mm diameter
185—185 mm spacing c/c

(*ii*) RT—10—200
RT—Ribbed torsteel
10—10 mm diameter
200—200 mm spacing c/c

(*iii*) 9—G_1—MR—25
9—9 numbers of bars
G_1 – Grade-1—Steel
MR—Mild steel round bars
25—25 mm diameter

(*iv*) 9—RT—20
9—9 number of bars
RT—Ribbed Torsteel
20—20 mm diameter.

I. CANTILEVER RETAINING WALLS

General Information 1

1·1. Definition

A wall designed to maintain the difference in elevations of the ground surfaces on each side of the wall is called a retaining wall.

These are extensively used in the construction of railways, highways, bridges, canals, basement walls in buildings, walls of underground reservoirs, swimming pools etc.

1·2. Types

(1) Solid gravity
(2) Semi gravity
(3) Reinforced concrete walls
  (*i*) Cantilever retaining walls
  (*ii*) Counterfort retaining walls
  (*iii*) Buttressed walls
  (*iv*) Crib retaining walls.

1·3. Stability

Factor of safety against overturning

$$= \frac{\text{Stabilizing moment}}{\text{Overturning moment}}$$

$$= \frac{W\bar{x}}{P\bar{y}},$$ should be greater than 1·5

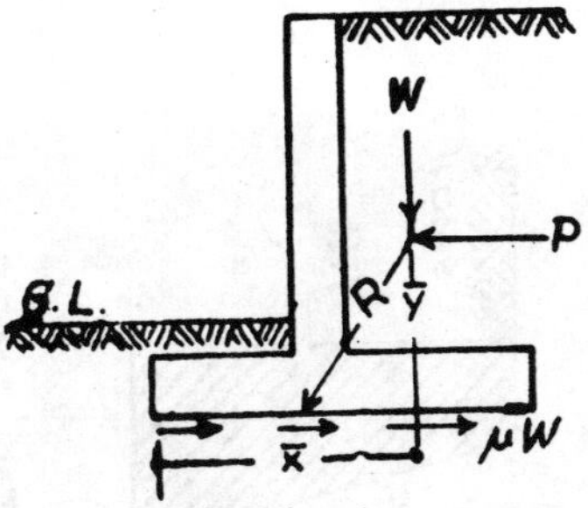

Factor of safety against sliding

$$= \frac{\mu W}{P},$$ should be greater than 1·5

Maximum compressive pressure below the base in the soil should not exceed the safe bearing capacity of the soil and the resultant should fall within the middle third of the base so that the pressures at toe and the heel will be compressive.

1·4. Required Data for Design

(*i*) Location of wall
(*ii*) Height of fill to be retained
(*iii*) Surcharge or any other additional loading on the wall

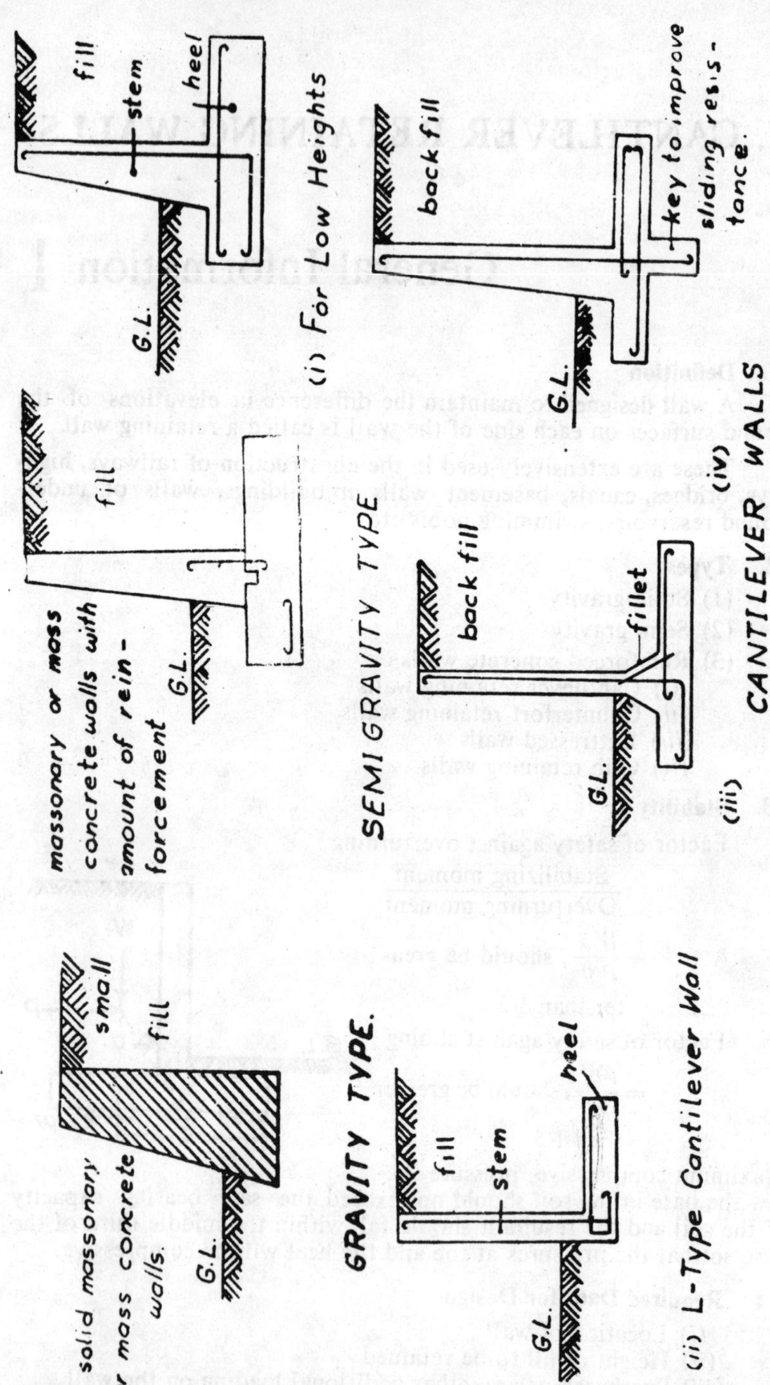
solid massonary
or mass concrete
walls
G.L.
fill
small
GRAVITY TYPE.
massonary or mass
concrete walls with
amount of rein-
forcement
G.L.
fill
SEMI GRAVITY TYPE
G.L.
fill
stem
heel
(i) For Low Heights
G.L.
fill
stem
heel
(ii) L-Type Cantilever Wall
G.L.
back fill
fillet
(iii)
G.L.
back fill
key to improve
sliding resis-
tance.
(iv)
CANTILEVER WALLS

(*iv*) Water table

(*v*) Safe bearing capacity of the soil

and other characteristics of the soil such as angle of internal friction, density of the soil etc.

1·5. Design

Design procedures for different types of structures are not exact but mostly based on experiments, judgements and experience. Conventional procedures have been developed through the years which have been widely accepted by structural engineers. Therefore such procedures are used in the design of R.C.C. retaining walls.

1·6. Earth Pressure Theories

Rankine's theory is mostly used for the walls whose backs are not plane, such as Cantilever and Counterforted walls. Coloumb's method is more applicable for the walls of solid gravity and semi-gravity walls whose backs are plane.

1·6·1. Substitute Cohesionless Soil

In computing the lateral earth pressure, it is assumed that the soil is cohesionless, but in most cases, it is known to possess cohesion. To account for the effects of omitting cohesion, the angle of internal friction is assumed to be larger than the angle of internal friction usually possessed by cohesive soils.

1·7. Trial Dimensions

Depth of foundation should not be less than 1 *m*. The other dimensions are as in the sketch.

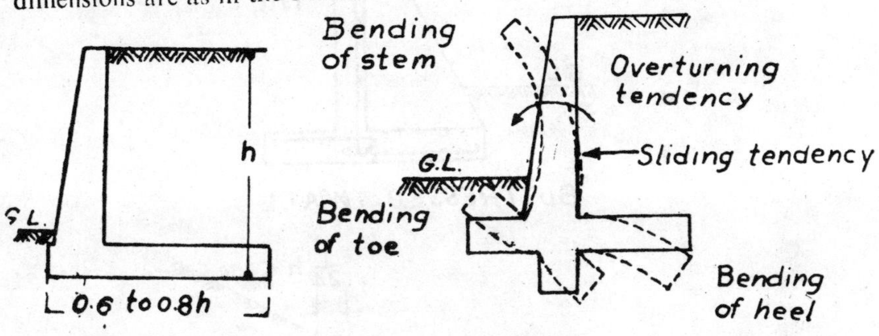

1·8. Drainage

Water which may permeate to the back of the wall may be encouraged to move towards the weep-holes by providing layers of gravel or loose stones suitably placed and directioned to facilitate efficient drainage.

1·9. Steps in Design

(*i*) Assume suitable stresses and obtain design constants.

(*ii*) Select trial dimensions for the wall.

(*iii*) Compute horizontal active earth pressure due to fill, surcharge or any other given loading conditions on the wall.

(*iv*) Design the stem as a cantilever slab fixed at the base.

(*v*) Calculate the pressure distribution under the base and check for bearing capacity.

(*vi*) Design the toe slab as a cantilever.

(*vii*) Design the heel slab as a cantilever

(*viii*) Check for stability of the wall for

(*a*) overturning

(*b*) sliding

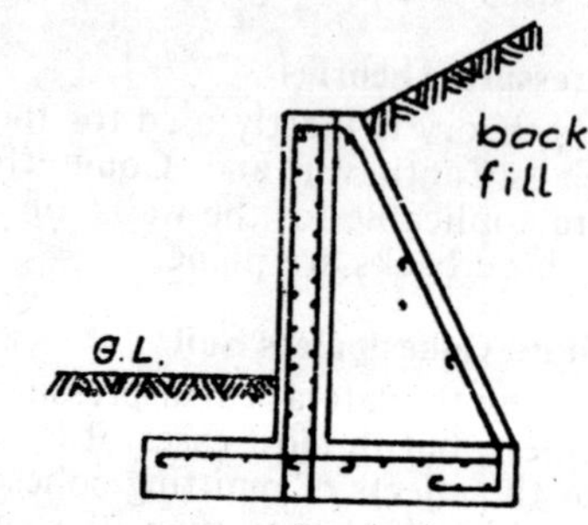

COUNTERFORT TYPE

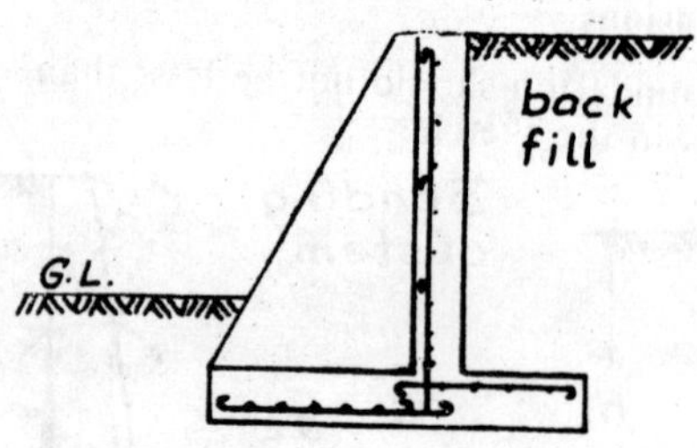

BUTTRESSED WALL

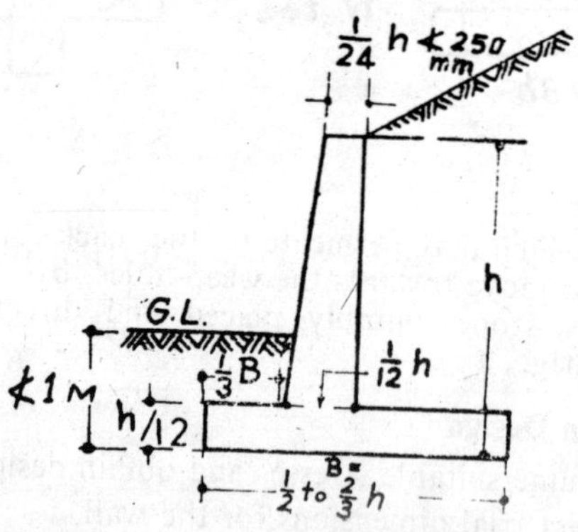

1·9·1. Values for K

δ°/φ°	25	30	35	40	45
10	0·430	0·353	0·287	0·225	0 175
15	0 465	0·376	0·300	0·240	0·180
20	0·570	0·420	0 225	0·253	0·190
25	0 900	0·560	0·355	0·278	0·210
30	—	0·870	0·450	0·315	0·240
35	—	—	0·820	0·400	0·276
40	—	—	—	0·760	0·350

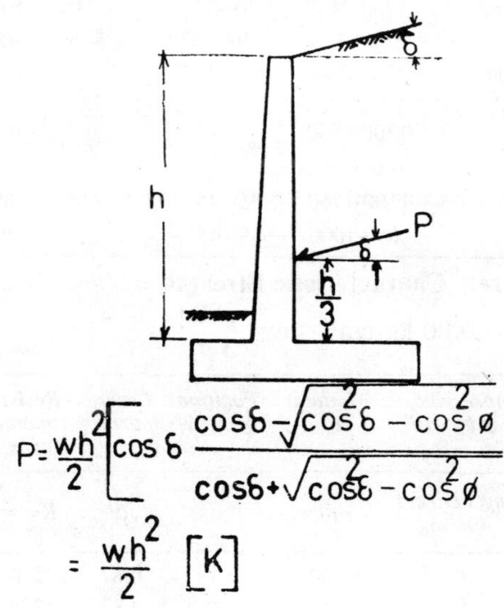

$$P = \frac{wh^2}{2}\left[\cos\delta\,\frac{\cos\delta - \sqrt{\cos^2\delta - \cos^2\phi}}{\cos\delta + \sqrt{\cos^2\delta - \cos^2\phi}}\right]$$

$$= \frac{wh^2}{2}\,[K]$$

1·9·2. Functions of Angle of Repose

angle φ	15°	20°	25°	30	35°	40°	45°
$\frac{1-\sin\phi}{1+\sin\phi}$	0·588	0·490	0·406	0·333	0·271	0·217	0·172
$\frac{1+\sin\phi}{1-\sin\phi}$	1·700	2·039	2·464	3·000	3·660	4·599	5·826
$\left(\frac{1-\sin\phi}{1+\sin\phi}\right)^2$	0·346	0·240	0·165	0·111	0·073	0·047	0·029
$\left(\frac{1+\sin\phi}{1-\sin\phi}\right)^2$	2·890	4·159	6·07	9·000	13·619	21·152	33·940
Sin φ	0·259	0·342	0·423	0·500	0·574	0·643	0·707
Cos φ	0·966	0·940	0·906	0·866	0·819	0·766	0·707
Tan φ	0·268	0·364	0·466	0·577	0·700	0 839	1·000

1·9·3. Particulars of Materials

Item	Weight kg/m³	Angle of internal friction φ	μ	Submerged weight kg/m³	Bearing capacity tonnes/m²
Clay (dry)	1450	—	0·1-0·6	—	5-10
Clay (damp)	1750	—		—	10 15
Earth (dry)	1500-1800	—	0·33-0·45	—	15-25
Earth (moist)	1600-2000	—	—	—	15-20
Sand (dry)	1550-1600	—	—	—	10-25
Sand (wet)	1750-2000	30°-35°	0·5-0·6	—	10-25
Gravel	1600-2000	35°-45°	·5-·6	1000-1300	45
Coarse to medium sand	1650-2100	35°-40°		1000-1300	25-45
Fine to salty sand	1750-2150	30°-35°		1000-1300	15
Granite and shale	1600-2100	30°-35°	·5-·6	1000-1300	45-90
Basalts and Dolerites	1750-2200	35°-45°		1100-1600	90
Lime stones and Sand stones	1300-1900	—		650-1300	45
Chalk	1000-1300	—		300-650	—
Broken brick	1100-1750	35°-45°		650-1000	—
Ashes	650-1000	35°-45°		300-500	—

1·9·4. Concrete Characteristic Strength

For $\sigma_{st}=1400$ kg/cm² $m=\dfrac{2800}{3\,\sigma_{cb}}$

Concrete grade	Compressive stress kg/cm²		Modular ratio	Position of N.A.	Lever arm	Resisting moment constant	q	σ_b
	Bending σ_{cb}	Direct σ_c	m	n/d	jd	$R=\dfrac{M}{bd^2}$	kg/cm²	
M_{100}	30	25	31·00	0·4	0·87	5·19	3	4
M_{150}	50	40	18·70	0·4	0·87	8·70	5	6
M_{200}	70	50	13·30	0·4	0·87	12·10	7	8
M_{250}	85	60	11·00	0·4	0·87	14·74	8	9
M_{300}	100	80	9·35	0·4	0·87	17·00	9	10
M_{350}	115	90	8·10	0·4	0·87	19·82	10	11
M_{400}	130	100	7·20	9·4	0·87	22·25	11	12

For σ_{st} - 1000 kg/cm²

Concrete grade	Tensile stress in direct	bending	Position of NA	Lever arm	Resisting moment constant
	σ_{ct}	σ_{cb}	n/d	Jd	$R=\dfrac{M}{bd^2}$
M_{200}	12	17	0·48	0·84	14·11
For $\sigma_{st}=1250$ kg/cm²					
M_{200}	12	17	0·42	0·86	12·65

Cantilever Retaining Wall with Constant Surcharge 2

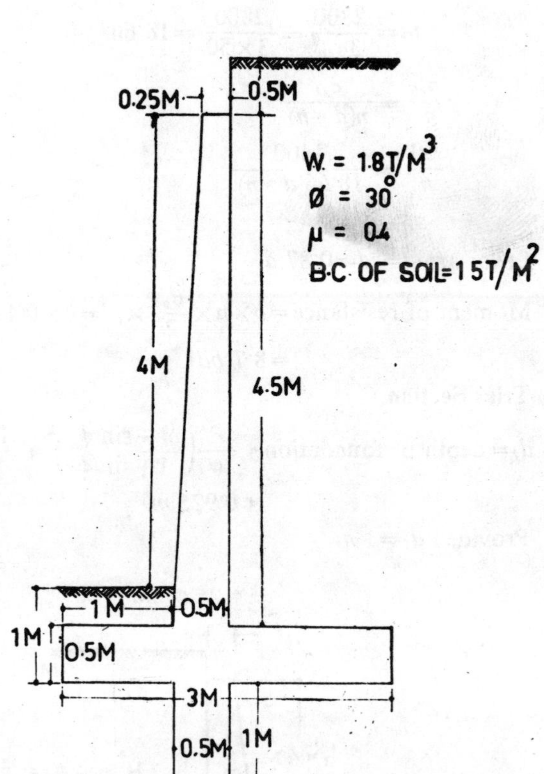

2·1. Data

Height of the fill to be retained by the wall = 4*m*
Surcharge = ½*m* constant
Density of soil = 1800 kg/m^3
Angle of internal friction = 30°
Coefficient of friction between the soil and concrete = 0·4
Bearing capacity of soil = 15 tonnes/m^2
IS—456—1964

2·2. Allowable Stresses

Concrete grade $M150$,
Compressive stress in bending, $\sigma_{cb}=50$ kg/cm²
In direct compression $\sigma_c = 40$ kg/cm²
Shear stress $q_s=5$ kg/cm²
Bond stress $\sigma_b=10$ kg/cm²
Steel-grade-1 $\sigma_{st}=1400$ kg/cm²

2·2·1. Characteristic Strengths

$$m=\frac{2800}{3\sigma_{cb}}=\frac{2800}{3\times 50}=18{\cdot}66$$

$$\frac{\sigma_{cb}}{n}=\frac{\sigma_{st}}{n(d-n)}$$

$$\frac{50}{n}=\frac{1400}{18{\cdot}66(d\ \ n)}$$

$$n=0{\cdot}4\ d$$

Lever arm $=jd=0{\cdot}87\ d$

$$\text{Moment of resistance}=b\times n\times\frac{\sigma_{cb}}{2}\times jd=b\times 0{\cdot}4\ d\times\frac{\sigma_{cb}}{2}\times 0{\cdot}87\ d$$

$$=8{\cdot}7\ bd^2$$

2·3. Trial Section

$$d_f=\text{depth of foundation} \nless \frac{P}{w}\left(\frac{1-\sin\phi}{1+\sin\phi}\right)^2 \nless \frac{15000}{1800}\times\frac{1}{9}$$

$$\nless 0{\cdot}925\ m$$

Provided $d_f=1\ m$

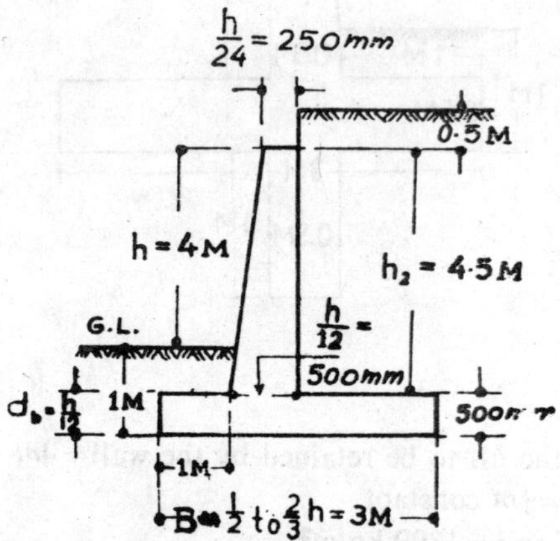

2·4. Earth Pressure Calculations

(1) Earth pressure due to surcharge$=wh_1\left(\frac{1-\sin\phi}{1+\sin\phi}\right)\times h_2$

$$=1800\times\tfrac{1}{2}\times\tfrac{1}{3}\times 4{\cdot}5=1350\text{ kg}$$

(2) Earth pressure due to backfill $= \frac{1}{2} \times wh_2 \left(\frac{1-\sin\phi}{1+\sin\phi} \right) \times h_2$

$= \frac{1}{2} \times 1800 \times 4{\cdot}5 \times \frac{1}{3} \times 4{\cdot}5 = 6075$ kg.

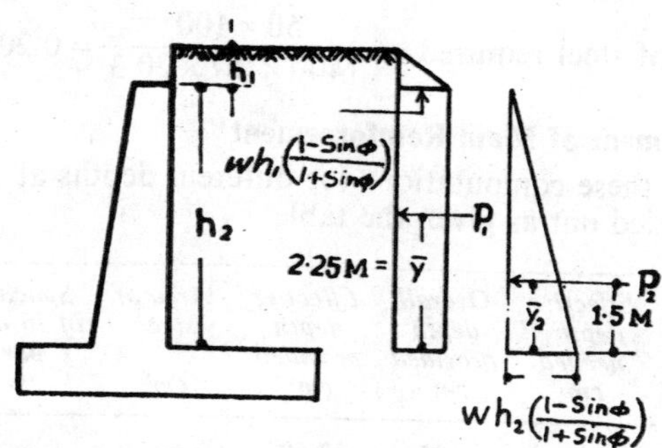

2·5. Stem Design

Maximum bending moment at the junction of the stem and the base $= P_1Y_1 + P_2Y_2 = 1350 \times 2{\cdot}25 + 6075 \times 1{\cdot}5 = 12150$ m kg

Effective depth of stem

$$d_e = \sqrt{\frac{12150 \times 100}{100 \times 8{\cdot}7}} = 37{\cdot}5 \text{ cm}$$

Overall depth $d = 37{\cdot}5 + \text{cover} = 37{\cdot}5 + 7{\cdot}5 = 45$ cm

However 50 cm assumed shall be used, therefore the effective depth = 42·5 cm

2·5·1. Shear Check

$$P = P_1 + P_2 = 1350 + 6075 = 7425 \text{ kg}$$

$$\text{Shear stress} = \frac{7425}{100 \times {\cdot}87 \times 42{\cdot}5} = 2{\cdot}01 \text{ kg/cm}^2 < 5 \text{ kg/cm}^2$$

2·5·2. Bending Moment Variation

The bending moment in the stem is maximum at the junction and reduces to zero at the top of the wall. Hence main reinforcement can be suitably reduced towards top and the wall can be tapered and the thickness reduced to achieve economy.

Bending of any depth h'

$$M = wh_1 \left(\frac{1-\sin\phi}{1+\sin\phi} \right) h' \times \frac{h'}{2} + wh' \left(\frac{1-\sin\phi}{1+\sin\phi} \right) \frac{h'}{2} \times \frac{h'}{3}$$

$$= 1800 \times \frac{1}{2} \times \frac{1}{3} \times \frac{h'^2}{2} + 1800 \times \frac{1}{3} \times \frac{h'^3}{6}$$

$$= 150(h')^2 + 100(h')^3$$

Calculations for $h' = 0{\cdot}5$ m

Bending moment $= 150 \times (\frac{1}{2})^2 + 100(\frac{1}{2})^3 = \frac{150}{4} + \frac{100}{8}$

$= 37{\cdot}5 + 12{\cdot}5 = 50$ m kg

Effective depth needed $d_{0.5}=\sqrt{\frac{50\times100}{100\times8{\cdot}7}}=2{\cdot}4$ cm

Overall depth provided $=25+\frac{25}{450}\times50=27{\cdot}8$ cm

Effective depth provided $=27{\cdot}8-7{\cdot}5$ (cover) $=20{\cdot}3$ cm

Amount of steel required $=\frac{50\times100}{1400\times{\cdot}87\times20{\cdot}3}=0{\cdot}202$ cm^2

2·5·3. Curtailment of Main Reinforcement

Similarly these computations for different depths at an interval of ½ m are carried out as given the table

h′ *m*	*BM* *m kg*	*Effective depth needed* *cm*	*Overall depth provided* *cm*	*Effective depth provided* *cm*	*Area of steel* *cm²*	*Spacing of 16 mm bar* *cm*	*Spacing adopted* *cm*
0	—	--	25	17·50	—	—	64
0·5	50·0	2·4	27·80	20·30	0·202	9·95	64
1·0	250	5·36	30·55	23·05	0·89	226	64
1·5	677	8·8	33·30	25·80	2·16	93	32
2·0	1400	12·7	36·10	28·60	4·01	50	32
2·5	2498	16·9	38·90	31·4	6·5	31	16
3·0	4050	21·6	41·65	34·15	9·73	20·6	16
3·5	6130	26·6	44·45	36·95	18·60	14·8	8
4·0	8800	31·8	47·20	39·70	18·40	10·9	8
4·5	12150	37·4	50·00	42·50	23·40	8·6	8

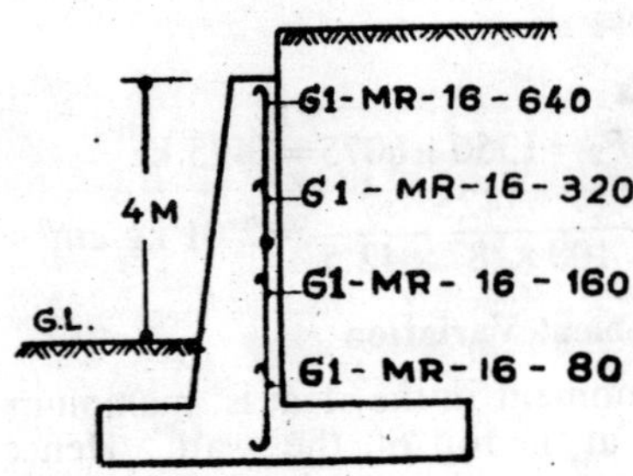

2·5·4. Secondary Reinforcement

Minimum reinforcement $=0{\cdot}15\%$ of gross cross sectional area of concrete

Concrete area $=\frac{1}{2}(25+50)\times450=16900$ cm^2

Area of steel $=\frac{16900\times{\cdot}15}{100}=25{\cdot}4$ cm^2

This reinforcement is distributed in the ratio of 1/3 on the earthface and 2/3 on the exposed face.

Using 10 mm ϕ spacing on the earth face $=\frac{0{\cdot}79}{8{\cdot}45}\times450=42$ cm

Spacing on the exposed face $= \dfrac{0{\cdot}79}{16{\cdot}9} \times 450 = 21$ cm

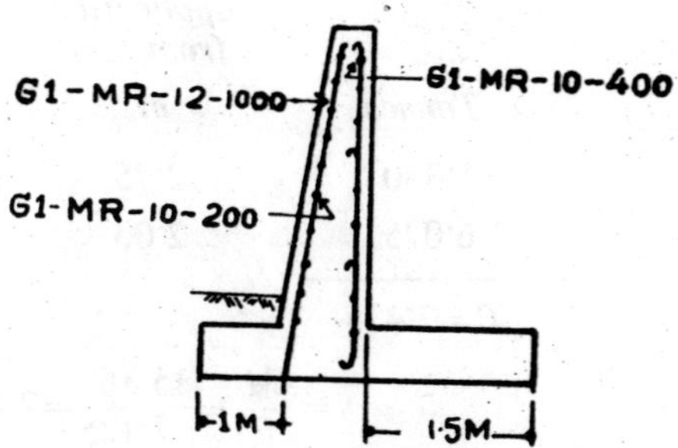

Use 10 mm ϕ, 20 cm c/c on the exposed face and 40 cm c/c on on the earth surfaae.

2·6. Base Design

Loads. Consider one metre length of wall

(1) weight of stem rectangular portion $= \frac{25}{100} \times 4{\cdot}5 \times 2400$

$W_1 = 2{\cdot}7$ tonnes

point of application from toe $x_1 = 1{\cdot}375$ m

(2) weight of stem-triangular portion $= \dfrac{25}{2 \times 100} \times 4{\cdot}5 \times 2400$

$W_2 = 1{\cdot}35$ tonnes

point of application from toe $x_2 = 1{\cdot}167$ m

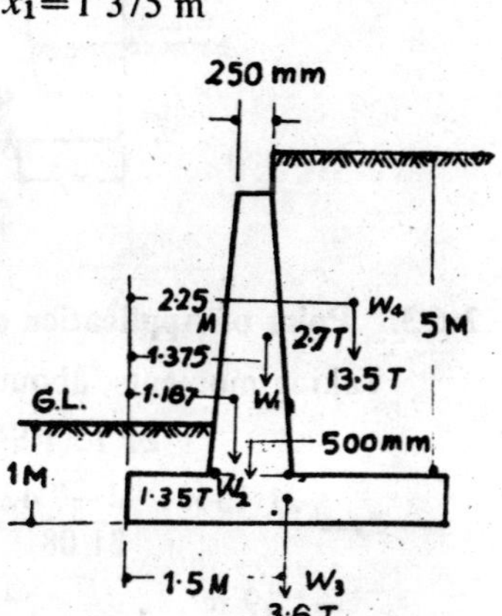

(3) weight of base $= \frac{1}{2} \times 3 \times 2400$

$W_3 = 3{\cdot}6$ tonnes

point of application from toe $x_3 = 1{\cdot}5$ m

(4) weight of backfill including surcharge $= 5 \times 1{\cdot}5 \times 1800$

W_4 13·5 tonnes

point of application from toe $x_4 = 2{\cdot}25$ m

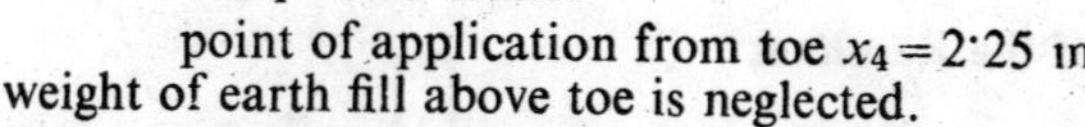

weight of earth fill above toe is neglected.

2·6·1. Point of Application of Resultant Vertical Load

	Load Tonnes	Pt. of application m	Moment mt
(1)	2·7 (W_1)	1·375 (x_1)	3·71
(2)	1·35 (W_2)	1·167 (x_2)	1·57
(3)	3·6 (W_3)	1·5 (x_3)	5·40
(4)	13·5 (W_4)	2·25 (x_4)	30·40
	$W = 21{\cdot}15$		$M = 41{\cdot}08$

$$\bar{x} = \frac{41{\cdot}08}{21{\cdot}15} = 1{\cdot}94 \text{ m}$$

2·6·2. Resultant of Horizontal Force on the Wall

Force due to	*Magnitude*	*Point of application from base*	*Moment*
	Tonnes	*m*	*mt*
Surcharge	1·350	2·75	3·70
Fill	6·075	2·00	12·16
	$P=7·425$		$M=15·86$

$$\bar{y}=\frac{M}{P}=\frac{15·86}{7·425}=2·14 \text{ m}$$

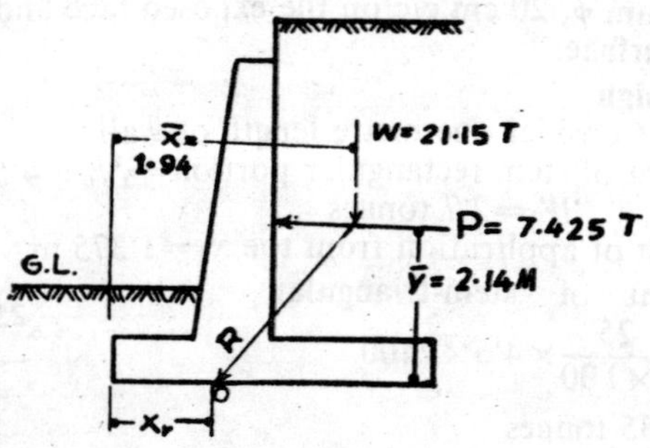

2·6·3. Point of Application of Resultant Force on the Base

Taking moments, about 0, $W(\bar{x}-\bar{x}_r)=P\bar{y}$

$$21·15(1·94-\bar{x}_r)=7·425\times 2·14$$

$$\bar{x}_r=\frac{21·15\times 1·94-7·425\times 2·14}{21·08}=\frac{41·08-15·86}{21·15}=\frac{25·22}{21·15}$$

$=1·19$ m

2·6·4. Eccentricity of the Resultant

$$e=\frac{B}{2}-\bar{x}_r$$

$=1·5-1·19$

$=0·31<\dfrac{B}{6}$

$=0·5$

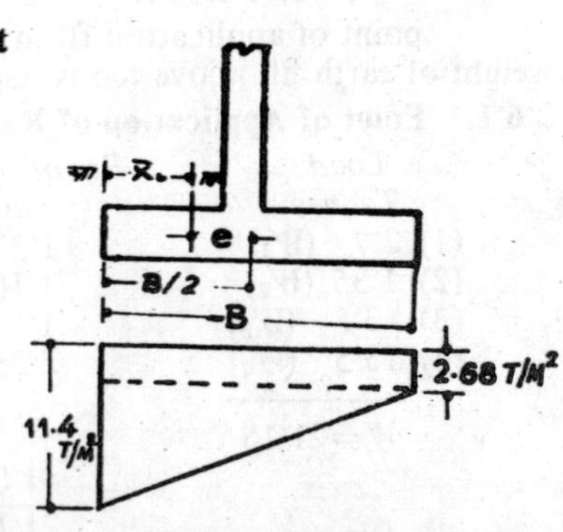

2·6·5. Pressure Distribution under the Base

Maximum pressure

$$= \frac{W}{B \times 1}\left(1 + \frac{6e}{B}\right)$$

$$= \frac{21{\cdot}150}{3 \times 1}\left(1 + \frac{6 \times 0{\cdot}31}{3}\right)$$

$= 11{\cdot}4$ tonnes/m^2 < 15 tonnes/m^2

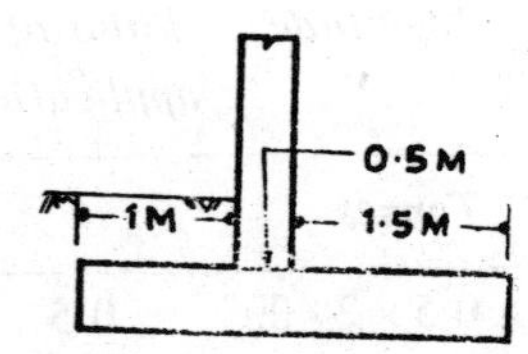

Minimum pressure

$$= \frac{W}{B \times 1}\left(1 - \frac{6e}{B}\right)$$

$$= \frac{21{\cdot}150}{3 \times 1}\left(1 - \frac{6 \cdot 0{\cdot}31}{3}\right)$$

$= 2{\cdot}68$ tonnes/m^2

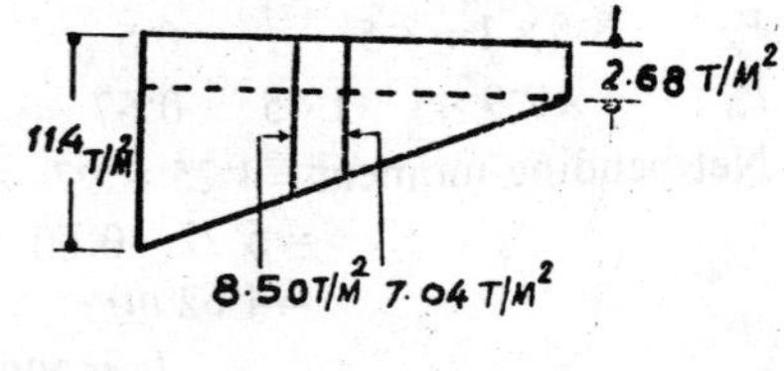

2·7. Heel Design

Forces acting on the heel

Force	*Magnitude* Tonnes	*Point of application* m	*Bending moment* mt
F_1	$1{\cdot}5 \times 5 \times 1{\cdot}800 = 13{\cdot}5$	0·75	+10·1
F_2	$1{\cdot}5 \times {\cdot}5 \times 2{\cdot}400 = 1{\cdot}8$	0·75	+1·35
F_3	$2{\cdot}68 \times 1{\cdot}5 - 4{\cdot}01$	0·75	−3·00
F_4	$\frac{1}{2} \times 1{\cdot}5 \times 4{\cdot}36 = 3{\cdot}28$	0·5	−1·64

Net bending moment $= 11{\cdot}45 - 4{\cdot}64 = 6{\cdot}81$ m tonnes

$$\text{Depth required} = \sqrt{\frac{681000}{100 \times 8{\cdot}7}} = 28 \text{ cm} < 50 \text{ cm}$$

Total depth provided = 50 cm

Effective depth provided $= 50 - 7{\cdot}5 = 42{\cdot}5$ cm

2·7·1. Main reinforcement

$$A_t = \frac{681000}{1400 \times {\cdot}87 \times 42{\cdot}5}$$

$= 13{\cdot}2$ cm^2

Use 16 mm ϕ @ 14 cm c/c

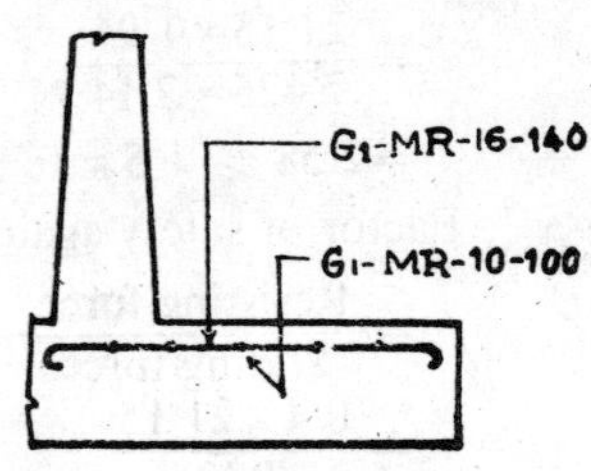

2·7·2. Secondary reinforcement

$$A_t = \frac{0{\cdot}15}{100} \times 100 \times 50$$

$= 7{\cdot}5$ cm^2

Use 10 cm ϕ @ 10 cm c/c

2.8. Toe design

Forces acting on the toe. Bending moment at the junction

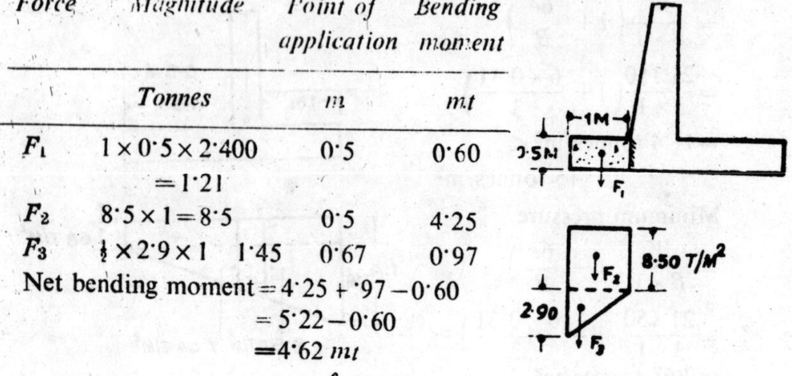

Force	*Magnitude*	*Point of application*	*Bending moment*
	Tonnes	*m*	*mt*
F_1	$1\times0.5\times2.400$ $=1.21$	0.5	0.60
F_2	$8.5\times1=8.5$	0.5	4.25
F_3	$\frac{1}{2}\times2.9\times1$ 1.45	0.67	0.97

$$\text{Net bending moment}=4.25+.97-0.60$$
$$=5.22-0.60$$
$$=4.62\ mt$$

$$\text{Depth required } d_e=\sqrt{\frac{462000}{100\times8.7}}=23.2\text{ cm} < 50\text{ cm}$$

Depth provided $=50$ cm

Effective depth $=42.5$ cm

2.8.1. Main reinforcement

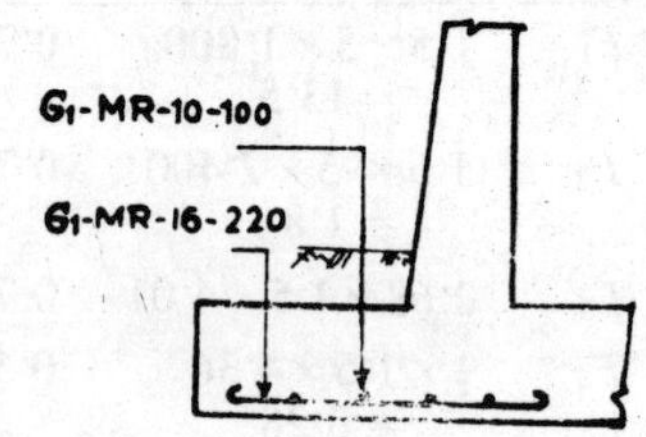

$$A_t=\frac{465000}{1400\times.87\times42.5}$$
$$=9\text{ cm}^2$$

Use 16 cm ϕ at 22 cm c/c

2.8.2. Secondary reinforcement

$$A_t=\frac{0.15}{100}\times100\times.50$$
$$=7.5\text{ cm}^2$$

Use 10 mm ϕ at 10 cm c/c

2.9. Stability Analysis

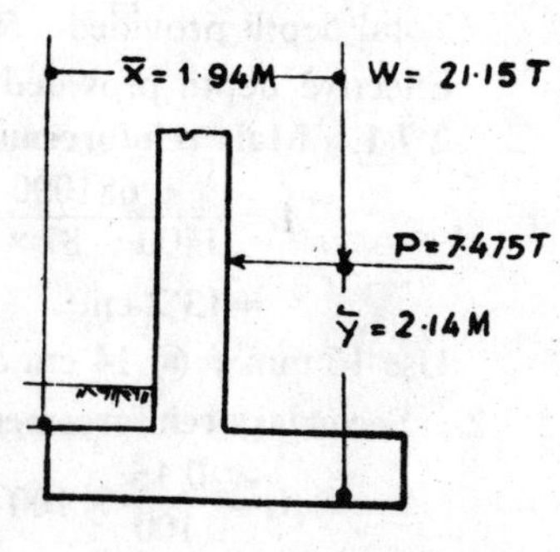

Factor of safety against overturning

$$=\frac{\text{Stabilizing moment}}{\text{overturning moment}}$$
$$=\frac{21.15\times1.94}{7.425\times2.14}$$
$$=2.58 > 1.5\text{ safe}$$

Factor of safety against sliding

$$=\frac{\text{Resisting force}}{\text{Sliding force}}=\frac{\mu W}{P}$$
$$=\frac{0.4\times21.15}{7.425}$$
$$=1.14 < 1.5\text{ unsafe}$$

2·10. Design of key

Provide key to prevent sliding failure
Resistance to sliding available = 21·15 × 0·4 = 8·45 tonnes
Minimum resistance required = 1·5 × 7·425 = 11·15 tonnes
Additional resistance to be provided by key = 11·15 8·45 tonnes
= 2·7 *t*

Assuming that key provides only 50% of passive resistance; magnitude of resistance to be developed by key = 2·7 × 2 = 5·4 tonnes

Passive pressure at level of base

$$= w \times h_1 \times \left(\frac{1+\sin}{1-\sin\phi}\right)$$

$$= 1800 \times 0{\cdot}5 \times 3 = 2700$$

P_1 $2700 \times d$,

$P_2 = 5400\, d \times \frac{1}{2} \times d = 2700\, d^2$

$P = P_1 + P_2 = 5400$

$2700\, d + 2700\, d^2 = 5400$

$d^2 + d = 2$

$d = 1\ m$

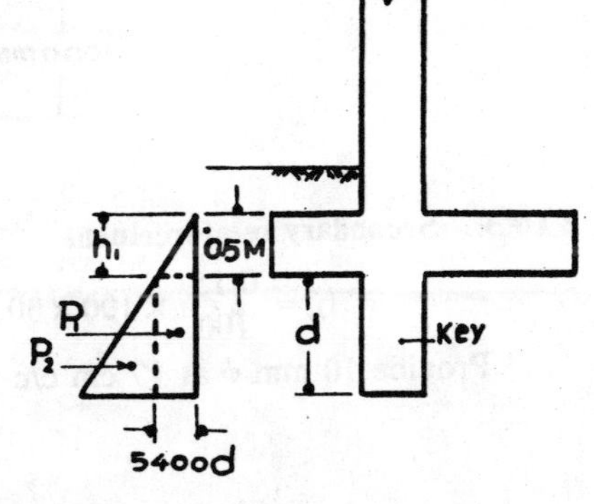

Revised factor of safety against sliding

$$= \frac{0{\cdot}4 \times 21{\cdot}14 + 5{\cdot}4}{7{\cdot}425} = \frac{13{\cdot}85}{7{\cdot}425} = 1{\cdot}865 > 1{\cdot}5 \text{ Safe}$$

2·10·1. Width of the key wall

Maximum Bending moment at the junction of the base and the key.

Force	*Magnitude*	*Point of application*	*Moment*
	T	*m*	*mt*
P_1	2·7 × 1	0·5	1·35
P_2	½ × 5·4 × 1	0·67	1·80

Maximum bending moment 3·15 *mt*

$$\text{width } b = \sqrt{\frac{315000}{100 \times 8{\cdot}7}} = 19{\cdot}2 \text{ cm}$$

Use 30 cm × 100 cm key
effective width = 22·5 cm

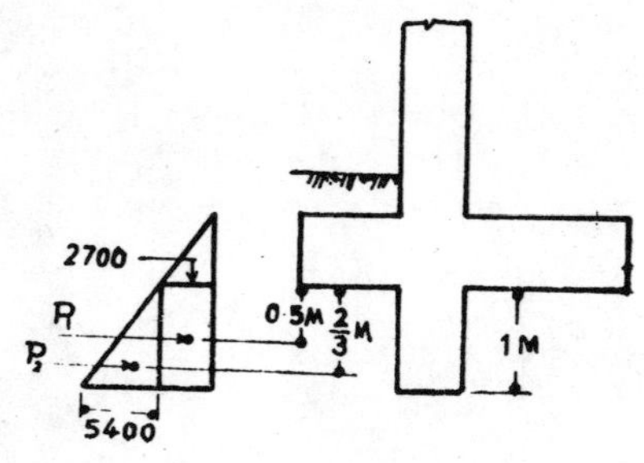

2·10·2. Main reinforcement

$$A_{st}=\frac{315000}{1400\cdot 0{\cdot}87\times 22{\cdot}5}=11{\cdot}5\ \text{cm}^2$$

rovide 16 mm ϕ
t 17 cm c/c

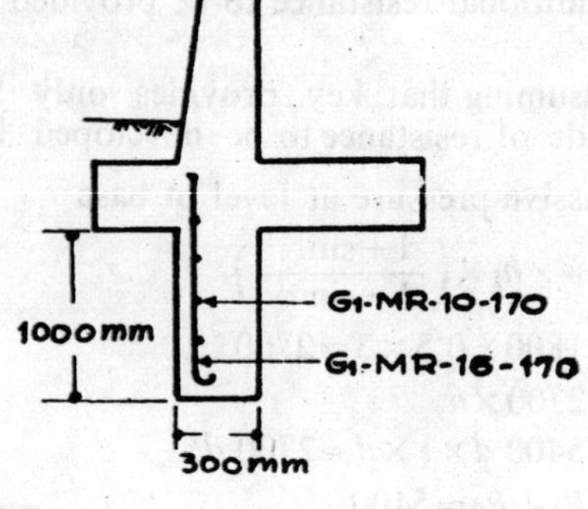

2·10·3. Secondary reinforcement

$$A_{st}=\frac{0{\cdot}15}{100}\times 100\times 30=4{\cdot}5\ \text{cm}^2$$

Provide 10 mm ϕ at 17 cm c/c

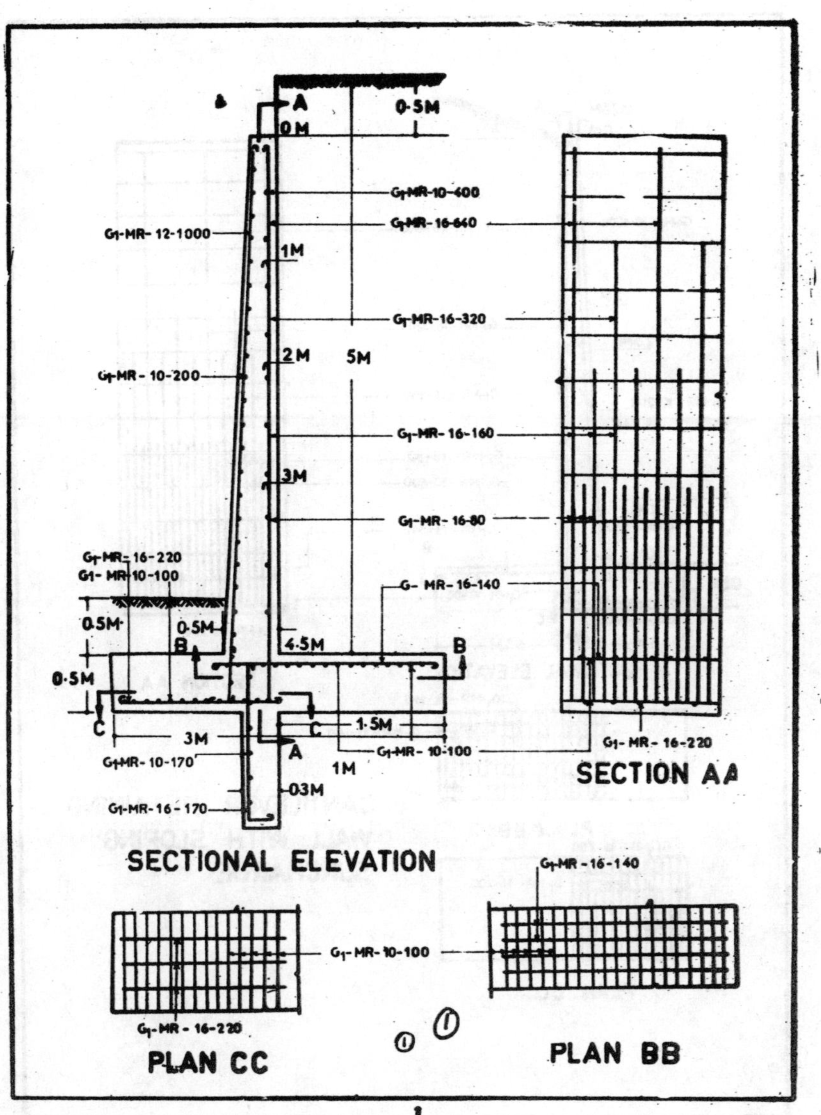

SECTION AA

SECTIONAL ELEVATION

PLAN CC

PLAN BB

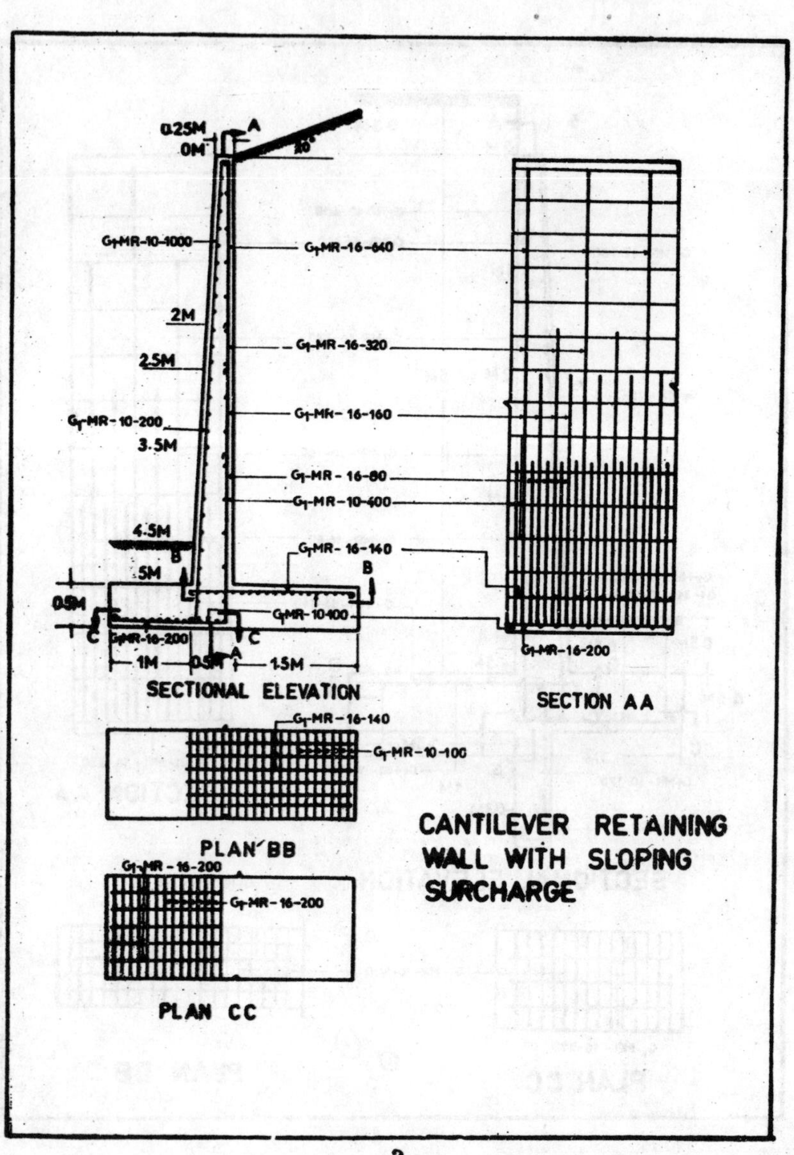
0.25M
A
0M
20°
G1-MR-10-1000
G1-MR-16-640
2M
G1-MR-16-320
2.5M
G1-MR-16-160
G1-MR-10-200
3.5M
G1-MR-16-80
G1-MR-10-400
4.5M
B
G1-MR-16-140
.5M
B
0.5M
G1-MR-10-100
C
G1-MR-16-200
C
A
1M
0.5M
1.5M
SECTIONAL ELEVATION
G1-MR-16-200
SECTION AA
G1-MR-16-140
G1-MR-10-100
PLAN BB
G1-MR-16-200
G1-MR-16-200
PLAN CC
CANTILEVER RETAINING WALL WITH SLOPING SURCHARGE

3 Cantilever Retaining Wall with Sloping Surcharge

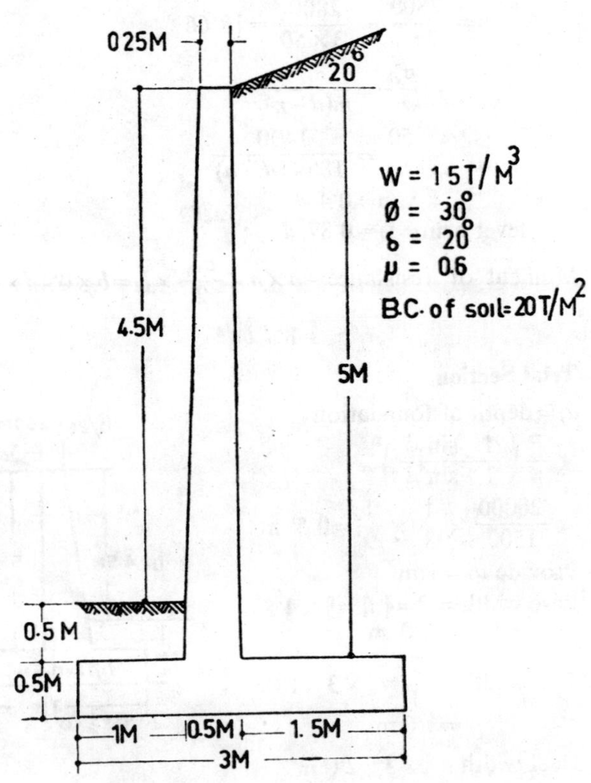

3·1. Data

Difference between ground levels 4·5m

Surcharge = δ = 20°

Density of soil $w = 1500$ kg/m³

Angle of internal friction = $\phi = 30°$

Coefficient of friction between the soil and concrete = 0·6

Bearing capacity of soil = 20 t/m²

Code – IS 456 – 1964

3·2. Allowable Stresses

Concrete—grade - *M* 150

Compressive stress in bending $=\sigma_{cb}=50$ kg/cm²

in direct compression $=40$ kg/cm²

shear stress $=5$ kg/cm²

bond stress $=10$ kg/cm²

Steel—grade—1 $\sigma_{st}=1400$ kg/cm²

3·2·1. Characteristic Strengths

$$m=\frac{2800}{3\sigma_{cb}}=\frac{2800}{3\times 50}=18{\cdot}66$$

$$\frac{\sigma_{cb}}{n}=\frac{\sigma_{st}}{m(d-n)}$$

$$\frac{50}{n}=\frac{1400}{18{\cdot}66(d-n)}$$

$$n=0{\cdot}4\ d$$

$$\text{lever arm}=j_d=0{\cdot}87\ d$$

$$\text{Moment of resistance}=b\times n\times\frac{\sigma_{cb}}{2}\times j_d=b\times 0{\cdot}4d\times\frac{\sigma_{cb}}{2}\times 0{\cdot}87\ d$$

$$=8{\cdot}7\ bd^2$$

3·3. Trial Section

d_f=depth of foundation

$$\nless \frac{P}{w}\left(\frac{1-\sin\phi}{1+\sin\phi}\right)^2$$

$$\nless \frac{20000}{1500}\times\frac{1}{3}\times\frac{1}{9}=0{\cdot}5\ \text{m}$$

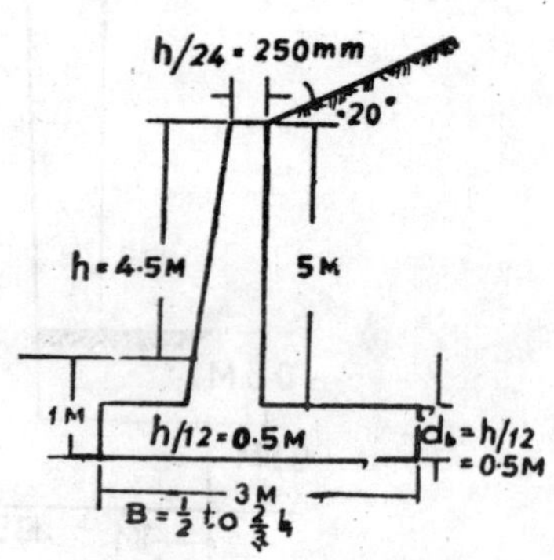

Provide $d_f=1$ m

Base width $=B=\frac{2}{3}h=\frac{2}{3}\times 4{\cdot}5$
$=3{\cdot}0$ m

Toe width $=\frac{B}{3}=\frac{1}{3}\times 3=1{\cdot}0$
$=1{\cdot}0$ m

Heel width $=\frac{2}{3}\times 3=2{\cdot}0$ m

$$\text{Stem width at the top}=\frac{4{\cdot}5\times 100}{24}=25{\cdot}00\ \text{cm}$$

$$\text{Stem width at bottom}=\frac{4{\cdot}5\times 100}{12}=50{\cdot}0\ \text{cm}$$

$$\text{Thickness of base slab}=\frac{4{\cdot}5\times 100}{12}=50{\cdot}0\ \text{cm}$$

3·3·1. Earth Pressure Calculations

Earth pressure due to backfill with surcharge

$$P=\frac{wh^2}{2}\cos\delta\ \frac{\cos\delta-\sqrt{\cos^2\delta-\cos^2\phi}}{\cos\delta+\sqrt{\cos^2\delta-\cos^2\phi}}$$

$$=1500\times\frac{25}{2}\times 0{\cdot}94\times\frac{0{\cdot}94-\sqrt{0{\cdot}88-0{\cdot}75}}{0{\cdot}94+\sqrt{0{\cdot}88-0{\cdot}75}}$$

$$=7800\text{ kg}=312\,(h')^2$$

$P\cos\delta=294{\cdot}6\,(h')^2=7340$ kg acts horizontally at a height $\frac{5}{3}$ m above base

$P\sin\delta=2660$ kg acts along the stem

3·4. Stem Design

Maximum bending moment at the junction of the stem and the base $=7340\times\frac{5}{3}=12{,}300$ m. kg

effective depth of stem $d=\sqrt{\dfrac{123000}{8{\cdot}7\times 100}}=37{\cdot}6$ cm

overall depth $=37{\cdot}6+\text{cover}=37{\cdot}6+7{\cdot}5=45{\cdot}1$ cm.

However 50 cm assumed shall be used therefore the effective depth $=42{\cdot}5$ cm

Shear check, Maximum force $=7340$ kg

shear stress $=\dfrac{7340}{100\times 0{\cdot}87\times 42{\cdot}5}=2{\cdot}12\text{ kg/cm}^2<5\text{ kg/cm}^2$ (safe)

3·4·1. Bending Moment Variation

The bending moment in the stem is maximum at the junction and reduces to zero at the top of the wall. Hence main reinforcement can be suitably reduced towards top and the wall can be tapered and the thickness reduced to achieve economy.

Bending moment at any depth h'

$$M=294{\cdot}6\,(h')^2\times\frac{h'}{3}=98{\cdot}2\,(h')^3$$

Calculations for $h'=0{\cdot}5$ m

Bending moment $=98{\cdot}2\,(h')^3=98{\cdot}2\times(\tfrac{1}{2})^3=98{\cdot}2\times\tfrac{1}{8}=12{\cdot}3$ m kg

Effective depth needed $d_{0.5}=\sqrt{\dfrac{12{\cdot}3\times 100}{100\times 8{\cdot}7}}=1{\cdot}23$ cm

Overall depth provided $=25+\dfrac{25}{450}\times 50=27{\cdot}8$ cm

Effective depth provided $=27{\cdot}8-7{\cdot}5$ (cover) $=20{\cdot}3$ cm

Amount of steel required $=\dfrac{12{\cdot}3\times 100}{1400\times 0{\cdot}87\times 20{\cdot}3}=0{\cdot}05\text{ cm}^2$

Similarly these computations for different depths at an interval of $\frac{1}{2}$ m are carried out as given in the table.

3·4·2. Curtailment of Main Reinforcement

h′	B.M.	Effective depth needed	Overall depth provided	Effective depth provided	Area of steel	Spacing of 16 mm bar	Spacing adopted
m	m kg	cm	cm	cm	cm²	cm	cm
0	—	—	25·0	17·5	—	—	64·00
0·5	12·3	1·23	27·5	20·30	0·05	4020·0	64·00
1·0	98·2	3·36	30·0	23·95	0·358	561·0	64·00
1·5	332·0	6·2	32·5	25·80	1·09	184·5	64·00
2·0	786·0	9·5	35·0	28·60	2·35	85·5	64·00
2·5	1532·0	13·3	37·5	31·40	4·2	48·0	32·00
3·0	2660·0	17·5	40·0	34·15	6·7	30·0	16·00
3·5	4220·0	22·0	42·5	36·95	9·9	20·3	16·00
4·0	6300·0	27·0	45·0	39·70	13·8	14·6	8'00
4·5	9000·0	32·2	47·5	42·50	18·45	10·9	8·00
5·0	12300·0	37·6	50·0	45·35	23·7	8·46	8·00

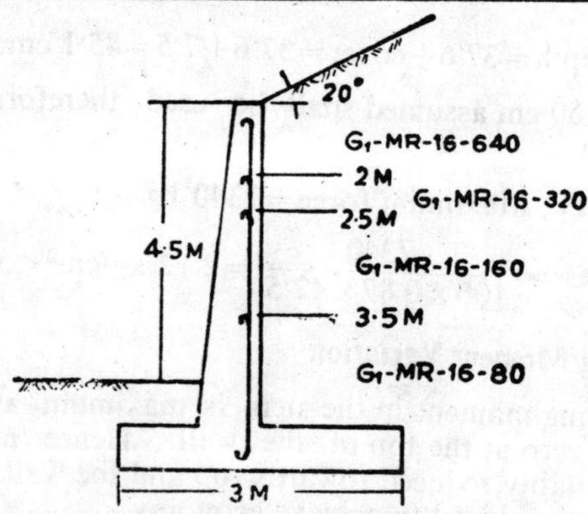

3·4·3. Secondary Reinforcement

Minimum reinforcement=0·15% of gross cross-sectional area of concrete

Concrete area

$=\frac{1}{2}(25+50)\times 500=18750 \text{ cm}^2$

Area of steel

$=18750\times\frac{\cdot 15}{100}=28\cdot 125 \text{ cm}^2$

This reinforcement is distributed in the ratio of $\frac{1}{3}$ on the earth face and $\frac{2}{3}$ on the exposed face.

Using 10 mm ϕ, spacing on the earth face

$=\frac{0\cdot 79}{9\cdot 375}\times 500=42\cdot 0$ cm

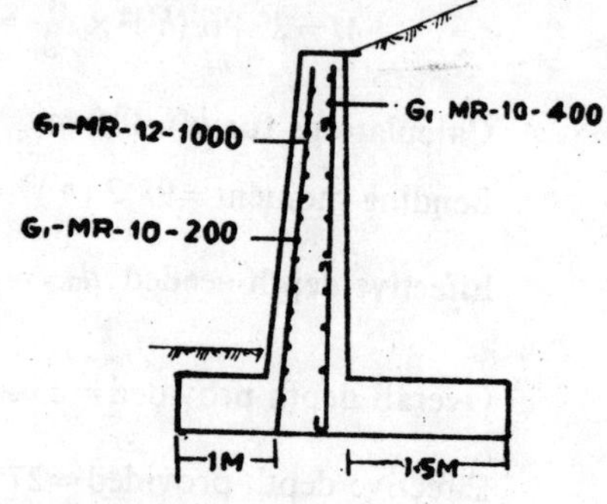

Spacing on the exposed face $=\frac{0\cdot 79\times 500}{18\cdot 75}=21\cdot 0$ cm

Use 10mm ϕ at 20 cm c/c on exposed face and 40 cm c/c on earth face.

3·5. Base Design

Loads-consider one metre length of wall

(1) weight of stem rectangular portion $=\frac{25}{100}\times 5\times\frac{2400}{1000}$

$W_1 = 3{\cdot}0$ tonnes

Point of application from toe $x_1 = 1{\cdot}375$ m

(2) weight of stem triangular portion

$=\frac{25}{2\times 100}\times 5\times\frac{2400}{1000}$, $W_2 = 1{\cdot}5$ tonnes

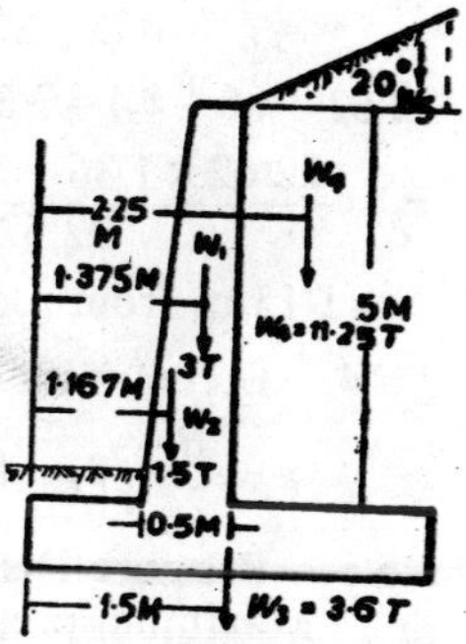

Point of application from toe

$x_2 = 1{\cdot}167$ m

(3) weight of base $=\frac{1}{2}\times 3\times\frac{2400}{1000}$

$W_3 = 3{\cdot}6$ tonnes

Point of application from toe $x_3 = 1{\cdot}5$ m

(4) weight of backfill $= 5\times 1{\cdot}5\times\frac{1500}{1000}$

$W_4 = 11{\cdot}25$ tonnes

Point of application from toe $x_4 = 2{\cdot}25$ m

(5) weight of surcharge $= \frac{1}{2}\times 1{\cdot}5\times \tan 20\times\frac{1{\cdot}5\times 1500}{1000}$

$W_5 = 0{\cdot}61$ tonnes

Point of application from toe $x_5 = 2{\cdot}5$ m

$P\sin\delta = 2{\cdot}66$ tonnes

Point of application from toe $= 1{\cdot}5$ m

3·5·1. Point of Application of Resultant Vertical Load

	Load Tonnes	Point of application m	Moment mt
(1)	3·0 (W_1)	1·375	4·13
(2)	1·5 (W_2)	1·17	1·756
(3)	3·6 (W_3)	1·5	5·4
(4)	11·25 (W_4)	2·25	25·3
(5)	0·61 (W_5)	2·5	1·525
(6)	2·66 ($P\sin\delta$)	1·5	4·00

$W = 22{\cdot}62$ tonnes $\qquad M = 42{\cdot}111$

$\bar{x} = \frac{42{\cdot}111}{22{\cdot}62} = 1{\cdot}86$ m from toe

3·5·2. Resultant of Horizontal Forces on the Wall

$\bar{y} = \frac{5}{3} + 0{\cdot}5 = 1{\cdot}667 + 0{\cdot}5$

$= 2{\cdot}167 = 2{\cdot}17$ m (Say)

3·5·3. Point of Application of Resultant Force on the Base

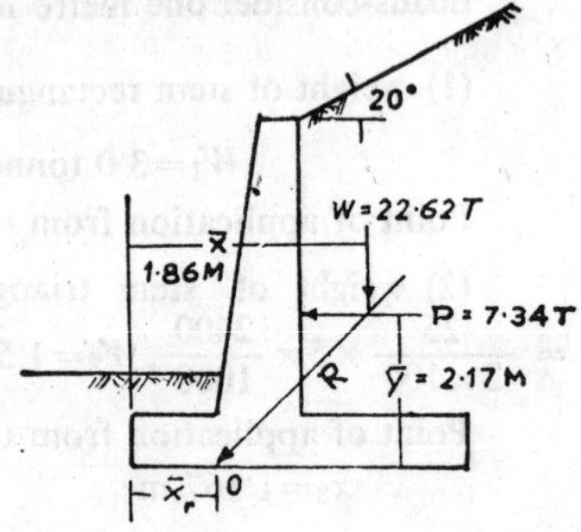

Taking moments about 0,

$$W(\bar{x} - \bar{x}_r) = P\bar{y}$$

$$22{\cdot}62(1{\cdot}86 - \bar{x}_r) = 7{\cdot}34 \times 2{\cdot}17$$

$$\bar{x}_r = \frac{22{\cdot}62 \times 1{\cdot}86 - 7{\cdot}34 \times 2{\cdot}17}{22{\cdot}62}$$

$= 1{\cdot}15$ m from toe

3·5·4. Eccentricity of the Resultant

$$e = \frac{B}{2} - \bar{x}_r = 1{\cdot}5 - 1{\cdot}15 = 0{\cdot}35 \text{ m} < \frac{B}{6} \quad 0{\cdot}5 \text{ m (Safe)}$$

3·5·5. Pressure Distribution under the Base

Maximum pressure

$$= \frac{W}{B \times 1}\left(1 + \frac{6e}{B}\right)$$

$$= \frac{22{\cdot}62}{3}\left(1 + \frac{6 \times 0{\cdot}35}{3}\right)$$

$$= \frac{22{\cdot}62}{3} \times 1{\cdot}7$$

$= 12{\cdot}8$ t/m$^2 < 20$ t/m^2

Minimum pressure

$$= \frac{W}{B \times 1}\left(1 - \frac{6e}{B}\right)$$

$$= \frac{22{\cdot}62}{3} \times 0{\cdot}3 = 2{\cdot}26 \text{ t/m}^2$$

(comp)

3·6. Heel Design

Forces acting on the heel. Bending moment at the junction.

Force	Magnitude	Point of application	B.M.
	tonnes	m	mt
F_1	$1{\cdot}5 \times 5 \times 1{\cdot}5$ $+0{\cdot}61$	0·75	+8·9
F_2	$1{\cdot}5 \times 0{\cdot}5 \times 2{\cdot}40$ $= 1{\cdot}8$	0·75	+1·35
F_3	$2{\cdot}26 \times 1{\cdot}5$ $= 3{\cdot}39$	0·75	−2·54
P_4	$\frac{1}{2} \times 1{\cdot}5 \times 5{\cdot}27$	0·5	1·95

Net B.M. $=10{\cdot}25-4{\cdot}49=5{\cdot}75$ m/t

depth required $=\sqrt{\dfrac{575000}{100\times 8{\cdot}7}}=25{\cdot}7$ cm <50 cm

depth provided $=50$ cm

effective depth provided $=50-7{\cdot}5=42{\cdot}5$ cm

3·6·1. Main Reinforcement

$$A_t=\frac{575000}{1400\times{\cdot}87\times 42{\cdot}5}=11{\cdot}1\text{ cm}^2$$

Provide 16 mm ϕ bars, spacing $=\dfrac{100}{11{\cdot}1}\times 2{\cdot}01=18{\cdot}1$cm c/c

Use 16 mm ϕ at 18 cm c/c

3·6·2. Secondary Reinforcement

$$A_t=\frac{0{\cdot}15}{100}\times 100\times 50$$

$=7{\cdot}5$ cm^2

Use 10 mm ϕ at 10 cm c/c

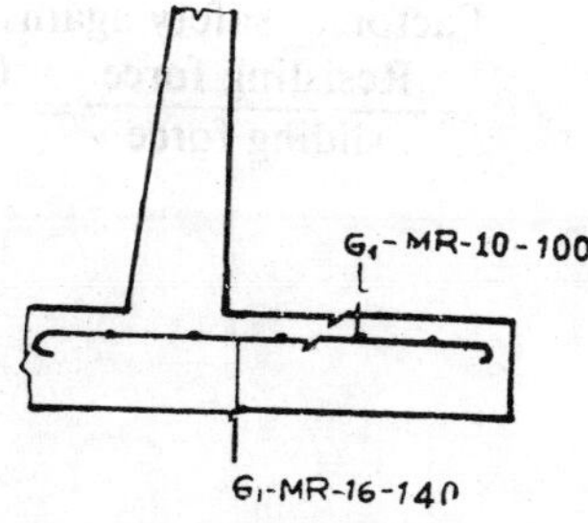

3·7. Toe Design

Forces acting on the toe. Bending moment at the junction.

Force	*Magnitude* tonnes	*Point of application* m	*B.M.* mt
F_1	$1\times 0{\cdot}5\times 2{\cdot}4 = 1{\cdot}2$	0·5	0·6
F_2	$9{\cdot}3\times 1=9{\cdot}3$	0·5	4·65
F_3	$\frac{1}{2}\times 3{\cdot}5\times 1=1{\cdot}75$	0·67	1·18

Net B.M. $=4{\cdot}65+1{\cdot}18-0{\cdot}6$
$=5{\cdot}83-0{\cdot}6=5{\cdot}23$ mt

depth required $d=\sqrt{\dfrac{523000}{100\times 8{\cdot}7}}=24{\cdot}5$ cm (say) 25 cm <50 cm

Total depth provided $=50$ cm

effective depth $=42{\cdot}5$ cm

3·7·1. Main Reinforcement

$$A_t=\frac{523000}{1400\times 0{\cdot}87\times 42{\cdot}5}=9{\cdot}97\text{ cm}^2$$

Provided 16 mm ϕ bars, spacing $= \frac{100}{9{\cdot}97} \times 2{\cdot}01 = 20{\cdot}1$ cm

Use 16 mm ϕ at 20 cm c/c

3·7·2. Secondary Reinforcement

$$A_t = \frac{0{\cdot}15}{100} \times 100 \times 50 = 7{\cdot}5 \text{ cm}^2$$

Use 10 mm ϕ at 10 cm c/c

3·8. Stability Analysis

Factor of safety against overturning

$$= \frac{\text{stabilizing moment}}{\text{overturning moment}} = \frac{22{\cdot}62 \times 1{\cdot}86}{7{\cdot}34 \times 2{\cdot}17} = 2{\cdot}64 > 1{\cdot}5$$

Factor of safety against sliding

$$= \frac{\text{Resisting force}}{\text{sliding force}} = \frac{0{\cdot}6 \times 22{\cdot}62}{7{\cdot}34} = 1{\cdot}85 > 1{\cdot}5 \text{ (Safe)}$$

Cantilever Retaining Wall With Track Loading 4

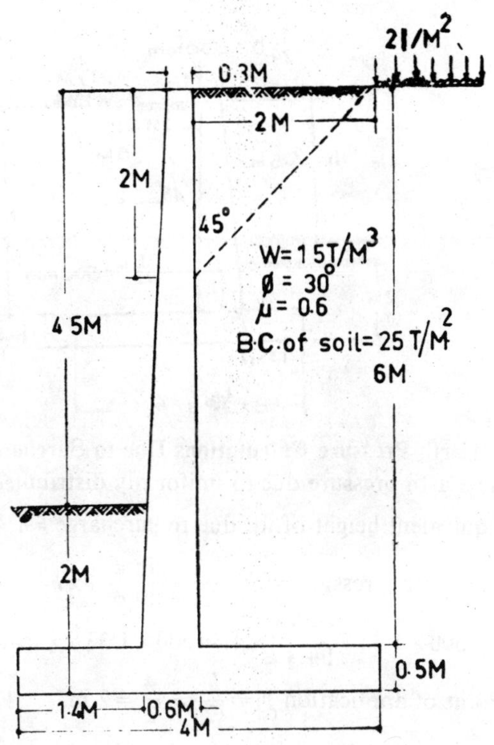

4.1. Data

Height of the fill to be retained by the wall = 4·5 m

Surcharge—uniformly distributed load of 2 t/m at a distance of 2 m from the edge of the wall

Density of soil = 1500 kg/m³

Angle of internal friction = 30°

Coefficient of friction between the soil and concrete = 0·6

Bearing capacity of soil = 25 t/m²

IS—456—1964

4·2. Allowable Stresses and Characteristic Strengths

Concrete — *M* 150 | Steel-grade — 1
$\sigma_{cb}=50$ kg/cm² | $R=8{\cdot}7$
$\sigma_{st}=1400$ kg/cm² | $j_d=0{\cdot}87$
$m=18$
$q=5$ kg/cm² | $\sigma_b=10$ kg/cm²

4·3. Trial Dimensions

$$d_f=\text{depth of foundation}=\frac{P}{w}\left(\frac{1-\sin\phi}{1+\sin\phi}\right)^2=\frac{25000}{1500}\left(\frac{1-\frac{1}{2}}{1+\frac{1}{2}}\right)^2$$

$=1{\cdot}8$ m say 2 m

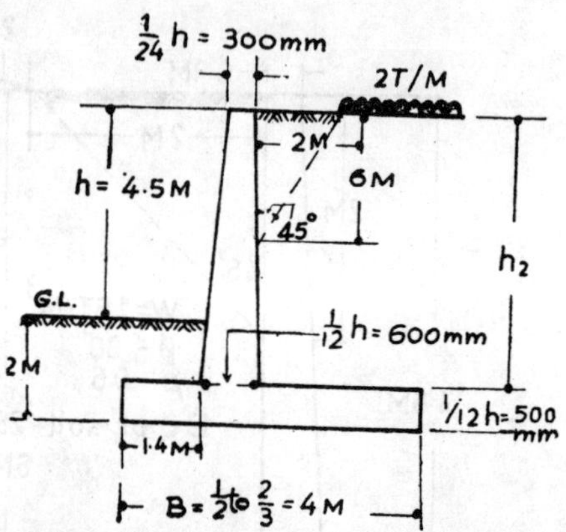

4·3·1. Earth Pressure Calculations Due to Surcharge

(*i*) Earth pressure due to uniformly distributed surcharge load

Equivalent height of fill due to surcharge $=h'=\dfrac{2000}{1500}=1{\cdot}33$ m

$$P_1=\text{earth pressure}=wh'\left(\frac{1-\sin\phi}{1+\sin\phi}\right)\times h_1$$

$$=1500\times1{\cdot}33\left(\frac{1-\frac{1}{2}}{1+\frac{1}{2}}\right)\times4=1500\times1{\cdot}33\times\tfrac{1}{3}\times4=665\times4=2660 \text{ kg}$$

Point of application $y_1=\dfrac{h_1}{2}=\dfrac{4}{2}=2$ m

4·3·2. Due to Backfill

(*ii*) Earth pressure due to backfill

$$P_2=\frac{1}{2}wh_2\left(\frac{1-\sin\phi}{1+\sin\phi}\right)\times h_2$$

$=\frac{1}{2}\times1500\times6(\frac{1}{3})\times6$
$=9000$ kg

Point of application

$=\dfrac{h_2}{3}=\dfrac{6}{3}=2$ m

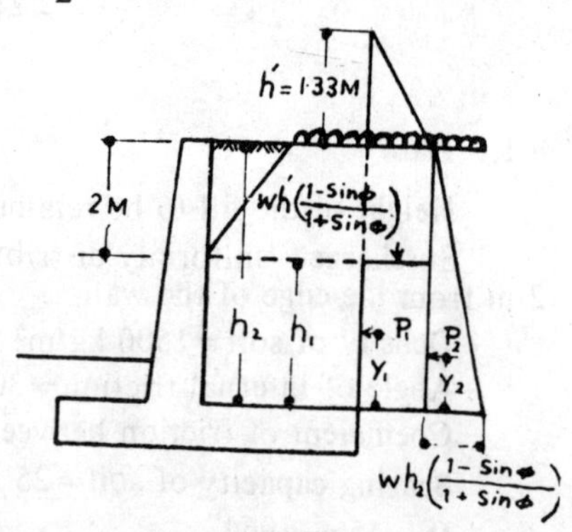

4·4. Stem Design

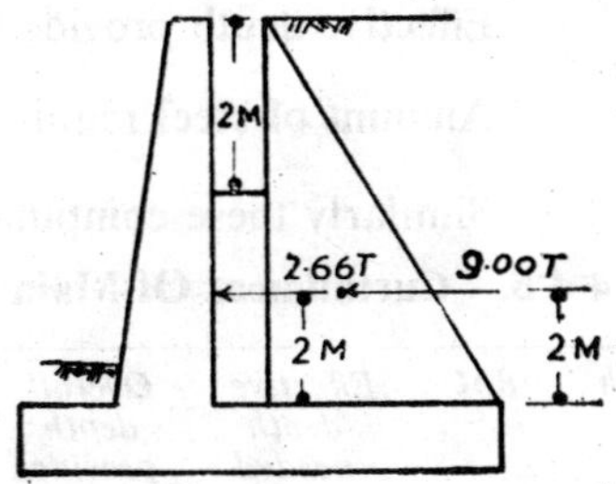

Maximum bending moment at the junction of the stem and the base

$=P_1Y_1+P_2Y_2$

$=2660\times 2+9000\times 2$

$=5320+18000$

$=23320$ m kg

4·4·1. Effective Depth

Effective depth $d_e=\sqrt{\dfrac{23320\times 100}{8{\cdot}7\times 100}}=51{\cdot}5$ cm <60 cm

Use an overall depth $=60$ cm

Effective depth provided $=60-7{\cdot}5=52{\cdot}5$ cm

4·4·2. Shear Check

Maximum shear at the junction $=P_1+P_2=2660+9000$
$=11660$ kg

Shear stress $=\dfrac{\text{shear force}}{b\times J_d}=\dfrac{11660}{100\times{\cdot}87\times 52{\cdot}5}$

$=2{\cdot}56$ kg/cm$^2<5$ kg/cm^2

4·4·3. Bending Moment Variation

The bending moment in the stem is maximum at the junction and reduces to zero at the top of the wall. Hence the main reinforcement can be suitably reduced towards top and the wall ean be tapered and the thickness reduced to achieve economy.

$M=$ Bending moment at any depth h from 0 to 2 m from top

$$=\tfrac{1}{2}\,wh_2\left(\frac{1-\sin\phi}{1+\sin\phi}\right)h_2\times\frac{h_2}{3}=\frac{1}{2}\times 1500\times\frac{1}{9}\,h_2^3=\frac{500}{6}\,h_2^3\,.$$

$$BM\text{ from 2 m to 6 m, }M=wh'\left(\frac{1-\sin\phi}{1+\sin\phi}\right)h_1\times\frac{h_1}{2}+\frac{500}{6}h_2^3\,.$$

$$=1500\times 1{\cdot}33\times\frac{1}{3}\times\frac{h_1^2}{2}+\frac{500}{6}=h_2^3\frac{666{\cdot}7}{2}\,h_1^2+\frac{500}{6}h_2^3$$

4·4·4. Sample Calculation

When $h_2=2$ m

$$M=\text{bending moment}=\frac{500}{6}\,h_2^3=\frac{500\times 2^3}{6}=\frac{500\times 8}{6}=\frac{4000}{6}$$

$=666{\cdot}7$ m kg

Effective depth $=\sqrt{\dfrac{666{\cdot}7\times 100}{8{\cdot}7\times 100}}=8{\cdot}6$ cm

Overall depth provided $=40$ cm

Effective depth provided $=30+\dfrac{30}{600}\times 200=40$ cm

Effective depth provided $=40-7{\cdot}5$ (cover) $=32{\cdot}5$ cm

Amount of steel required $=\dfrac{66670}{1400\times{\cdot}87\times32{\cdot}5}=1{\cdot}69\text{ cm}^2$

Similarly these computations are made for other depths

4·4·5. Curtailment Of Main Reinforcement

h	*BM*	*Effective depth needed cm*	*Overall depth provided cm*	*Effective depth provided cm*	*Area of steel cm²*	*Spacing 20 cm φ required*	*Spacing of 20mm φ used*
1	83 3	3·1	35	27·5	0·19	16·50	96 cm
2	666·7	8·6	40	32·5	1·69	18·5	96 cm
3	2583	15·2	45	37·5	5.65	55·5	48 cm
4	6667	27·8	50	42·5	12·9	24·2	24 cm
5	13400	39·2	55	47·5	23·2	13·5	8 cm
6	23320	51·8	60	52·5	35·5	8·8	8 cm

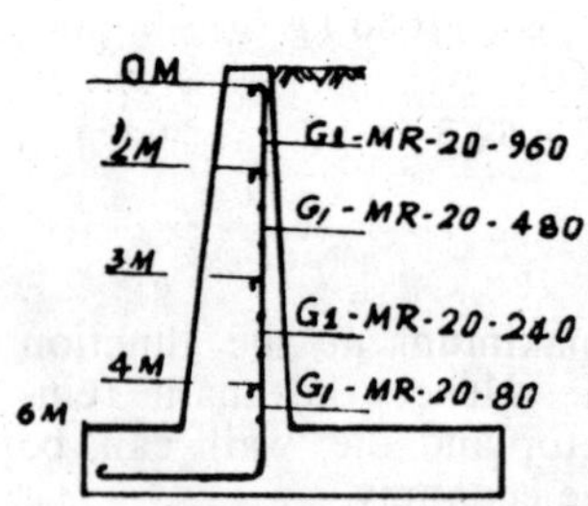

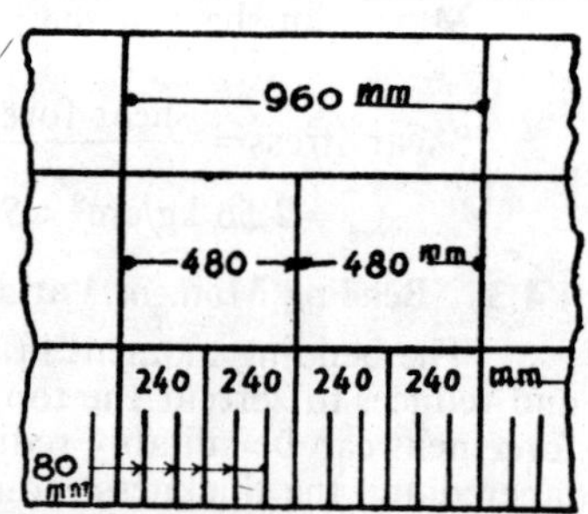

4·4·6. Secondary Reinforcement

Minimum reinforcement $=0{\cdot}15\%$ of cross-sectional area of concrete

Concrete area $=\frac{1}{2}(30+60)\times600=45\times60=27000\text{ cm}^2$

Area of secondary steel $=\dfrac{27000\times0{\cdot}15}{100}=40{\cdot}5\text{ cm}^2$

$\frac{1}{3}$ of this reinforcement is placed on the earthface and $\frac{2}{3}$ on the exposed face.

A_t on the earth face

$=\dfrac{40{\cdot}5}{3}=13{\cdot}5\text{ cm}^2$

Using 10 mm ϕ spacing

$=\dfrac{0{\cdot}79}{13{\cdot}5}\times600=35$ cm

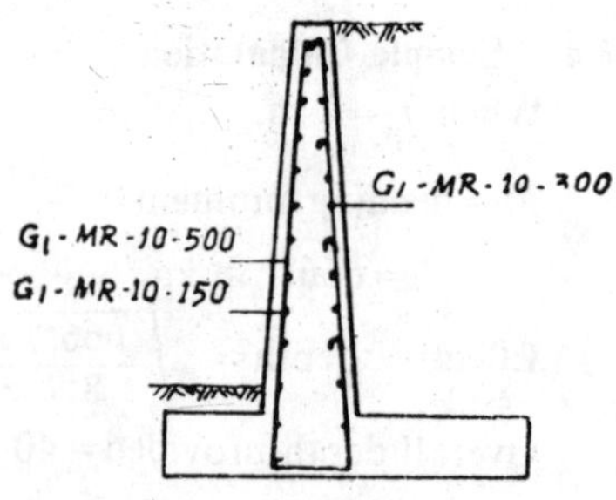

Use 10 mm ϕ at 30 cm c/c

A_t on the exposed face $=\frac{2}{3}\times40{\cdot}5=27\text{ cm}^2$

Use 10 mm ϕ at 15 cm c/c.

4·5. Base Design Vertical Loads

Consider 1 m length of wall

(*i*) weight of stem-rectangular portion $W_1 = \frac{30}{100} \times 1 \times 6 \times 2400$

$= 4320$ kg $= 4·32$ tonnes

Point of application from toe $x_1 = 1·4 + ·3 + \frac{·3}{2} = 1·85$ m

(*ii*) weight of stem-triangular portion

$W_2 = \frac{1}{2} \times \frac{30}{100} \times 6 \times 1 \times 2400$

$= 2160$ kg $= 2·16$ tonnes

Point of application from toe $x_2 = 1·40 + \frac{2}{3} \times 0·3$

$x_2 = 1·60$ m

(*iii*) weight of base

$W_3 = \frac{1}{2} \times 4 \times 1 \times 2400 = 4800$ kg

$= 4·8$ tonnes

Point of application from toe $x_3 = \frac{4}{2} = 2$ m

(*iv*) weight of backfill above heel slab $W_4 = 2 \times 6 \times 1 \times 1500$
$= 18000$ kg $= 18$ tonnes

Point of application from toe $x_4 = 2 + \frac{2}{2} = 3$ m

(*v*) weight of soil above toe slab $W_5 = 1·5 \times 1 \times 1·4 \times 1500$
$= 3150$ kg $= 3·15$ tonnes

Point of application $x_5 = \frac{1·4}{2} = 0·7$ m

4·5·1. Point Of Application of Resultant Vertical Loads

Load tonnes		*Point of application* m	*Momnet* mt
W_1	4·32	1·85	8·00
W_2	2·16	1·60	3·46
W_3	4·80	2·00	9·60
W_4	18·0	3·00	54·00
W_5	3·15	0·70	2·105
$\Sigma W = 32·43$			$M = 77·165$

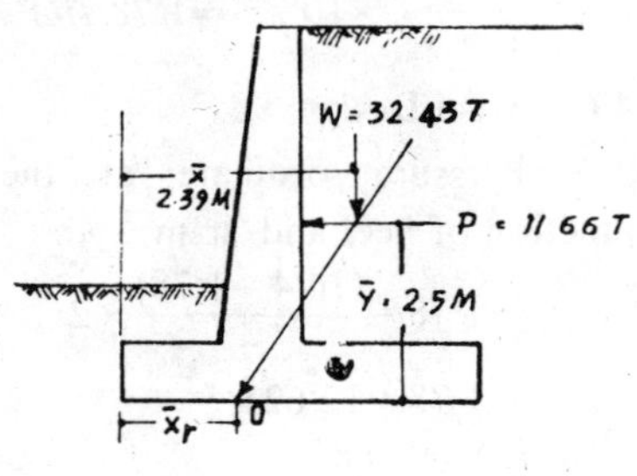

$$\bar{x} = \frac{77·165}{32·43} = 2·39 \text{ m}$$

4·5·2. Point of Application of Horizontal Forces

Force due to	*Magnitude*	*Point of application from base*	*Moment*
	tonnes	*m*	*mt*
Surcharge load P_1	2·66	2·5	6.65
Backfill P_2	9·00	2·5	22·50
ΣP 11·66			$\Sigma M = 29·15$

$\bar{y} = 2·5$ m

4·5·3. Point of Application of the Resultant

Taking moments about 0, $W(\bar{x} - \bar{x}_r) \quad P\,\bar{y}$

$32·43(2·39 - \bar{x}_r) \quad 11·66 \times 2·5 = 29·15$

$77·165 - 32·43\,\bar{x}_r = 29·15$

$32·43\,\bar{x}_r = 77·165 - 29·15 = 48·015$

$$\bar{x}_r = \frac{48·015}{32·43} = 1·48 \text{ m}$$

4·5·4. Eccentricity of the Resultant

$$e \quad \frac{B}{2} - \bar{x}_r = \frac{4}{2} - 1·48 = 2 - 1·48 = 0·52 < \frac{B}{6} = 0·66 \text{ m}$$

4·5·5. Pressure Distribution Below the base

$$\text{Maximum pressure} = \frac{W}{B \times 1}\left(1 + \frac{6e}{B}\right) = \frac{32·43}{4 \times 1}\left(1 + \frac{6 \times 0·52}{4}\right)$$

$$= \frac{32·43}{4}(1 \cdot 0·78) = 14·4 \; t/m^2 < 25 \; t/m^2$$

$$\text{Minimum pressure} = \frac{W}{B \times 1}\left(1 - \frac{6e}{B}\right) = \frac{32·43}{4}(1 - 0·78)$$

$$= \frac{32·43}{4} \times (0·22) = 1·78 \; t/m^2$$

4·6. Heel Design

Pressure ordinate at the junction of heel and stem

$$= 1·78 + \frac{(14·4 - 1·78)}{4} \times 2$$

$$= 1·78 + 12·62 \times \tfrac{2}{4}$$

$$= 8·09 \; t/m^2$$

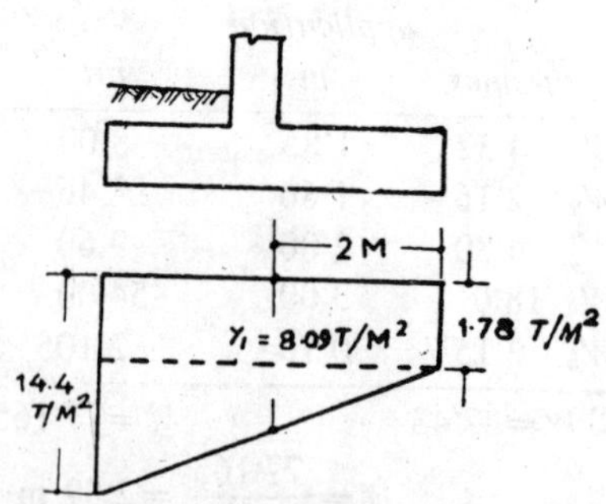

Bending moment at the junction of the stem

(*i*) due to upward forces

Forces due to	*Magnitude* tonnes	*Point of application* m	*Bending moment* mt
F_1=Rectangular pressure	$1{\cdot}78 \times 2 = 3{\cdot}56$	1·0	3·56
F_2=Triangular pressure	$\frac{1}{2} \times 6{\cdot}31 \times 2 = 6.31$	2/3	4·22
		Total upward moment=	7·78

(*ii*) due to downward forces

F_3 Backfill soil	$6 \times 2 \times 1{\cdot}5 = 18$	1·0	18·00
F_4 Self weight of heel	$0.50 \times 2 \times 2{\cdot}4 = 2{\cdot}4$	1·0	2 40
		Total downward moment=	20·40

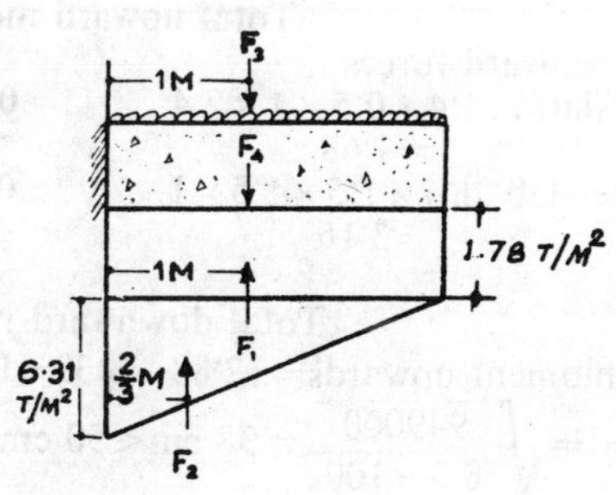

Net bending moment downwards $= 20{\cdot}4 - 7{\cdot}78 = 12{\cdot}62$ mt
$= 1262000$ cm kg

depth required $d = \sqrt{\dfrac{1262000}{8{\cdot}7 \times 100}} = 38{\cdot}2$ cm < 50 cm assumed

Use $d = 50$ cm and effective depth$=42{\cdot}5$ cm with a cover of 7·5 cm

4·6·1. Main Reinforcement

$$A_t = \frac{1262000}{1400 \times {\cdot}87 \times 42{\cdot}5} = 24\ 4 \text{ cm}^2$$

Use 20 mm ϕ at 12·5 cm c/c

4·6·2. Secondary Reinforcement

$$A_t = \frac{0{\cdot}15}{100} \times 100 \times 50 = 7{\cdot}5 \text{ cm}^2$$

Use 10 mm ϕ at 10 mm c/c

4·7. Toe Design

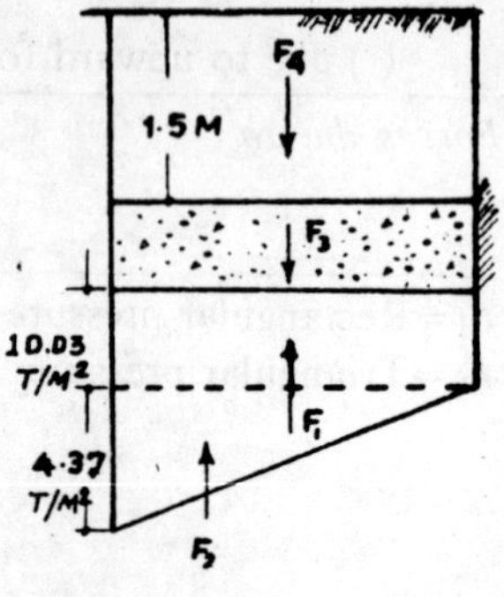

Pressure ordinate at the junction of toe and stem

$$=1{\cdot}78+12{\cdot}62\times\frac{2{\cdot}6}{4}$$

$$=1{\cdot}78+8{\cdot}25$$

$$=10{\cdot}03\ \text{t/m}^2$$

Bending moment at the junction of of the toe and stem

(*i*) due to upward forces

Force due to	*Magnitude* tonnes	*Point of application* m	*Moment* mt
F_1 rectangular pressure	$10{\cdot}03\times1{\cdot}4$ 14·14	0·7	10·05
F_2=triangular pressure	$\frac{1}{2}\times4{\cdot}37\times1{\cdot}4=3{\cdot}02$	$\frac{2}{3}\times1{\cdot}4$	2·83
		Total upward moment=	12·88 mt
(*ii*) due to downward forces			
F_3=self weight of slab	$1{\cdot}4\times0{\cdot}5\times1\times2{\cdot}4=1{\cdot}68$	0·7	1·18
F_4=earth above toe slab	$1{\cdot}4\times1{\cdot}5\times1{\cdot}5\times1=3{\cdot}16$	0·7	2·21
		Total downward moment	3·39 mt

Net bending moment upwards=12·88 – 3·39=9·49 mt

$$\text{depth required}=\sqrt{\frac{949000}{8{\cdot}7\times100}}=33\ \text{cm}<50\ \text{cm}$$

Use d = 50 cm and effective depth=42·5 cm with a cover of 7·5 cm.

4·7·1. Main Reinforcement

$$A_t\ \text{required}=\frac{949000}{1400\times{\cdot}87\times42{\cdot}5}=18{\cdot}2\ \text{cm}^2$$

Use 16 mm ϕ at 11 cm c/c

4·7·2. Secondary Reinforcement

$$A_t=\frac{{\cdot}15}{100}\times50\times100=7{\cdot}5\ \text{cm}^2$$

Use 10 mm ϕ at 10 cm c/c

4·8. Stability Investigation

$$\text{Factor of safety against overturning}=\frac{\text{stabilizing moment}}{\text{overturning moment}}=\frac{W\bar{x}}{P\bar{y}}$$

$$=\frac{32{\cdot}43\times2{\cdot}39}{11{\cdot}66\times2{\cdot}5}=\frac{77{\cdot}165}{29{\cdot}15}=2{\cdot}66>1{\cdot}5$$

$$\text{Factor of safety against sliding}=\frac{\text{Resisting force}}{\text{sliding force}}=\frac{\mu W}{P}$$

$$=\frac{0{\cdot}6\times32{\cdot}43}{11{\cdot}66}=1{\cdot}66>1{\cdot}5$$

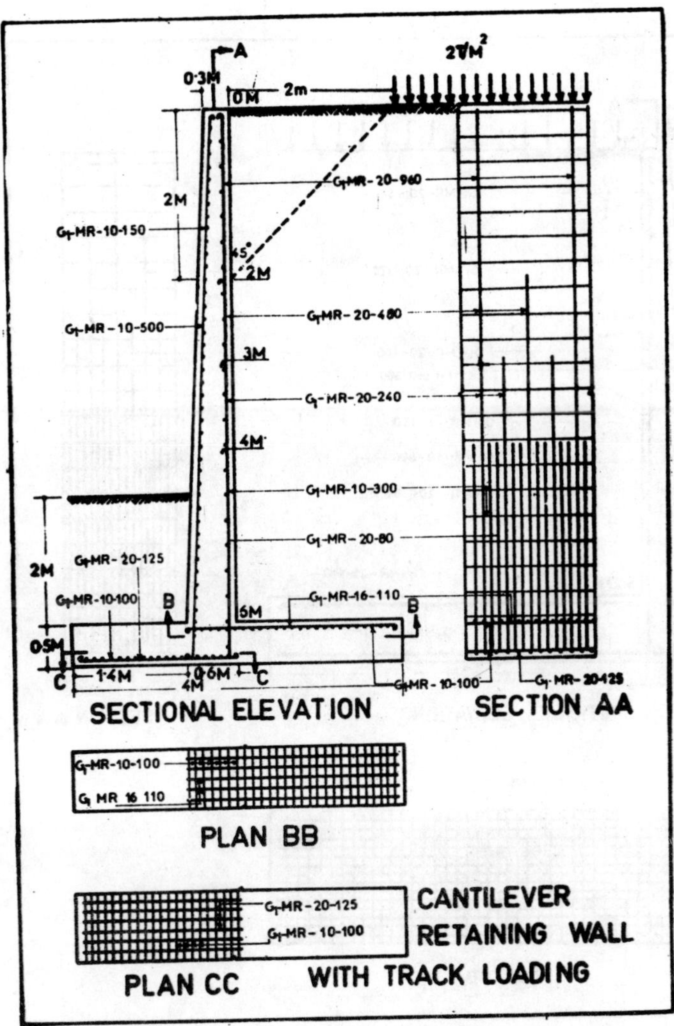

CANTILEVER RETAINING WALL WITH TRACK LOADING

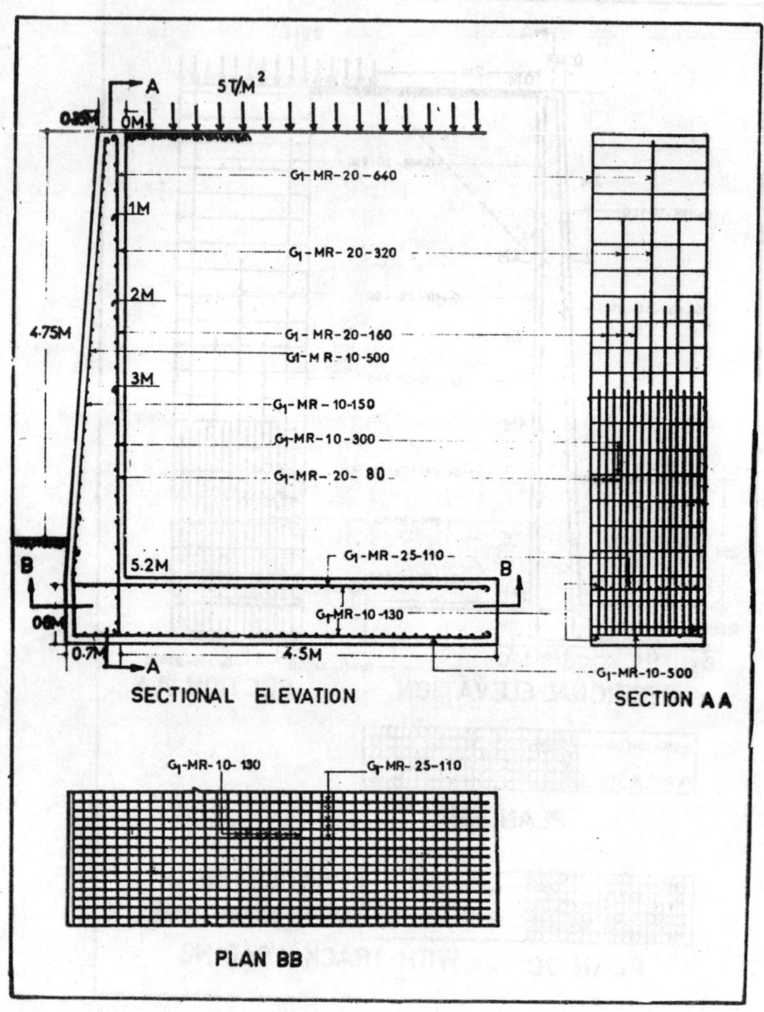

A
5T/M²
0.35M
0M
G1-MR-20-640
1M
G1-MR-20-320
2M
4.75M
G1-MR-20-160
G1-MR-10-500
3M
G1-MR-10-150
G1-MR-10-300
G1-MR-20-80
B
5.2M
G1-MR-25-110
B
G1-MR-10-130
0.6M
0.7M
4.5M
A
G1-MR-10-500
SECTIONAL ELEVATION
SECTION AA
G1-MR-10-130
G1-MR-25-110
PLAN BB

5 L-Shaped Retaining Wall with Uniformly Distributed Superimposed Load

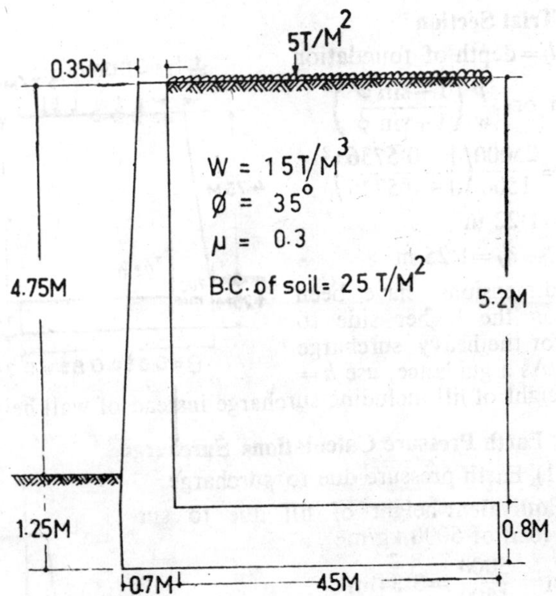

5·1. Data

Height of the fill to be retained by the wall = 4·75 m
Surcharge = 5 t/m² uniformly distributed.
Density of soil = 1500 kg/m³
Angle of internal friction = 35°
Coefficient of friction between the soil and concrete = 0·3
Bearing capacity of soil = 25 t/m²
IS—456—1964

5·2. Allowable Stresses

Concrete—grade *M* 150
Compressive stress for concrete in bending $\sigma_{cb} = 50$ kg/cm²
In direct compression = 40 kg/cm²
Shear stress, $q = 5$ kg/cm²
Bond stress $\sigma_b = 10$ kg/cm²
Steel-grade—1, $\sigma_{st} = 1400$ kg/cm²

5·2·1. Characteristic Strengths

$$\text{Modular ratio} = m = \frac{2800}{3\ \sigma_{cb}} = \frac{2800}{3 \times 50} = 18{\cdot}66$$

$$\frac{\sigma_{cb}}{n}=\frac{\sigma_{st}}{m(d-n)},\quad \frac{50}{n}=\frac{1400}{18{\cdot}66(d-n)}$$

Neutral axis $n=0{\cdot}4\ d$

Lever arm $=j_d=0{\cdot}87\ d$

Moment of resistance $=b\times n\times\frac{\sigma_{cb}}{2}\times j_d=b\times 0{\cdot}4\ d\times\frac{50}{2}\times 0{\cdot}87\ d$

$=8{\cdot}7\ bd^2$

5·3. Trial Section

d_f = depth of foundation

$$\nless 1\text{ m or }\frac{p}{w}\left(\frac{1-\sin\phi}{1+\sin\phi}\right)^2$$

$$=\frac{25000}{1500}\left(\frac{1-0{\cdot}5736}{1+0{\cdot}5736}\right)^2$$

$=1{\cdot}22$ m

Use $d_f=1{\cdot}25$ m

Trial dimensions have been taken on the higher side to allow for the heavy surcharge load. As a guidance, use h = total height of fill including surcharge instead of wall height h_2

5·3·1. Earth Pressure Calculations Surcharge

(1) Earth pressure due to surcharge

Equivalent height of fill due to surcharge load of 5000 kg/m²

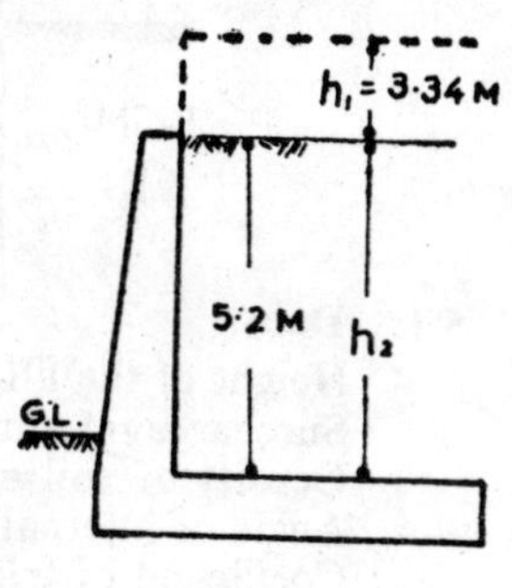

$$h_1=\frac{5000}{1500}=3{\cdot}34\text{ m}$$

$$P_1=\text{earth pressure}=wh_1\left(\frac{1-\sin\phi}{1+\sin\phi}\right)h_2$$

$=1500\times 3{\cdot}34\times 0{\cdot}271\times 5{\cdot}2=7060$ kg

Point of application $=\frac{5{\cdot}2}{2}=2{\cdot}6$ m

5·3·2. Back Fill

(2) Earth pressure due to backfill

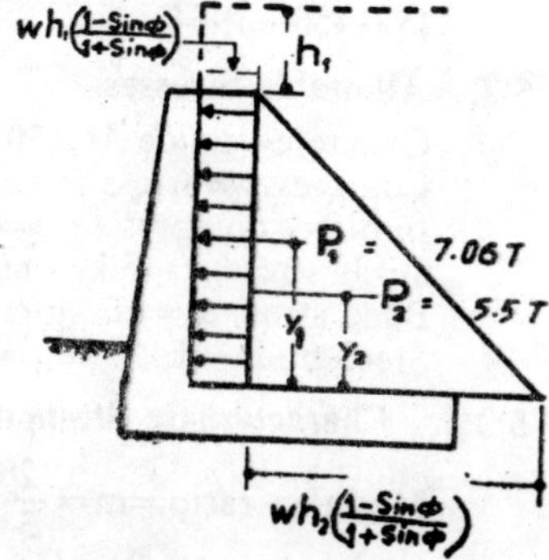

$$P_2=\tfrac{1}{2}\ wh_2\left(\frac{1-\sin\phi}{1+\sin\phi}\right)\times h_2$$

$=\frac{1}{2}\times 1500\times 5{\cdot}2\times 0{\cdot}271\times 5{\cdot}2$

$=5500$ kg

Point of application $=\frac{5{\cdot}2}{3}$

$=1{\cdot}75$ m

5·4. Stem Design

Maximum bending moment at the junction of the stem and the base$=P_1y_1+P_2y_2=7060\times\frac{5\cdot2}{2}+5500\times\frac{5\cdot2}{3}=18300+9500$

$=27{,}800$ m kg

5·4·1. Effective Depth

Effective depth $d_e=\sqrt{\frac{27800\times100}{8\cdot7\times100}}=56\cdot5$ cm<70 cm assumed

Use an overall depth of 70 cm and effective depth$=70-7\cdot5$ $=62\cdot5$ cm

5·4·2. Shear Check

Maximum shear at the junction$=P_1+P_2=7060+5500$
$=12560$ kg

Shear stress$=\frac{12560}{100\times\cdot87\times62\cdot5}=2\cdot31$ kg/cm$^2<5$ kg/cm

5·4·3. Bending Moment Variation

The bending moment in the stem is maximum at the junction and reduces to zero at the top of the wall. Hence, the main reinforcement can be suitably reduced towards top and the wall can be tapered and the thickness reduced to achieve economy.

Bending moment at any depth h'

$$M=wh_1\left(\frac{1-\sin\phi}{1+\sin\phi}\right)h'\times\frac{h'}{2}+wh'\left(\frac{1-\sin\phi}{1+\sin\phi}\right)\frac{h'}{2}\times\frac{h'}{3}$$

$$=1500\times3\cdot34\times0\cdot271\times\frac{(h')^2}{2}+1500\times\cdot271\times\frac{(h')^3}{6}$$

$$M=675\,(h')^2+67\cdot5\,(h')^3$$

5·4·4. Sample Calculations

Calculations for $h'=1$ m

Bending moment$=675\times1^2+67\cdot5\times1^3=675+67\cdot5=732\cdot5$ m kg

Effective depth needed $d_1=\sqrt{\frac{732\cdot5\times100}{100\times8\cdot7}}=9\cdot18$ cm

Overall depth provided $=30+\frac{40}{520}\times100=30+7\cdot7=37\cdot7$ cm

Effective depth provided$=37\cdot7-7\cdot5$ (cover)$=30\cdot2$ cm

Amount of steel required$=\frac{732\cdot5\times100}{1400\times\cdot87\times30\cdot2}=1\cdot99$ cm^2

Similarly these computations for different depths at an interval of 1 m are carried out as given in the table.

5·4·5. Curtailment of Main Reinforcement

depth	B.M.	Effective depth needed	Overall depth provided	Effective depth provided	Area of steel	Spacing of 20 mm	Spacing adopted
m	m kg	cm	cm	cm	cm²	cm	cm
0	0	0	30	22·5	—	—	64 cm
1	732·5	9·18	37·70	30·2	1·99	157	64 cm
2	3242·0	19·6	45·40	37·9	7·01	44·5	32 cm
3	7890	29·9	53·00	45·5	14·20	22·0	16 cm
4	15100	41·6	60·80	53·3	23·2	14·0	8 cm
5	26380	54·0	68·4	60·9	31·0	10·1	8 cm
5·2	27800	56·5	70·00	62·5	36·5	8·8	8 cm

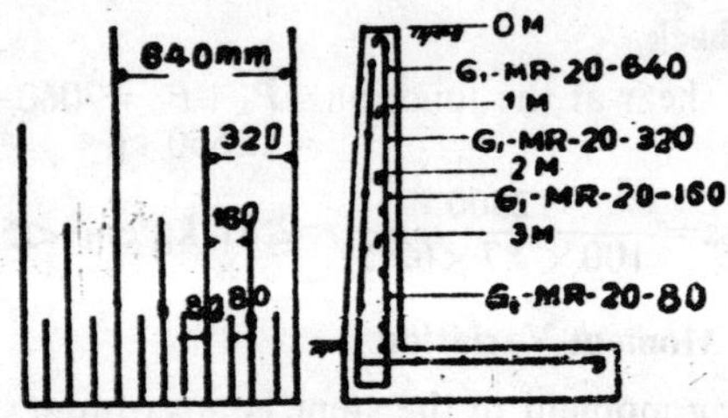

5·4·6. Secondary Reinforcement

Minimum reinforcement = 0·15% of cross-sectional area of concrete

Concrete area

$= \frac{1}{2} \times (30+70) \times 520$

$= 50 \times 520 = 26000$ cm²

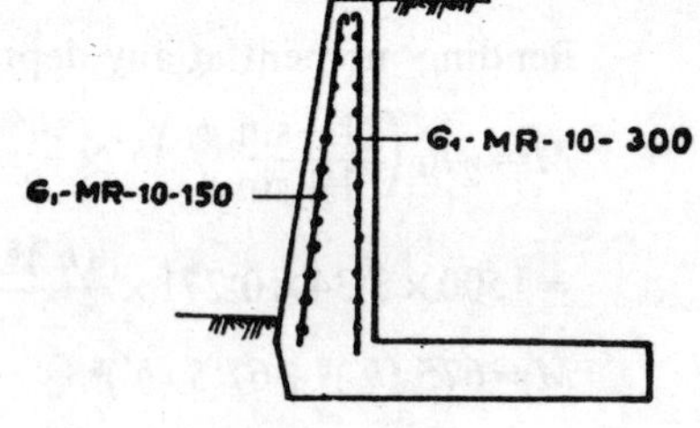

Area of secondary steel

$= 26000 \times \frac{0{\cdot}15}{100}$

$= 39{\cdot}0$ cm²

This reinforcement is distributed in the ratio of $\frac{1}{3}$ on the earth face and $\frac{2}{3}$ on the exposed face

Using 10 mm ϕ, spacing on the earth face $= \frac{0{\cdot}79}{13} \times 520 = 31{\cdot}6$cm

Use 30 cm c/c

Spacing on the exposed face $= \frac{0{\cdot}79}{26} \times 520 = 15{\cdot}8$ cm

Use 15 cm c/c

5·5 Base Design

Vertical loads—Consider one metre length of wall

(1) weight of stem rectangular portion

$= \frac{30}{100} \times 5{\cdot}2 \times 2400 = 3750$ kg

Point of application from 0,

$=4{\cdot}65$ m

(2) weight of stem triangular portion

$$=\frac{1}{2}\times\frac{40}{100}\times 5{\cdot}2\times 2400=2500 \text{ kg}$$

Point of application from 0

$=4{\cdot}93$ m

(3) weight of base $=5{\cdot}2\times\frac{80}{100}\times 2400$

$=10{,}000$ kg

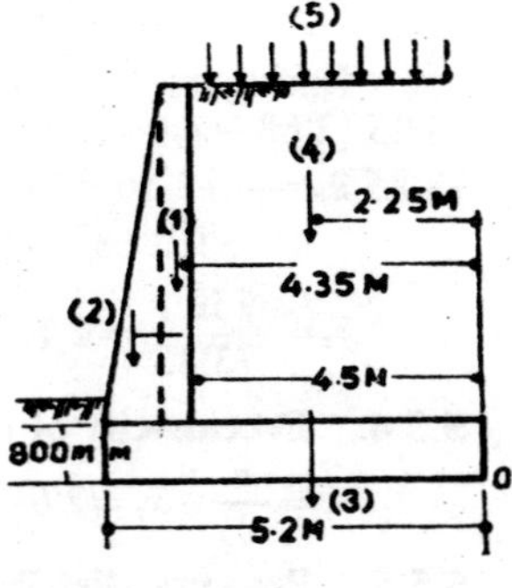

Point of application from $0=2{\cdot}6$ m

(4) weight of back fill $=1500\times 4{\cdot}5\times 5{\cdot}2=35100$ kg

Point of application from $0=2{\cdot}25$ m

(5) weight due to surcharge $=5000\times 4{\cdot}5=22500$ kg

Point of application $=2{\cdot}25$ m

5·5·1. Point of Application of Resultant Vertical Load

	Load Tonnes	*points of application* m	*Moment* mt
(1)	3.75	4.65	17·5
(2)	2·50	4·93	12·3
(3)	10·00	2.60	26·0
(4)	35·10	2·25	79·0
(5)	22·50	2·25	51·0
	$W=73{\cdot}85$		$M_v=185{\cdot}8$

$$\bar{x}_1=\frac{185{\cdot}8}{73{\cdot}85}=2{\cdot}52 \text{ m}$$

5·5·2. Point of Application of Horizontal Forces

Force due to	*Magnitude* Tonnes	*Point of application from base* m	*Moment* mt
Surcharge load P_1	7·060	3·4	24·0
Fill P_2	5·500	2·53	13·9
	$P=12{\cdot}56$		$M_h=37{\cdot}9$

$$\bar{y}=\frac{37{\cdot}9}{12{\cdot}56}=3{\cdot}01 \text{ m}$$

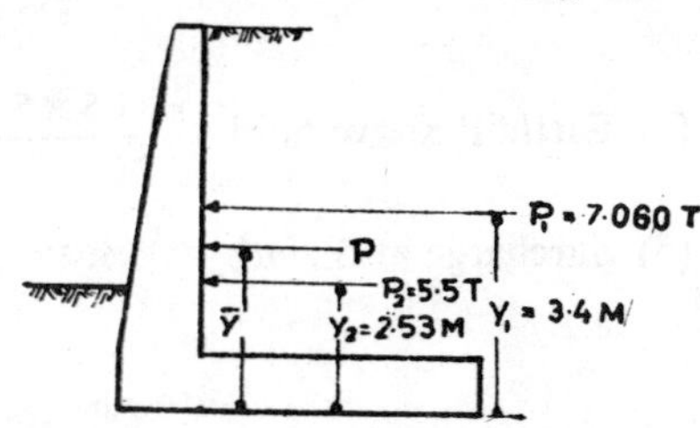

5·5·3. Point of Application of Resultant Force on the Base

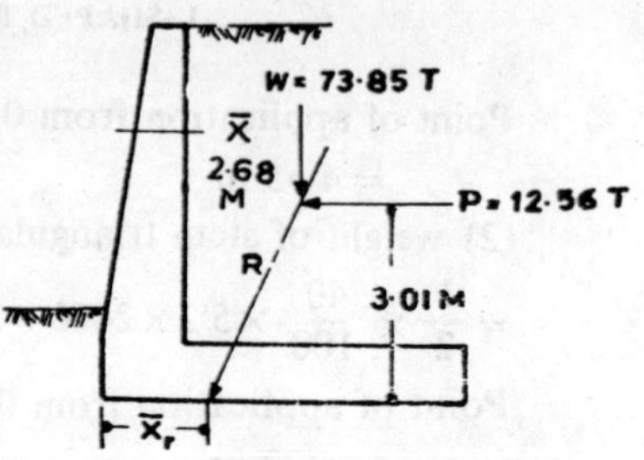

Taking moments about 0′

$73{\cdot}85\,(2{\cdot}68-\bar{x}_r)=12{\cdot}56\times3{\cdot}01$

$73{\cdot}85\,\bar{x}_r=-12{\cdot}56\times3{\cdot}01+73{\cdot}85\times2{\cdot}68-37{\cdot}9+197{\cdot}0$

$\bar{x}_r=\dfrac{159{\cdot}1}{73{\cdot}85}=2{\cdot}16$ m

5·5·4. Eccentricity of the Resultant

$$e=\frac{B}{2}-\bar{x}_r=2{\cdot}6-2{\cdot}16=0{\cdot}44<\frac{B}{6}=\frac{5{\cdot}2}{6}=0{\cdot}87\text{ m}$$

5·5·5. Pressure Distribution Under the Base

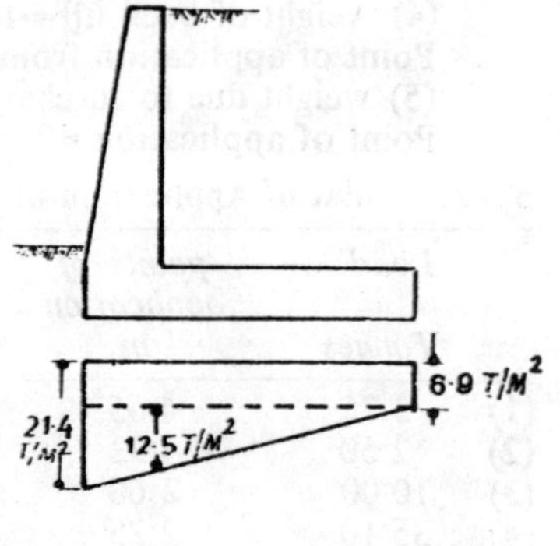

Maximum pressure

$$=\frac{W}{B\times1}\left(1+\frac{6e}{B}\right)$$

$$=\frac{73{\cdot}85}{5{\cdot}2\times1}\left(1+\frac{6\times0{\cdot}44}{5{\cdot}2}\right)$$

$=21{\cdot}4\ t/m^2$

Minimum pressure

$$=\frac{W}{B\times1}\left(1-\frac{6e}{B}\right)$$

$$=\frac{73{\cdot}85}{5{\cdot}2\times1}\left(1-\frac{6\times0{\cdot}44}{5{\cdot}2}\right)$$

$=6{\cdot}9\ t/m^2$

5·6. Heel Design

Forces acting on the heel :—

Bending moment at the junction of the stem due to

(*i*) Upward Forces

Force due to	*Magnitude tonnes*	*Point of application m*	*Bending moment mt*
(1) Rectangular pressure	$6{\cdot}9\times4{\cdot}5=31{\cdot}1$	2·25	68·0
(2) Triangular pressure	$\frac{1}{2}\times12{\cdot}5\times4{\cdot}5=28{\cdot}0$	1·5	42·0
		Total moment upward	=110·0
(*ii*) downward forces			
(3) Self weight heel	$0{\cdot}8\times4{\cdot}5\times\frac{2400}{1000}=8{\cdot}65$	2·25	19·50
(4) Earthfill above heel	$\frac{4{\cdot}5\times5{\cdot}2\times1\times1500}{1000}=35$	2·25	79·00
(5) Surcharge at 5 t/m^2	$5\times4{\cdot}5\times1=22{\cdot}5$	2·25	50·75
		Total downward moment	=149·25

Net bending moment downwards $=149{\cdot}25-110=39{\cdot}25$ *mt*

$$\text{Depth required } d=\sqrt{\frac{39{\cdot}25\times1000\times100}{100\times8{\cdot}7}}$$

$$=\sqrt{4520}=67{\cdot}5 \text{ cm} < 80 \text{ cm assumed}$$

Use a total depth of 80 cm and an effective depth of 72·5 cm allowing a cover of 7·5 cm.

5·6·1. Main Reinforcement

$$A_t=\frac{39{\cdot}25\times1000\times100}{1000\times{\cdot}87\times72{\cdot}5}=44{\cdot}4 \text{ cm}^2$$

Use 25mm ϕ at 11 cm c/c

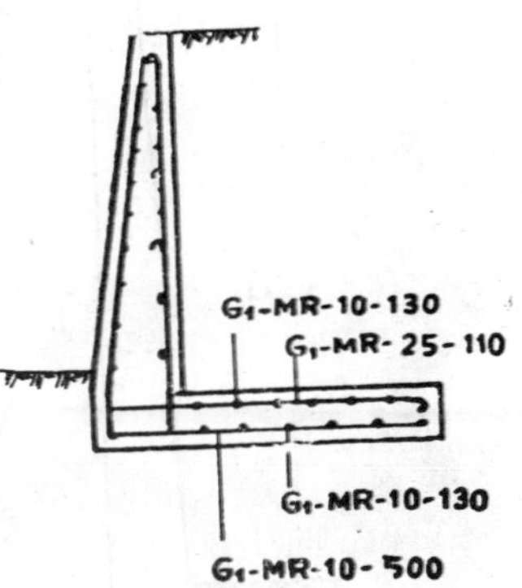

5·6·2. Secondary Reinforcement

$$A_t=\frac{0{\cdot}15}{100}\times100\times80$$

$$=12 \text{ cm}^2$$

10mm ϕ at 13 cm c/c on both faces

5·7. Stability Analysis

Factor of safety against overturning

$$=\frac{\text{Stabilizing moment}}{\text{Overturning moment}}=\frac{W\bar{x}}{P\bar{y}}$$

$$=\frac{73{\cdot}85\times2{\cdot}68}{12{\cdot}56\times3{\cdot}01}=\frac{197}{37{\cdot}9}=5{\cdot}2>1{\cdot}5$$

$$\text{Factor of safety against sliding}=\frac{\text{Resisting force}}{\text{Sliding force}}=\frac{\mu W}{P}$$

$$=\frac{0{\cdot}3\times73{\cdot}85}{12{\cdot}56}=1{\cdot}78>1{\cdot}5$$

L-Shaped Retaining Wall With Sloping Surcharge 6

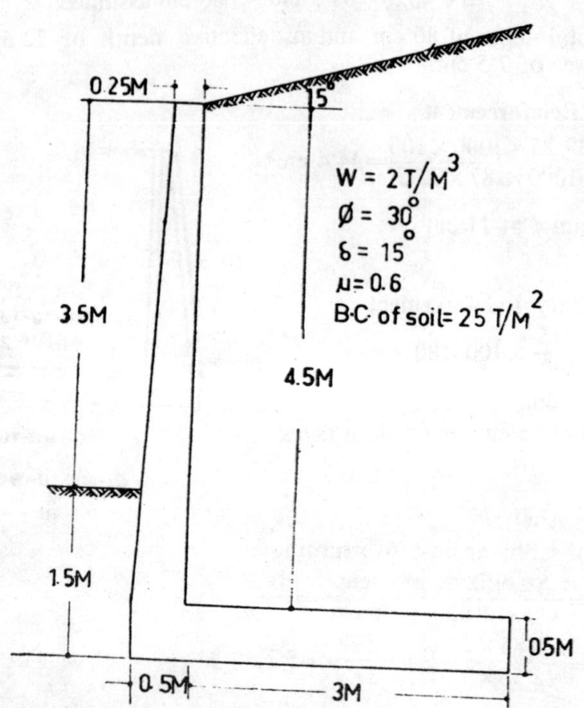

6·1. Data

Height of the fill to be retained by the wall = 3·5 m
Surcharge angle $\delta = 15°$
Density of the soil = 2000 kg/m³
Angle of internal friction = $\phi = 30°$
Coefficient of friction between the soil and concrete = 0·6
Bearing capacity of soil = 25 t/m²
I.S.—456—1964

Allowable Stresses

Concrete—Grade *M* 150
Compressive stress in bending = $\sigma_{cb} = 50$ kg/cm²
In direct compression = 40 kg/cm²
Shear stress $q = 5$ kg/cm²
Bond stress $\sigma_b = 10$ kg/cm²
Steel—Grade 1, $\sigma_{st} = 1400$ kg/cm²

6·2·1. Characteristic strengths

$$m=\frac{2800}{3\,\sigma_{cb}}=\frac{2800}{3\times50}=18{\cdot}66$$

$$\frac{\sigma_{cb}}{n}=\frac{\sigma_{st}}{m(d-n)},\quad \frac{50}{n}=\frac{1400}{18{\cdot}66(d-n)}$$

$$n=0{\cdot}4\,d$$

Lever arm $=j_d=0{\cdot}87\,d$

Moment of resistance $=b\times n\times\sigma_{cb}/2\times j_d=b\times0{\cdot}4d\times\sigma_{cb}/2\times0{\cdot}87d$

$$=8{\cdot}7\,bd^2$$

6·3. Trial Section

d_f=depth of foundation

$$\nless\frac{p}{W}\left(\frac{1-\sin\phi}{1+\sin\phi}\right)^2$$

$$\nless\frac{25000}{2500}\times\frac{1}{9}$$

$$\nless 1{\cdot}39\text{ m}$$

Provide $d_f=1{\cdot}5$ m

6·3·1. Earth Pressure Calculations

Earth Pressure due to Surcharge and back fill.

$$P=\frac{Wh^2}{2}\left[\cos\delta\cdot\frac{\cos\delta-\sqrt{\cos^2\delta-\cos^2\phi}}{\cos\delta+\sqrt{\cos^2\delta-\cos^2\phi}}\right]$$

$$=\frac{2000\times4{\cdot}5^2}{2}\times0{\cdot}376=7600\text{ kg.}$$

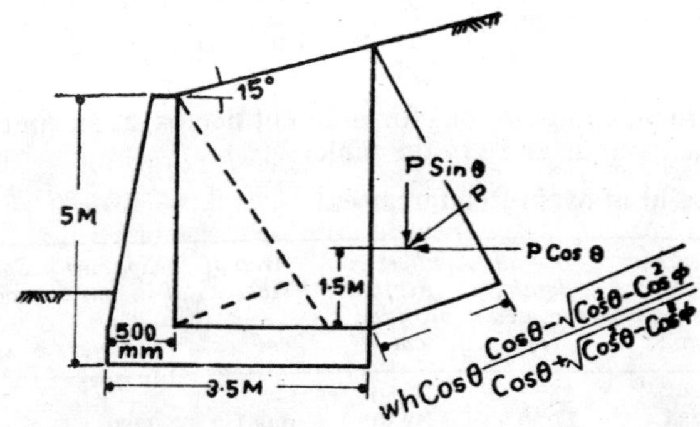

This pressure 'P' acts at an angle of 15° to the vertical face of retaining wall at a height of 1·5 m

Horizontal Component $=P\cos\delta=7600\times0{\cdot}965=7350$ kg

Vertical Component $=P\sin\delta=7600\times0{\cdot}259=1970$ kg

6·4. Stem Design

Maximum bending moment at the junction of the stem and the base $\quad P\cos\theta\times\bar{y}=7350\times1{\cdot}5=11000$ m kg

Effective depth of stem $= d_e = \sqrt{\dfrac{11000 \times 100}{100 \times 8.7}} = 35.8$ cm

Over all depth $= 35.8 + \text{cover} = 35.8 + 7.5 = 43.3$ cm

However 50 cm assumed shall be used.

Therefore the effective depth $= 42.50$ cm

6·4·1. Check For Shear

Maximum Horizontal Pressure $= 7350$ kg.

Shear Stress $= \dfrac{7350}{100 \times 0.87 \times 42.5} = 1.98$ kg/cm² $\lessdot$ 5 kg/cm²

6·4·2. Bending Moment Variation

The bending moment in the stem is maximum at the junction and reduces to zero at the top of the wall. Hence, main reinforcement can be suitably reduced towards top and the wall can be tapered and the thickness reduced to achieve economy.

Bending moment at any depth 'h'

$$M = Wh\left[\cos\delta \frac{\cos\delta - \sqrt{\cos^2\delta - \cos^2\phi}}{\cos\delta + \sqrt{\cos^2\delta - \cos^2\phi}}\right] \times \frac{h}{2} \times \cos\delta \times \frac{h}{3}$$

$$= \frac{2000 \times h^3}{2 \times 3} \times 0.376 \times 0.965 = 121\ h^3$$

Calculations for 'h' $= 0.5$ m

Bending Moment $= 121 \times 0.5^3 = 15.1$ m kg

Effective depth needed $= \sqrt{\dfrac{15.1 \times 100}{100 \times 8.7}} = 1.34$ cm

Over all depth provided $= 25 + \dfrac{25}{450} \times 50 = 27.8$ cm

Effective depth provided $= 27.8 - 7.5 = 20.3$ cm

Amount of steel required $A_t = \dfrac{BM}{t \times jd} = \dfrac{15.1 \times 100}{1400 \times 0.87 \times 20.3}$

$= 0.061$ cm²

Similarly these computations for different depths at an interval of ½ m are carried out as given in the table.

6·4·3. Curtailment of Main Reinforcement

'h' m	B.M. mkg	Effective depth needed cm	overall depth provided cm	Effective depth provided cm	Area of steel cm²	Spacing of 16 mm ϕ Bar cm	Spacing adopted cm
0	—	—	25	17·50	—	—	54
0·5	15·1	1·34	27·80	20·30	0·061	3280	,,
1·0	121·0	3·73	30·55	23·05	0·43	465	,,
1·5	408	6·85	33·30	25·80	1·30	154	,,
2·0	968	10·55	36·10	28·60	2·76	72·5	,,
2·5	1880	14·70	38·90	31·40	4·92	40·6	18
3·0	3262	19·45	41·65	34·15	7·86	25·45	,,
3·5	5200	24·40	44·45	36·95	11·55	17·30	9
4·0	7750	29·80	47·20	39·70	16·05	12·50	,,
4·5	11000	35·80	50·00	42·50	21·20	9·50	,,

6·4·4. Secondary Reinforcement

Minimum Reinforcement=0·15% of gross cross sectional area of concrete

Concrete area$=\frac{1}{2}(25+50)\times 450=16900$ cm²

Area of steel$=16900\times\frac{0\cdot 15}{100}=25\cdot 4$ cm²

This reinforcement is distributed in the ratio of $\frac{1}{3}$ on the earth face and $\frac{2}{3}$ on the exposed face.

Using 10 mm ϕ bars Spacing on earth face$=\frac{0\cdot 79}{8\cdot 45}\times 450$
$=42$ cm

Spacing on exposed face$=\frac{0\cdot 79}{16\cdot 9}\times 450=21$ cm

6·5. Base Design

Vertical loads—Consider one metre length of wall

1. Weight of stem rectangular portion

$$=\frac{25}{100}\times 4\cdot 5\times\frac{2400}{1000}=W_1=2\cdot 7 \text{ tonnes}$$

Point of application from front face$=x_1=0\cdot 25+\frac{0\cdot 25}{2}$
$=0\cdot 375$ m

2. Weight of stem triangular portion$=\frac{25}{2\times 100}\times 4\cdot 5\times\frac{2400}{1000}$
$=W_2=1\cdot 35$ tonnes

Point of application from front face $x_2=\frac{0\cdot 25}{3}\times 2=0\cdot 16$ m

3. Weight of base$=\frac{1}{2}\times 3\cdot 5\times\frac{2400}{1000}=W_3=4\cdot 2$ tonnes

Point of application from front face $x_3=1\cdot 75$ m

4. Weight of backfill excluding

surcharge $=4\cdot 5\times 3\times\frac{2000}{1000}$
$=W_4=27$ tonnes

Point of application from front face
$x_4=2$ m

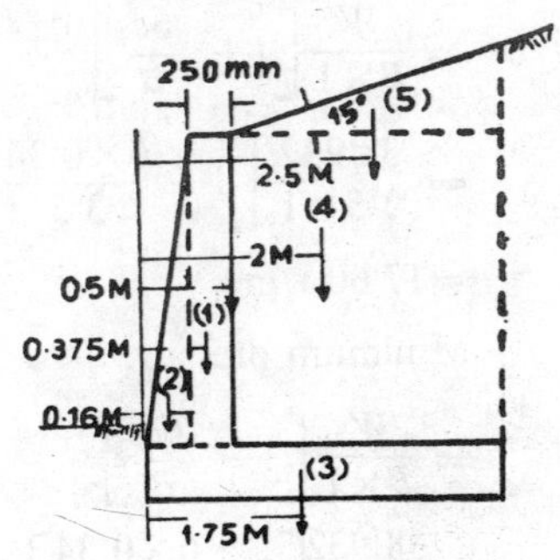

5. Weight of surcharge

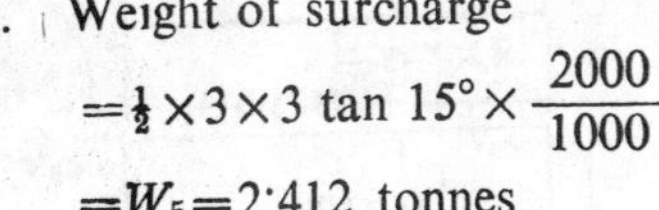

$$=\frac{1}{2}\times 3\times 3\tan 15^\circ\times\frac{2000}{1000}$$

$=W_5=2\cdot 412$ tonnes

Point of application from front face $x_5=0\cdot 5+3\times\frac{2}{3}=2\cdot 5$ m

6. Vertical component of pressure $P=W_6=P\sin\theta=1\cdot 97$ tonnes

This acts along the vertical face of the wall $x_6=0\cdot 5$ m

6·5·1. Point of Application of Resultant

	Load tonnes	Point of application m	Moment mt
1.	$W_1=2·7$	0·375	1·013
2.	$W_2=1·35$	0·16	0·216
3.	$W_3=4·20$	1·75	7·350
4.	$W_4=27·0$	2·00	54·000
5.	$W_5=2·412$	2·5	6·030
6.	$W_6=1·97$	0·5	0·985
	$\Sigma W=38·932$ t		$\Sigma M=69·584$ mt

$$\bar{x}=\frac{69·584}{38·932}=1·79 \text{ m}$$

6·5·2. Point of Application of the Resultant Force on the Base

Taking moments about '0'

$W(\bar{x}-\bar{x}_r)=P\bar{y}$

$38·932(1·79-\bar{x}_r)=7·35\times 2$

$$\bar{x}_r=\frac{38·932\times 1·79-7·35\times 2}{38·932}$$

$$=\frac{69·60-14·70}{38·932}$$

$$=\frac{54·90}{38·932}$$

$=1·41$ m

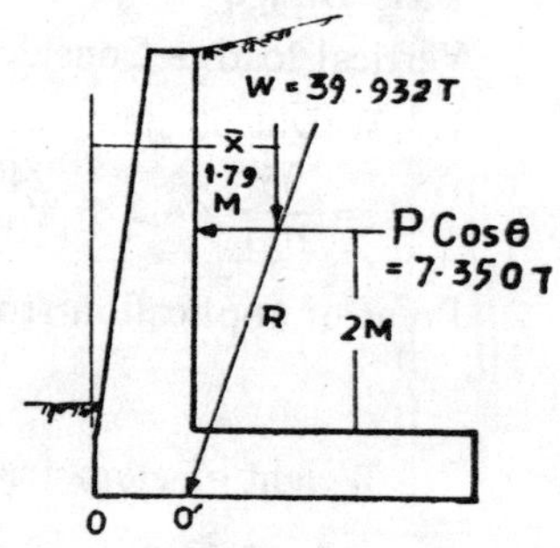

6·5·3. Eccentricity of Resultant

$$e=\frac{B}{2}-\bar{x}_r=\frac{3·5}{2}-1·41=1·75-1·41=0·34 \text{ m}<\frac{B}{6}=0·6 \text{ m}$$

6·5·4. Pressure Distribution under the Base

Maximum pressure

$$=\frac{W}{B\times 1}\left[1+\frac{6e}{B}\right]$$

$$=\frac{38·932}{3·5\times 1}\left[1+\frac{6\times 0·34}{3·5}\right]$$

$=17·60$ t/m²

Minimum pressure

$$=\frac{W}{B\times 1}\left[1-\frac{6e}{B}\right]$$

$$=\frac{38·932}{3·5}\left[1-\frac{6\times 0·34}{3·5}\right]$$

$=4·63$ t/m² (comp.)

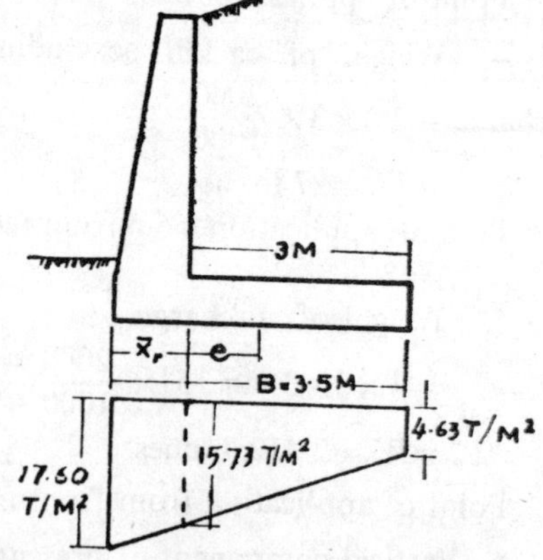

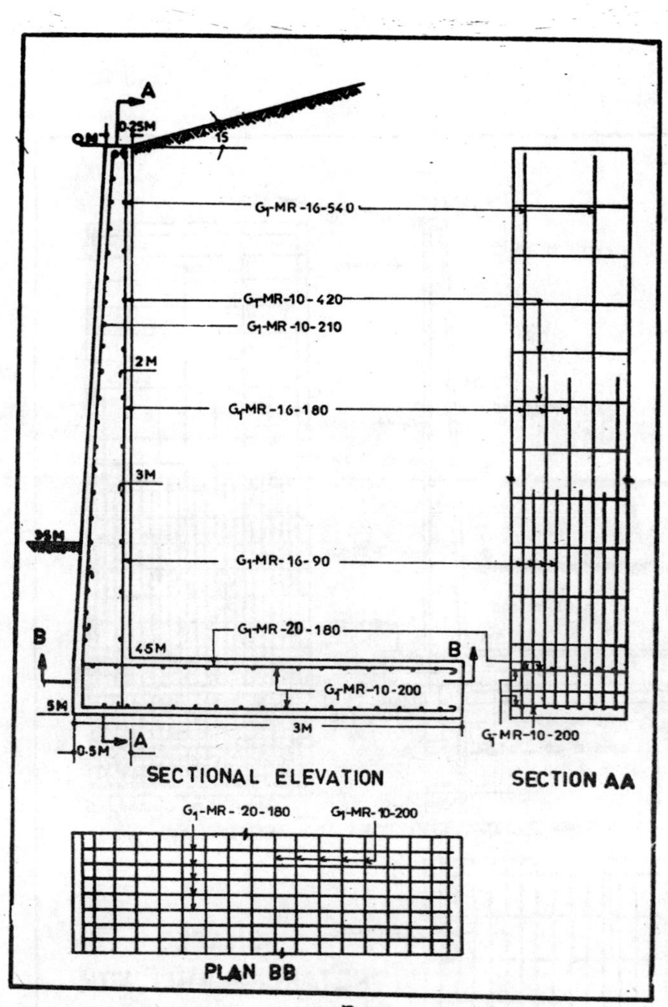
A
0.25M
0M
15
G1-MR-16-540
G1-MR-10-420
G1-MR-10-210
2M
G1-MR-16-180
3M
3.5M
G1-MR-16-90
G1-MR-20-180
B
4.5M
B
G1-MR-10-200
5M
3M
0.5M
A
G1-MR-10-200
SECTIONAL ELEVATION
SECTION AA
G1-MR-20-180
G1-MR-10-200
PLAN BB

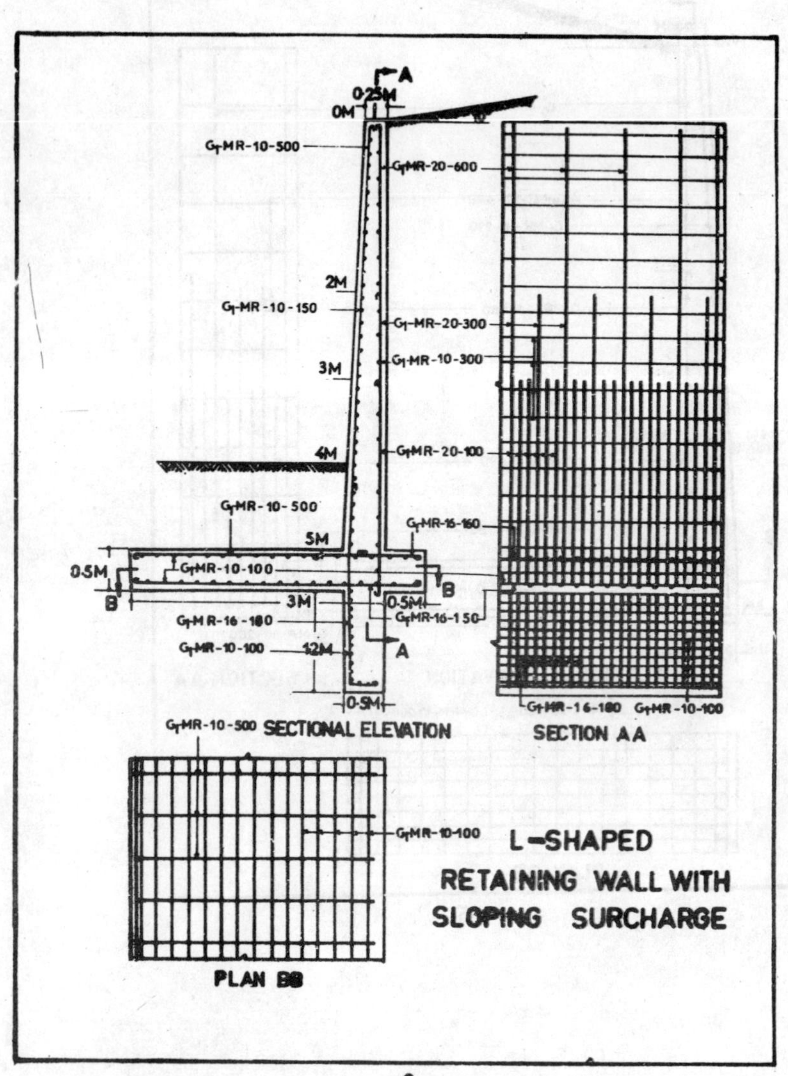

L-SHAPED RETAINING WALL WITH SLOPING SURCHARGE

6·6. Heel Design

Forces acting on the Heel. Bending moment at the junction

Force	*Magnitude tonnes*	*Point of application m*	*Moment mt*
F_1	$3\times4.5\times2000=27{\cdot}0$	1·5	+40·50
F_2	$3\times0{\cdot}5\times2400=3{\cdot}6$	1·5	+5·40
F_3	$4{\cdot}63\times3=13.89$	1·5	−20·84
F_4	$3\times\frac{1}{2}\times11{\cdot}10=16{\cdot}65$	1	−16·65

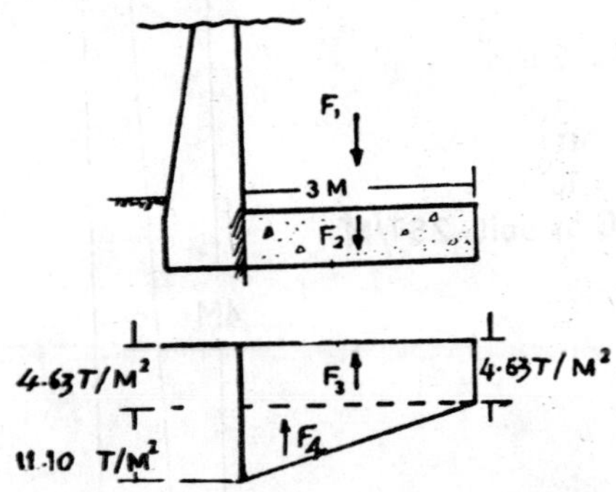

Net bending moment$=45{\cdot}9-37{\cdot}49=8{\cdot}51$ mt

$$\text{Depth required}=\sqrt{\frac{8510\times100}{100\times8{\cdot}7}}=31{\cdot}3\text{ cm}<50\text{ cm}$$

Depth provided$=50$ cm and Effective depth provided$=50-7{\cdot}5$ $=42{\cdot}5$ cm

6·6·1. Main Reinforcement

$$A_t=\frac{8510\times100}{1400\times0{\cdot}87\times42{\cdot}5}=16{\cdot}4\text{ cm}^2$$

Use 20 mm ϕ at 18 cm c/c

6·6·2. Secondary Reinforcement

$$A_t=\frac{0{\cdot}15}{100}\times100\times50=7{\cdot}5\text{ cm}^2$$

Use 10 mm ϕ at 20 cm c/c on both faces.

6·7. Stability Analysis

$$\text{Factor of safety against overturning}=\frac{\text{Stabilizing moment}}{\text{Overturning moment}}$$

$$=\frac{W\bar{x}}{P\bar{y}}=\frac{38{\cdot}932\times1{\cdot}79}{7{\cdot}35\times2}=\frac{69{\cdot}60}{14{\cdot}70}=4{\cdot}74>1{\cdot}5$$

$$\text{Factor of safety against sliding}=\frac{\text{Resisting force}}{\text{Sliding force}}=\frac{\mu W}{P}$$

$$=\frac{0{\cdot}6\times38{\cdot}932}{7{\cdot}35}=3{\cdot}18>1{\cdot}5$$

L-Shaped Retaining Wall With Sloping Surcharge 7

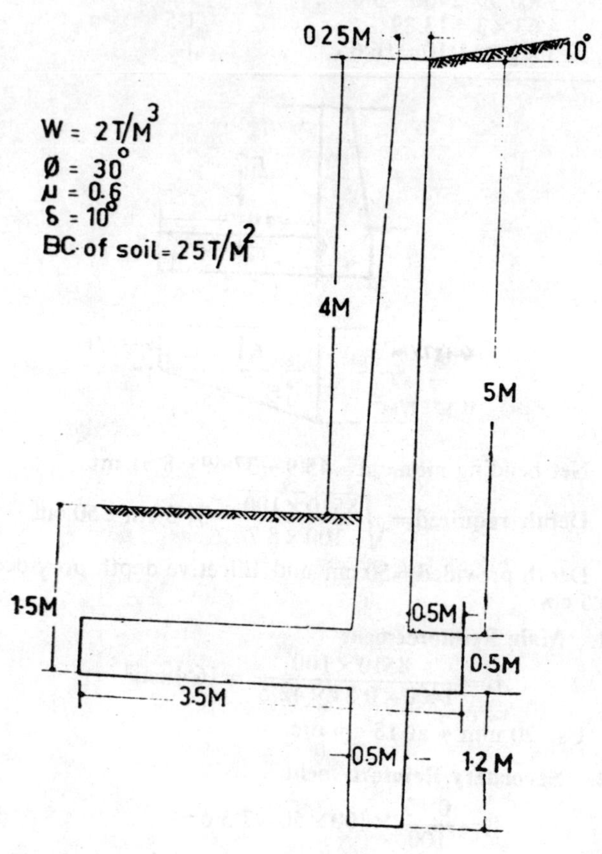

7·1. Data

Height of the fill to be retained by the wall = 4 m
Angle of Surcharge = $\delta = 10°$
Density of soil = 2 t/m³
Angle of internal friction = 30°
Coefficient of friction between the soil and concrete = $\mu = 0{\cdot}6$
Bearing capacity of soil = 25 t/m²
I.S. – 456 – 1964

7·2. Allowable Stresses

Concerte grade = M 150
$\sigma_{cb} = 50$ kg/cm², $q = 5$ kg/cm², $\sigma_b = 10$ kg/cm²
Steel grade – 1, $\sigma_{st} = 1400$ kg/cm²

7·2·1. Characteristic Strengths

Modular ratio$=m=\dfrac{2800}{3\times\sigma_{cb}}=\dfrac{2800}{3\times150}=18{\cdot}66$

$\dfrac{\sigma_{cb}}{n}=\dfrac{\sigma_{st}}{m(d-n)}$, $\dfrac{50}{n}=\dfrac{1400}{18{\cdot}66(d-n)}$

$n=0{\cdot}4\ d$

$j_d=$lever arm $\left(d-\dfrac{n}{3}\right)=0{\cdot}87\ d$

Moment of resistance of the balanced section

$=b\times n\times\dfrac{\sigma_{cb}}{2}\times j_d=b\times0{\cdot}4\ d\times\dfrac{50}{2}\times0{\cdot}87\ d=8{\cdot}7\ bd^2$

7·3. Trial Dimensions

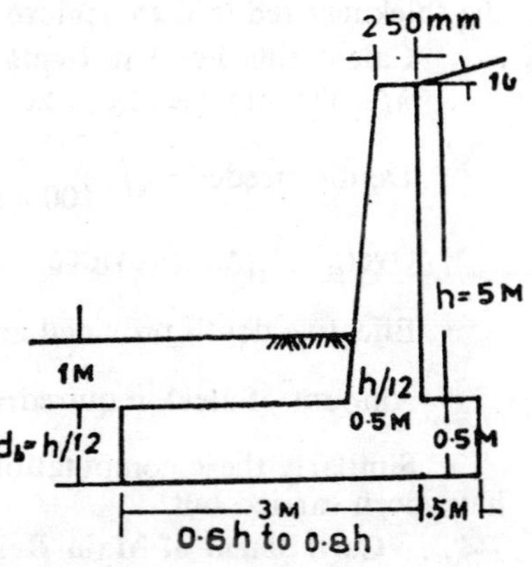

Depth of foundation

$\nless \dfrac{P}{w}\left(\dfrac{1-\sin\phi}{1+\sin\phi}\right)^2$

$=\dfrac{25}{2}\left(\dfrac{1}{3}\right)^2=1{\cdot}39$ m

Use $d_f=1{\cdot}5$ m

Thickness of base slab$\simeq\dfrac{h}{12}$

$=50$ cm (say)

Height of stem$=h=5$ m

Base width$=0{\cdot}6$ to $0{\cdot}8\ h$

try a width of 3·5 m

Stem thickness at top $\dfrac{h}{24}=25$ cm

Stem thickness at the junction$=\dfrac{h}{12}=50$ cm

7·3·1. Earth Pressure

Active earth pressure ordinate at any depth

$=p=wh\cos\delta\left[\dfrac{\cos\delta-\sqrt{\cos^2\delta-\cos^2\phi}}{\cos\delta+\sqrt{\cos^2\delta-\cos^2\phi}}\right]$

$=0{\cdot}353\ wh$

Total horizontal pressure$=P$

$=0{\cdot}353\ wh\times\frac{1}{2}\ h\times\cos\delta$

$={\cdot}353\times2000\times5\times\frac{5}{2}\times\cos 10^\circ$

$=8700$ kg$=8{\cdot}7$ tonnes

7·4. Stem Design

Maximum bending moment at any depth h'

$=2000\times0{\cdot}353\times h'\times h'/2\times\cos\delta\times h'/3=118\ h'^3$

Maximum bending moment at the junction of the stem and base

$=118\times5^3=14700$ m kg

7·4·1. Effective Depth of Stem

d_e at 5 m depth$=\sqrt{\dfrac{14700}{100\times 8{\cdot}7}}=41$ cm

Use $d_e=42{\cdot}5$ cm and overall depth$=42{\cdot}5+7{\cdot}5$ (cover)$=50$ cm

Shear stress at the junction$=\dfrac{8700}{100\times{\cdot}87\times 42{\cdot}5}=2{\cdot}23<5$ kg/cm^2

7·4·2. Bending Moment Variation

The bending moment in the stem is maximum at the junction and reduces to zero at the top of the wall. Hence main reinforcement can be suitably reduced towards top and the wall can be tapered and the thickness reduced to achieve economy.

Calculation For 1 m Depth

$M=118\times(1)^3=118$ m kg

Depth needed$=\sqrt{\dfrac{11800}{100\times 8{\cdot}7}}=3{\cdot}66$ cm

Overall depth provided$=25+\dfrac{25}{500}\times 100=30$ cm

Effective depth provided at 1 m depth$=30-7{\cdot}5=22{\cdot}5$ cm

Amount of steel required$=\dfrac{11800}{1400\times{\cdot}87\times 22{\cdot}5}=0{\cdot}43$ cm^2

Similarly these computations for different depths at 1 m intervals have been carried out.

7·4·3. Curtailment of Main Reinforcement

Depth from top *m*	*B.M.* *m kg*	*Effective depth needed* *cm*	*Effective depth provided* *cm*	*Area of steel* *cm*	*Spacing of 20 mm φ bars* *cm*	*Spacing used* *cm*
0	—	—	17·50	—		60 cm
1	118	3·66	22·50	0·43	743	60 cm
2	944	10·32	27·50	2·8	113	30 cm
3	3130	19·00	32·50	7·9	39·8	30 cm
4	9425	29·20	37·50	16·3	19·3	10 cm
5	14700	41·00	42·5	28·5	11	10 cm

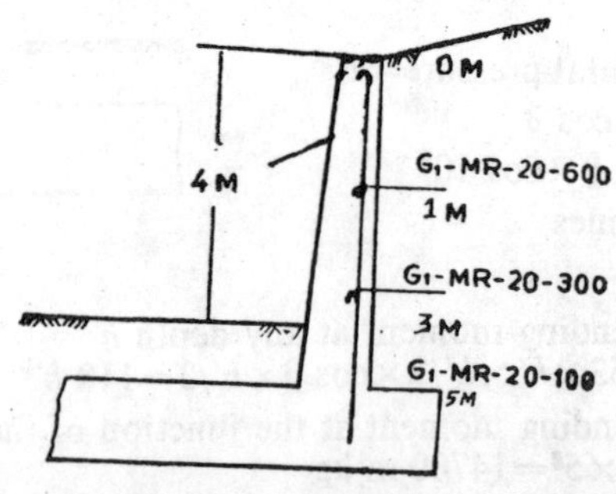

7·4·4. Secondary Reinforcement

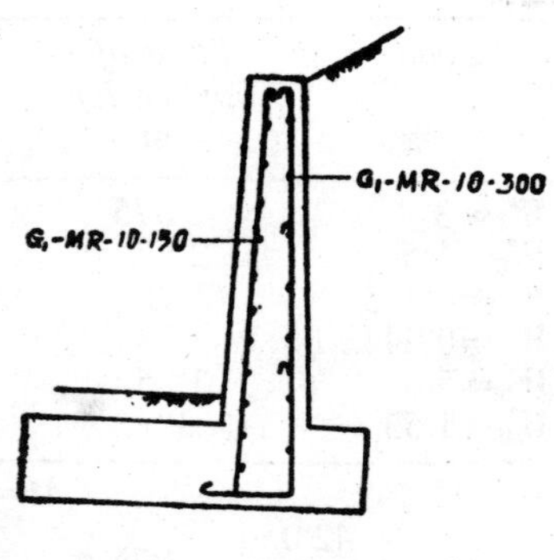

Minimum reinforcement needed

$=0{\cdot}15\%$ of concrete area

$$=\frac{0{\cdot}15}{100}\times500\times\left(\frac{25+50}{2}\right)$$

$=28{\cdot}1\ \text{cm}^2$

Provide 1/3 of this on earth face and 2/3 on the exposed face

Use 10 mm ϕ bars

Spacing on the earth face

$$=\frac{0{\cdot}79}{9{\cdot}36}\times500=33{\cdot}6\ \text{cm}$$

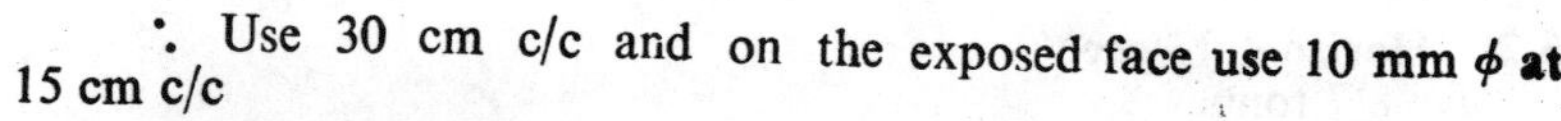

$\therefore$ Use 30 cm c/c and on the exposed face use 10 mm ϕ at 15 cm c/c

7·5. Base Design

Loads : Consider 1 m length of wall

(*i*) Weight of stem, rectangular portion

$$=\frac{25}{100}\times1\times5\times2400=3000\ \text{kg},\quad W_1=3\ \text{tonnes}$$

Point of application from the point 0, $x_1=2{\cdot}875$ m

(*ii*) Weight of stem triangular portion

$$=\frac{1}{2}\times\frac{25}{100}\times1\times5\times2400=1500\ \text{kg},\quad W_2=1{\cdot}5\ \text{tonnes}$$

Point of application from 0, $x_2=2{\cdot}67$ m

(*iii*) Weight of base slab

$$=3{\cdot}5\times\frac{50}{100}\times1\times2400=4375\ \text{kg},\quad W_2=4{\cdot}375\ \text{tonnes}$$

Point of application from toe$=x_3=1{\cdot}75$ m

(*iv*) Backfill soil weight triangular portion $=\frac{1}{2}\times0{\cdot}5\tan10^\circ\times2000\times0{\cdot}5$

$W_4=44{\cdot}2$ kg $=0{\cdot}0442$ tonnes

Point of application from toe

$x_4=3{\cdot}33$ m

(*v*) Backfill soil weight rectangular portion

$$W_5=5\times\frac{50}{100}\times1\times2000=5000\ \text{kg}$$

$=5$ tonnes

Point of application from toe$=3{\cdot}25$ m

(*vi*) Vertical component of earth

$$\text{pressure}=P\sin\ \delta=0{\cdot}353\times\frac{200\times5^2}{2}\times\sin10^\circ$$

$W_6=1530$ kg $=1{\cdot}53$ tonnes

Point of application$=x_6=3{\cdot}5$ m

7·5·1. Point of Application of Resultant Vertical Load

Load tonnes	Point of application m	Moment mt
$W_1=3$	2·875	8·625
$W_2=1·5$	2·67	4·025
$W_3=4·375$	1·75	7·660
$W_4=0·0442$	3.33	0·146
$W_5=5$	3·25	16·250
$W_6=1·53$	3·5	5·350
$W=15·449$		$M=42·056$

$$\bar{x}=\frac{42·056}{15·449}=2·74 \text{ m}$$

7·5·2. Horizontal Force

$P=8·7$ tonnes

Point of application from base $\frac{5}{3}+0·5=2·17$ m

7·5·3. Point of Application of the Resultant on the Base

$W(\bar{x}-\bar{x}_r)=P\bar{y}$

$15·449(2·74-\bar{x}_r)=8·7\times 2·17$

$$\bar{x}_r=\frac{2·74\times 15·449-8·7\times 2·17}{15·449}=\frac{42·056-18·9}{15·449}=\frac{23·156}{15·449}=1·52 \text{ m}$$

7·5·4. Eccentricity

$$e=\frac{B}{2}-\bar{x}_r=1·75-1·52=0·23 \text{ m}$$

7·5·5. Pressure Distribution Under Base

Maximum pressure

$$=\frac{W}{B\times 1}\left(1+\frac{6e}{B}\right)$$

$$=\frac{15·449}{3·5}\left(1+6\times\frac{0·23}{3·5}\right)$$

$=6·16<25$ t/m²

Minimum pressure

$$=\frac{W}{B\times 1}\left(1-\frac{6e}{B}\right)$$

$$=\frac{15·449}{3·5}\left(1-\frac{6\times 0·23}{3·5}\right)$$

$=2·68$ t/m²

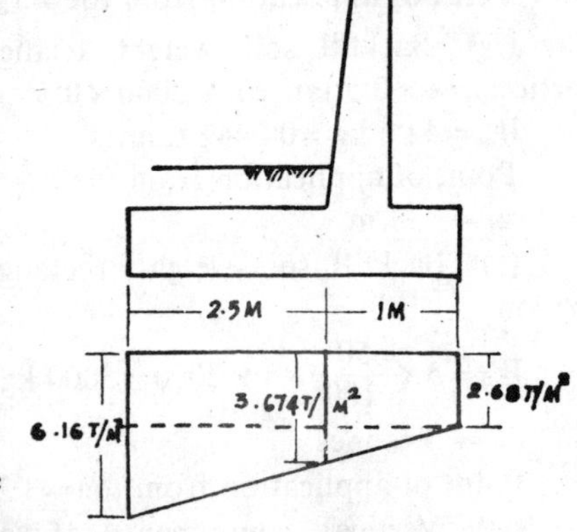

7·6. Design of Toe

Pressure ordinate at the junction

$$=2·68+(6·16-2·68)\frac{1}{3·5}=2·68+0·994=3·674 \text{ t/m}^2$$

Forces acting on the toe :

Net bending moment at the junction

Force	*Magnitude tonnes*	*Pt. of application m*	*Moment mt*
F_1	$1\times2{\cdot}5\times1\times2=5$	1·25	+6·25
F_2	$0{\cdot}5\times2{\cdot}5\times1\times2{\cdot}4=3$	1·25	+3·75
F_3	$3{\cdot}674\times2{\cdot}5\times1=9{\cdot}2$	1·25	−11·5
F_4	$\frac{1}{2}\times2{\cdot}5\times2{\cdot}486\times1=3{\cdot}1$	1·66	−5·1

Net BM downward = 6·7 mt

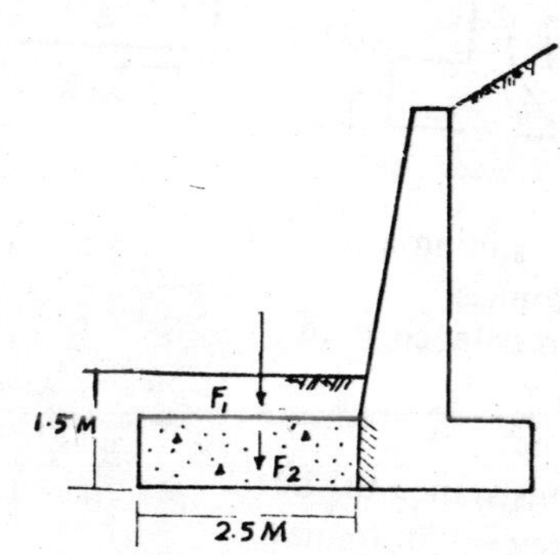

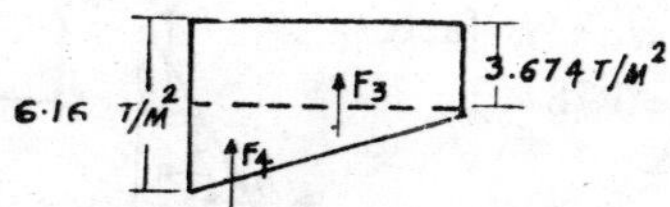

depth required $d_e=\sqrt{\dfrac{670000}{100\times8{\cdot}7}}=27{\cdot}8$ cm

effective depth provided = 50 − 7·5 = 42·5 cm

7·6·1. Main and Secondary Reinforcement

$$A_t=\frac{670000}{1400\times{\cdot}87\times42{\cdot}5}=13\text{ cm}^2$$

Use 16 mm ϕ at 15 cm c/c

$$A_t=\frac{0{\cdot}15}{100}\times100\times50=7{\cdot}5\text{ cm}^2$$

Use 10 cm ϕ at 10 cm c/c

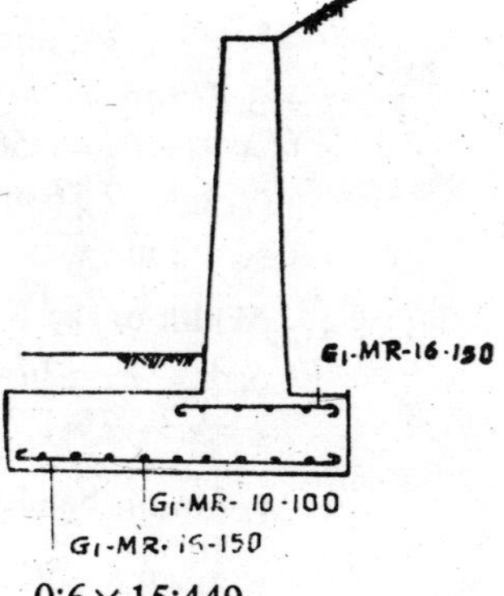

7.7· Stability Analysis

Factor of safety against overturning

$$=\frac{\text{stabilising moment}}{\text{overturning moment}}=\frac{42{\cdot}056}{18{\cdot}9}$$

$=2{\cdot}22>1{\cdot}5$

Factor of safety against sliding $=\dfrac{\mu W}{P}=\dfrac{0{\cdot}6\times15{\cdot}449}{8{\cdot}7}$

$=1{\cdot}12<1{\cdot}5$ (unsafe)

Therefore key is to be provided.

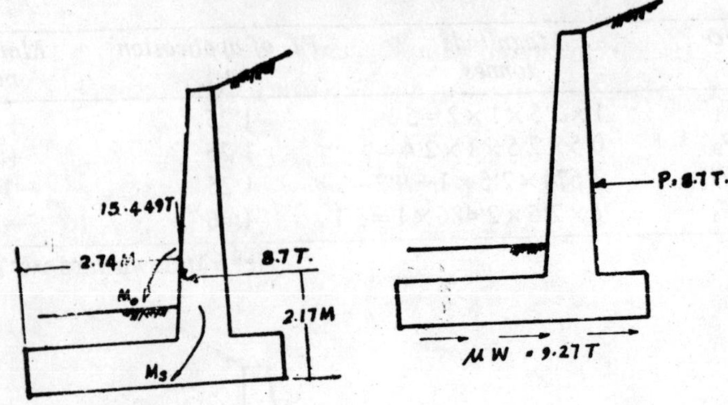

7·8. Key Design

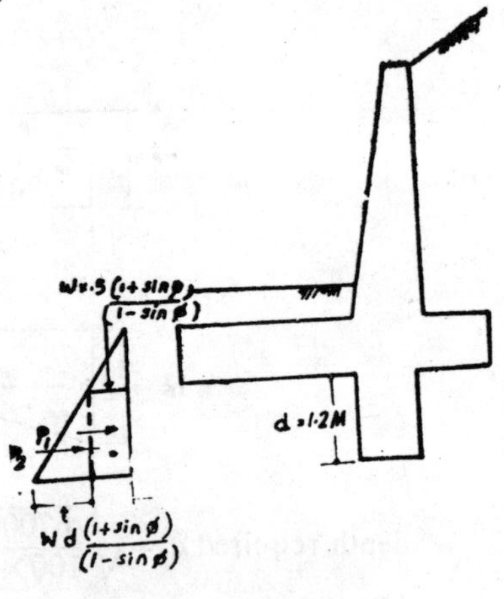

Resistance to sliding available $= \mu W = 9{\cdot}2$ tonnes

Minimum resistance e required $1{\cdot}5\ P = 1{\cdot}5 \times 8{\cdot}7$
$= 13{\cdot}05$ tonnes

Additional resistance to be provided by the key $= 3{\cdot}78$ tonnes

Magnitude of passive earth pressure required with a factor of safety of $2 = P_p = 3{\cdot}78 \times 2$
$= 7{\cdot}56$ tonnes

Passive earth pressure

$$P_1 = wh\left(\frac{1+\sin\phi}{1-\sin\phi}\right)\times d$$

$= 2000 \times 0{\cdot}5 \times 3 \times d$

$= 3000\ d = 3\ d$ tonnes

$$P_2 = \tfrac{1}{2} \times wd\left(\frac{1+\sin\phi}{1\ \ \sin\phi}\right)\times d = \tfrac{1}{2} \times 2000 \times d \times 3 \times d = 3000\ d^2$$

$= 3\ d^2$ tonnes

$P_1 + P_2 = P_p = 7{\cdot}56$ tonnes

$3d + 3d^2 = 7{\cdot}56$ and $d = 1{\cdot}15$ m

Use $1{\cdot}2$ m

7·8·1. Width of the Key Wall

Loads : $P_1 = 3d = 3 \times 1{\cdot}2 = 3{\cdot}6$ tonnes

$P_2 = 3d^2 = 3 \times 1{\cdot}2 \times 1{\cdot}2 = 4{\cdot}32$ tonnes

Maximum bending moment $= P_1 \times \frac{1{\cdot}2}{2} + P_2 \times \frac{1{\cdot}2}{3}$

$= 3{\cdot}6 \times 0{\cdot}6 + 4{\cdot}32 \times 0{\cdot}8 = 2{\cdot}16 + 3{\cdot}456 = 5{\cdot}616$ mt

Effective depth required $= \sqrt{\frac{561600}{8{\cdot}7 \times 100}} = 25{\cdot}6$ cm

Use 50 cm overall with 42·5 cm as used for stem

7·8·2. Main Reinforcement and Secondary Reinforcement

$$A_t \text{ needed} = \frac{561600}{1400 \times \cdot 87 \times 42 \cdot 5} = 10 \cdot 9 \text{ cm}^2$$

Use 16 mm ϕ at 18 cm c/c

$$A_t = 0 \cdot 15\% \text{ of concrete area} = 0 \cdot 15 \times \frac{50}{100} \times 100 = 7 \cdot 5 \text{ cm}^2$$

Use 10 mm ϕ at 10 cm c/c

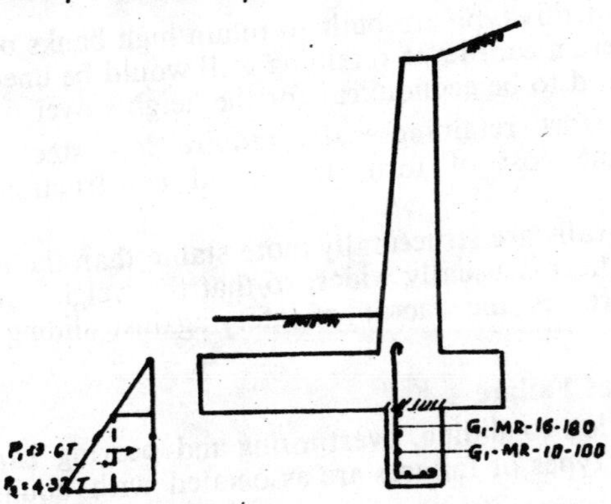

II. COUNTERFORT RETAINING WALLS

General Information 1

1. General Information

Walls of this type are built to retain high banks of earth or other sub-soils where a cantilever retaining wall would be uneconomical. The walls are found to be economical for the heights over 6 m.

Counterfort retaining walls require less steel than cantilever walls, but the cost of form work and construction is appreciably greater

These walls are structurally more stable than the cantilever walls because, the heel is usually wider, so that the weight of the retained material increases the factor of safety against sliding and tilting or overturning.

1.1. Types of Failure

In addition to sliding, overtnrning and bearing capacity failures the following types of failures are associated with counterfort retaining walls.

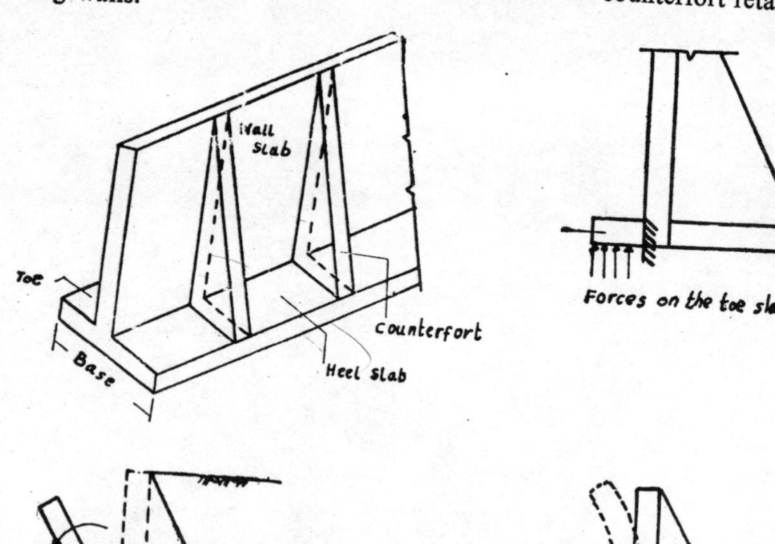

Forces on the toe slab

Tearing of wall slab

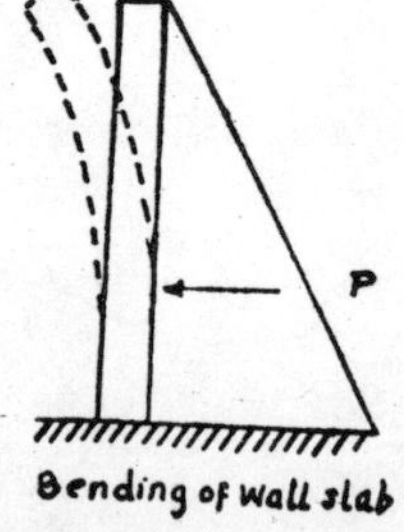

Bending of wall slab

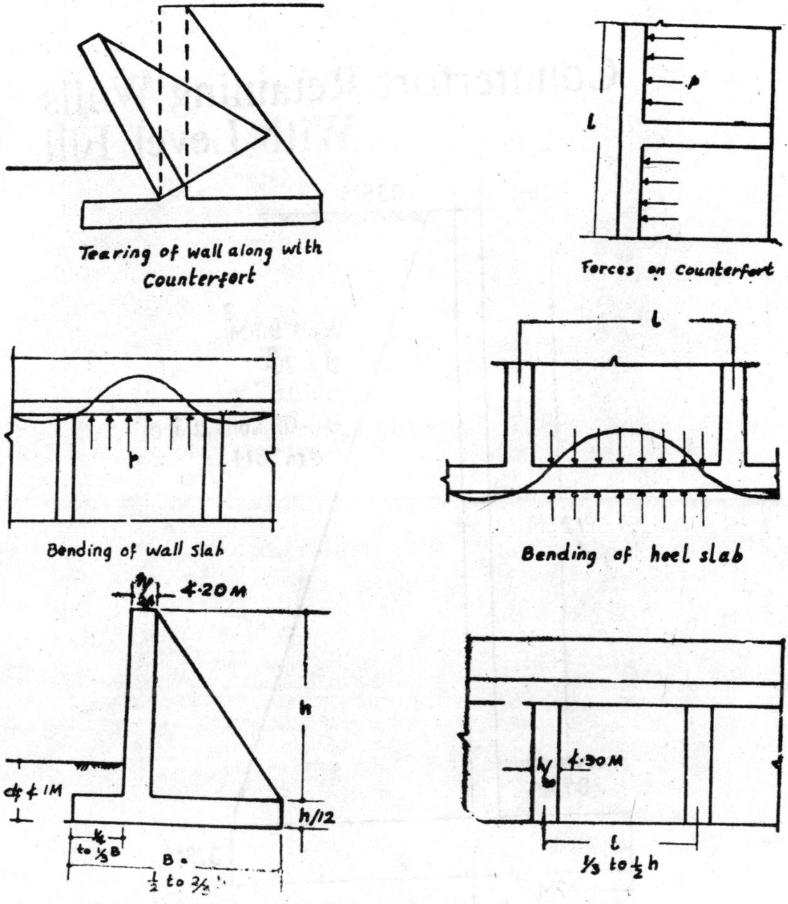

Tearing of wall along with counterfort

Forces on counterfort

Bending of wall slab

Bending of heel slab

1·3. Components of the Wall

(*i*) wall slab (*ii*) Counterfort (*iii*) Base slab consisting of heel and toe

1·4. Design

The wall slab is designed as a continuous slab spanning over a number of counterforts. The counterfort is designed as a cantilever beam fixed to the base.

The heel slab is designed as a continuous slab spanning across the counterforts. The toe slab is designed as a cantilever slab fixed at the junction of the wall slab and toe.

1·5. Trial Dimensions

Depth of foundation shall not be less than 1 m or depth required by Rankine's formula

$$d_f = \frac{p}{w}\left(\frac{1-\sin\phi}{1+\sin\phi}\right)^2$$

Spacing of counterforts, $l = \frac{1}{3}$ to $\frac{1}{2}\,h$

Width of counterfort$=\frac{h}{24}$ or $\nless$ 20 cm

Counterfort Retaining Walls With Level Fill 2

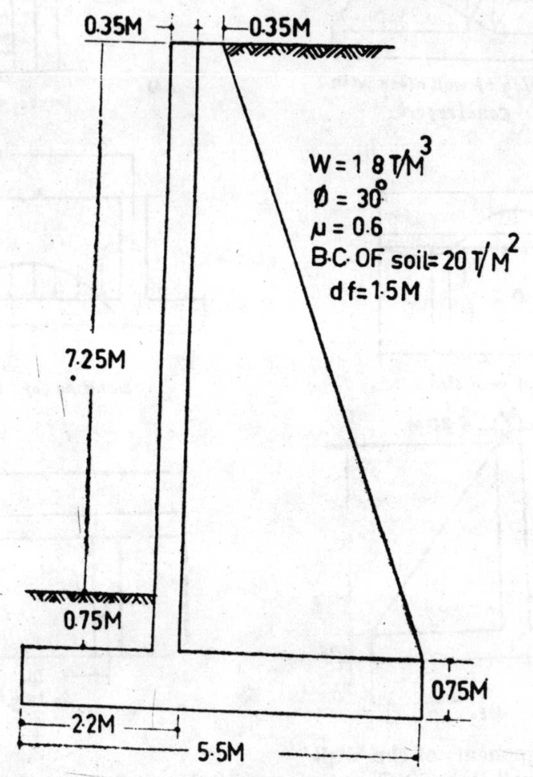

2·1. Data

Height of the fill to be retained by the wall = 7·25 m
No surcharge and the fill is level. Weight of earth = 1800 kg/m³
Angle of internal friction = 30°
Coefficient of friction between soil and base slab = 0·6
Bearing capacity of soil = 20 t/m²
Depth of foundation = 1·5 m
Material available *M* 150 concrete and Grade-I steel
I.S. 456—1964

2·2. Allowable Stresses

Concrete—grade *M* 150
Compressive stress of concrete in bending = σ_{cb} = 50 kg/cm²
shear stress = q = 5 kg/cm²
bond stress = 10 kg/cm²
Steel—grade—I σ_{st} = 1400 kg/cm²

2·2·1. Characteristic Strengths

Modular ratio $= m = \dfrac{2800}{3\sigma_{cb}} = \dfrac{2800}{3\times50} = 18{\cdot}66$

$\dfrac{\sigma_{cb}}{n} = \dfrac{\sigma_{st}}{m(d-n)}, \quad \dfrac{50}{n} = \dfrac{1400}{18{\cdot}66(d-n)}, \; n = 0{\cdot}4d$

Lever arm $= jd = 0{\cdot}87d$

Moment of resistance $= b\times n\times\dfrac{c}{2}\times jd = b\times0{\cdot}4d\times\dfrac{50}{2}\times0{\cdot}87d$

$= 8{\cdot}7bd^2$

2·3. Trial Dimensions

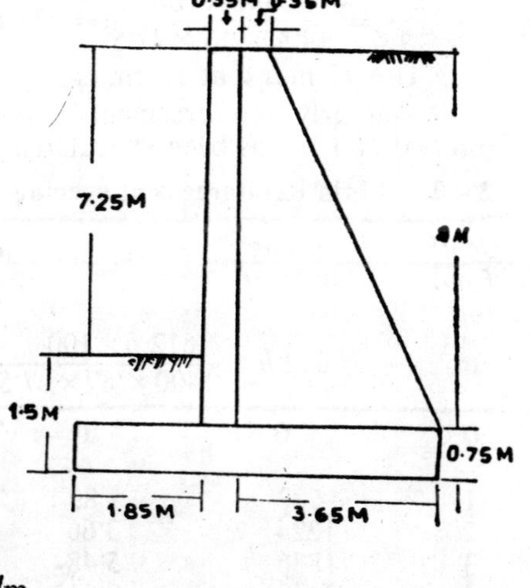

Base width $= B = \frac{1}{2}$ to $\frac{2}{3}h$
$= 5{\cdot}5$ m

Toe width $= \frac{1}{3}$ of $B = 1{\cdot}85$ m

Heel width $= \frac{2}{3}$ of $B = 3{\cdot}65$ m

Wall slab thickness $= \frac{1}{24}h$
$= 35$ cm

Base slab thickness $= \frac{1}{12}h$
$= 75$ cm

Spacing of counterforts $= \frac{1}{2}$ to $\frac{1}{3}h = 3{\cdot}5$ m

2·3·1. Earth Pressure Calculations

Max. earth pressure due to level fill $= p = wh\left(\dfrac{1-\sin\phi}{1+\sin\phi}\right)$

$= 1800\times8\times\frac{1}{3} = 600h = 4800$ kg/m

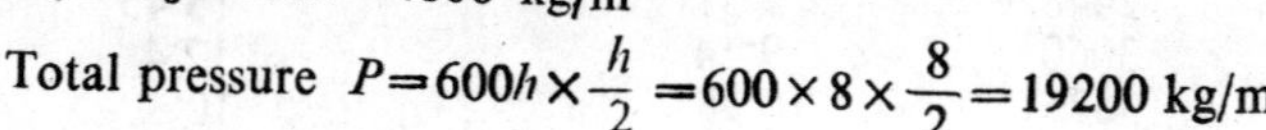

Total pressure $P = 600h\times\dfrac{h}{2} = 600\times8\times\dfrac{8}{2} = 19200$ kg/m

Point of application of this force P from the base $= \bar{y} = (\frac{8}{3}+0{\cdot}75)$
$= 2{\cdot}67+0{\cdot}75 = 3{\cdot}42$ m

2·4. Wall Slab Design

Maximum bending moment in the wall slab occurs at the base where the pressure ordinate of the pressure triangle is the maximum $p = 600h$

Wall slab is designed as a continuous slab spanning over the counterforts

Span of the slab $= 3{\cdot}5$m

Maximum bending moment

$= \dfrac{pl^2}{12} = \dfrac{600h\times3{\cdot}5\times3{\cdot}5}{12}$

$= \dfrac{600\times8\times3{\cdot}5\times3{\cdot}5}{12} = 4900$ mkg

600 h

3·5M

depth of wall slab $= \sqrt{\dfrac{490000}{100\times8{\cdot}7}} = 23{\cdot}6$ cm

Use 27·5 cm effective depth with 7·5 cm cover

2·4·1. Bending Moment Variation

The bending moment in the wall slab reduces towards the top of the slab. Therefore spacing of the main reinforcement can be increased towards the top.

$$\text{Bending moment at any depth } h' = \frac{600\,h' \times l^2}{12} = \frac{600\,h' \times 3{\cdot}5 \times 3{\cdot}5}{12}$$

$= 612\,h'$

Sample Calculations

Calculations for $h' = 4$ m

$M_4 = 612 \times 4 = 2448$ m/kg

$$A_t = \frac{2448 \times 100}{1400 \times .87 \times 27{\cdot}5} = 7{\cdot}31 \text{ cm}^2$$

Use 12 mm ϕ at 15 cm c/c

Similarly reinforcement required at different depths at an interval of 1 m has been calculated as in the table

2·4·2. Main Reinforcement spacing

depth from top	*bending moment*	A_t	*diameter of bar*	*spacing required*	*spacing adopted*
m	$612\,h$	$\frac{612\,h \times 100}{1400 \times {\cdot}87 \times 27{\cdot}5}$	mm	cm	
0	0	0	12	60 cm (< 3d)	60
1	612	1·83	,,	61·9	60
2	1224	3·66	,,	31·0	30
3	1836	5·48	,,	20·6	20
4	2448	7·31	,,	15·45	15
5	3060	9·14	,,	12·4	12
6	3675	10·98	,,	10·3	10
7	4300	12·80	,,	8·85	8·5
8	4900	14·6	,,	7·75	7·5

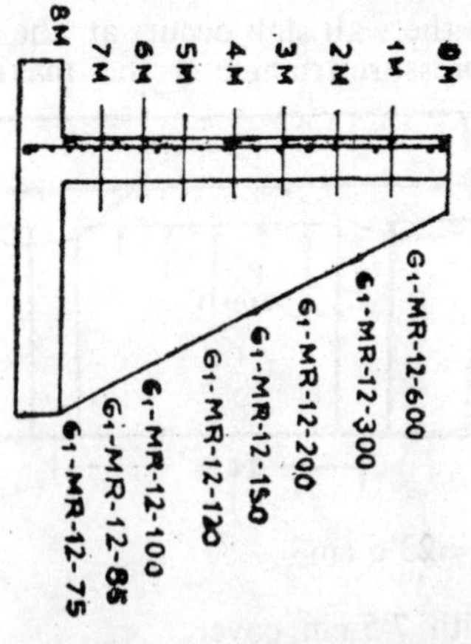

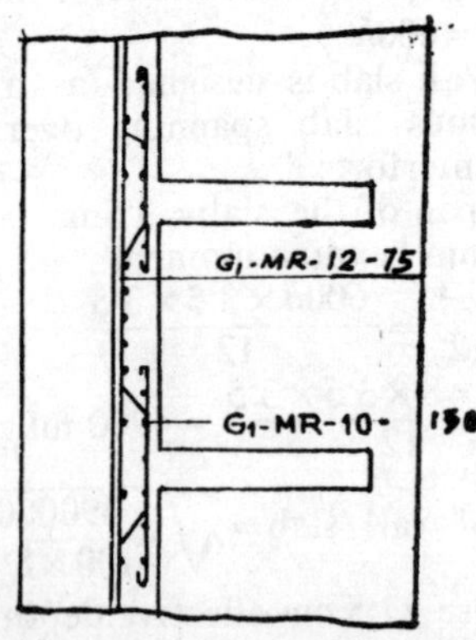

2·4·3. Secondary Reinforcement

Minimum reinforcement = 0·15% of concrete area

$$=\frac{0{\cdot}15\times 3{\cdot}5\times 100}{100}=5{\cdot}25 \text{ cm}^2$$

Use 10 mm ϕ at 15 cm c/c

2·5. Design of Base Slab

component	*Load*	*Magnitude tonnes*	*Pt. of application from 0*	*B.M. mt*
Wall slab	W_1	$8\times 0{\cdot}35\times 2{\cdot}4=6{\cdot}71$	2·025	13·6
Base slab	W_2	$5{\cdot}5\times {\cdot}0{\cdot}75\times 2{\cdot}4=9{\cdot}9$	2·75	27·2
Earth fill above heel	W_3	$3{\cdot}3\times 8\times 1\times 1{\cdot}8=47{\cdot}5$	3·85	183·0
	ΣW	64·11 tonnes		$\Sigma M=223{\cdot}8$ mt

$$\bar{x}=\frac{223{\cdot}8}{64{\cdot}11}=3{\cdot}48 \text{ m}$$

Horizontal force P due to earth-fill = 19200 kg = 19·2 tonnes

Point of application of the force P above base $=\frac{8}{3}+0{\cdot}75=3{\cdot}42$ m

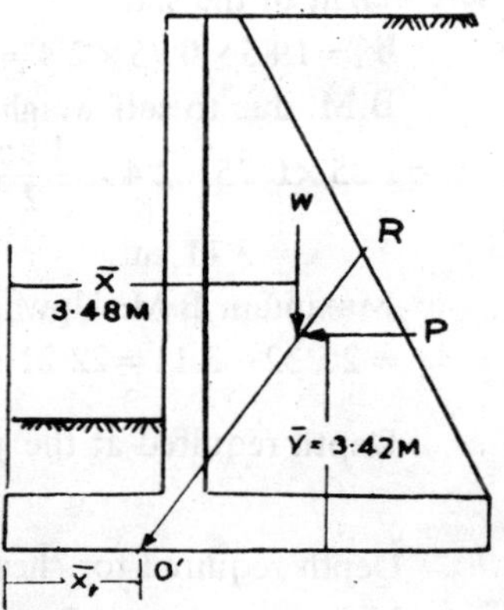

2·5·1. Point of Application of the Resultant

Taking moments about 0′

$W(\bar{x}-\bar{x}_r)=P\,\bar{y}$

$64{\cdot}11\,(3{\cdot}48-\bar{x}_r)=19{\cdot}2\times 3{\cdot}42$

$\bar{x}_r=2{\cdot}46$ m

2·5·2 Eccentricity of the Resultant

$e=B/2-\bar{x}_r=5{\cdot}5/2-2{\cdot}46$

$=2{\cdot}75-2{\cdot}46=0{\cdot}29<B/6$

The resultant falls within the middle third

2·5·3. Pressure Distribution Under the Base

Maximum pressure at toe end $=\frac{W}{B}\left(1+\frac{6e}{B}\right)$

$$=\frac{64{\cdot}11}{5{\cdot}5}\left(1+\frac{6\times 0{\cdot}29}{5.5}\right)=15{\cdot}35<20 \text{ t/m}^2$$

Minimum pressure

$$=\frac{64{\cdot}11}{5{\cdot}5}\left(1-\frac{6e}{B}\right)$$

$$=\frac{64{\cdot}11}{5{\cdot}5}\left(1-\frac{6\times 0{\cdot}29}{5{\cdot}5}\right)$$

$=8$ t/m²

2·6. Toe Design

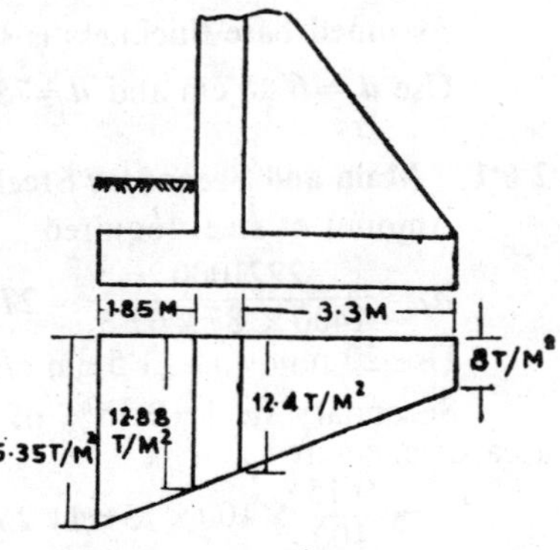

Pressure ordinate at the junction of wall slab and toe

$$=8+(15{\cdot}35-8)\times\frac{3{\cdot}65}{5{\cdot}5}$$

$$=8+7{\cdot}35\times\frac{3{\cdot}65}{5{\cdot}5}$$

$=8+4{\cdot}88=12{\cdot}88$ t/m²

Bending moment due to forces acting on the toe

(*i*) due to upward soil pressure

Force	*Magnitude tonnes*	*Point of application m*	*Moment mt*
P_1	$12.88\times1.85\times1=23.8$	$\frac{1.85}{2}=0.925$	22.10
P_2	$\frac{1}{2}\times1.85\times(15.35-12.88)$ $=\frac{1}{2}\times1.85\times2.47=2.28$	$\frac{2}{3}\times1.85=1.234$	3.22

Total upward moment=25.32mt

(*ii*) due to downward force self weight of the slab

$W_t=1.85\times0.75\times2.4=3.34$

B.M. due to self weight of slab

$$=1.85\times0.75\times2.4\times\frac{1.85}{2}$$

$$=3.11 \text{ mt}$$

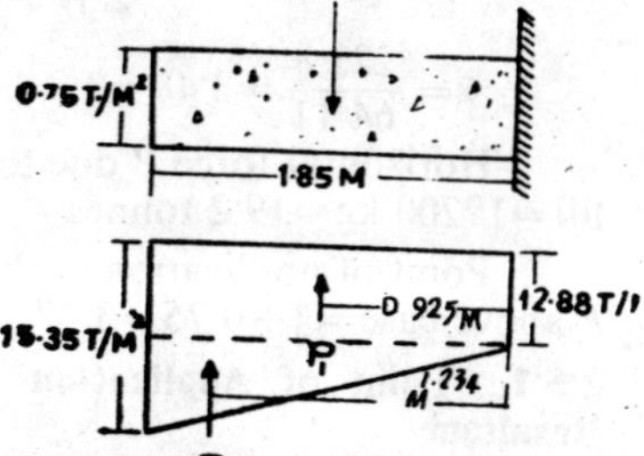

Maximum B.M. Upwards

$=25.32-3.11=22.21$ mt

Depth required at the junction$=\sqrt{\dfrac{2221000}{8.7\times100}}$

$=50.25$ cm <75 cm provided

Depth required for shear

Maximum shear $=P_1+P_2-W_t=23.8+2.28-3.34=22.74$ tonnes

$$5=\frac{22740}{100\times.87\times d_e}$$

$$d_e=\frac{22740}{100\times.87\times5}=53 \text{ cm}$$

$d=53+7.5=60.5$ cm <75 cm

Assumed base thickness is sufficient

Use $d_e=6\!:\!.5$ cm and $d=75$ cm

2.6.1. Main and Secondary Steel

Amount of steel required

$$A_t=\frac{2221000}{1400\times.87\times67.5}=27.0 \text{ cm}^2$$

Use 20 mm ϕ at 11.5 cm c/c

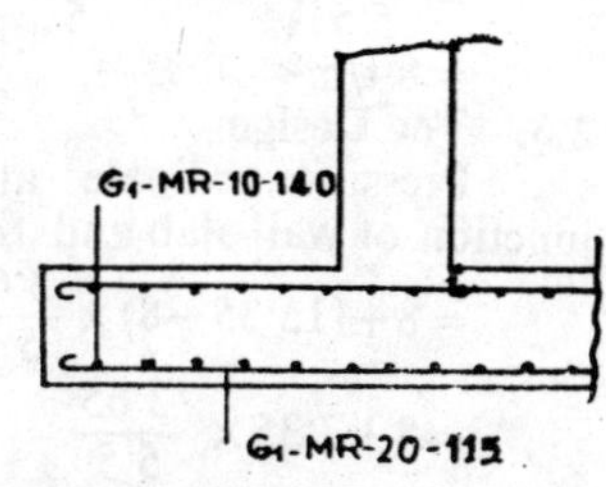

Secondary steel $=0.15\%$ of gross area of concrete

$$=\frac{0.15}{100}\times100\times75=11.25 \text{ cm}^2$$

10 mm ϕ at 14 cm c/c on both faces

2·6·2. Heel Design

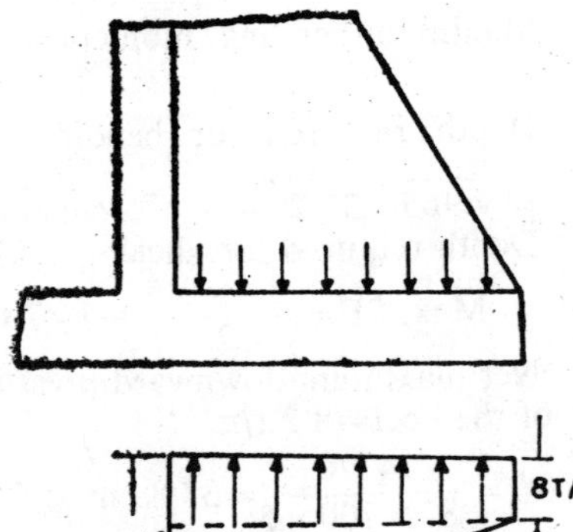

The heel slab is designed as a continuous slab spanning between the counterforts similar to wall slab.

Forces acting on the heel slab.

(*i*) Average soil pressure

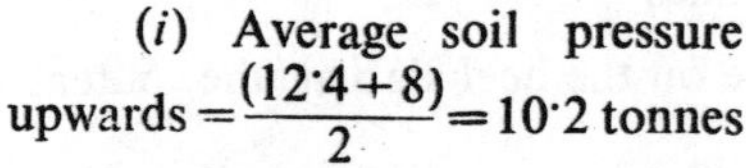

upwards $= \frac{(12·4+8)}{2} = 10·2$ tonnes

(*ii*) Forces due to soil fill acting downwards $= 1·8 \times 8$

$= 14·4$ t/m

(*iii*) Self weight of heel acting downwards $= 0·75 \times 1 \times 2·4$

$= 1·8$ t/m

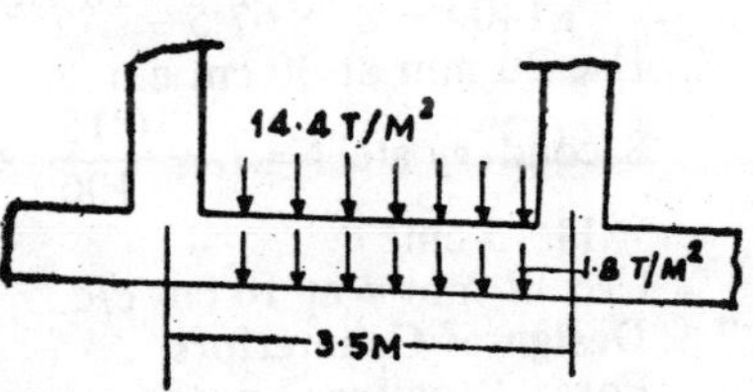

Net downwards load uniformly distributed $= 14·4 + 1·8$

$= 16·2$ t/m

Net pressure acting on the heel slab at the end (*b*) downwards

$= 16·2 - 8 = 8·2$ t/m

Net pressure acting on the heel slab at the end (*a*) downwards

$= 16·2 - 12·4 = 3·8$ t/m

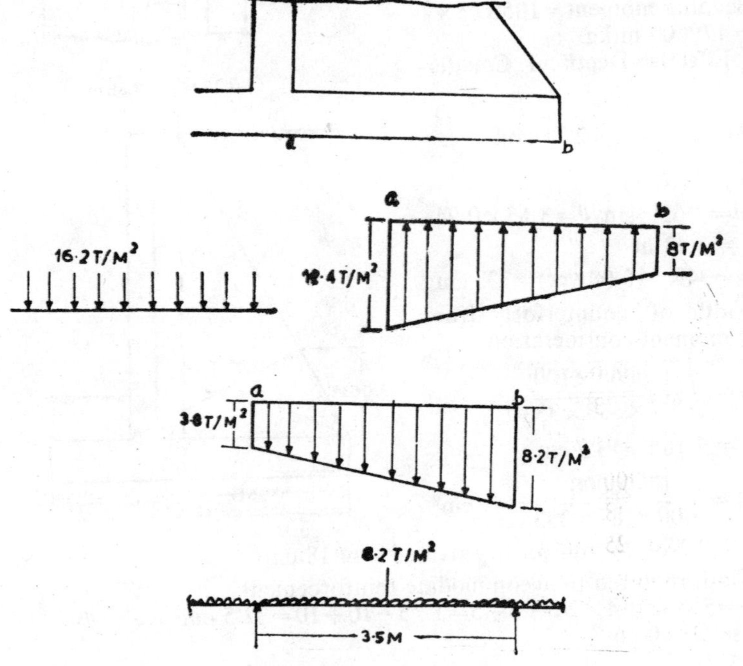

Maximum bending moment $= \frac{pl^2}{12} = \frac{8{\cdot}2 \times 3{\cdot}5^2}{12} = 8{\cdot}4$ mt

Depth required for bending moment $d_e = \sqrt{\frac{840000}{100 \times 8{\cdot}7}}$

$= \sqrt{965} = 31{\cdot}2$ cm $<$ 75 cm assumed

Depth required for shear

Max. SF $= \frac{8{\cdot}4 \times 3{\cdot}5}{2} = 14{\cdot}7$ tonnes

Net maximum downward pressure on the heel slab at the outer edge of the heel $= 8{\cdot}2$ t/m

$d_e = \frac{14700}{{\cdot}87 \times 100 \times 5} = 33{\cdot}8$ cm $<$ 75 cm

Use $d = 75$ cm and $d_e = 67{\cdot}5$ cm

2·6·3. Main and Secondary Steel

$A_t = \frac{840000}{1400 \times {\cdot}87 \times 67{\cdot}5} = 10{\cdot}1$ cm²

Use 20 mm at 30 cm c/c

Secondary steel $= A_t = \frac{0{\cdot}15}{100} \times 75 \times 100$

$= 11{\cdot}15$ cm²

Use 12 mm ϕ at 10 cm c/c

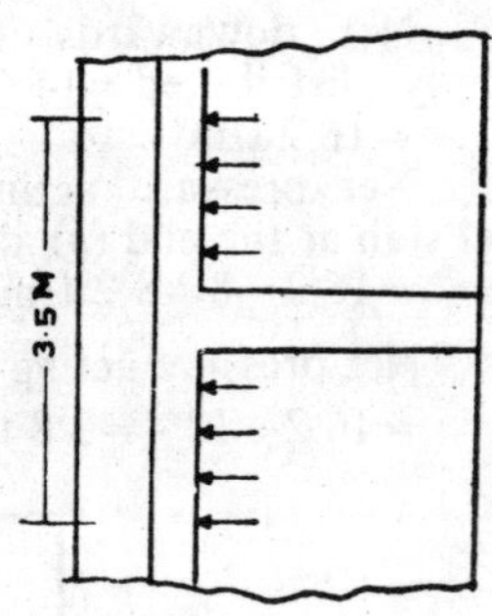

2·7. Design of Counterfort

Force P acting on the counterfort per metre width $= \frac{1}{2} \times 600\, h \times h = 300\, h^2$

Total force on the counterfort $= P \times l$

$= 300\, h^2 \times 3{\cdot}5 = 1050\, h^2$ kg $= 67250$ kg

Bending moment $= 1050\, h^2 \times \frac{8}{3}$

$= 179000$ m kg

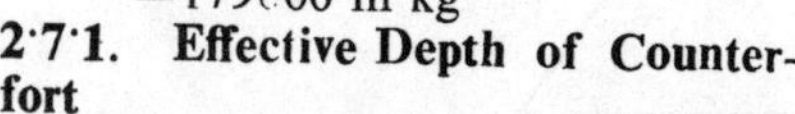

2·7·1. Effective Depth of Counterfort

$\tan\theta = \frac{8}{2{\cdot}95} = 2{\cdot}71$ and

$\sin\theta = 0{\cdot}94$

$d = 3{\cdot}65 \times \sin\theta = 3{\cdot}65 \times 0{\cdot}94$

$= 3{\cdot}43$ m

$d_e = 343 \quad 10$ (cover) $= 333$ cm

Width of counterfort from bending moment consideration

$b = \sqrt{\frac{179000 \times 100}{8{\cdot}7 \times 333 \times 333}}$

$= \sqrt{186} = 13{\cdot}7$ cm

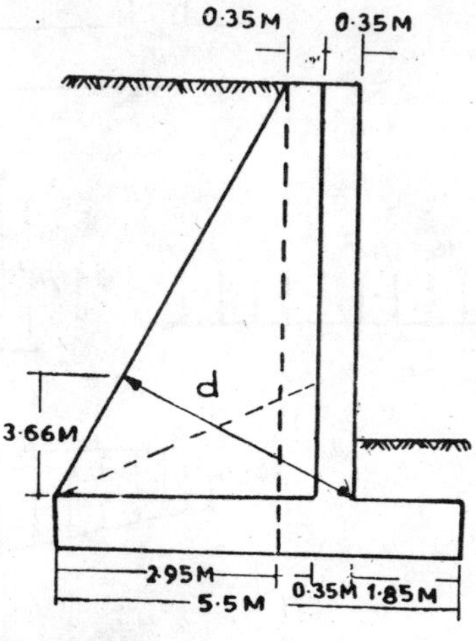

$A_t = \frac{17900000}{1400 \times {\cdot}87 \times 333} = 44$ cm²

Use 9 Nos 25 mm ϕ; A_t provided $= 44{\cdot}18$ cm²

Width required to accommodate reinforcement

$b = 5 \times 2{\cdot}5 + 4 \times 2{\cdot}5 + 5 + 5 = 12{\cdot}5 + 10 + 10 = 32{\cdot}5$ cm

Use $b = 40$ cm

2·7·2. Width of Counterfort on Shear Consideration

Total shear at the junction of base and the counterfort

$= P - T \cos \theta = \frac{1}{2} \times 600 \times h^2 \times 3·5 - 1400 \times 44·18 \times 0·345$

$= 67250 - 21350 = 45900$ kg

Shear stress $= 5 = \dfrac{45900}{0·87 \times b \times 333}$

Width required to resist shear $b = \dfrac{45900}{5 \times ·87 \times 333} = 31·7$ cm

Use $b = 40$ cm

2·7·3. Curtailment of Main Steel

9 bars of 25 mm diameter are required at the lowest level of the counterfort, since the maximum bending moment occurs at that level. Towards the top, the bending moment reduces and the reinforcement can be curtailed at different levels suitably.

Sample calculations.

Bending moment at 2 m depth from top. BM at any depth

$h_1 = \frac{1}{2} \times 600\, h_1 \times h_1 \times 3·5 \times \dfrac{h_1}{3} = 350\, (h_1)^3$ m kg

BM at 2 m depth $= 350 \times 2^3 = 2800$ m kg

Effective depth provided at 2 m = horizontal depth $\times \sin\theta$ − cover

$= \left(35 + 35 + \dfrac{295}{8} \times 2\right) \times \sin\theta - 10 = (70 + 73·75) \sin\theta - 10$

$= 143·75 \times 0·94 \quad 10 = 135 - 10 = 125$ cm.

Amount of steel required at 2 m depth

$A_t = \dfrac{280000}{·87 \times 125 \times 1400} = 1·84$ cm²

Use 2 Nos of 25 mm ϕ

Similarly calculations for different levels are calculated as in the Table.

Depth from top	*Effective depth provided*	*Bending moment ($350\, h_1^3$) m kg*	A_t	*No. of bars*
0	—	—	—	2
2	125	2800	1·84	2
4	195	22400	5·2	2
6	265	75600	23·3	5
8	333	179000	44·0	9

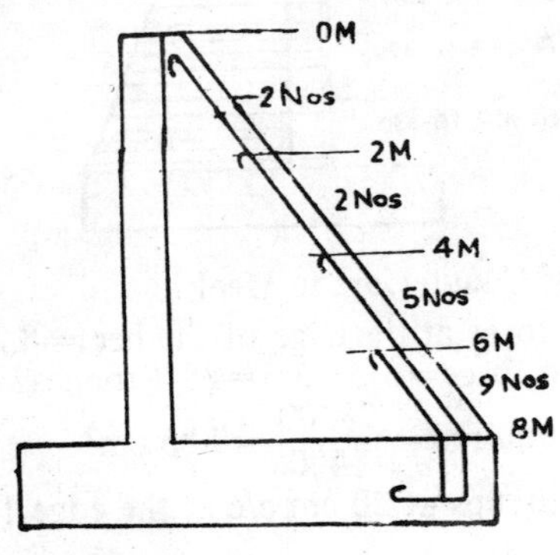

2·7·4. Connection of Counterfort With Wall Slab

Horizontal pressure acting over a length of 3·5 m on the wall might tear off the wall from counterfort. Therefore to prevent such failure the main reinforcement in the wall slab is tied with the main reinforcement in the counterfort.

Maximum horizontal pressure occurs at the junction of the base and counterfort and the pressure reduces towards the top. Therefore the reinforcement required to resist this pull is calculated on the basis of the pressure at different levels and the reinforcement is provided in the form of closed stirrups tied firmly to main reinforcement in the wall slab and counterfort.

Maximum horizontal forces at any depth h_1

$$=wh_1\times\left(\frac{1-\sin\phi}{1+\sin\phi}\right)\times l=1800\times h_1\times\tfrac{1}{3}\times l=600\ h_1\ l$$

At 1 m depth $=600\times1\times3{\cdot}5=2100$ kg

Amount of steel required $=\dfrac{2100}{1400}=1{\cdot}5\ \text{cm}^2$

From top to 1 m depth use 2 strirrups of 6 mm ϕ

Similarly the horizontal stirrups required are calculated

Depth of plank from top *m*	*Horizontal force* *600 h_1 l* *kg*	A_t *cm*²	*Spacing of 2 legged stirrups* *cm*
1	2100	1·5	6 mm—33
2	4200	3	8 mm—30
3	6300	4·5	8 mm—20
4	8400	6	8 mm—16
5	10500	7·5	10 mm—20
6	12600	9	10 mm—17
7	14700	10·5	10 mm—15
8	16800	12·0	10 mm—13

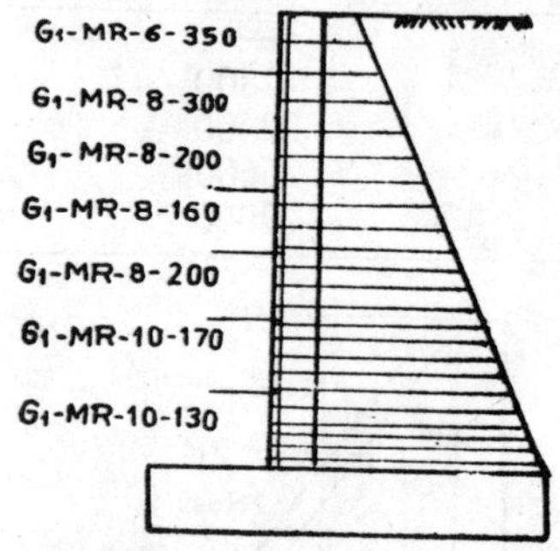

2·7·5. Connection of Counterfort to Heel

Net downward force at the edge of the heel $=8{\cdot}2$ t/m

Total downward force $=8{\cdot}2\times3{\cdot}5=28{\cdot}7$ tonnes

Area of steel required $=\dfrac{28700}{1400}=20{\cdot}5\ \text{cm}^2$

Use 16 mm ϕ stirrups at 20 cm c/c at the edge 1 m

Pressure at the junction oı wall & slab = 3·8 t/m

Total downward load which tries to separate heel from counterfort = 3800 × 3·5 = 10900 kg

$$A_t = \frac{10900}{1400} = 7{\cdot}8 \text{ cm}^2$$

Use 16 mm ϕ at 40 cm c/c

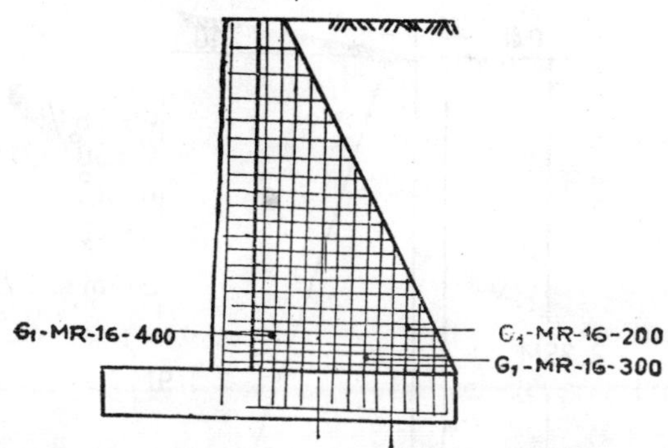

2·7·6. Stability

Factor of safety against overturning $= \frac{w\,\bar{x}}{P\,\bar{y}} = \frac{64{\cdot}11 \times 3{\cdot}48}{19{\cdot}2 \times 3{\cdot}42}$

$= 3{\cdot}42 > 1{\cdot}5$

Factor of safety against sliding $= \frac{\mu W}{P} = \frac{0{\cdot}6 \times 64{\cdot}11}{19{\cdot}2} = 2 > 1{\cdot}5$

Counterfort Retaining Wall With Sloping Surcharge 3

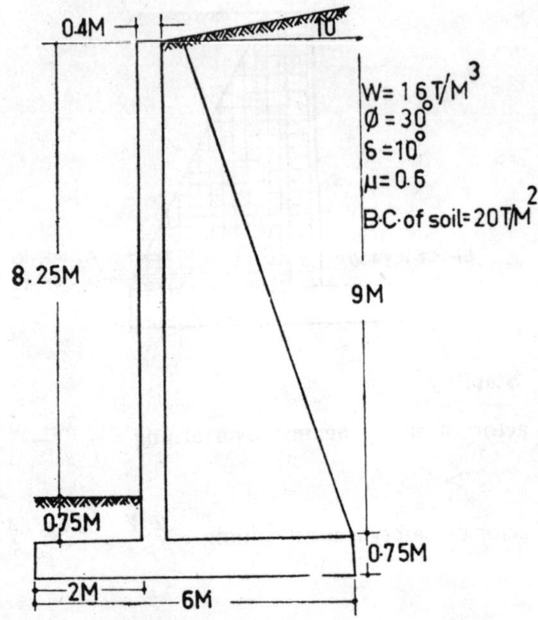

3·1. Data

Height of the fill to be retained by the wall = 9 m
Surcharge = δ = earth sloping at 10°
Density of the soil = 1600 kg/m³
Angle of internal friction $\phi = 30°$
Coefficient of friction between soil and base slab = 0.6
Bearing capacity of soil = 20 t/m²
Materials available M 150 concrete and grade – 1 steel

3·2. Allowable Stresses

Concrete-Grade M 150

Compressive stress of concrete in bending $= \sigma_{cb} = 50\text{kg/cm}^2$, shear stress = 5kg/cm², bond stress = 10kg/cm², Steel grade-I $\sigma_{st} = 1400$ Kg/cm²

3·2·1 Characteristic Strengths

$$\text{Modular ratio} = m = \frac{2800}{3 \times \sigma_{cb}} = \frac{2800}{3 \times 50} = 18{\cdot}66$$

$$\frac{\sigma_{cb}}{n} = \frac{\sigma_{st}}{n\,(d\text{-}n)}$$

$$\frac{50}{n}=\frac{1400}{18{\cdot}66\,(d\text{-}n)}$$

$n = 0{\cdot}4\text{d}$

Lever arm $jd = 0{\cdot}87\ d$

$$\text{Moment of resistance} = b \times n \times \frac{\sigma_{cb}}{2} \times jd$$

$$= b \times 0{\cdot}4d \times \frac{50}{2} \times 0{\cdot}87\ d = 8{\cdot}7\ bd^2$$

3·3. Trial Section

d_f= depth of foundation 1 m or depth as given by

$$\frac{20000}{1600}\left(\frac{1-\frac{1}{2}}{1+\frac{1}{2}}\right)^2 = \frac{20{,}000}{1600} \times \frac{1}{9} = 1{\cdot}39\text{m}$$

Use a depth = 1·5m

Spacing of counter forts = $\frac{1}{2}$ to $\frac{1}{3}h = 4m$ (say)

3·3·1. Earth Pressure Calculations

Horizontal active earth pressure ordinate at any depth h' from top

$$P_h = wh' \cos^2\delta \left[\frac{\cos\delta - \sqrt{\cos^2\delta - \cos^2\phi}}{\cos\delta + \sqrt{\cos^2\delta - \cos^2\phi}}\right]$$

$$= wh' \cos^2 \delta K = 1600\ h' \times 0{\cdot}985 \times 0.353 = 550\ h'$$

$$\text{Horizontal active earth pressure } P_h = \tfrac{1}{2} \times 550\ h' \times h = \frac{550 \times 9 \times 9}{2}$$

$$= 22500 \text{ kg}$$

$$\text{Point of application} = h/3 = \frac{9}{3} = 3 \text{ m}$$

3·4. Wall Slab Design

Maximum bending moment in the wall slab occurs at the base where the pressure ordinate of the pressure triangle is the maximum $= 550\ h$

Wall slab is treated as a continuous slab spanning over the counterforts = span of the slab = 4 m

$$\text{Maximum bending moment} = \frac{pl^2}{12} = \frac{550h \times 4 \times 4}{12}$$

$$= \frac{550 \times 9 \times 4 \times 4}{12} = 6600 \text{ mkg}$$

$$\text{Depth of wall slab} = \frac{\sqrt{6600 \times 100}}{8{\cdot}7 \times 100} = 27{\cdot}6 \text{ cm} < 40 \text{ cm assumed}$$

Provide effective depth = 32·5 cm

3·4·1. Bending Moment Variation

The bending moment in the wall slab reduces towards the top of the slab. Therefore spacing of the main reinforcement can be increased towards the top.

$$\text{Bending moment at any depth } h' = \frac{pl^2}{12},\ M = \frac{550h' \times 4^2}{12}$$

$$= 735\ h'$$

Calculations for $h'=2$ m

M = 735×2 = 1470 mkg.

$$A_t=\frac{1470\times100}{1400\times\cdot87\times32\cdot5}=3\cdot71 \text{ cm}^2$$

Use 16 mm ϕ at 54 cm c/c

3·4·2. Main Steel Spacing

Similarly reinforcement required at different depths at an interval of 2m is calculated as in the table

depth from top	*Bending moment*	A_t	*Spacing of 16 mm* ϕ
m	$735h'$	$\frac{735h'}{1400\times\cdot87\times32\cdot5}$	cm
0 – 2	1470	3·71	54
4	2940	7·43	27
6	4400	11.1	18
8	5860	14·8	13·5
9	6600	16·7	12

34·3. Secondary Reinforcement

Minimum reinforcement = 0·15% of concrete area

$$=\frac{0\cdot15}{100}\times40\times100=6 \text{ cm}^2$$

Use 10mm ϕ at 13 cm c/c

3·5. Design of Base Slab

Vertical loads—Consider 1 metre length of wall.

Component	*Load*	*Magnitude*	*Point of application from toe – 0*	*Bending moment*
		tonnes	*m*	*mt*
Wall slab	W_1	$9\times0\cdot4\times1\times2\cdot4=8\cdot65$	2·2	19·00
Base slab	W_2	$6\times0\cdot75\times1\times2\cdot4=10\cdot8$	3·0	32·40
Earth fill above heel rectangular portion	W_3	$3\cdot6\times9\times1\times1\cdot6=51\cdot75$	4·2	217·50
Triangular surcharge portion	W_4	$\frac{1}{2}\times3\cdot6\times3\cdot6\times0.1763\times1\cdot6=1\cdot82$	4·8	8·75
Vertical component of pressure	P_V	$1600h\times K\times\sin\delta\times\frac{9}{2}=\frac{1600}{1000}\times9\times\cdot353\times\cdot1736\times\frac{9}{2}=3\cdot97$	6·0	23·80
		$\Sigma W_1=76\cdot99$ tonnes		$\Sigma M_V=301.45$ *mt*

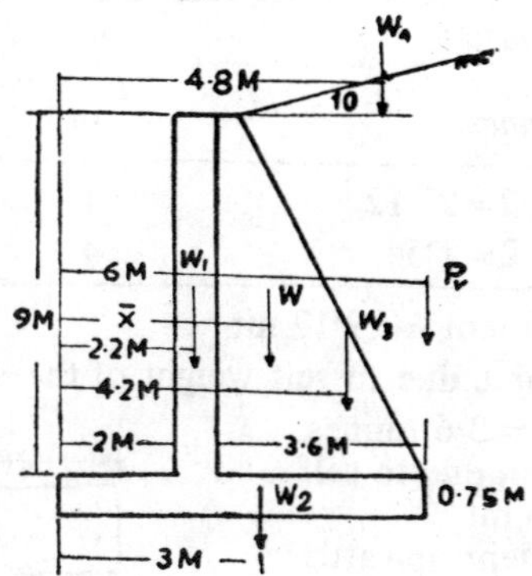

3·5·1. Point of Application of the Resultant Force

$\bar{x} = \frac{301{\cdot}45}{76{\cdot}99} = 3{\cdot}94$ m

$P_h = 22{\cdot}5$ tonnes

Point of application $\bar{y} = 3{\cdot}75$ m

Taking moments about 0',

$W(\bar{x} - \bar{x}_r) = P_h \bar{y}$

$76{\cdot}9\ (3{\cdot}94 - \bar{x}_r) = 22{\cdot}5 \times 3{\cdot}75$

$$\bar{x}_r = \frac{76{\cdot}9 \times 3{\cdot}94 - 22{\cdot}5 \times 3{\cdot}75}{76{\cdot}99}$$

$$= \frac{301{\cdot}45 - 84{\cdot}5}{76{\cdot}99}$$

$$= \frac{216{\cdot}95}{76{\cdot}99} = 2{\cdot}82 \text{ m}$$

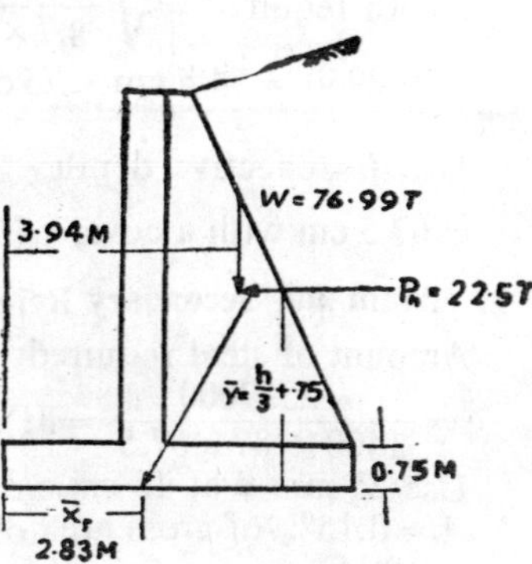

3·5·2. Eccentricity of the Resultant

$$e = \frac{B}{2} - \bar{x}_r = 3 - 2{\cdot}82 = 0{\cdot}18 \text{ m} < \frac{B}{6}$$

3·5·3. Pressure Distribution Under the Base

Maximum pressure at the toe end $= \frac{W}{B}\left(1 + \frac{6e}{B}\right)$

$$= \frac{76{\cdot}99}{6}\left(1 + \frac{6 \times 0{\cdot}18}{6}\right) = 15.1 \text{ t/m}^2$$

Minimum pressure at heel end $= \frac{W}{B}\left(1 - \frac{6e}{B}\right)$

$$= \frac{76{\cdot}99}{6}\left(1 - \frac{6 \times {\cdot}18}{6}\right) = \frac{76{\cdot}99}{6} \times {\cdot}82 = 10{\cdot}5 \text{ t/m}^2$$

3·6. Toe Design

Pressure ordinate at the junction of the toe and wall slab

$$= 10{\cdot}5 + \frac{(15{\cdot}1 - 10{\cdot}5) \times 4}{6} = 10{\cdot}5 + 4{\cdot}6 \times \frac{4}{6} = 13{\cdot}56 \text{ t/m}^2$$

Bending moment due to forces acting on the toe

(*i*) due to upward soil pressure

Force	*Magnitude* tonnes	*Point of application* m	*Moment* mt
P_1	$13{\cdot}56\times2=27{\cdot}12$	1	27·12
P_2	$\frac{1}{2}\times1{\cdot}54\times2=1{\cdot}54$	$\frac{4}{3}$	2·05

Total upward moment $=29{\cdot}17$ mt

(*ii*) Downward force due to self weight of the slab

$P_3=2\times0{\cdot}75\times2{\cdot}4=3{\cdot}6$ tonnes

Downward moment due to self weight $P_3=3{\cdot}6\times1=3{\cdot}6$ mt

Net bending moment upwards
$=29{\cdot}17-3{\cdot}60=25{\cdot}57$ mt

$$\text{Depth required}=\sqrt{\frac{557000}{8.7\times100}}$$

$=\sqrt{2950}=54{\cdot}5$ cm <75 cm provided

Use d_e=effective depth

$=67{\cdot}5$ cm with a cover of 7·5 cm

3·6·1. Main and Secondary Reinforcement

Amount of steel required

$$=\frac{2557000}{1400\times{\cdot}87\times67.5}=31{\cdot}5\text{ cm}^2$$

Use 25 mm ϕ at 15 cm c/c

A_t=0.15% of gross area of concrete

$$=\frac{0{\cdot}15}{100}\times100\times75=11{\cdot}25\text{ cm}^2$$

Use 12 mm ϕ at 10 cm c/c on bottom face or 12 mm ϕ at 20 cm c/c on both the faces of the slab.

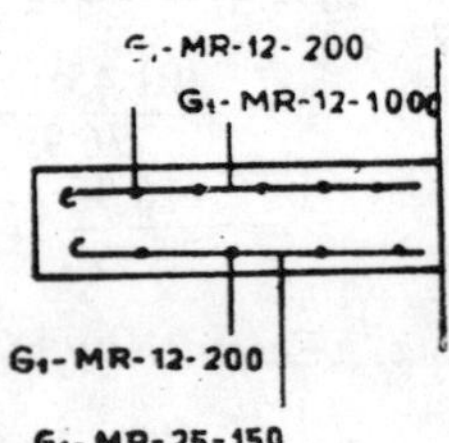

3·6·2. Shear at the Junction of Toe and Wall Slab

Maximum shear force $=P_1+P_2-P_3=27{\cdot}17+1{\cdot}54-3{\cdot}6$
$=25{\cdot}06$ tonnes.

$$\text{Shear stress}=\frac{25{\cdot}060}{100\times{\cdot}87\times67{\cdot}5}=4{\cdot}34\text{ kg/cm}^2<5\text{kg/cm}^2$$

3·7. Design of Heel Slab

The heel is designed as a continuous slab spanning between the counterforts similar to wall slab.

Forces acting on the heel.

$$\text{Average soil pressure upwards}=\frac{13{\cdot}25+10{\cdot}5}{2}=11{\cdot}875\text{ t/m}^2$$

Forces due to soil fill acting downwards forces due to level fill $=9\times1{\cdot}6=14{\cdot}4$ t/m²

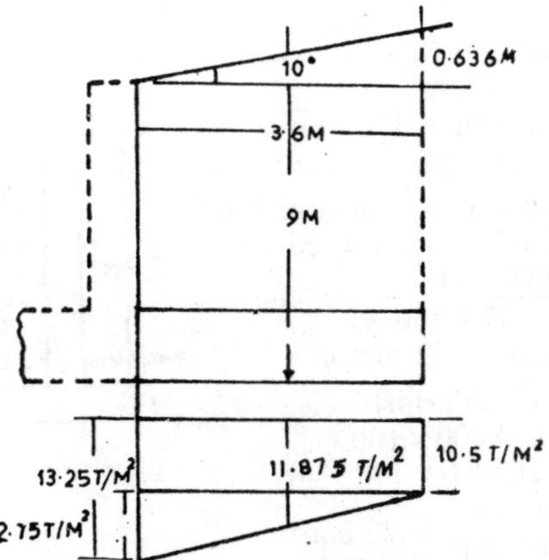

Surcharge $= \frac{\cdot 636}{2} \times 1 \cdot 6 = 0 \cdot 51$ t/m²

Self weight of slab $= 0 \cdot 75 \times 2 \cdot 4 \times 1 = 1 \cdot 8$ t/m²

Total downward load

= 16·71 tonnes

Net force downwards

$= 16 \cdot 710 - 11 \cdot 875 = 4 \cdot 835$ t/m

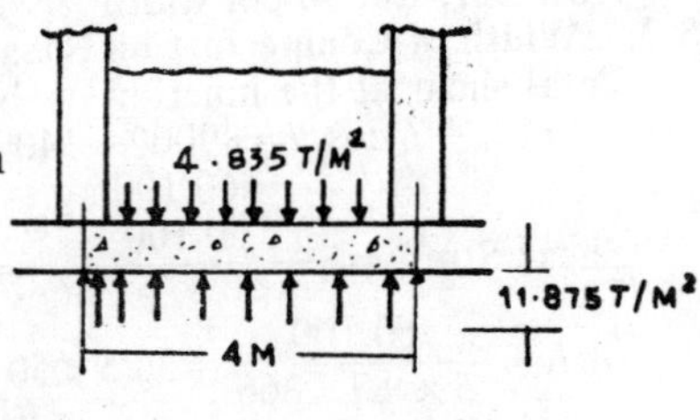

Bending moment $= \frac{wl^2}{12}$

$= \frac{4 \cdot 835 \times 4 \times 4}{12} = 6 \cdot 42$ mt

= 6420 mkg

3·7·1. Main and Secondary Reinforcement

$A_t = \frac{642000}{1400 \times \cdot 87 \times 67 \cdot 5} = 7 \cdot 82$ cm²

Use 12 mm ϕ at 14 cm c/c

$A_t = \frac{0 \cdot 15}{100} \times 75 \times 100 = 11 \cdot 25$ cm²

Use 12 mm ϕ at 10 cm c/c

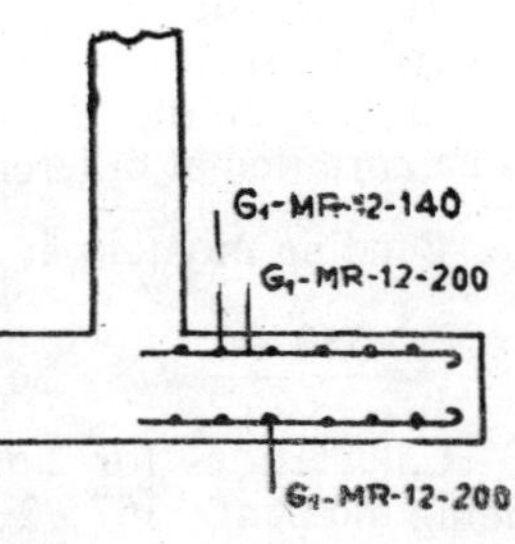

3·8. Design of Counterfort, Effective Depth of Counterfort

Total horizontal force acting on the counterfort

$P_h = 550\, h \times \frac{h}{2} \times l$

$= 550 \times 9 \times \frac{9}{2} \times 4 = 89000$ kg

Maximum bending moment at the junction of counterfort and base

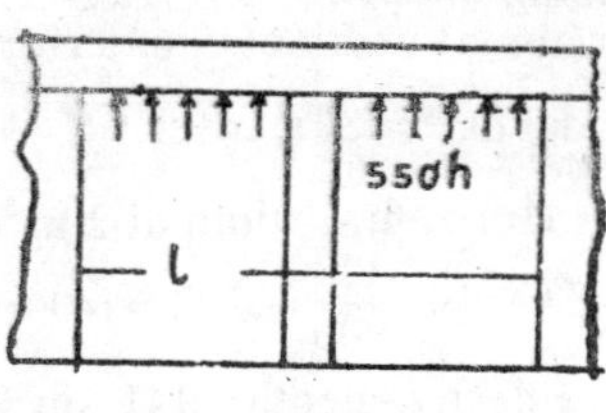

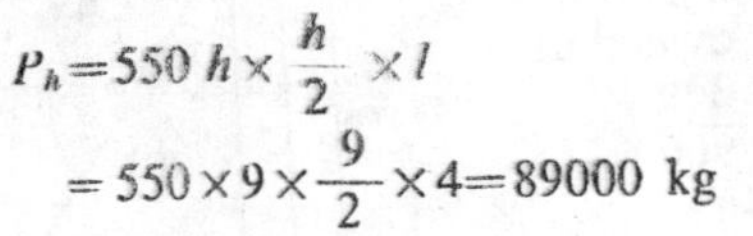

$= 8\text{9}000 \times \frac{h}{3} = 89000 \times \frac{9}{3}$

268,000 mkg $=268$ mt

$\text{Tan }\theta=\frac{9}{3{\cdot}2}=2{\cdot}81$ and $\theta=70^\circ-30'$

$d=4\times\sin\theta=4\times0{\cdot}9436$
$=3{\cdot}76$ m $=376$ cm

d_e = effective depth $=376-$cover
$=376-10=366$ cm

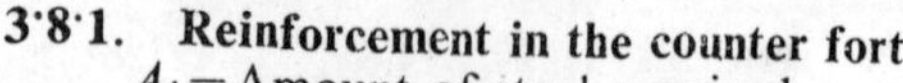

3·8·1. Reinforcement in the counter fort

A_t = Amount of steel required

$$=\frac{268000\times100}{1400\times{\cdot}87\times366}=60\text{ cm}^2$$

Provide 16 bars of 22 mm ϕ

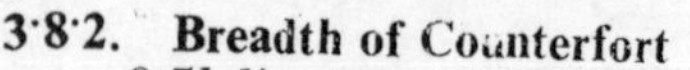

3·8·2. Breadth of Counterfort

$8{\cdot}7bd^2=M=268000\times100$

$8{\cdot}7b\times366^2=268000\times100$

$$b=\frac{26800000}{8{\cdot}7\times366\times366}=23\text{ cm}$$

To accommodate 16 bars in two layers width required $=5$ cm cover on either side $+8$ bar diameters $+7$ spacings of bar diameter or aggragate size

$=5+5+8\times2{\cdot}2+7\times2{\cdot}2=10+17{\cdot}6+15{\cdot}4=43$ cm.

However, use 50 cm width

3·8·3. Width of Counterfort on Shear Consideration

Total shear at the junction of base and the counterfort

$=p_h-T\cos\theta=89000-1400\times60{\cdot}82\times0{\cdot}3338=89000-28900$
$=60{\cdot}100$ kg

$$\text{Shear Stress}=\frac{60.100}{b\times0{\cdot}87\times366}=5$$

$$\text{or } b=\frac{60{,}100}{5\times{\cdot}87\times366}=36{\cdot}5<50\text{ cm, provided}$$

3·8·4. Curtailment Reinforcement in the Counterfort

Sample Calculation

16 bars of 22mm ϕ are required at the lowest level of the counterfort, since maximum bending moment occurs at that level. Towards the top the bending moment reduces and the reinforcement can be curtailed at different levels.

$$\text{Bending moment at any depth}=550h\times\frac{h}{2}\times l\times\frac{h}{3}=\frac{550h^3l}{6}$$

$$=\frac{550\times4}{6}\times h^3=367h^3$$

Calculations for 2 m depth

Bending moment $=367\times2^3=2940$ mkg

Horizontal width of counterfort provided at any depth $h'=80-\text{cover}+x=70+\frac{320}{900}h'$

Horizontal width at 2 m depth

$$=70+\frac{320}{900}\times200=141\text{ cm}$$

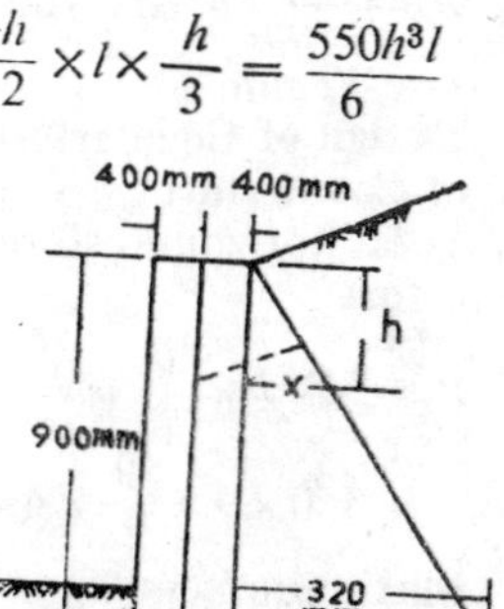

Effective depth $=141\sin\theta=141\times0{\cdot}9426=133$cm

Amount of steel required $A_t = \frac{BM}{\sigma_{st}\, jd} = \frac{2940 \times 100}{1400 \times \cdot 87 \times 133}$

$= 1.82\ cm^2$

Similarly amount of steel required at defferent depths, is calculated as in the table

depth h' from top m	bending moment mkg	horizontal width d $=70+\frac{320h'}{900}$ cm	effective depth d_e =d sine θ cm	Ast cm²	No. of tars of 22mm φ required	No of bars used
2	2940	141	133	1.82	1	2
4	23450	212	200	9·65	3	4
6	79320	283·5	267	24.5	7	8
8	188000	354·0	333	46·4	13	—
9	268000	390·0	366	60	16	16

3·8·5. Connections to Counterfort and Wall Slab

The main reinforcement in the counterfort is tied to the main reinforcement in the wall slab through 2-legged stirrups. Total horizontal pressure to which the counterfort is subjected varies from a spacing the base to a minimum at the top of the wall. Therefore the max. at of these links can be increased towards the top.

Sample calculations.

p = horizontal pressure at any depth $h' = 550h'$

Total horizontal pressure carried by each counterfort$=550h' \times l$ $=550h' \times 4 = 2200h'$

when $h' = 9m$

Total horizontal pressure$=2200 \times 9 = 19800$ kg

Amount of reinforcement required for bottom 1 metre height

$= \frac{19800}{4\ 100} = 14{\cdot}2\ \text{cm}^2$

Using 2 legged stirrups of 10mm φ spacing$= \frac{0{\cdot}79 \times 2}{14{\cdot}2} \times 100$

$= 11{\cdot}1$ cm

Use 2 legged 10mm φ ties at 11 cm c/c

Similarly spacing of 10mm φ bars as ties at different depths has been calculated as in the table.

Plank	*Location m*	*size area m²*	*pressure on the plank per m²*	*Total force kg*	A_t	*Spacing adopted cm*
1	1	1×4	550	2200	1·57	50
2	2	1×4	1100	4400	3·00	50
3	3	1×4	1650	6600	4·71	33
4	4	1×4	2200	8800	6·3	25
5	5	1×4	2750	11000	7·85	20
6	6	1×4	3300	13200	9·43	16
7	7	1×4	3850	15400	11·0	14
8	8	1×4	4400	17600	12·6	12
9	9	1×4	4950	19800	14·2	11

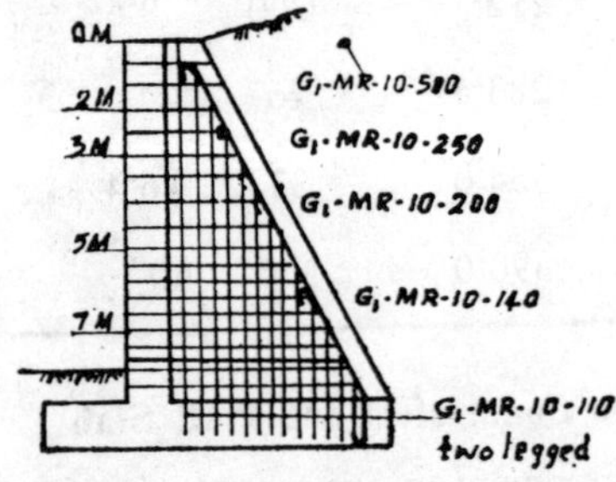

3·8·6. Connection of Counterfort and Heel Slab

Net downward load acting on the heel slab = 4·835t/m

Total downward load/m = 4×4·835 = 19·30 tonnes

Amount of steel required = $\dfrac{19300}{1400}$ = 13·8 cm²

Use 10mm ϕ – 2 legged ties 11 cm c/c

3·9. Stability Investigation

Factor of safety against overturning

$$= \frac{\text{stabilizing moment}}{\text{overturning moment}} = \frac{W\bar{x}}{P\bar{y}} = \frac{76{\cdot}99\times3{\cdot}94}{22{\cdot}5\times3{\cdot}75}$$

$$= \frac{301{\cdot}45}{84.5} = 3{\cdot}6 > 1{\cdot}5$$

Factor of safety against sliding = $\dfrac{\text{Resisting force}}{\text{Sliding force}}$

$$= \frac{\mu W}{P} = \frac{0{\cdot}6\times76{\cdot}99}{22{\cdot}5} = 2{\cdot}04 > 1{\cdot}5$$

Counterfort Retaining Wall with Superimposed Load and Stratified Back Fill 4

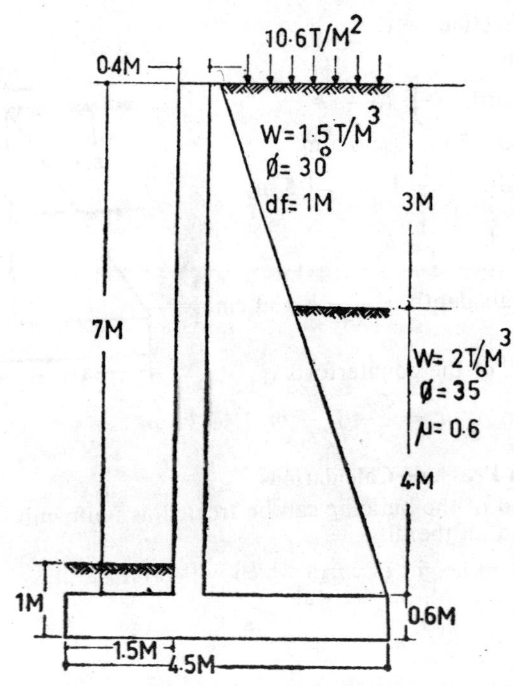

4·1. Data

Fill material characteristics. Soil fill upto top 3 m depth

Density of soil $= w_1 = 1500$ kg/cu m, Angle of internal friction $\phi_1 = 30°$

Soil fill below 3 m level

Density of soil $= w_2 = 2000$ kg/cu m, Angle of internal friction $\phi_2 = 35°$

Loading on the Retaining wall—a building which distributes a load of 10·6 t/m² from the edge of the wall.

Coefficient of friction $\mu = 0·6$

Material available—Concrete—M 150, Steel—grade—I

4·2. Allowable Stresses

Compressive Stress of concrete in bending $= \sigma_{cb} = 50$ kg/cm², shear stress $= q = 5$ kg/cm², Bond stress $= \sigma_b = 10$ kg/cm²

Steel grade—I—$\sigma_{st} = 1400$ kg/cm²

4·2·1. Characteristic Strengths

Modular ratio $=n=\frac{2800}{3\sigma_{cb}}=\frac{2800}{3\times 50}=18{\cdot}66$

$\frac{\sigma_{cb}}{n}=\frac{\sigma_{st}}{m(d-n)}, \quad \frac{50}{n}=\frac{1400}{18{\cdot}66\,(d-n)}$

$n=0{\cdot}4\,d$ and Lever arm $jd=0{\cdot}87\,d$

Moment of resistance $=b\times\frac{\sigma_{cb}}{2}\times jd\times n=8{\cdot}7\,bd^2$

4·3. Trial Section

$h=7$ m

Base width $=\frac{1}{2}$ to $\frac{2}{3}\,h$

$=3{\cdot}5$ to $4{\cdot}7$ m, Use $4{\cdot}5$ m

Toe width $=\frac{1}{3}$ base $=1{\cdot}5$ m,

Base slab depth $=\frac{1}{12}\,h=60$ cm

Wall slab depth $=\frac{1}{24}\,h=30$ cm,

Use 40 cm

Spacing of the counterforts

$=\frac{1}{3}$ to $\frac{1}{2}\,h=\frac{7}{3}$ to $\frac{7}{2}$ m, Use $3{\cdot}5$ m

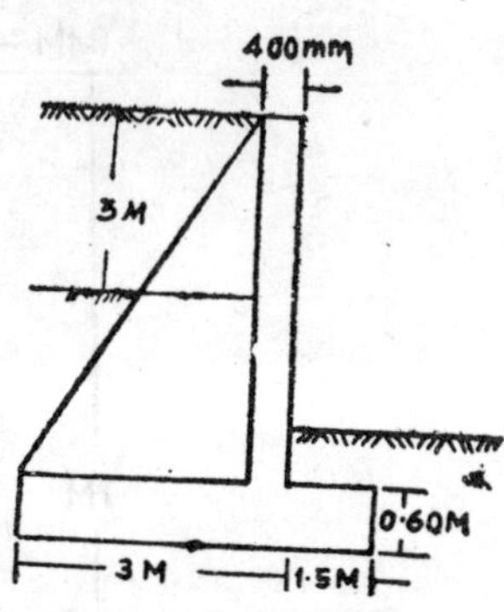

4·3·1. Earth Pressure Calculations

The load of the building can be treated as uniformly distributed surcharge load on the fill.

Equivalent height of earth which will produce same intensity of pressure due to u.d.l. on the wall

$h'=\frac{q}{w_1}=\frac{10600}{1500}=7{\cdot}06$ m

Active earth pressure intensity due to surcharge

$P_1=wh'\left(\frac{1-\sin\phi_1}{1+\sin\phi_1}\right)=1500\times 7{\cdot}06\times\frac{1}{3}=3540$ kg/m^2

$P_2=$ pressure intensity due to upper strata of soil $=w_1h_1\left(\frac{1-\sin\phi_1}{1+\sin\phi_1}\right)$

$P_2=1800\times 3\times\frac{1}{3}=1500$ kg/m^2

Pressure intensity due to lower strata of soil fill

$P_3=w_2h_2\left(\frac{1-\sin\phi_2}{1+\sin\phi_2}\right)$

$=2000\times 4\left(\frac{1-\sin 35^\circ}{1+\sin 35^\circ}\right)$

$=2000\times 4\times{\cdot}271=2216$ kg/m^2

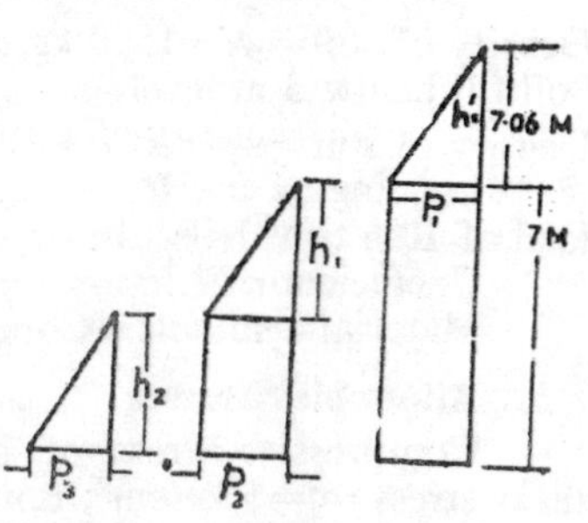

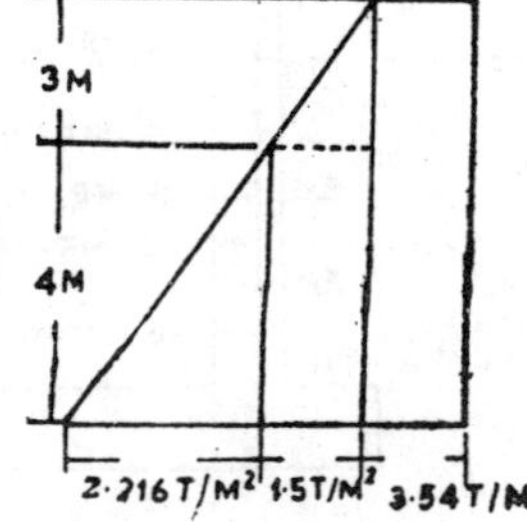

Maximum intensity of earth pressure at the junction of the base and wall slab

$p = p_1 + p_2 + p_3 = 3540 + 1500 + 2216$
$= 7266 \text{ kg/m}^2$

4·4. Wall Slab Design

Maximum bending moment in the wall slab occurs at the junction of wall slab and base

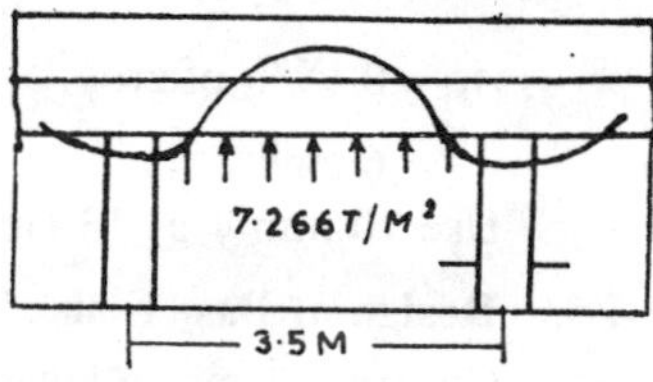

Maximum bending moment

$$M = \frac{pl^2}{12} = \frac{7266 \cdot 3{\cdot}5^2}{12}$$

$= 7380$ mkg

depth of wall slab

$$d = \sqrt{\frac{7380 \times 100}{100 \times 8{\cdot}7}}$$

$= 29{\cdot}2 \text{ cm} < 40$ cm assumed

Provide 40 cm total depth and an effective depth of 32.5 cm

4·4·1. Bending Moment Variation and Main Steel Spacing

The bending moment in the wall slab is maximum at the lowest level and reduces to a minimum at the top of the wall. Therefore the amount of reinforcement required varies from bottom to top.

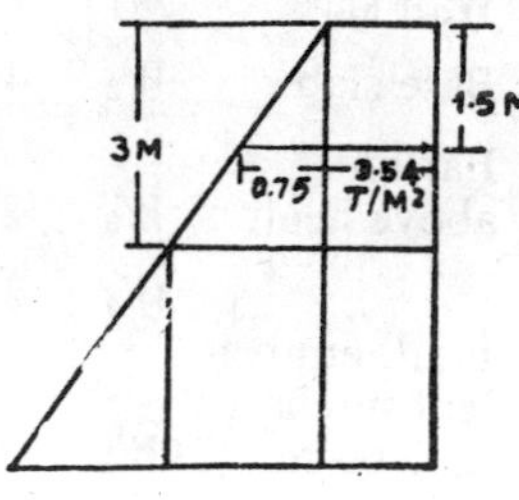

Calculations for 1·5 m depth pressure

intensity $= p_1 + p_2 \times \frac{x}{3} = 3540 + 1500 \times \frac{1{\cdot}5}{3}$

$= 3540 + 750 = 4290 \text{ kg/m}^2$

Bending moment at 1·5 m level from

top $= \frac{pl^2}{12} = \frac{4290 \times 3{\cdot}5^2}{12} = 4380$ mkg

A_t required $= \frac{438000}{1400 \times {\cdot}87 \times 32{\cdot}5}$

$= 11{\cdot}1 \text{ cm}^2$

Use 16 mm ϕ at 18 cm c/c

Similarly, the amount of reinforcement required and the spacing of 16 mm ϕ are calculated as in table

Depth	*Pressure*	*Bending moment*	*A_t required*	*Spacing of 16 mm ϕ*
m	*p* *kg/m²*	$\frac{pl^2}{12}$ *mkg*	*cm²*	*cm*
1·5	4290	4380	11·1	18
3	5040	5150	13·0	15
5	6128	6310	16·0	12
7	7216	7380	18·6	10

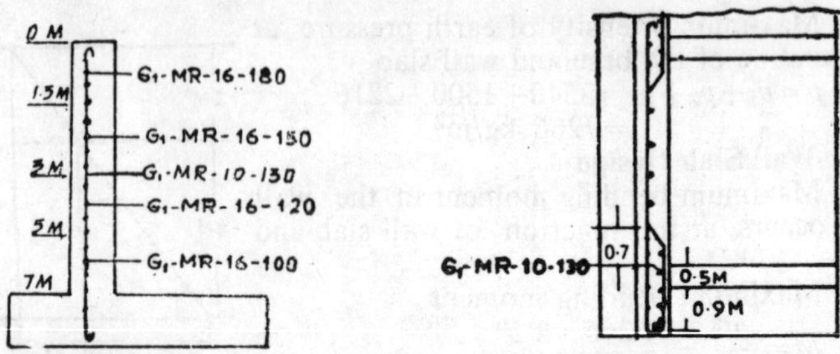

4·4·2. Secondary Steel

A_{st} = 0·15% of gross area of concrete $= \frac{0.15}{100} \times 40 \times 100$

= 6 cm²

Use 10 mm ϕ at 13 cm c/c

4·5. Design of Base Slab

Consider 1 metre length wall. Vertical loads

Component	*Load*	*Magnitude tones*	*Point of application from toe point, 0 m*	*Bending moment mt*
Wall slab	W_1	$7 \times 1 \times 0.4 \times 2.4 = 6.71$	1·7	11·4
Base slab	W_2	$4.5 \times .6 \times 1 \times 2.4 = 6.48$	2·25	14·6
Earthfill above heel	W_3	$2.6 \times 3 \times 1 \times 1.5 = 11.7$	3·20	37·5
	W_4	$2.6 \times 4 \times 1 \times 2 = 20.8$	3·20	66·5
u.d.l. equivalent surcharge of 7·06 m	W_5	$7.06 \times 1 \times 2.6 \times 1.5 = 27.5$	3·20	88·5

$\Sigma W = 73.19$ tonnes $\qquad \Sigma M = 218.5$ mt

$$\bar{x} = \frac{218.5}{73.19} = 2.96 \text{ m}$$

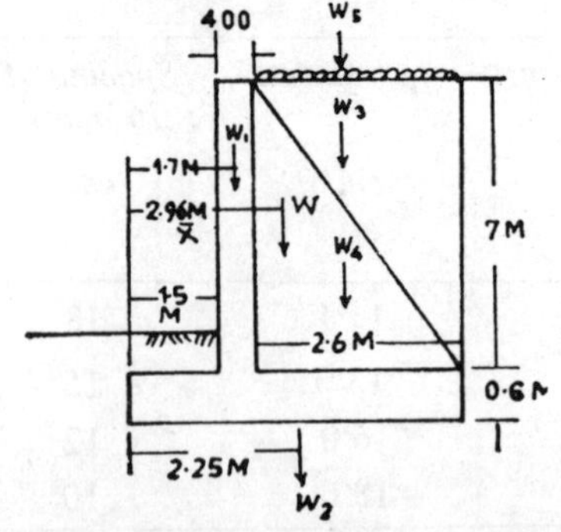

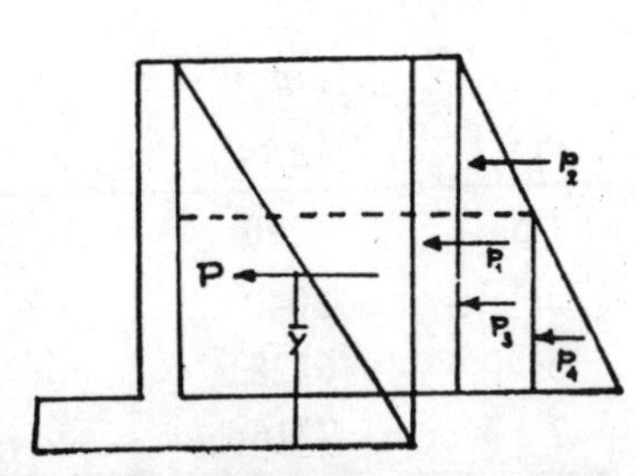

Horizontal forces on one meter width of the wall

	Force due to	Magnitude (tonnes)	Point of application from base (m)	Bending moment (mt)
P_1	u.d.l. from building equivalent to a surcharge of 7·06 m	$p_1 \times 7 = 3.54 \times 7 = 24{\cdot}7$	4·1	99·0
P_2	Earthfill 0 to 3 m triangular portion	$\frac{1}{2} \times p_2 \times 3 = \frac{1}{2} \times 1{\cdot}5 \times 3 = 2{\cdot}25$	5·6	12.6
P_3	Rectangular portion	$p_2 \times 4 = 1{\cdot}5 \times 4 = 6$	2·6	15·3
P_4	Earthfill 3m to 7 m	$\frac{1}{2} \times p_3 \times 4 = \frac{1}{2} \times 2{\cdot}216 \times 4 = 4{\cdot}432$	1·93	8·6

$\Sigma P = 37{\cdot}382$ tonnes $\qquad \Sigma M_h = 135{\cdot}5$ mt

$\bar{y} = 135{\cdot}5/37{\cdot}382 = 3{\cdot}64$ m

4·5·1. Point of Application of the Resultant

Taking moments about 0′,

$W(\bar{x} - \bar{x}_r) = P\bar{y}$

$73{\cdot}19 \times 2{\cdot}96 - 73{\cdot}19 \times \bar{x}_r = 135{\cdot}5$

$73{\cdot}19\bar{x}_r = 218{\cdot}5 - 135{\cdot}5$

$73{\cdot}19\bar{x}_r = 83$

$\bar{x}_r = \dfrac{83}{73{\cdot}19} = 1{\cdot}14$ m

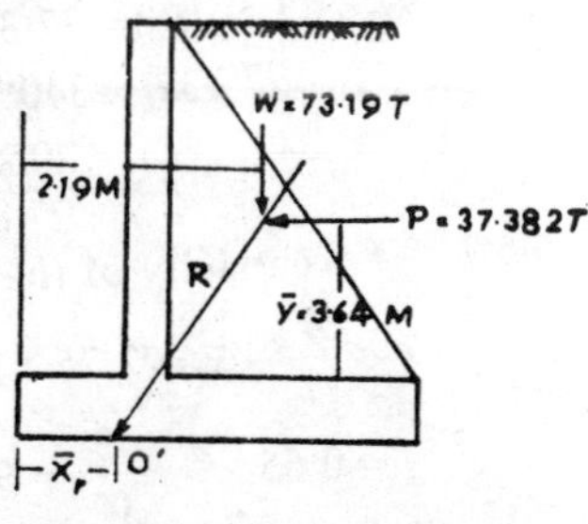

4·5·2. Eccentricity of the Resultant

$e = \dfrac{B}{2} - \bar{x}_r = \dfrac{4{\cdot}5}{2} - 1{\cdot}14$

$= 2{\cdot}25 - 1{\cdot}14 = 1{\cdot}11$ m $\therefore \dfrac{B}{6} = \dfrac{4{\cdot}5}{6} = 0{\cdot}75$ m

The Resultant falls beyond middle third therefore unsafe. Due to heavy surcharge the trial dimension used for the base width is not sufficient. Therefore when a wall is subjected to heavy surcharge the base width may be increased to start with to avoid repetition of calculations for stability.

Let base width be increased to 5·5 m

Component	Load	Magnitude (tonnes)	Point of application from the point 0	Bending moment (mt)
Wall slab	W_1	$7 \times 1 \times 0{\cdot}4 \times 2{\cdot}4 = 6{\cdot}71$	1·7	11·4
Base slab	W_2	$5{\cdot}5 \times {\cdot}6 \times 1 \times 2{\cdot}4 = 7{\cdot}92$	2·75	21·8
Earthfill above heel top 0–3m height	W_3	$3.6 \times 3 \times 1 \times 1{\cdot}5 = 16.2$	3·7	60·0
3m 7m fill	W_4	$3{\cdot}6 \times 4 \times 1 \times 2{\cdot}0 = 28{\cdot}8$	3·7	106·5
u.d.l. equivalent to 7·06 m surcharge	W_5	$7{\cdot}06 \times 1 \times 3{\cdot}6 \times 1{\cdot}5 = 38{\cdot}2$	3·7	141·0

$\Sigma W = 97{\cdot}83$ tonnes $\qquad \Sigma M_v = 340{\cdot}7$ mt

$\bar{x} = 340{\cdot}7/97{\cdot}83 = 3{\cdot}5$ m

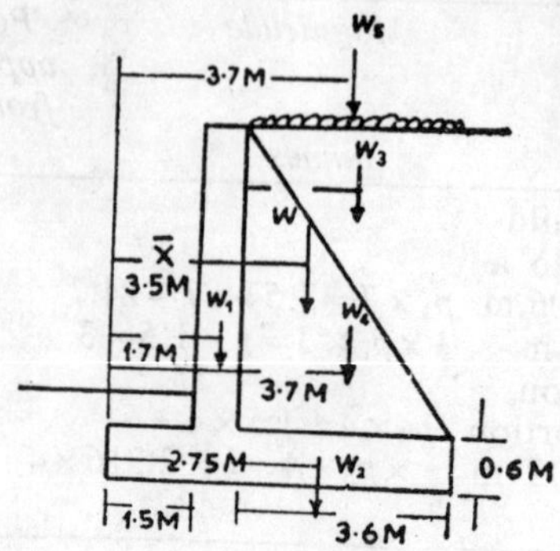

4·5·1. Point of Application of the Resultant

Taking moments 0', $W(\bar{x}-\bar{x}_r)=P\bar{y}$

$$97{\cdot}83(3{\cdot}5 \quad \bar{x}_r)=37{\cdot}382\times3{\cdot}64$$

$$97{\cdot}83\bar{x}_r=340{\cdot}7-135{\cdot}5$$

$$\bar{x}_r=\frac{205{\cdot}2}{97{\cdot}83}=2{\cdot}1\text{ m}$$

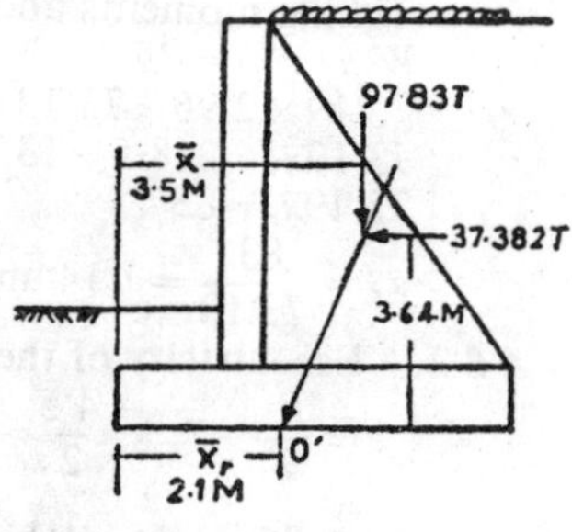

4·5·2. Eccentricity of the Resultant

$$e=\frac{B}{2}-\bar{x}_r=2.75 \quad 2{\cdot}1$$

$$=0{\cdot}65<\frac{5{\cdot}5}{6}={\cdot}916\text{ m}$$

Resutant falls within the middle third.

4·5·3. Pressure Distribution Under the Base

Maximum pressure at the toe end $=\dfrac{W}{B}\left(1+\dfrac{6e}{B}\right)$

$$=\frac{97{\cdot}83}{5{\cdot}5}\left(1+\frac{6\times0{\cdot}65}{5{\cdot}5}\right)$$

$$=\frac{97{\cdot}83}{5{\cdot}5}\times1{\cdot}71$$

$$=30{\cdot}4\text{ t/m}^2$$

Minimum pressure

$$=\frac{W}{B}\left(1-\frac{6e}{B}\right)$$

$$=\frac{97{\cdot}83}{5{\cdot}5}\left(1-\frac{6\times0{\cdot}65}{5{\cdot}5}\right)$$

$$=\frac{97{\cdot}83}{5{\cdot}5}(1 \quad 0{\cdot}71)$$

$$=\frac{97{\cdot}83}{5{\cdot}5}\times0{\cdot}29=5{\cdot}15\text{ t/m}^2$$

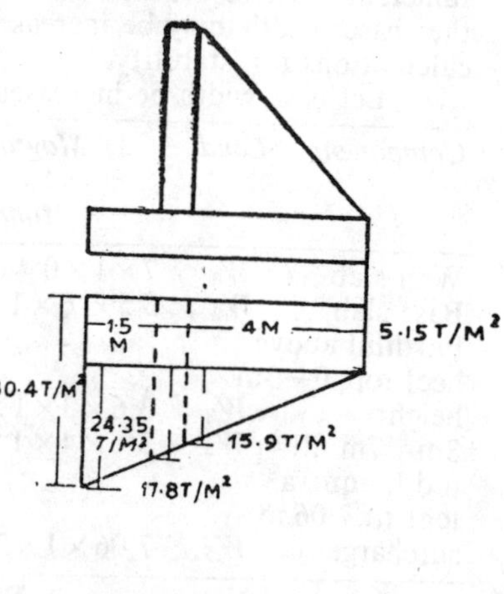

4·6. Toe Design

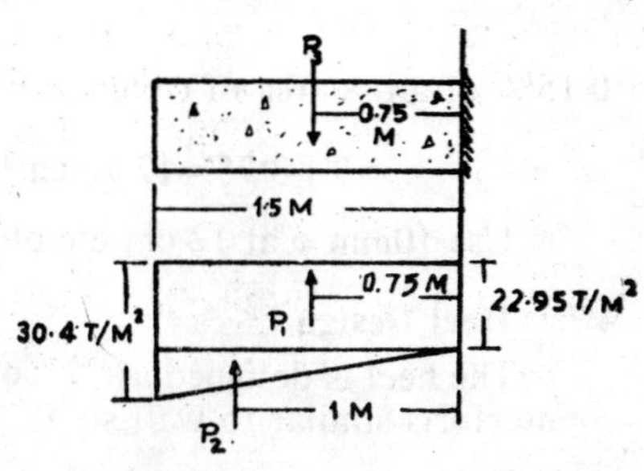

Pressure ordinate at the junction of wall slab and toe

$=5{\cdot}15+(30{\cdot}4-5{\cdot}15)\times\dfrac{4}{5{\cdot}5}$

$=5{\cdot}15+17.80=22{\cdot}95$ t/m²

Bending moment due to forces acting on the toe

(*i*) due to upward soil pressure

Force	*Magnitude tonnes*	*Point of application m*	*Moment mt*
P_1	$22{\cdot}95\times1{\cdot}5=34{\cdot}5$	0·75	25·80
P_2	$\frac{1}{2}\times(30{\cdot}4-22{\cdot}95)\times1{\cdot}5$ $=\frac{1}{2}\times7{\cdot}45\times1{\cdot}5$ $=5{\cdot}6$	1·0	5·60

Total upward moment $=31{\cdot}40\,mt$

(*ii*) due to downward force

self weight of the slab $=1{\cdot}5\times0{\cdot}6\times2{\cdot}4=2{\cdot}16$ tonnes

bending moment $=1{\cdot}5\times0{\cdot}6\times2{\cdot}4\times0{\cdot}75=1{\cdot}62\,mt$

Net bending moment upwards $=31{\cdot}40-1{\cdot}62=29{\cdot}78\,mt$

depth required at the junction $=\dfrac{\sqrt{29780\times100}}{8{\cdot}7\times100}=\sqrt{3410}$

$=58{\cdot}7$ cm

depth required for shear.

Maximum shear $=P_1+P_2-P_3=34{\cdot}5+5{\cdot}6-2{\cdot}16=37{\cdot}94$ tonnes

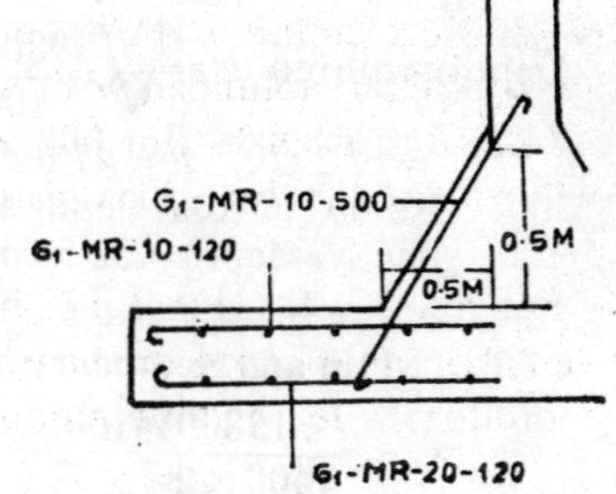

$q_s=5=\dfrac{37940}{100\times{\cdot}87\times d_e}$

$d_e=\dfrac{37940}{100\times{\cdot}87\times5}=87{\cdot}5$ cm

$d=87{\cdot}5+7{\cdot}5$ (cover) $=95{\cdot}0$ cm

assumed thickness of 60 cm for toe slab is not sufficient on both bending moment and shear considerations

Hence provide fillet of 50 cm × 50 cm

depth at the junction = 110 cm

effective depth = 100 cm (say)

4·6·1. Main and Secondary Steel

Amount of steel required

$A_t=\dfrac{29780\times100}{1400\times{\cdot}87\times100}=24{\cdot}5$ cm²

Use 20 mm ϕ at 12 cm c/c

0·15% of gross area of concrete $= \frac{\cdot 15}{100} \times 100 \times 60 + \frac{\cdot 15}{100} \times 50 \times 50$

$= 9 + 3\cdot 75 = 12\cdot 75 \text{ cm}^2$

Use 10mm ϕ at 12 cm c/c on both faces

4·7. Heel Design

The heel is designed as a continuous slab spanning between the counterforts similar to wall slab

Forces acting on the heel

(*i*) Average soil pressure upwards $= \frac{(21\cdot 05 + 5\cdot 15)}{2} = \frac{26\cdot 25}{2}$

$= 13\cdot 125 \text{ t/m}^2$

(*ii*) Forces due to soil fill acting downwards

Top 3m depth of soil $= 3 \times 1 \times 1\cdot 5 = 4\cdot 5 \text{ t/m}^2$

3m to 7m depth of soil
$= 4 \times 1 \times 2 = 8 \text{ t/m}^2$

Surcharge of 7·06 m
$= 7\cdot 06 \times 1 \times 1\cdot 5 = 10\cdot 6 \text{ t/m}^2$

self weight of slab
$= \cdot 60 \times 1 \times 2\cdot 4 = 1\cdot 44 \text{ t/m}^2$

Total downward load
$= 4\cdot 5 + 8 + 10\cdot 6 + 1\cdot 44$
$= 24\cdot 54 \text{ t/m}^2$

Maximum net load downwards at the edge of the heel

$= 24\cdot 54 - 5\cdot 15 = 19\cdot 39 \text{ t/m}^2$

Bending moment at the edge of the heel slab

$\frac{wl^2}{12} = \frac{19\cdot 39 \times 3\cdot 5^2}{12} = 19\cdot 80 \text{ mt}$

Depth required $d_e = \sqrt{\frac{19800 \times 100}{100 \times 8\cdot 7}}$

$= \sqrt{2270}, \quad d_e = 47\cdot 6 \text{ cm}$

Use 60 cm total depth and effective depth $= 52\cdot 5$ cm

4·7·1. Main and Secondary Steel

$A_t = \frac{19800 \times 100}{1400 \times \cdot 87 \times 52\cdot 5}$

$= 31 \text{ cm}^2$

Use 22mm ϕ at 12 cm c/c

$A_t = \frac{\cdot 15}{100} \times 60 \times 100$

$= 9 \text{ cm}^2$

Use 12 mm ϕ 12 cm c/c

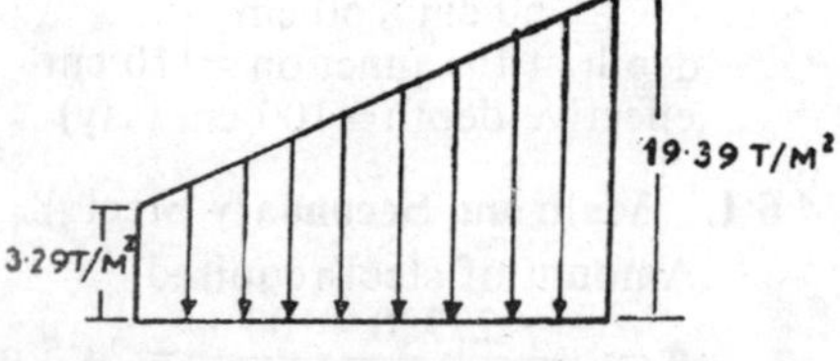

4·8. Design of Counterfort

Maximum Bending Moment

	Force	*Magnitude tonnes*	*Point of application m*	*Moment mt*
Surcharge pressure	P_1	$3{\cdot}54\times7\times3{\cdot}5 = 84{\cdot}0$	3·5	294·0
Upper strata of soil	P_2	$\frac{1}{2}\times3\times1{\cdot}5\times3{\cdot}5 = 7{\cdot}89$	5	39·4
	P_3	$1{\cdot}5\times4\times3{\cdot}5=21{\cdot}0$	2	42·0
Lower strata of soil	P_4	$\frac{1}{2}\times2{\cdot}216\times4\times3{\cdot}5 = 15{\cdot}5$	1·33	20·6

$P_h = 128{\cdot}39$ tonnes $\qquad\qquad \Sigma M_h = 396{\cdot}0$ mt

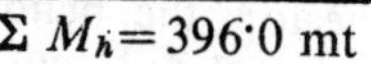
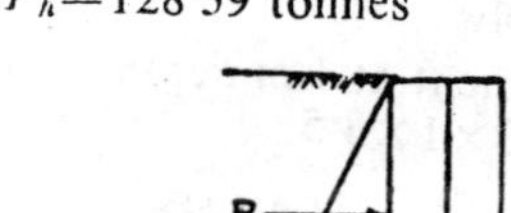
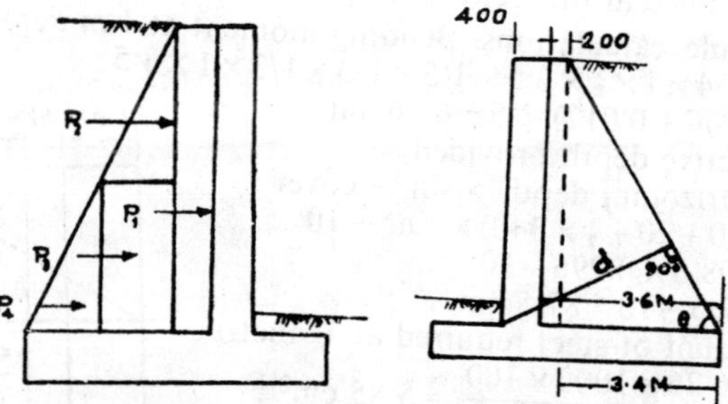

$$\tan\theta = \frac{7}{3{\cdot}4} = 2{\cdot}06 \text{ and } \theta = 64^\circ - 6'$$

$d = 4{\cdot}0\times\sin\theta = 4\times{\cdot}8996\times100 = 360$ cm

$d_e = 360 - 10$ (cover) $= 350$ cm

Width of counterfort from bending moment consideration

$$b = \frac{396\times1000\times100}{8{\cdot}7\times350\times350} = 38{\cdot}6 \text{ cm}$$

4·8·1. Main Reinforcement

A_t = Amount of steel required

$$= \frac{396\times1000\times100}{1400\times{\cdot}87\times350} = 95{\cdot}5 \text{ cm}^2$$

No. of 32 mm ϕ required $= \frac{95{\cdot}5}{8{\cdot}04} = 12$ Nos.

Using 12 Nos of 32 mm ϕ in two Layers, ($A_t = 112{\cdot}5$ cm^2) width required to accommodate 7 bars

= 5 cm cover on either side + 6 bar diameters + 5 spacings each equal to bar diameter or maximum size of the aggregate

$= 10 + 6\times3{\cdot}2 + 5\times3{\cdot}2$

$= 10 + 19{\cdot}2 + 16{\cdot}0 = 45.2$ cm

4·8·2. Width of counterfort on Shear Consideration

Total shear at the junction of base and the counterfort

$$= P_h - T\cos\theta = 128{\cdot}39 - 112{\cdot}5\times0{\cdot}4368\times\frac{1400}{1000}$$

$= 128{\cdot}39 - 69{\cdot}0 = 59{\cdot}39$ tonnes

$$\text{Shear stress} = \frac{59390}{b\times{\cdot}87\times310} = 5$$

$$b = \frac{59390}{5\times{\cdot}87\times310} = 44 \text{ cm}$$

Width of counterfort is governed by the arrangement of steel that is b 45·2 cm

Use $b=50$ cm $> \frac{h}{20}$ so that the shear stress would be less than 5 kg/cm².

4·8·3. Curtailment of Main Reinforcement in the Counterfort

12 bars of 32 mm ϕ are required at the lowest level of the counterfort, since the maximum bending moment occurs at that level. Towards the top, the bending moment reduces and the reinforcement can be curtailed at different levels suitably.

Sample calculations. Bending moment at 1 m depth from top
$= 3{\cdot}54\times1\times\frac{1}{2}\times3{\cdot}5+1{\cdot}5\times1/3\times1/3\times1\times3{\cdot}5$
$=(1{\cdot}77+0{\cdot}017)\ 3{\cdot}5 = 6{\cdot}26$ mt
Effective depth provided
$=$horizontal depth $\times \sin\theta -$ cover
$=(40+20+\frac{1}{7}\times340)\times\sin\theta - 10$
$=108{\cdot}5\times{\cdot}8996-10$
$=97{\cdot}5-10=87{\cdot}5$ cm
Amount of steel required at 1 metre

$$\text{depth} = \frac{6{\cdot}26\times1000\times100}{{\cdot}87\times87{\cdot}5\times1400} = 5{\cdot}85 \text{ cm}^2$$

No. of bars of 32 mm ϕ required = 1

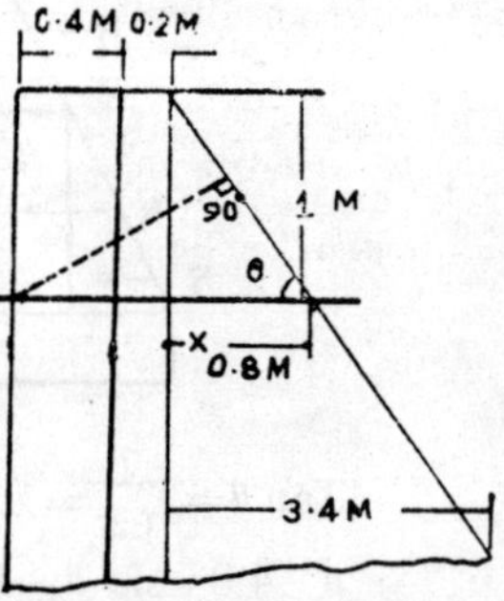

Similarly calculations for different levels are calculated as in Table

depth m	Bending moment mt	Horizontal depth	Effective depth	A_t	No. of bars of 32 mm ϕ required	No. of bars used
1	6·26	108·5	87·5	5·85	1	1
2	27·25	157·0	131	17·1	3	3
3	63·5	204·0	173	30·0	4	
4	118	254	217	44·0	6	6
5	191	303·0	262	59·8	8	9
6	287·5	352·0	305	77·1	10	
7	396	400·0	350	95·5	12	12

4·8·4. Connection of Counterfort with Wall Slab

Horizontal pressure acting over a length of 3·5 m on the wall might tear off the wall from counterfort. Therefore to prevent such failures the main reinforcement in the wall slab is tied with the main reinforcement in the counterfort.

Maximum horizontal pressure occurs at the junction of the base and counterfort and the pressure reduces towards the top. Therefore the reinforcement required to resist this pull is calculated on the basis of the pressure at different levels and the reinforcement is provided in the form of closed stirrups tied firmly to main reinforcement in the wall slab and counterfort.

Depth	*Maximum horizontal pressure*	*Tensile force*	A_t	*Spacing of 12 mm φ stirrups*
m	*kg/m²*	*kg*	*cm²*	*cm*
1·5	4290	15000	10 7	20
3	5040	17600	12·6	18
5	6128	24900	17 8	12
7	7216	25300	18·1	12

4·8 5. Connection of Base and Counterfort

In order to prevent wall slab along with counterfort tearing off from the base, vertical stirrups are used to tie the main reinforcement in the counterfort and the base slab.

Main pressure at the edge of the heel
$=19{\cdot}39\ t/m^2$

Pull acting downwards over a length of 3·5 m
$=3\ 5\times 19{\cdot}39=67{\cdot}8$ tonnes

$$A_t \text{ required} = \frac{67800}{1400} = 48{\cdot}5\ cm^2$$

Pull at the juction of the wall slab $=3290\times 3{\cdot}5=11{,}500$ kg

$$A_t \text{ required} = \frac{11500}{1400} = 8{\cdot}2\ cm^2$$

Spacing of two legged 16 mm φ stirrups at the edge $=8$ cm
at the junction $=40$ cm

Use 16 mm φ 2-legged stirrups at the edge with a minimum spacing of 8 cm c/c at the edge gradually increasing to a maximum spacing 40 cm c/c towards the junction.

4·9. Stability

$$\text{Factor of safety against overturning} = \frac{\text{stabilizing moment}}{\text{overturning moment}}$$

$$=\frac{W\bar{x}}{P\bar{y}}=\frac{97{\cdot}83\times 3.5}{37{\cdot}382\times 3{\cdot}64}=2\ 52>1{\cdot}5$$

$$\text{Factor of safety against sliding}=\frac{\text{Resisting force}}{\text{Sliding force}}=\frac{\mu W}{P}$$

$$=\frac{0{\cdot}6\times 97{\cdot}83}{37{\cdot}382}=1\ 57>1{\cdot}5$$

5 Counterfort Retaining Wall for Bridge Abutment

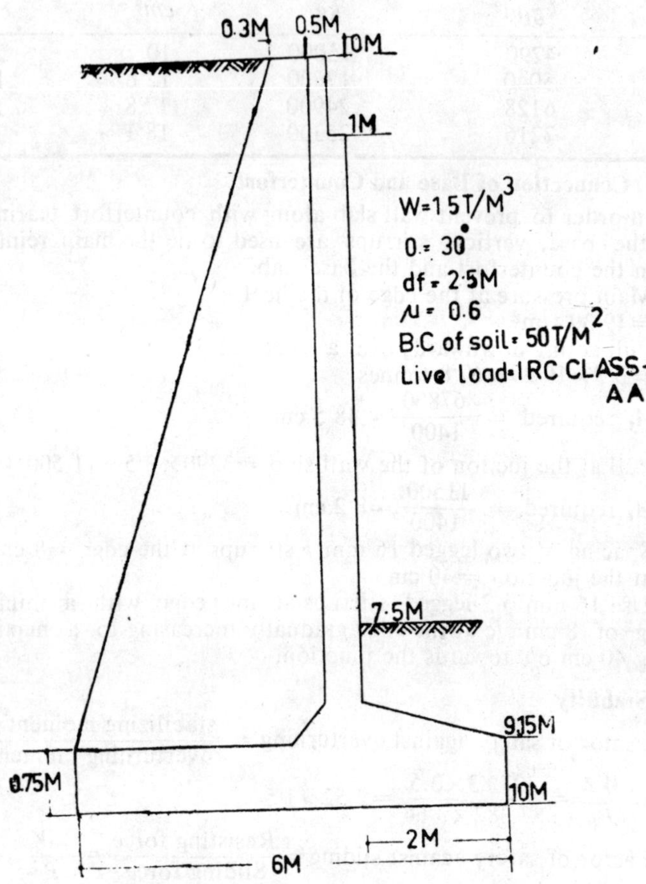

5.1. Data

Loading = I R C Class AA
Type = Multilane – T – beam bridge
Height of the abutment above Bed level = 7.5 m
Length of abutment = 7.6 m for multilane bridges
Depth of foundation = 2.5 m
Weight of fill material = 1.5 t/m³
Angle of internal friction = 30°
Bearing capacity of soil = 50 t/m²
Coefficient of friction between soil and concrete = 0.6
Materials Grade – I – steel, Concrete M 150

5·2. Allowable Stresses

Concrete compressive strength in bending=50 kg/cm², in shear =5 kg/cm², in bond=10 kg/cm²

σ_{st}=tensile strength of steel=1400 kg/cm²

$$\text{Modular ratio} = m = \frac{2800}{3\,\sigma_{cb}}$$

$$= \frac{2800}{3\times50} = 18{\cdot}66$$

Lever arm=jd=0·87 d and Moment of resistance =8·7 bd^2

5·3. Trial Section

Spacing of counterforts $l=\frac{1}{2}$ to $\frac{1}{3}$ h=3·75 m to 2·5 m

Use three counterforts over a length of 7·6 m of Abutment, say l=3·5 m

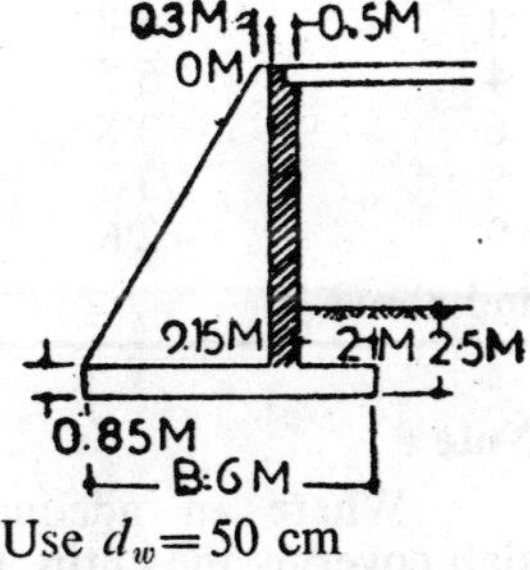

As the wall is subjected to heavy surcharge the trial dimensions may be suitably increased.

Base width=$\frac{1}{2}$ to $\frac{2}{3}$ h=3·75 m to 5 m
Use B=6 m

$$\text{Thickness of base slab} = \frac{h}{12} = \frac{750}{12}$$

=62 5 cm Use d_b=85 cm

$$\text{Width of wall slab} = \frac{h}{12} = 31{\cdot}25 \text{ cm} \quad \text{Use } d_w = 50 \text{ cm}$$

The wall slab should be wide enough to accommodate the bridge girder

Width of toe slab=$\frac{1}{3}B$=2 m

5·3·1. Earth Pressure Calculations

Horizontal earth pressure ordinate due to backfill.

$$= wh\left(\frac{1-\sin\phi}{1+\sin\phi}\right) = 1500\times h\left(\frac{1-\sin 30^\circ}{1+\sin 30^\circ}\right) = 1500h\times\tfrac{1}{3} = 500\,h$$

Earth pressure due to class AA loading

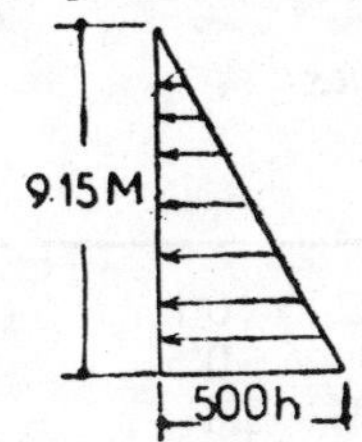

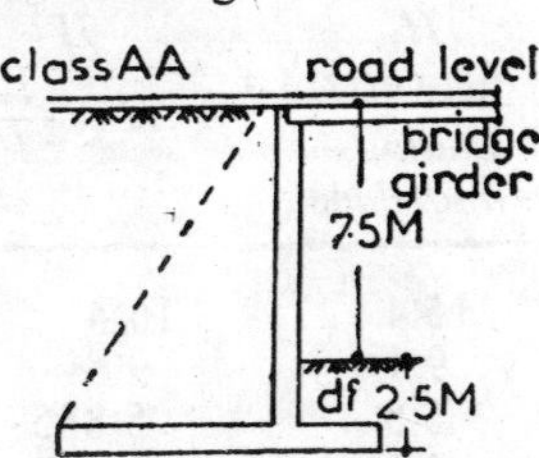

For bridge abutments, the concentrated surface loads due to the wheel or track loads of any of the IRC standard vehicles or trains on the backfill, shall be considered to have the same effect as equivalent heights of surcharges as shown in the table, which are based on the Spangler's equation.

Table of Equivalent Heights and Surcharge of Earth

Depth of abutment below the road level in m	$w=1600$ kg/m^3 $\phi=30°$ *H in metre for the concentrated surface loads due to the wheel or track loads of the following IRC standard loadings*					
	IRC CLASS-AA		*IRC CLASS-A*		*IRC CLASS-B*	
	single lane bridge	*Multilane bridge*	*Single lane*	*Multi-lane*	*Single lane*	*Multi-lane*
0·2	26	15·4	14·3	17 2	8·3	10·0
1	15	9·1	8·5	10·0	5·1	5·8
2	8	5·5	5·1	6·1	3·0	3·7
3	6·8	4·1	3·8	4·6	2·3	2·7
4	5·5	3·3	3·0	3·5	1·8	2·1
6	3·8	2·8	2·2	2·6	1·3	1·5
8	3	1·8	1·7	2·0	1·0	1·2
10 and above	2·6	1.5	1·4	1·7	0·9	1·0

Note :

Where an adequately designed reinforced concrete approach slab covering the entire width of the roadway, with one end resting on the structure designed to retain earth and extending for a length of not less than 3·5 m into the approach is provided, no live load surcharge need be considered in the design.

5·3·2. Earth Pressure Ordinates due to LL Surcharge

Earth pressure ordinates due to L.L. surcharge				*due to back fill tonnes*	*Total pressure ordinate*
depth of abutment below the road level m	*Equivalent height of surcharge H for class AA loading multilane*	*Modified height of surcharge H* $=H\times\frac{1.6}{1·5}$	*Pressure ordinate per Metre width tonnes*		
0·2	15·4	16·4	8·22	0·1	8·32
1	9·1	9·8	4·9	0·5	5·4
2	5·5	5·885	2·94	1	3·94
3	4·1	4·387	2·19	1·5	3·69
4	3·3	3·531	1·76	2	3·76
6	2·3	2·461	1·23	3	4·23
7·5	1·9	2·033	1·02	3·75	4·95
8	1·8	1·926	·96	4	4·96
9·15	1·63	1·74	·87	4·575	5·345
10	1·5	1·6	·80	—	—

Sample calculation for 0·2 m depth
$H=15·4$ m for $w=1·6$ t/m³ and $\phi=30°$
But in the problem $w=1·5$ t/m³ and $\phi=30°$
Therefore modified equivalent height

$$=H'=15·4\times\frac{1·6}{1·5}=16·4 \text{ m}$$

Pressure ordinate due to this surcharge per metre width

$$=wh'\left(\frac{1-\sin\phi}{1+\sin\phi}\right)=1·5\times16·4\times\tfrac{1}{3}=8·225 \text{ tonnes}$$

Similarly these ordinates are calculated for various depths

5·4. Wall Slab Dessign

Sample calculation for 0·2 m depth
Pressure ordinate$=p=8·32$ tonnes

Maximum bending moment (only two spans)$=\dfrac{pl^2}{10}$

$$=\frac{8·32}{10}\times3·5^2=10·4 \text{ mt}=1040{,}000 \text{ cm kg}$$

$$\text{Depth of slab required}=\sqrt{\frac{10{,}40{,}000}{100\times8·7}}=34·6 \text{ cm}$$

Use $d=50$ cm and $d_e=42·5$ cm

At other depths the pressure ordinates are less than the ordinate at 0·2 m depth. Hence the same depth can be provided throughout. However spacing of the reinforcement can be changed.

5·4·1. Main and Secondary Reinforcement

$$A_t \text{ required at 0·2 m depth}=\frac{10{,}40{,}000}{·87\times42·5\times1400}=20·2 \text{ cm}^2$$

Use 20 mm ϕ at 15 cm c/c

depth below from top m	*pressure ordinate tonnes*	*BM* $\frac{pl^2}{10}$ *mt*	A_t *cm²*	*bar dia mm*	*spacing mm*
0·2	8·32	10·4	20·2	20	150
1	5·4	6·61	12·8	16	155
2	3·94	4·83	9·35	12	120
3	3·69	4·52	8·75	12	125
4	3·76	4·6	8·90	12	125
6	4·23	5·18	10·00	16	200
7·5	4·95	6·05	11·70	16	170
8	4·96	6·06	11·7	16	170
9·15	5·345	6·55	12·7	16	155

$$A_t=\frac{0·15\times50\times100}{100}=7·5 \text{ cm}^2$$

Use 10 mm ϕ at 10 cm c/c

5·5. Base Design

Verticle loads. Consider one metre length of wall

Component	*Load*	*Magnitude* tonnes	*Point of application from 0* m	*Moment* mt
Wall slab	W_1	9·15×·5×2·4=11	2·25	24·75
Base slab	W_2	6×·85×2·4=12·25	3	36·75
Back fill	W_3	9·15×3·5×1·5=48	4·25	204·00
Live load	W_4	70	4·3	300·00
		ΣW=141·25 tonnes		ΣM=565·5 mt

$$\bar{x}\ \frac{\Sigma M}{\Sigma W}=\frac{565{\cdot}5}{141{\cdot}25}=4 \text{ m}$$

Horizontal loads (*i*) due to earth fill

$$P_1=\frac{1}{2}\times\frac{h}{2}\times h$$

$$=\frac{1}{2}\times\frac{9{\cdot}15}{2}\times 9{\cdot}15=20{\cdot}75 \text{ tonnes}$$

Point of application above base

$$=\frac{h}{3}=\frac{9{\cdot}15}{3}=3{\cdot}05 \text{ m}$$

$\bar{y}_1=3{\cdot}9$ m (from bottom of the base)

(*ii*) due to L.L. surcharge

Force	*Magnitude* tonnes	*Point of application from bottom of the base* m	*Moment* mt
1	2	3	4
Force			
P_1	4·9×·8=3·92	9·4	37
P_2	½×(8·22−4·9)×·8=½×3·32×·8 =1·328	9·54	12·5
P_3	2·94×1	8·5	25·0
P_4	½(4·90 2·94)×1=·98	8·67	8·45
P_5	2 19×1	7 5	16·4
P_6	½(2·94 2·19)=0.375	7·67	2·87
P_7	1·76×2=3·52	6·5	22·9
P_8	½(2·19 1·76)×2=0·43	6·67	2·86
P_9	1·23×2=2·46	5	12·20
P_{10}	½(1·76 1·23)×2=0·53	5·33	2·84
P_{11}	0·96×2=1·92	3	5·76
P_{12}	½(1·23−0·96)×2=0·27	3·33	0·91
P_{13}	0·80×2=1·60	1	1·60
P_{14}	½(·96−·80)×2=·16	1·33	0·21
	ΣP_2=22·623 tonnes		ΣM_h=151·50 mt

$$\bar{y}_2 = \frac{151{\cdot}50}{22{\cdot}623} = 6{\cdot}7 \text{ m}$$

Resultant of P_1 and P_2

$$y = \frac{P_1\bar{y}_1 + P_2\bar{y}_2}{P+P_2}$$

$$= \frac{20{\cdot}75 \times 3{\cdot}9 + 22{\cdot}623 \times 6{\cdot}7}{20{\cdot}75 + 22{\cdot}623}$$

$$= \frac{81+151{\cdot}5}{43{\cdot}373} = \frac{232{\cdot}5}{43{\cdot}373} = 5{\cdot}35 \text{ m}$$

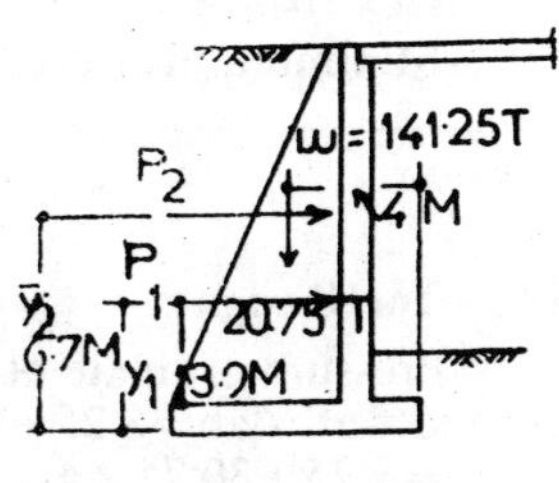

5·5·1. Point of Application of the Resultant Force R

Taking moments about 0′

$W(\bar{x} - \bar{x}_r) = Py$

$141{\cdot}25(4 - \bar{x}_r) = 43{\cdot}373 \times 5{\cdot}35$

$565 - 141{\cdot}25\bar{x}_r = 232{\cdot}5$

$$\bar{x}_r = \frac{565 - 232{\cdot}5}{141{\cdot}25} = \frac{332{\cdot}5}{141{\cdot}25} = 2{\cdot}35 \text{ m}$$

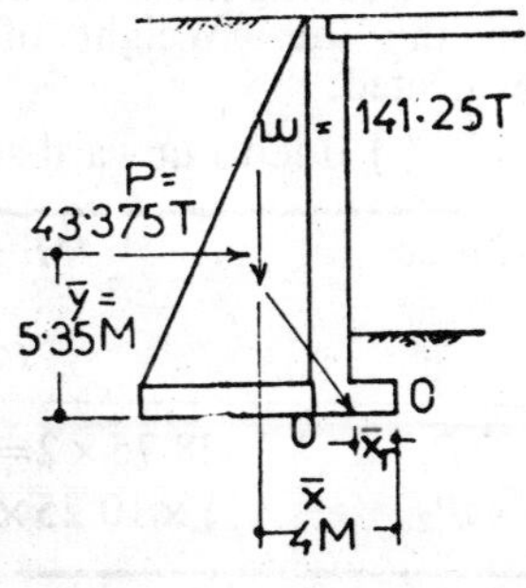

5·5·2. Eccentricity of the Resultant

$$e = \frac{B}{2} - \bar{x}_r = \frac{6}{2} - 2{\cdot}35 = 0{\cdot}66 \text{ m} < \frac{B}{6}$$

Case (ii) When Live Load is not Present

Total vertical load $= 141{\cdot}25 - 70$
$= 71{\cdot}25$ tonnes

Moment of vertical loads
$= 565{\cdot}6 - 300 = 265{\cdot}6$ mt

$$\bar{x} = \frac{265{\cdot}6}{71{\cdot}25} = 3{\cdot}73 \text{ m}$$

Horizontal force $P = 20{\cdot}75$ tonnes

$\bar{y} = 3{\cdot}9$ m

Point of Application of the Resultant

$W(\bar{x} - \bar{x}_r) = P\bar{y}$

$71{\cdot}25(3{\cdot}73 - \bar{x}_r) = 20{\cdot}75 \times 3{\cdot}9$

$$\bar{x}_r = \frac{266-81}{71{\cdot}25} = \frac{185}{71{\cdot}25} = 2{\cdot}6 \text{ m}$$

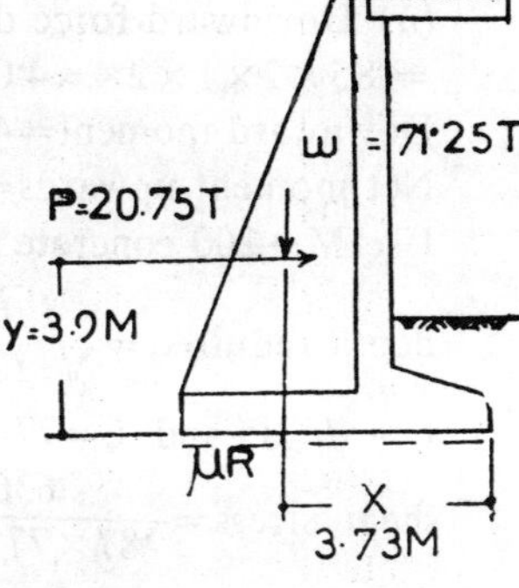

Eccentricity

$$e = \frac{B}{2} - \bar{x}_r = 3 - 2{\cdot}6 - 0{\cdot}4 \text{ m} < 1 \text{ m}$$

Since the eccentricity is greater in case (*i*) when the live load is present, base is to be designed for case (*i*) loading.

5·5·3. Pressure Distribution under the Base

$$\text{Maximum pressure at toe point} = \frac{W}{B}\left(1 + \frac{6e}{B}\right)$$

$$= \frac{141{\cdot}25}{6}\left(1 + \frac{6 \times 0{\cdot}65}{6}\right) = \frac{141{\cdot}25}{6} \times 1{\cdot}65 = 39 \text{ tm}^2 < 50 \text{ t/m}^2$$

Minimum pressure $=\frac{W}{B}\left(1-\frac{6e}{B}\right)=\frac{141\cdot25}{6}\left(1-\frac{6\times\cdot065}{6}\right)$.

$$=\frac{141\cdot25}{6}\times0\cdot35=8\cdot25\ \text{t/m}^2$$

5·6. Toe Design

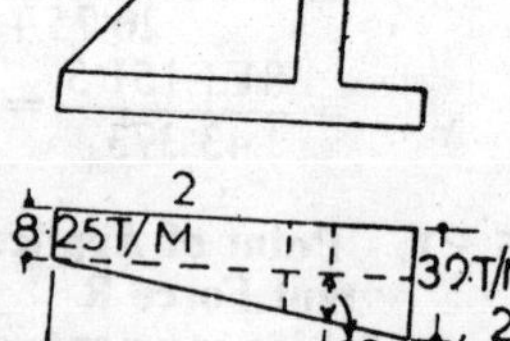

Pressure ordinate at the junction of the toe and wall slab $=8\cdot25+(39-8\cdot25)\times\frac{4}{6}$
$=8\cdot25+30\cdot75\times\frac{4}{6}$ - $8\cdot25+20\cdot5$
$=28\cdot75$ t/m²

Bending moment due to force acting on the toe. Weight of fill above toe is neglected.

(*i*) due to upward soil pressure

Force	*Magnitude* tonnes	*Point of application* m	*Moment* mt
P_1	$28\cdot75\times2=57\cdot50$	1	57·50
P_2	$\frac{1}{2}\times10\cdot25\times2=10\cdot25$	1·33	13·40

Total upward moment = 70·9 mt

(*ii*) Downward force due to self weight of the toe slab
$=\cdot85\times2\times1\times2\cdot4=4\cdot06$ tonnes

Downward moment $=4\cdot06\times1=4\cdot06$ mt

Net moment upwards $=70\cdot9-4\cdot06=66\cdot84$ mt

Use M$-$200 concrete for the base slab

$$\text{depth required}=\sqrt{\frac{6684000}{12\cdot1\times100}}=\sqrt{550}=74\ \text{cm}$$

Use $d=85$ and $d_e=77\cdot5$ cm

$$\text{shear stress}=\frac{63690}{\cdot87\times77\cdot5\times100}=9\cdot4\ \text{kg/cm}^2>7\ \text{kg/cm}^2$$

The depth is not sufficient on shear consideration. Therefore increase the depth.

$$\text{depth required to resist shear } d=\frac{63690}{\cdot87\times7\times100}=105\ \text{cm}$$

Use $d=115$ cm and $d_e=107\cdot5$ cm at the jucntion and 85 cm at the end

5·6·1. Main and Secondary Steel

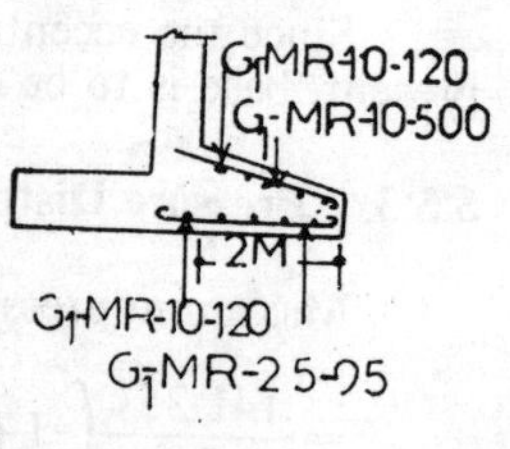

$$A_t=\frac{6684000}{1400\times\cdot87\times107\cdot5}=51\ \text{cm}^2$$

Use 25 cm at 9·5 cm c/c

$$A_t=\frac{0\cdot15}{100}\times100\times85=12\cdot75\ \text{cm}^2$$

Use 10 mm at 12 cm c/c both faces.

5·7. Design of Heel Slab

The heel is designed as a continuous slab spanning between the counterforts similar to wall slab.

Forces acting on the heel, pressure ordinate at the junction of heel slab and wall slab.

$$= 8{\cdot}25 + (39 - 8{\cdot}25) \times \frac{3{\cdot}5}{6}$$

$$= 8{\cdot}25 + 30{\cdot}75 \times \frac{3{\cdot}5}{6} = 8{\cdot}25 + 17{\cdot}9$$

$= 26{\cdot}15$ t/m²

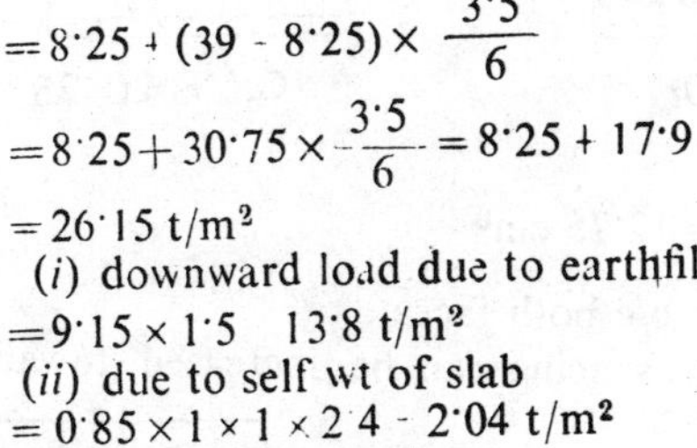

(*i*) downward load due to earthfill

$= 9{\cdot}15 \times 1{\cdot}5 \quad 13{\cdot}8$ t/m²

(*ii*) due to self wt of slab

$= 0{\cdot}85 \times 1 \times 1 \times 2{\cdot}4 - 2{\cdot}04$ t/m²

(*iii*) due to live load

$$= \frac{70}{3{\cdot}6 \times 2 \times {\cdot}85} = 11{\cdot}4 \text{ t/m}^2$$

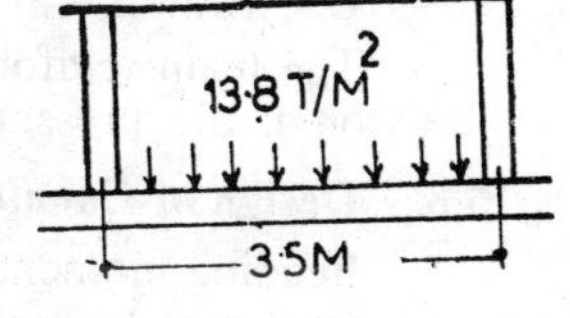

Total downward uniformly distributed load
$= 13{\cdot}8 + 2{\cdot}04 + 11{\cdot}4 = 27{\cdot}24$ tonnes

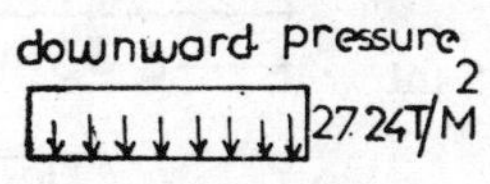

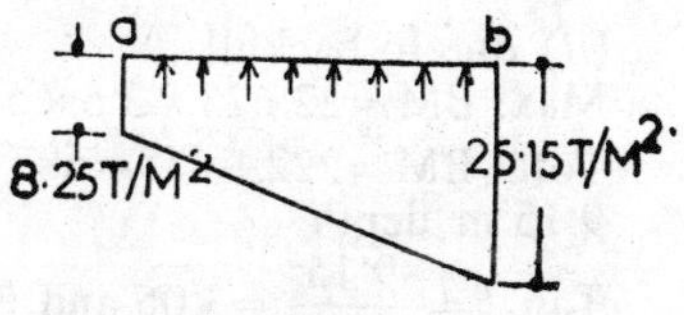

Pressure ordinate at heel point (*a*) $27{\cdot}24 - 8{\cdot}25 = 18{\cdot}99$ t/m²

Pressure ordinate at junction point (*b*) $= 27{\cdot}24 - 26{\cdot}15 = 1{\cdot}09$ t/m²

Maximum downward pressure at the edge of the heel
$= 18.99$ t/m²

$$\text{Bending moment} = \frac{wl^2}{10} = \frac{18{\cdot}99 \times 3{\cdot}5^2}{10} = 23{\cdot}2 \text{ mt}$$

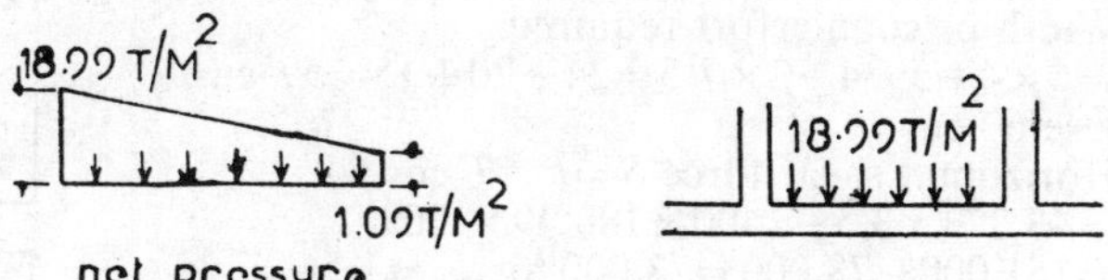

$$\text{depth required} = \sqrt{\frac{2320000}{12{\cdot}1 \times 100}} = \sqrt{1920} = 44 \text{ cm} < 85 \text{ cm}$$

Use $d = 85$ and $d_e = 77{\cdot}5$ cm

5·7·1. Main and Secondary Reinforcement

$$A_t = \frac{2320000}{1400 \times {\cdot}87 \times 77{\cdot}5} = 24{\cdot}5 \text{ cm}^2$$

Use 20 mm ϕ at 12·5 cm c/c

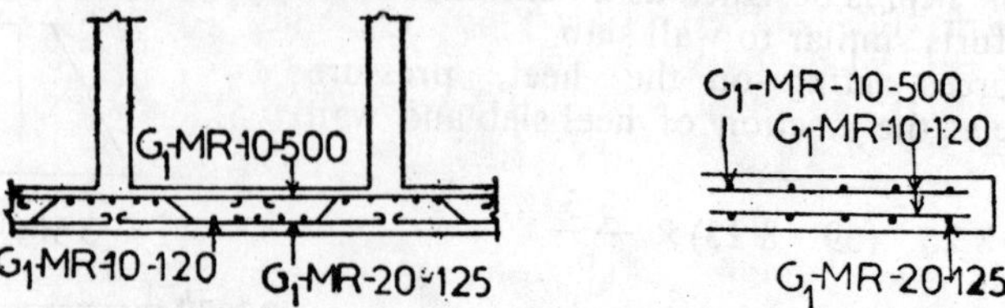

$$A_t = \frac{0\cdot15}{100} \times 85 \times 100 = 12\cdot75 \text{ cm}^2$$

Use 10 mm ϕ at 12 cm c/c both faces

The main reinforcement spacing can be increased towards the junction as the pressure decreases.

5·8. Design of Counterfort

Bending moments due to horizontal forces

(*i*) due to back fill

Max. BM$=P_1 \times l \times y_1$

$=20\cdot75 \times 3\cdot5 \times 3\cdot05 = 222$ mt

(*ii*) due to backfill

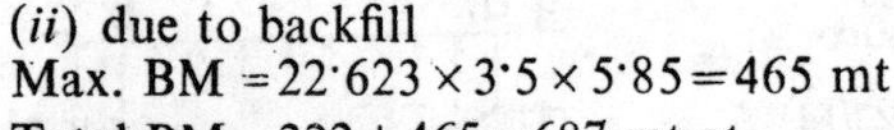

Max. BM $=22\cdot623 \times 3\cdot5 \times 5\cdot85 = 465$ mt

Total BM$=222+465=687$ mt at 9·15 m depth

$\tan\theta = \dfrac{9\cdot15}{3} = 3\cdot05$ and $\theta = 72°$

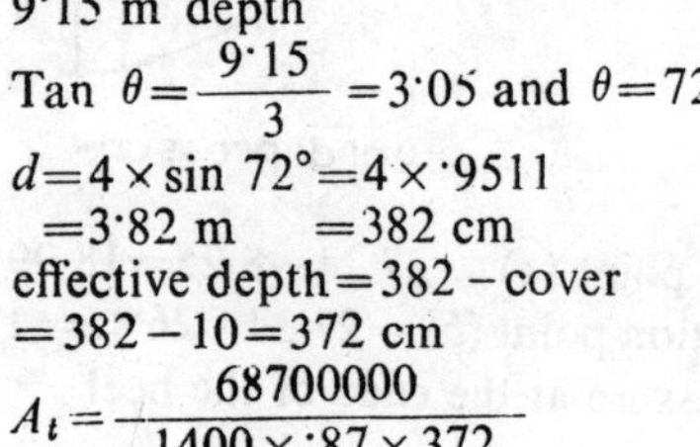

$d = 4 \times \sin 72° = 4 \times \cdot9511$

$=3\cdot82$ m $=382$ cm

effective depth$=382-$cover

$=382-10=372$ cm

$$A_t = \frac{68700000}{1400 \times \cdot87 \times 372}$$

$=151$ cm^2

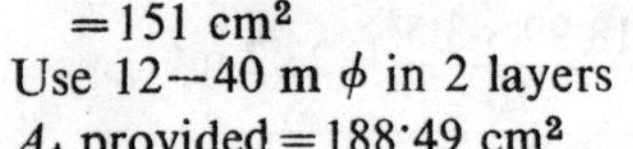

Use 12—40 m ϕ in 2 layers

A_t provided$=188\cdot49$ cm^2

Width of counterfort required

$=6\times4+5\times4+2\times7\cdot5=24+20+15=59$ cm.

Use 60 cm

Horizontal shear force $S=P \quad T\cos\theta$

$=43{,}373\times3\cdot5-1400\times188\cdot49\times\cdot309$

$=151{,}000-78{,}000=73{,}000$ kg

$$\text{Shear stress} = \frac{73{,}000}{60\times\cdot87\times400} = 3\cdot5 \text{ kg/cm}^2 < 5 \text{ kg/cm}^2$$

5·8·1. Curtailment of Main Reinforcement

The horizontal pressure diagram due to earth fill and live load is not a regular one. Therefore the bending moment at any depth from top is calculated considering the pressure ordinates at different levels. The total pressures and their moments are calculated

Sample calculation. Bending moment at 2 m depth from top

$= [5{\cdot}4 \times {\cdot}8 \times 3{\cdot}5] \times 1{\cdot}4$
$+ [\tfrac{1}{2}(8{\cdot}32 - 5{\cdot}40) \times 3{\cdot}5 \times {\cdot}8] \times 1{\cdot}53$
$+ [3{\cdot}94 \times 1 \times 3{\cdot}5] \times 0{\cdot}5$
$+ [\tfrac{1}{2}(5{\cdot}40 - 3{\cdot}94) \times 3{\cdot}5 \times {\cdot}8] \times 0{\cdot}67$
$= 15{\cdot}1 \times 1{\cdot}4 + 4{\cdot}1 \times 1{\cdot}53 + 13{\cdot}8 \times 0{\cdot}5$
$+ 2{\cdot}56 \times 0{\cdot}67$
$= 21{\cdot}2 + 6{\cdot}25 + 13{\cdot}8 + 1{\cdot}71 = 42{\cdot}96$ mt

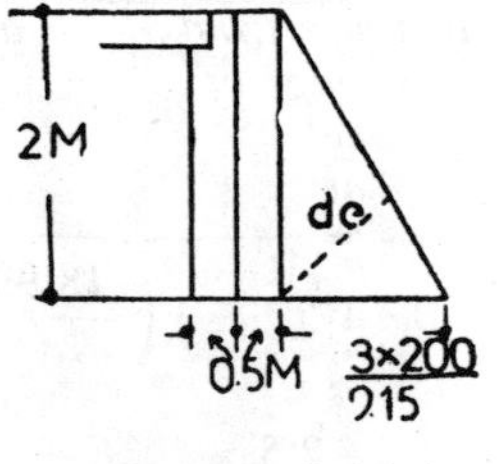

Horizontal depth provided at 2 m depth

$= (100 - \text{cover}) + \dfrac{3}{9{\cdot}15} \times 200$

$= (100 - 10) + 67 = 157$ cm

effective depth $d_e = 157 \sin \theta$

$= 157 \times {\cdot}9511 = 150$ cm

Amount of reinforcement required

$= \dfrac{4296000}{1400 \times {\cdot}87 \times 150} = 23{\cdot}4 \text{ cm}^2$

Use Two – 40 mm ϕ

Similarly the computations are made for 4, 6 and 8 m depths.

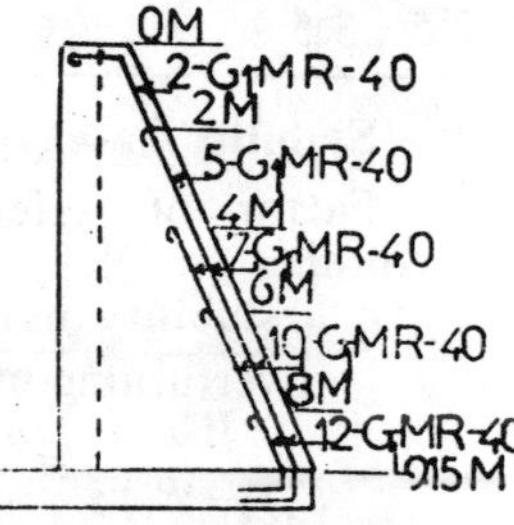

depth from top m	*B.M. mt*	*Horizontal depth cm*	*Effective depth cm*	*A_t cm²*	*No. of 40 mm bar*
2	42·96	157	150	23·4	2
4	134·12	224	214	51·6	5
6	285·80	291	278	85·0	7
8	497·85	358	340	120·0	10
9·15	687·0	382	372	151·0	12

5·8·2. Connection to Counterfort and Wall Slab

depth from top m	*Average horizontal pull – p tonnes*	*A_t p/1400 cm²*	*Spacing 2 legged stirrups cm*
0–1	$\left(\dfrac{8{\cdot}32+5{\cdot}4}{2}\right) \times 3{\cdot}5 \times {\cdot}8$ = 19·2	13·8	12 mm – 16
1 – 2	16·3	11·6	12 mm – 19.5
2 – 3	13·4	9·6	10 mm – 16
3 – 4	13·1	9·6	10 mm – 16
4 – 6	28	20	16 mm – 20
6 – 8	32	22·8	16 mm – 17·5

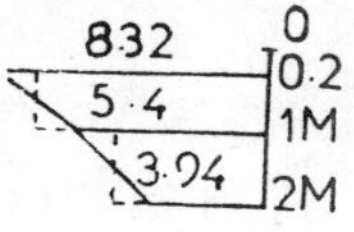

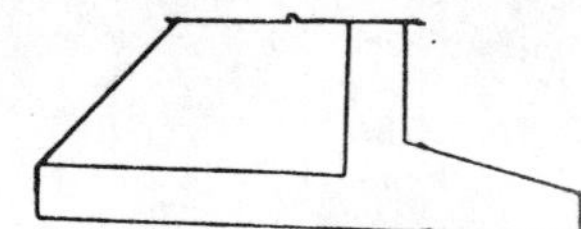

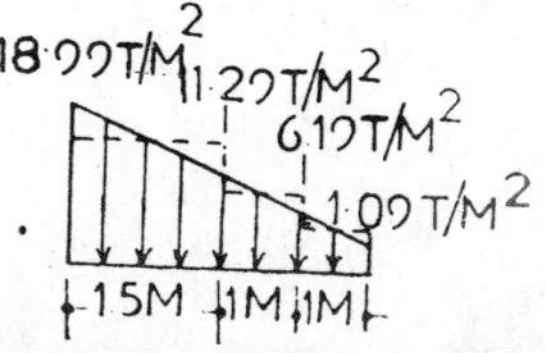

5·8·3. Connection to Counterfort and Heel Slab

distance from the edge of heel m	average downward pull = P tonnes	$A_t = P/1400$ cm²	spacing of two legged stirrups cm
0 – 1·5	$\left(\frac{18{\cdot}99+11{\cdot}79}{2}\right)\times 3{\cdot}5\times 1{\cdot}5$ = 80·5	56·2	22 – 13·5
1·5 - 2·5	30.6	21·8	16 - 18
2·5 3·5	12·7	9·15	10 – 16 5

5·9. Stability Case (i) LL is Present

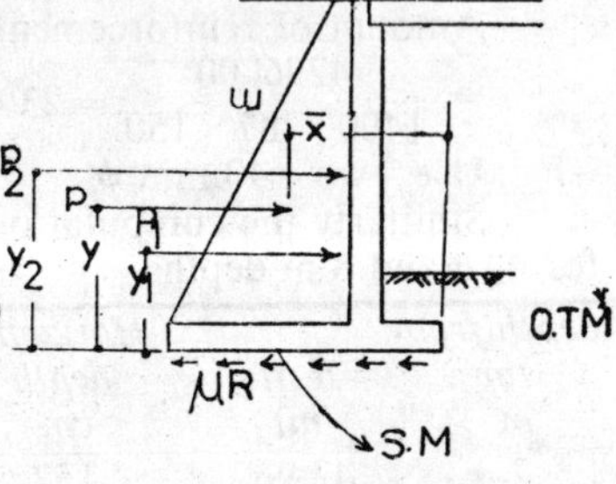

Factor of safety against overturning

$$= \frac{\text{stability moment}}{\text{overturning moment}}$$

$$= \frac{W\bar{x}}{P_1 y_1 + P_2 y_2} = \frac{W\bar{x}}{Py}$$

$$= \frac{141{\cdot}25\times 4}{43{\cdot}373\times 5{\cdot}35}$$

$$= 2{\cdot}45 > 1{\cdot}5$$

Factor of safety against sliding $= \dfrac{\mu W}{P} = \dfrac{0{\cdot}6\times 141{\cdot}25}{43{\cdot}373}$

$$= 1{\cdot}86 > 1{\cdot}5$$

Case (ii) When LL is not Present

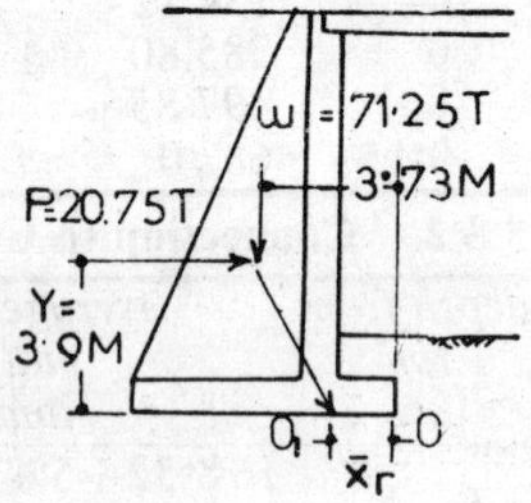

Factor of safety against overturning

$$= \frac{71{\cdot}25\times 3{\cdot}73}{20{\cdot}75\times 3{\cdot}9} = 3{\cdot}3 > 1{\cdot}5$$

Factor of safety against sliding

$$= \frac{0{\cdot}6\times 71{\cdot}25}{20{\cdot}75} = 2{\cdot}06 > 1{\cdot}5$$

III. DOMED ROOFS

General Information 1

1·1. Types

(*i*) Spherical (*ii*) Conical (*iii*) Segmental (*iv*) Elliptical

1·2. Loading

Wind loads as per IS –875, self weight of dome, snow loads if any, screeding if any

1·3. Materials Allowable Stresses

Steel grade – 1, $\sigma_{st}=1400$ kg/cm²
Concrete—*M* 150, $\sigma_{cb}=50$ kg/cm²

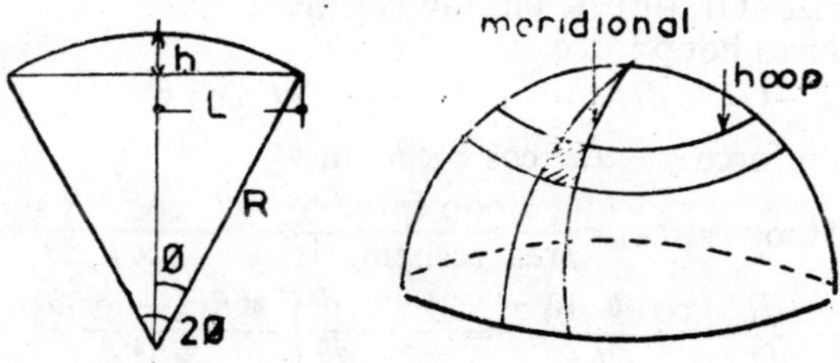

1·4. General Information

Thickness of dome=7·5 to 15 cm
Rise of the dome=1/5 to 1/7 span
Minimum reinforcement ≮ 2% of gross cross-section of concrete
Stresses (*i*) meridional stress, (*ii*) hoop stress
h = rise of the dome, R=Radius of the dome, 2ϕ=angle subtended at the centre

1·5. Spherical Domes

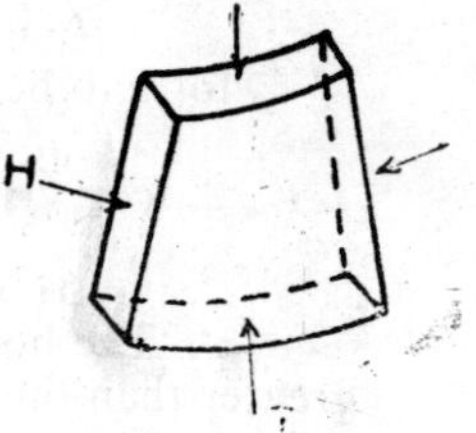

Let w be the intensity of load due to wind and accidental loads, self weight of dome etc. per square metre of surface of the dome.

Load acting on plane $AB = Q'$=surface area of the segment of a circle of radius $R\times$intensity of uniformly distributed load w

=Surface area of cylinder of radius R and a height $h\times$ intensity of load w

$$=2\pi R\times h'\times w=2\pi R(R-R\cos\theta)w=2\pi R^2(1-\cos\theta)w$$

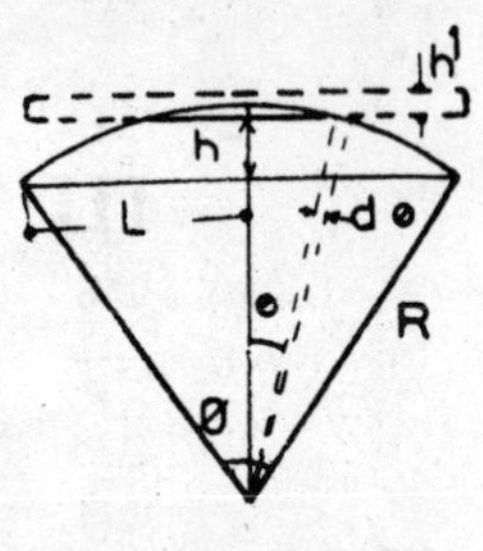

Vertical load per unit length along the circumference of the segment AB

$$V'=\frac{Q'}{2\,\pi l}=\frac{Q'}{2\,\pi R\sin\theta}$$

$$=\frac{2\,\pi R^2(1-\cos\theta)\,w}{2\,\pi R\sin\theta}$$

$$=wR\,\frac{(1-\cos\theta)}{\sin\theta}$$

Meridional thrust acting on the dome

$$T'=\frac{V'}{\sin\theta}=\frac{wR(1-\cos\theta)}{\sin^2\theta}$$

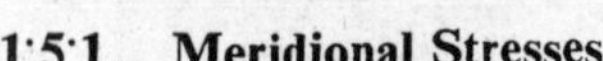

1·5·1. Meridional Stresses

Compressive stress along the meridion at any section defined by θ

$$c=\frac{T'}{t\times 1}=\frac{wR(1-\cos\theta)}{t\sin^2\theta}$$

$$=\frac{w\,R(1-\cos\theta)}{t\,(1+\cos^2\,)}=\frac{w\,R}{t}\quad[K_1]$$

1·5·2. Hoop Stresses

Horizontal thrust on the element that produces hoop force

$$=H'-(H'+dH')=-dH'=-d\,TH'\cos\theta$$

$$\text{Hoop force}=-dT'\cos\theta\times R\sin\theta$$

$$s=\text{Hoop stress}=\frac{\text{Hoop force}}{\text{area resisting it}}=\frac{dT'\cos\theta\,R\sin\theta}{t\times R\,d\theta}$$

$$s=\frac{-dT'}{d\theta}\;\frac{\cos\theta\sin\theta}{t}=\frac{1}{-t}\;\frac{d}{d\theta}\left[\frac{wR(1-\cos\theta)}{\sin^2\theta}\right]\cos\theta\sin\theta$$

$$=-\frac{wR}{t}\;\frac{d}{d\theta}\left[\frac{(1-\cos\theta)}{\sin^2\theta}\right]\cos\theta\;\sin\theta$$

$$=-\frac{wR}{t}\;\frac{d}{d\theta}\left[\frac{\cos\theta\sin\theta}{1+\cos\theta}\right]=-\frac{wR}{\text{t}}\;\frac{d}{d\theta}\left[\cot\theta-\operatorname{cosec}\theta+\sin\theta\right]$$

$$=-\frac{wR}{t}\left[(-\operatorname{cosec}^2\theta+\cot\theta\operatorname{cosec}\theta+\cos\theta)\right]$$

$$=-\frac{wR}{t}\left[(1-\cos\theta)-\cos\theta\,(1-\cos^2\theta)\right]$$

$$=-\frac{wR}{t}\left[\frac{1}{1+\cos\theta}-\cos\theta\right]=-\frac{wR}{t}\left[\frac{(1-\cos\theta-\cos^2\theta)}{(1+\cos\theta)}\right]$$

for s to be zero $1-\cos\theta-\cos^2\theta=0$

$$\therefore\;\cos\theta=\frac{1\pm\sqrt{1+4}}{t}=0{\cdot}618$$

$\therefore\;\theta=51°-48'$

$\therefore$ The hoop stress is compressive upto $\theta=51°\,48'$, and if θ is greater than this angle the hoop stress is tensile.

$$s=-\frac{wR}{t}\left(\frac{1-\cos\theta-\cos^2\theta}{1+\cos\theta}\right)=-\frac{wR}{t}\quad[K_2]$$

1·5·3. Stresses Under Lantern Loads

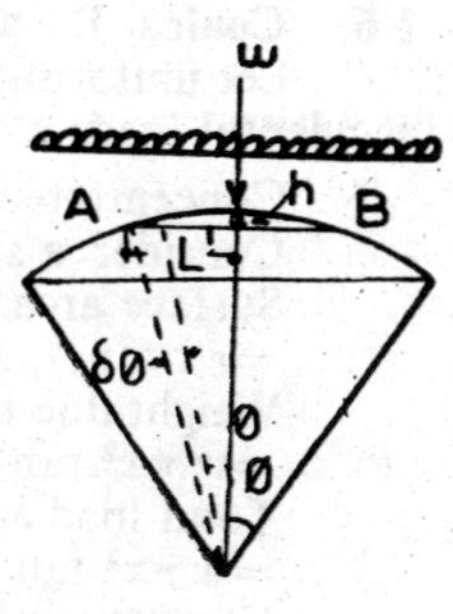

Load acting on any plane AB is

$$Q'=2\pi R^2(1-\cos\theta)\,w+W$$

$$V'=\frac{2\pi R^2(1-\cos\theta)\,w+W}{2\pi R\sin\theta}$$

$$T'=\frac{WR(1-\cos\theta)}{\sin^2\theta}-\frac{W\operatorname{cosec}^2\theta}{2\pi R}$$

1·5·4. Meridional Stress

$$c=\frac{WR}{t}[K_1]-\frac{W}{Rt}\left[\frac{\operatorname{cosec}^2\theta}{2\pi}\right]$$

$$=\frac{WR}{t}[K_1]\quad\frac{W}{Rt}[K_3]$$

That is, meridional thrust is due to uniformly distributed loads w, + due to concentrated load W

Horizontal thrust acting on the element $R\,d\theta$

$$(H'-dH')-H'=-dH'$$

Hoop force $=-dH'=-d(T'\cos\theta)\times R\sin\theta$

$$=-d\left[\frac{WR(1-\cos\theta)}{\sin^2\theta}+\frac{W}{2\pi R\sin^2\theta}\right]\times R\sin\theta\cos\theta$$

1·5. Hoop Stress

Hoop Stress $s=\dfrac{-dH'}{Rd\theta\times t}$

$$=\frac{-d}{d\theta}\left[\frac{WR}{t}\;\frac{(1-\cos\theta)}{\sin^2\theta}+\frac{W}{2\pi Rt\sin^2\theta}\right]\sin\theta\cos\theta$$

$$-s=\left[\frac{wR(1-\cos\theta-\cos^2\theta)}{t(1+\cos\theta)}+\frac{d}{d\theta}\left\{\frac{W\sin\theta\cos\theta}{2\pi Rt\sin^2\theta}\right\}\right]$$

$$=-\left[s_w+\frac{d}{d\sigma}\left\{\frac{W\cot\theta}{2\pi Rt}\right\}\right]=-\left[s_w+\frac{W(-\operatorname{cosec}^2\theta)}{2\pi Rt}\right]$$

$$s=-s_w-\frac{w\operatorname{cosec}^2\theta}{2\pi Rt}=-\frac{WR}{t}[K_2]-\frac{W}{Rt}[K_3]$$

VALUES OF K_1, K_2, K_3.

$\left(\dfrac{1-\cos\theta}{\sin^2\theta}\right)$	$\left(\dfrac{1-\cos\theta-\cos^2\theta}{1+\cos\theta}\right)$	$\dfrac{\operatorname{cosec}^2\theta}{2\pi}$	θ
0·5	−0·5	∝	0
0·5	0·496	21·0	5
0·505	−0·48	5·3	10
0·516	−0·425	1·37	20
0·537	−0·33	0·64	30
3·566	−0·20	0·38	40
0·608	−0·034	0·27	50
0·618	0·0	0·26	51°−48′
0·667	+0·167	0·21	60
0·747	+0·402	0·18	70
0·838	+0·68	0·16	80
1·00	+1·0	0·16	90
K_1	K_2	K_3	θ

1·6. Conical Domes

Let uniformly distributed load due to self weight, wind and accidental loads etc. $= w$/sq. m

Concentrated lantern load $= W$

Consider a strip δx at any depth x,

Surface area of dome $= \pi r \times$ slant height

$= r\, x \tan\theta,\ x \sec\theta = \pi x^2 \tan\theta \sec\theta$

Weight due to uniformly distributed load

$= w\, \pi x^2 \tan\theta \sec\theta$

Total load acting on the dome

$= w\, \pi x^2 \tan\theta \sec\theta + W$

$V =$ vertical load per metre

$$= \frac{w\, \pi x^2 \tan\theta \sec\theta}{2\, \pi x \tan\theta} + \frac{W}{2\, \pi x \tan\theta}$$

$$= \frac{wx \sec\theta}{2} + \frac{W}{2\, \pi x \tan\theta}$$

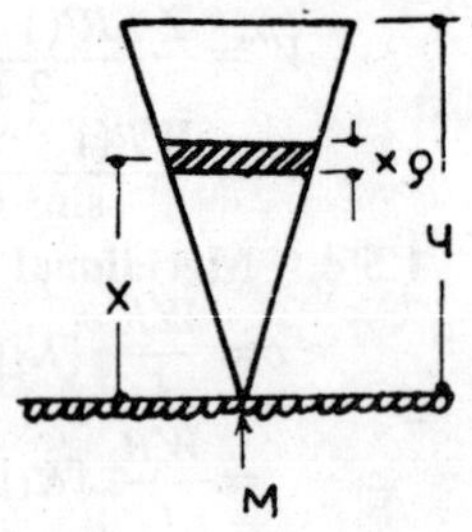

1·6·1. Meridional Stress

Meridional thrust $T = \frac{V}{\cos\theta} = \frac{wx \sec\theta}{2\cos\theta} = \frac{W}{2\, \pi x \tan\theta \cos\theta}$

$$= \frac{wx \sec^2\theta}{2} + \frac{W}{2\, \pi x \sin\theta}$$

$c =$ meridional stress $= \frac{T}{t \times 1} = \frac{wx \sec^2\theta}{2\, t} + \frac{W}{2\, \pi x\, t \sin\theta}$

1·6·2. Hoop Stresses

Horizontal thrust causing hoop compressive force in the dome at any section $= dH$

$= dT \sin\theta$

Hoop force $= dT \sin\theta \times \tan\theta$

$s =$ hoop stress $= \frac{dT \sin\theta \times \tan\theta}{t \times dx \sec\theta}$

$$= \frac{dT}{dx} \frac{x \sin^2\theta}{t}$$

$$s = \frac{d}{dx}\left[\frac{wx \sec^2\theta}{2} + \frac{W \sec\theta}{2\, \pi x \tan\theta}\right] \frac{x \sin^2\theta}{t}$$

$$= \frac{d}{dx}\left[\frac{wx^2 \sec^2\theta \sin^2\theta}{2} + \frac{wx \sec\theta \sin^2\theta}{2\, \pi x \tan\theta}\right]$$

$$= \frac{wx \tan^2\theta}{t}$$

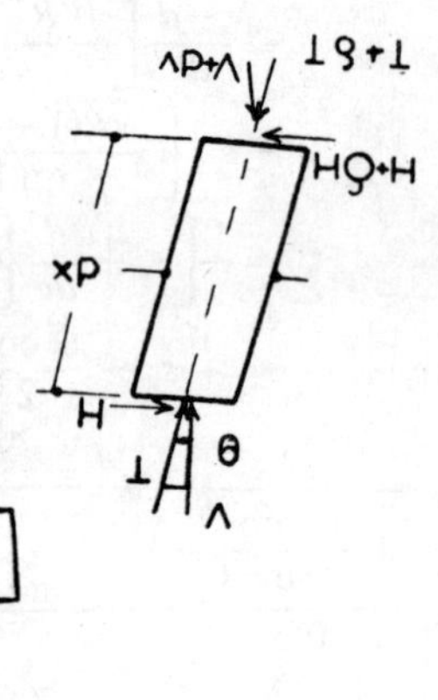

The hoop stress is independent of the lantern load W.

Meridional stress $c = \frac{wx \sec^2\theta}{2\, x\, t} + \frac{W \sec\theta}{2\, \pi x\, t \tan\theta}$

Hoop stress $s = \frac{wx \tan^2\theta}{t}$

Circular Roof Dome for a Reservoir 2

2·1. Data

Diameter of the dome = 20 m
Concrete = M 150, Steel = grade I, Lantern load = nil,
Type-spherical.

2·2. Trial Dimensions

Rise of dome = $\frac{1}{5}$ span = $\frac{20}{5}$ = 4 m

Thickness of the dome = t = 20 to 15 cm
Assume t = 15 cm

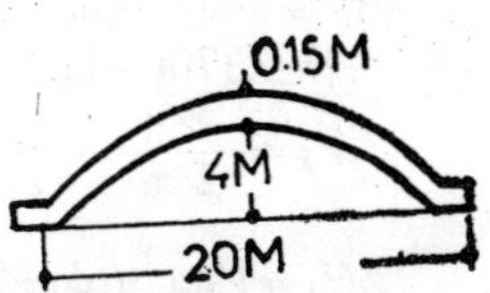

2·2·1. Loading

Self weight of dome (15 cm) = $\frac{15}{100} \times 1 \times 1 \times 2400 = 360$ kg/m²

Live load and accidental load due to wind etc. = 200 kg/m²

water proofing say = 90 kg/m²

Total load = w = 650 kg/m².

2·2·2. Characteristic Strengths of Materials

$\sigma_{cb} = 50$ kg/m², $\sigma_{st} = 1400$ kg/cm²
$q_s = 5$ kg/cm², σ_{ct} = cracking stress in concrete = 11 kg/cm²

2·2·3. Geometry of the Dome

$$m = \frac{2800}{3 \times 50} = 18{\cdot}66$$

Radius = R, Rise = h, l = 10

$$(R-h)^2 + l^2 = R^2$$

$$R = \frac{l^2 + h^2}{2h} = \frac{10^2 + 4^2}{2 \times 4} = \frac{116}{8} = 14{\cdot}5 \text{ m}$$

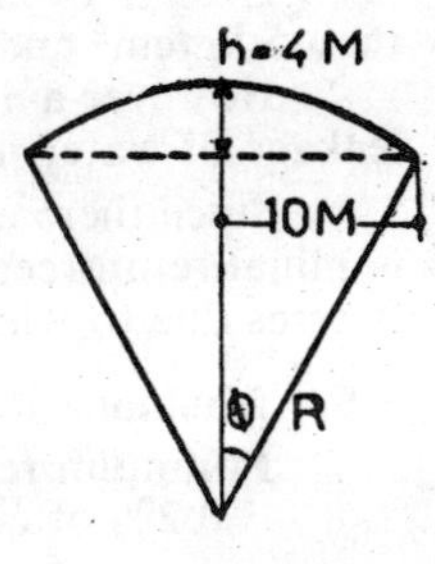

$$\sin\phi = \frac{10}{14{\cdot}5} = 0{\cdot}67, \quad \phi = 44° < 51°\ 48'$$

The entire dome is in compression.

2·4. Stresses in the Dome

Hoop stresses s = Hoop stress at crown, $\phi = 0$

$$s = -\frac{wR}{2t}\left(\frac{1-\cos\theta-\cos^2\theta}{1+\cos\theta}\right)$$

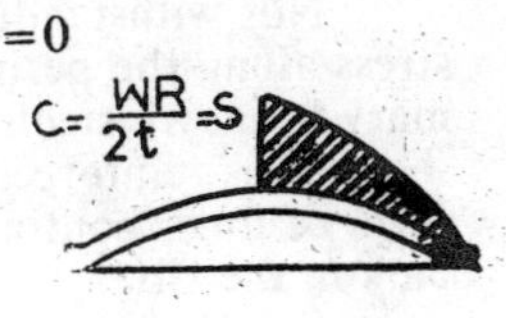

Maximum hoop stress occurs at $\phi = 0$

$$s = \frac{wR}{2t} = \frac{650 \times 14{\cdot}5}{2 \times 0{\cdot}15} = 31500 \text{ kg/m}^2$$

$$= 3{\cdot}15 \text{ kg/cm}^2$$

and this reduces as the angle increases.

Meridional stresses. Maximum meridional stress occurs at $\phi=44°$.

$$c=\frac{wk}{t(1+\cos\theta)}=\frac{650\times14\cdot5}{(14\cdot7193)\,0\cdot15}=36200\text{ kg/cm}^2=3\cdot62\text{ kg/cm}^2$$

at $\phi=0$, $c=3\cdot15$ kg/cm².

The compressive stress in the dome is very low and a very thin slab could be used, but there is not much advantage in reducing the thickness t below 10 cm.

2·4·1. Force Components at the Springing

Surface area$=2\,\pi Rh=2\times\pi\times14\cdot5\times4=365$ sq. m

Total load on the dome,

$W=w\times$surface area$=650\times365$

$=237000$ kg$=237$ tonnes

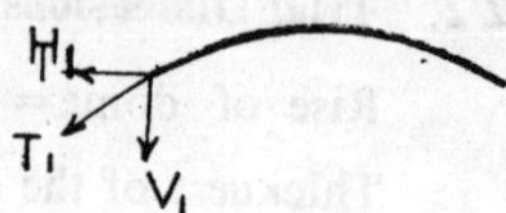

$$V_1=\frac{W}{2\,\pi l}=\frac{237}{2\times\pi\times10}=3\cdot8\text{ tonnes}$$

$$H_1=V_1\cot\phi=V_1\frac{(R-h)}{l}=\frac{3\cdot8(14\cdot5-4)}{10}=4\text{ tonnes}$$

$$T_1=\frac{V_1}{\sin\phi}=\frac{V_1R}{l}=\frac{3\cdot8\times14\cdot5}{10}=5\cdot51\text{ tonnes}$$

2·4·2. Shear Stress

Shear stress along the perimeter of the dome$=\dfrac{V_1}{\text{area}}=\dfrac{3802}{100\times15}$

$=2\cdot52<5$ kg/cm².

Even if 10 cm thick dome is used the shear stress will be less than 5 kg/cm² and $=3\cdot8$ kg/cm².

However a 15 cm thick slab provides better protection against leakage of water into the tank..

Since there are no tensile stresses anywhere in the dome, only a nominal reinforcement is required, to allow for such indeterminate stresses due to wind, distortion, shrinkage and temperature etc.

2·5. Minimum Reinforcement in the Dome

Minimum reinforcement

$=0\cdot2\%$ of concrete area

$$=\frac{0\cdot2\times15\times100}{100}=3\text{ cm}^2$$

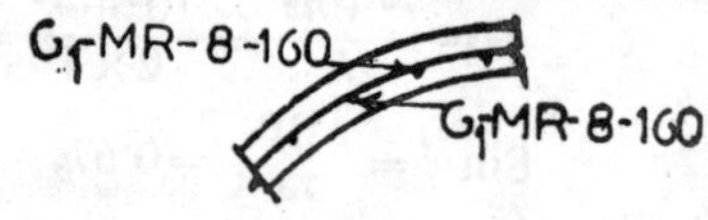

8 mm ϕ 16 cm c/c in both the directions.

2·5·1. Reinforcement for Shear at the Junction of the Rib and Dome

Not withstanding the low shear stress along the perimeter, it is customary to insert steel to resist all the shear, the reinforcement being in the form of links continuous around main bars in the rib.

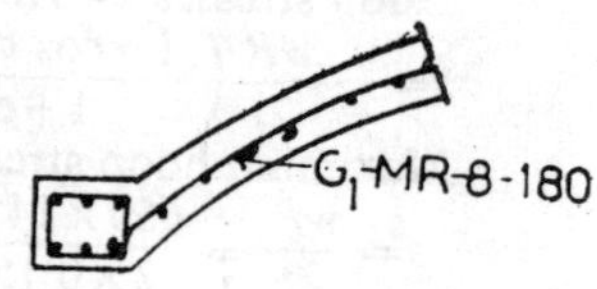

Steel required for shear

$$=\frac{V_1}{1400}=\frac{3800}{1400}=2\cdot7\text{ cm}^2.$$

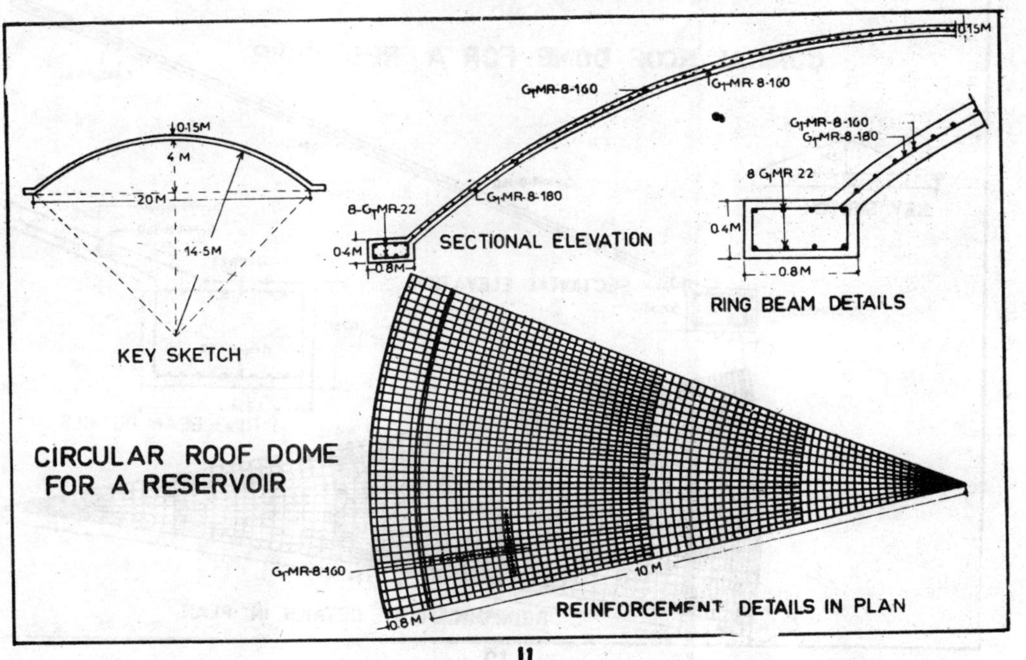
0·15M
4 M
20M
14·5M
KEY SKETCH
CIRCULAR ROOF DOME
FOR A RESERVOIR
G1MR-8-160
G1MR-8·160
0·15M
8-G1MR-22
G1MR-8-180
0·4M
0·8M
SECTIONAL ELEVATION
G1MR-8-160
G1MR-8-180
8 G1MR 22
0·4M
0·8M
RING BEAM DETAILS
G1MR-8-160
10 M
0·8M
REINFORCEMENT DETAILS IN PLAN

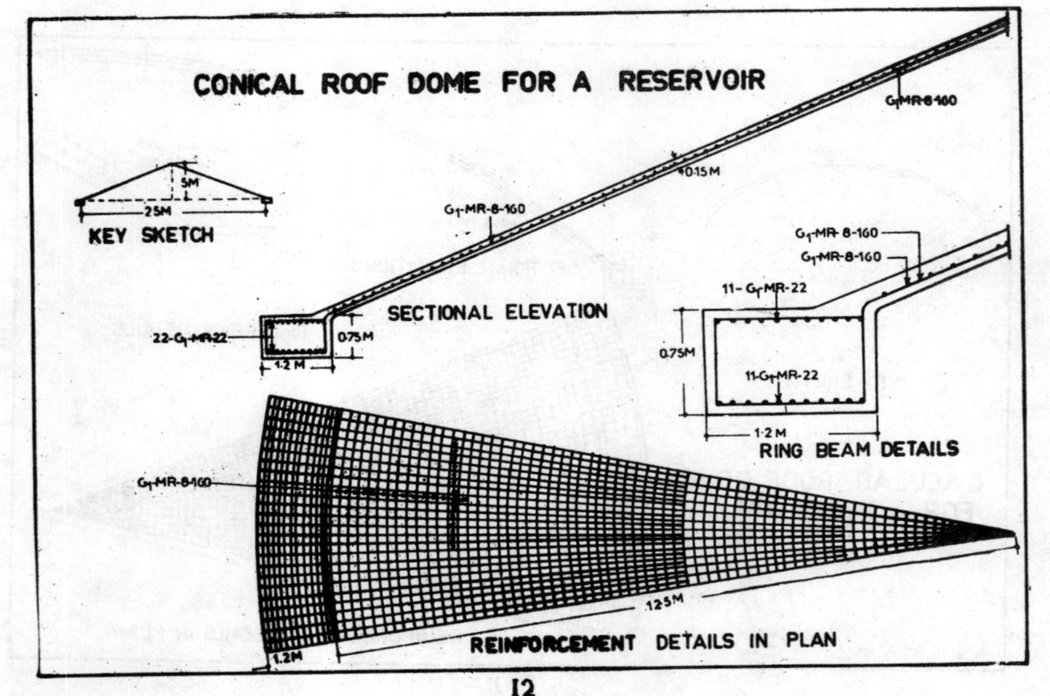
CONICAL ROOF DOME FOR A RESERVOIR
5M
25M
KEY SKETCH
G1-MR-8-160
0.15M
G1-MR-8-160
22-G1-MR-22
0.75M
1.2 M
SECTIONAL ELEVATION
G1-MR-8-160
G1-MR-8-160
11-G1-MR-22
0.75M
11-G1-MR-22
1·2 M
RING BEAM DETAILS
G1-MR-8-160
12.5M
1.2M
REINFORCEMENT DETAILS IN PLAN

2·6. Rib Design

Maximum hoop tension $T = H_1 \times \frac{D}{2} = \frac{4000 \times 20}{2} = 40{,}000$ kg

Amount of steel required$= \frac{40{,}000}{1400} = 28{\cdot}5$ cm^2

22 mm ϕ 8 Nos. A_{st} provided$=30{\cdot}41$ cm^2

2·6·1. Rib Dimension

Cracking stress in concrete

$$\sigma_{ct} = 11 \text{ kg/cm}^2 = \frac{T}{A_c + (m-1)\,A_{st}}$$

$$11 = \frac{40000}{A_c + 17{\cdot}66 \times 30{\cdot}41}$$

$$11A_c + 17{\cdot}66 \times 30{\cdot}41 \times 11 = 40{,}000$$

$$A_c = \frac{40000 - 5900}{11} = \frac{34100}{11}$$

$$= 3100 \text{ cm}^2$$

Use 80 cm $\times$ 40 cm rib.

Conical Roof Dome for a Reservoir 3

3·1. Data

Diameter of the reservoir = 25 m
Lantern load = nil.
Materials available, Concrete = M 200 Steel = grade—1
Type of dome desired = Conical

3·2. Trial Dimensions

Diameter of the dome = 25 m
Rise of the dome = $1/5 \times 25 = 5$ m
t = thickness of the dome = 10 to 15 cm
say $t = 15$ cm

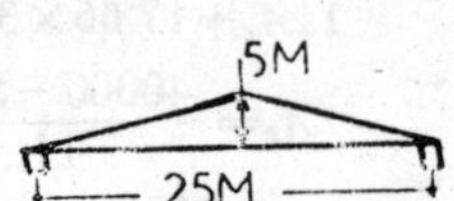

3·3. Loading

Self weight of the dome (t=15 cm) $= \dfrac{15}{100} \times 1 \times 2400$

$= 360$ kg/m^2

wind and accidental loads = 150 kg/m^2
Screeding or water proofing = 40 kg/m^2
Total load = w = 550 kg/m^2

3·4. Characteristic Strength of Materials

$\sigma_{cb} = 70$ kg/cm^2 $\sigma_{ct} = 12$ kg/cm^2
$\sigma_{st} = 1400$ 7g/cm^2 $m = 13{\cdot}33$
$q = 7$ kg/cm^2

3·4·1. Geometry of the Dome

$\tan\phi = \dfrac{12{\cdot}5}{5} = 2{\cdot}5,\ \phi = 68^\circ - 12'$

$\sec\phi = 2{\cdot}6927$

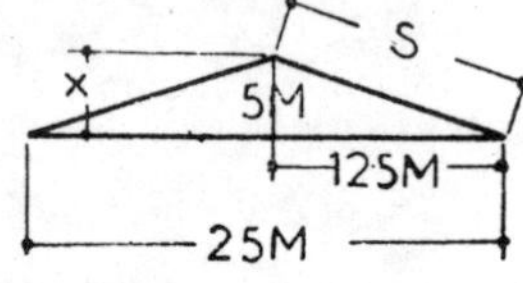

3·4·2. Force Components

Surface area of the dome = S_a
$\pi R \times$ slant height $= \pi R \times S$
$= \pi x \tan\phi \times x \sec\phi = \pi x^2 \tan\phi \sec\phi$
$= \pi \times 5 \times 5 \times 2{\cdot}5 \times 2{\cdot}6927 = 530$ Sq. m.

Total load = $W = w \times$ surface area $= 550 \times 530$
$= 291500$ kg $= 291{\cdot}5$ tonnes

$$V_1 = \frac{wx \sec\phi}{2} = \frac{w \times 5 \times 2{\cdot}6927}{2}$$

$$= \frac{550 \times 5 \times 2{\cdot}6927}{2} = 3690 \text{ kg}$$

$$T_1 = \frac{w \times x \sec^2\phi}{2} = V_1 \sec\theta = 3690 \times 2{\cdot}6927$$

$= 9950$ kg $= 9{\cdot}95$ tonnes

Horizontal component of the thrust$=H_1$
$=T_1 \sin\theta = 9{\cdot}95\times{\cdot}9285 = 9{\cdot}25$ T $=9250$ kg

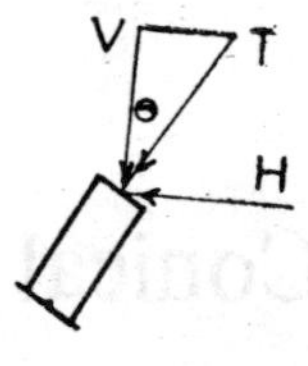

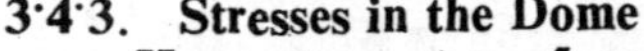

3·4·3. Stresses in the Dome

Hoop stress at $x=5$ m

$$s=\frac{H_{max}}{\text{area}}=\frac{wx\tan^2\phi}{100\times 15}=\frac{550\times 5\times(2{\cdot}5)^2}{150}$$

$$=\frac{17100}{15}=11{\cdot}4 \text{ kg/cm}^2 \text{ (comp)}$$

$$\text{meridional stress } c=\frac{T_1}{\text{area}}=\frac{9950}{100\times 15}=6{\cdot}65 \text{ kg/cm}^2 \text{ (comp)}$$

3·5. Minimum Reinforcement

The entire dome is in compression and the compressive stresses are small. However a minimum reinforcement at 0·2% of concrete area shall be provided.

$$A_{st}=\frac{0{\cdot}2}{100}\times 15\times 100=3 \text{ cm}^2$$

8 mm ϕ—16 cm c/c in both the directions

G₁MR-8-160

3·6. Ring Beam or Rib Design

Horizontal component of the thrust

$H_1=9100$ kg

Hoop tension at the edge of the dome

$$T'=\frac{9250\times 25}{2}=116{,}000 \text{ kg}$$

$$A_{st}=\frac{116{,}000}{1400}=83 \text{ cm}^2$$

22 mm—22 Nos

A_t provided $=83{\cdot}62$ cm²

G₁MR-8-160
11-G₁MR-22
0.75M
1.2M
11-G₁MR-22

3·6·1. Size of theR ib

$$\sigma_{ct}=\frac{T'}{A_c-(m-1)A_{st}},\quad 12=\frac{116{,}000}{A_c+12{\cdot}33\times 83{\cdot}62}$$

$$12\,A_c+12\times 12{\cdot}33\times 83{\cdot}62=116{,}000$$

$$A_c=\frac{103{,}600}{12}=8{,}600 \text{ cm}^2$$

Use 120×75 cm

Conical Roof Dome for a Temple 4

4·1. Data

Diameter of dome = 20 m
Rise of the dome = 14 m
Lantern load = 6000 kg
Type – Conical
Materials available, Concrete – M 150, Steel – grade I

4·2. Trial Dimensions

Diameter $D = 20$ m
Rise $h = 14$ m
Thickness of the dome (assume) = $t = 15$ cm

4·3. Loads

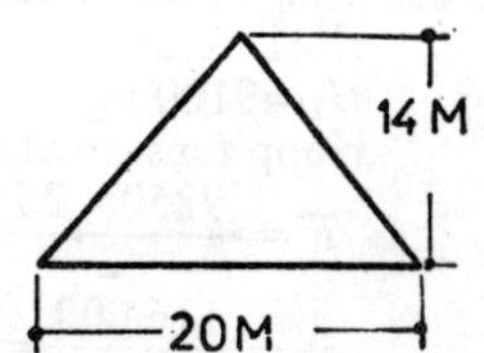

Self weight of dome (15 cm)

$= \frac{15}{100} \times 1 \times 1 \times 2400 = 360$ kg/cm²

Wind and accidental loads = 140 kg/cm²
Total load $w = 500$ kg/cm²
Lantern load = $W_1 = 6000$ kg/cm²

4·3·1. Characteristic Strength of Materials

$\sigma_{cb} = 50$ kg/cm² $\sigma_{ct} = 11$ kg/cm²
$\sigma_{st} = 1400$ kg/cm², $m = 18{\cdot}66$
$q = 5$ kg/cm²

4·4. Geometry of the Dome

$\tan \phi = \frac{10}{14} = 0{\cdot}714$, $\phi = 35° - 30'$,

$\sec \phi = 1{\cdot}2283$

4·4·1. Force Components

Surface area of the dome

$S_a = \pi R \times \text{slant height} = \pi R \times s$

$= \pi x \tan \phi \times x \sec \phi = \pi h \tan \phi \times h \sec \phi$

$= \pi \times 14 \times 0{\cdot}714 \times 14 \times 1{\cdot}2283 = 540$ sq. m

Total vertical load $W = S_a \times w + W_1 = 500 \times 540 + 6000$

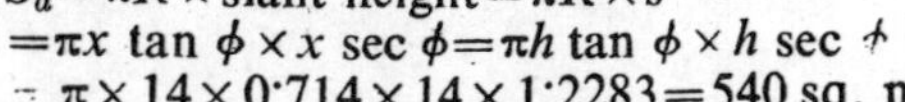

$= 276000$ kg $= 276$ tonnes

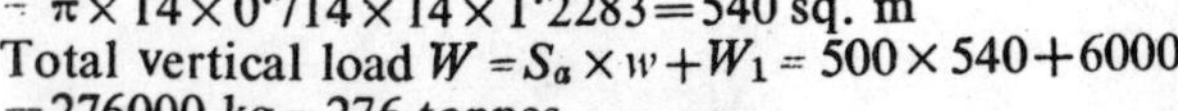

$$V_1 = \frac{w \times x \sec \phi}{2} + \frac{W_1}{2 \pi x \tan \phi}$$

$$= \frac{500 \times 14 \times 1{\cdot}228}{2} + \frac{6000}{2 \pi \times 14 \times {\cdot}714}$$

$= 4500 + 950 = 5450$ kg

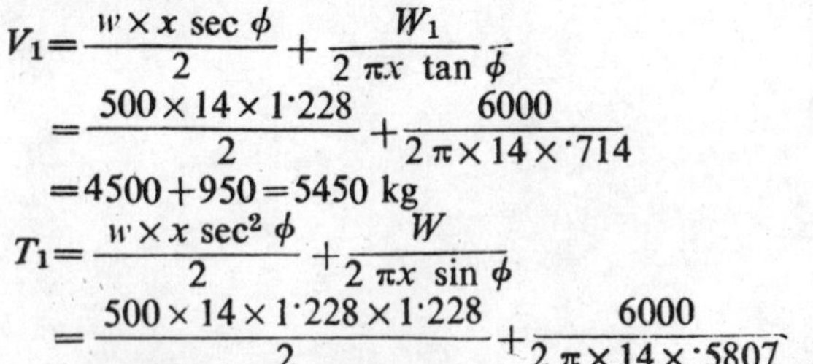

$$T_1 = \frac{w \times x \sec^2 \phi}{2} + \frac{W}{2 \pi x \sin \phi}$$

$$= \frac{500 \times 14 \times 1{\cdot}228 \times 1{\cdot}228}{2} + \frac{6000}{2 \pi \times 14 \times {\cdot}5807}$$

$= 5800 + 117 = 5917$ kg

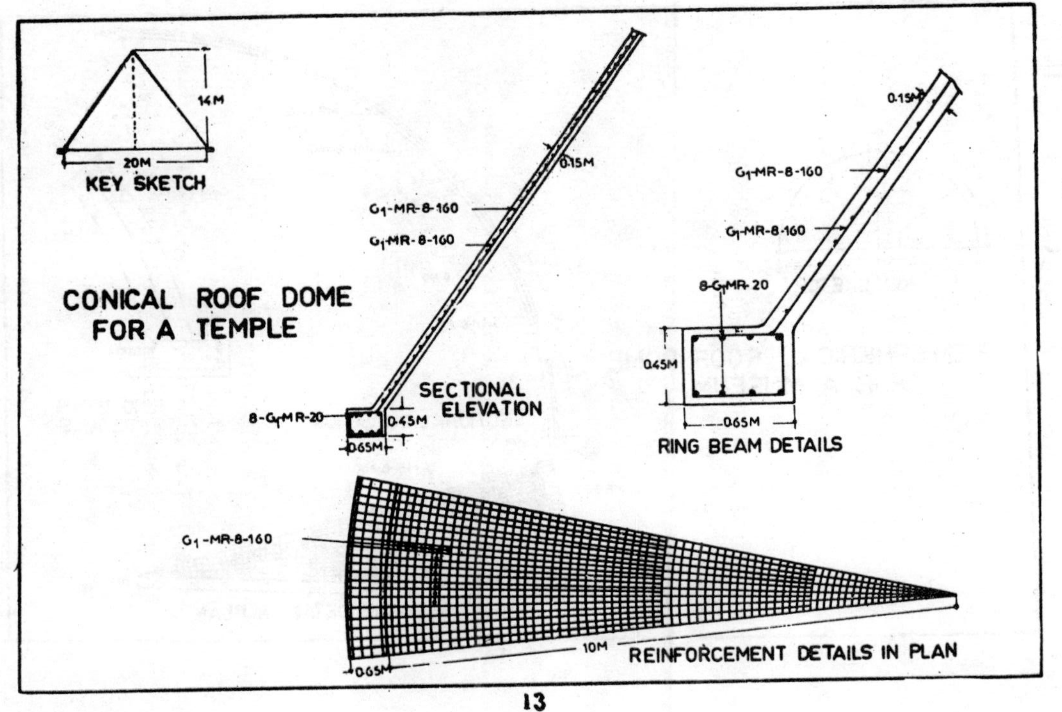
14M
20M
KEY SKETCH
CONICAL ROOF DOME
FOR A TEMPLE
0.15M
G1-MR-8-160
G1-MR-8-160
8-G1MR-20
0.45M
0.65M
SECTIONAL
ELEVATION
0.15M
G1-MR-8-160
G1-MR-8-160
8-G1MR-20
0.45M
0.65M
RING BEAM DETAILS
G1-MR-8-160
10M
0.65M
REINFORCEMENT DETAILS IN PLAN

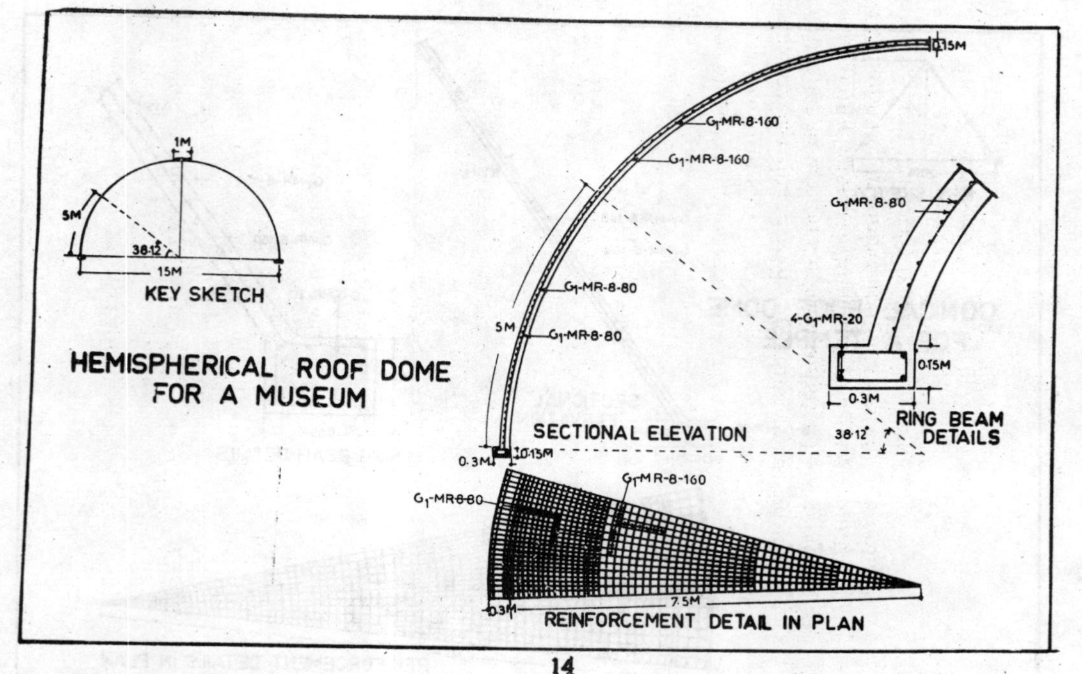
1M
5M
38·12°
15M
KEY SKETCH
HEMISPHERICAL ROOF DOME
FOR A MUSEUM
0·15M
G1-MR-8-160
G1-MR-8-160
G1-MR-8-80
5M
G1-MR-8-80
0·3M
0·15M
SECTIONAL ELEVATION
G1-MR-8-80
4-G1-MR-20
0·15M
0·3M
RING BEAM
DETAILS
38·12°
G1-MR-8-160
G1-MR-8-80
0·3M
7·5M
REINFORCEMENT DETAIL IN PLAN

14

$H_1 = T_1 \sin\phi = 5917 \times {\cdot}5807 = 3450$ kg

Hoop tension at springing $T = H_1 \dfrac{D}{2} = 3450 \times 10 = 34500$ kg

4·4·2. Stresses In the Dome

Hoop stresses at springing $= \dfrac{H_{max}}{\text{area}} = \dfrac{w \times x \tan^2 \phi}{15 \times 100}$

$= \dfrac{500 \times 14 \times {\cdot}714 \times {\cdot}714}{1500} = 2{\cdot}38$ kg/cm²

meridional stresses $C = \dfrac{T_1}{\text{area}} = \dfrac{5917}{15 \times 100} = 3{\cdot}95$ kg/cm² (comp)

4·5. Minimum Reinforcement

The entire dome is in compression and the compressive stresses are small. However a minimum reinforcement at 0·2% of concrete area shall be provided.

$A_{st} = \dfrac{0{\cdot}2}{100} \times 15 \times 100 = 3$ cm²

8 mm ϕ - 16 cm c/c in both the directions.

4·6. Ring Beam or Rib Design

$T = 34500$ kg

$A_t = \dfrac{34500}{1400} = 24{\cdot}6$ cm²

A_t provided 20 mm ϕ—8 Nos (25·14 cm²)

$\sigma_{ct} = \dfrac{T}{A_c + (m-1) A_t}, \ 11 = \dfrac{34500}{A_c + 17{\cdot}33 \times 25{\cdot}14}$

$11 A_c + 11 \times 17{\cdot}33 \times 25{\cdot}14 = 34500$

$A_c = \dfrac{34500 - 4950}{11} = \dfrac{29550}{11} = 2700$ cm²

Provide 45 × 65 cm rib

Hemispherical Dome to Cover a Central Circular Hall of a Museum 5

5·1. Data

Diameter of the dome = 15 m

Lantern load due to ornamental work = 1000 kg distributed over 1 m diameter at crown.

Materials available, Concrete—$M150$, Steel—grade—I

Type—Hemispherical.

5·2. Trial Dimension

Diameter of the dome $= D = 15$ m

Rise of the dome $h = R = 7{\cdot}5$ m

Thickness of the dome $= t = 15$ cm.

5·3. Loading

Self weight of the dome ($t = 15$ cm)

$$= \frac{15}{100} \times 1 \times 1 \times 2400 = 360 \text{ kg/cm}^2$$

Wind and accidental loads $= 200$ kg/m²

Water proofing etc. $= 90$ kg/m²

Total load $= w = 650$ kg/m²

Lantern load distributed over 1 m diameter $= W_1 = 1000$ kg.

5·4. Characteristic Strengths of Materials

$\sigma_{cb} = 50$ kg/cm², $\sigma_{st} = 1400$ kg/cm², $q_s = 5$ kg/cm², $\sigma_{ct} = 11$ kg/cm² and $m = 18{\cdot}66$.

5·4·1. Geometry of the Dome

$\phi = 90°$ since hemispherical and ϕ' = angle subtended at the centre by the distribused load on 1 m diameter.

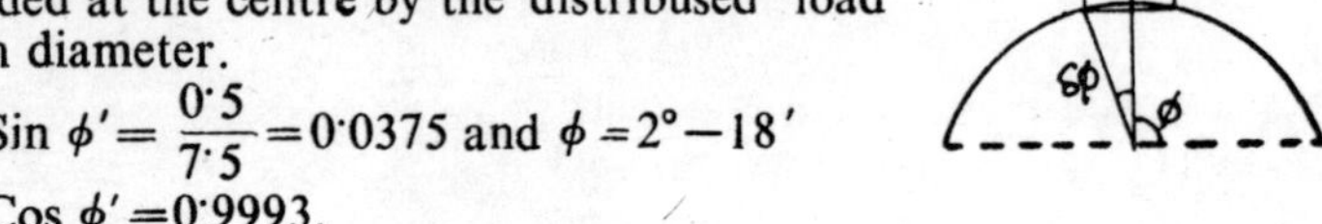

$$\text{Sin } \phi' = \frac{0{\cdot}5}{7{\cdot}5} = 0{\cdot}0375 \text{ and } \phi = 2° - 18'$$

$$\text{Cos } \phi' = 0{\cdot}9993.$$

5·4·2. Stresses in the Dome

Hoop stresses at crown $\phi = 2° - 12'$

$$S = -\frac{wR}{t}\left(\frac{1 - \cos\phi - \cos^2\phi}{1 + \cos\phi}\right) - \frac{W}{Rt}\left(\frac{\text{Cosec}^2\,\phi}{2}\right)$$

$$S = -\frac{650 \times 7{\cdot}5}{0{\cdot}15}\left(\frac{1 - 0{\cdot}9993 - {\cdot}99}{1 + 0{\cdot}9993}\right) - \frac{1000}{7{\cdot}5 \times 15 \times 2\,\pi \times ({\cdot}0375)^2}$$

$$= \frac{-650 \times 7{\cdot}5}{0{\cdot}15}\left(\frac{1 - 1{\cdot}9893}{1 + 0{\cdot}9993}\right) - \frac{1000}{7{\cdot}5 \times {\cdot}15 \times 2\,\pi \times {\cdot}0014}$$

$$= \frac{650 \times 7{\cdot}5}{0{\cdot}15} \times \frac{0{\cdot}9893}{1{\cdot}9993} - 101{,}000$$

$$= 16{,}000 - 101{,}000 = -85000 \text{ kg/m}^2 = -8{\cdot}5 \text{ kg/cm}^2$$

at $\phi=90°$, $s=\frac{-650\times7{\cdot}5}{0{\cdot}15}\;\frac{(1-0-0)}{1+0}-\frac{1000}{7{\cdot}5\times{\cdot}15\times2\pi\times(1)^2}$

$=-32500-1420=-33920$ kg/m² $=-3{\cdot}392$ kg/cm²

at $\phi=51°-48'$, $s=0$.

Stresses will be α, under the concentrated load at crown that is at $\phi=0°$. Therefore the load should be distributed over an area.

Intensity of load $w'=650+\frac{1000}{\frac{\pi\,1^2}{4}}=650+1280=1930$ kg

Hoop stresses at $\phi=0$

$s=-\frac{w'R}{t}\left(\frac{1-\cos\phi-\cos^2\phi}{1+\cos\phi}\right)=+\frac{w'R}{2t}=\frac{1930\times7{\cdot}5}{0{\cdot}15\times2}$

$=4825$ kg/m² $=4{\cdot}825$ kg/cm² (comp.)

meridional stress $c=\frac{w'R}{t(1+\cos\phi)}=4{\cdot}825$ kg/cm² (comp.)

The dome is partly in tension. The tensile stresses, however, are within the limits of allowable tensile stress in concrete, *i.e.* 11 kg/cm².

5·4·3. Force Components

Surface area $=S_a=2\pi R^2$

$=2\pi\times7{\cdot}5\times7{\cdot}5=354$ sq. m

Total load on the dome

$=w\times$ surface area $+W_1$

$=650\times354+1000=230000+1000$

$=231000$ kg.

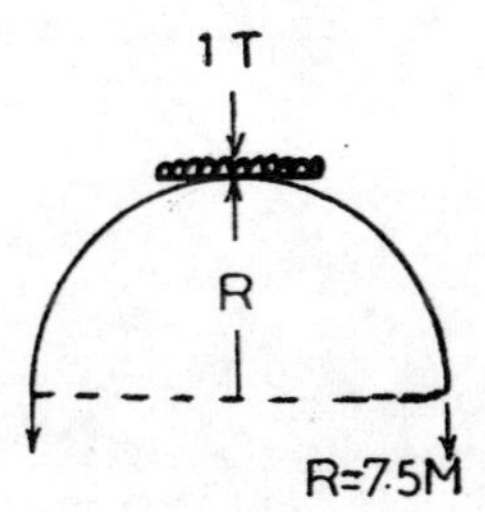

$V_l=\frac{W}{2\pi l}=\frac{W}{2\pi R}=\frac{231}{2\pi\times7{\cdot}5}$

$=4{\cdot}93$ tonnes

$H_1=V_1\cot\phi,=V_1\left(\frac{R-h}{l}\right)=0$

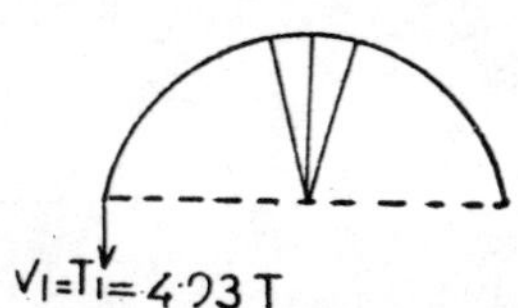

$T_1=\frac{V_1}{\sin\phi}=\frac{V_1R}{R}=V_1$

$=4{\cdot}93$ tonnes

5·4·4. Shear Stress

Shear stress along the perimeter of the dome

$=\frac{V_1}{\text{area}}=\frac{4930}{100\times15}=3{\cdot}27$ kg/cm² <5 kg/cm²

5·5. Reinforcement

Tensile stresses are present at the crown under lantern load

maximum tensile stress $=8{\cdot}5$ kg/cm²

tensile force $=8{\cdot}5\times15\times100=12750$ kg

A_{st} required $=\frac{12750}{1400}=9{\cdot}1$ cm²

Use 12 mm ϕ at 12 cm c/c in both the directions at crown 1·5 m diameter area.

5·5·1. Minimum Reinforcement

Minimum reinforcement$=0{\cdot}2\%$ of concrete area

$$=\frac{0{\cdot}2\times15}{100}\times100=3\ \text{cm}^2$$

Use 8 mm ϕ at 16 cm c/c in both directions.

$$R\theta=7{\cdot}5\times38{\cdot}12\times\frac{\pi}{180}=5\ \text{m}$$

The tensile stresses produced over a length of 5·0 m from the springing are small.

Hoop stress$=3{\cdot}392$ kg/cm²

Tensile force $=3{\cdot}392\times15\times100$

$=5010$ kg

$$A_t\ \text{required}=\frac{5010}{1400}=3{\cdot}6\ \text{cm}^2$$

Provide 8 mm at 8 cm c/c over a length of 6·5 m from springing.

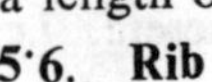

5·6. Rib

The hoop force is zero at the springing. Use a nominal size of 30 cm × 15 cm with 4 – 20 mm ϕ.

IV. OPEN WATER TANKS ON GROUND

R.C.C. Tanks—General Information 1

1·1. Types

(*i*) Tanks resting on ground

(*a*) Water reservoirs (*b*) Settling tanks etc.

These may be circular, square or rectangular in shape and may be open or closed at top.

(*ii*) Underground tanks

(*a*) Purification tanks (*b*) Septic tanks (*c*) Gas holders (*d*) Imhoff tanks. etc.

(*iii*) Swimming pools.

(*iv*) Elevated tanks

(*a*) Tanks on roofs of buildings (*b*) Elevated tanks on staging

These may be Intze type or circular, square or rectangular in shape and are usually closed

1·2. Allowable Stresses

Concrete : *Tensile stress* σ_{ct}

grade of concrete	*direct tension*	*in bending*	*in shear*
M 150	11	15	15
M 200	12	17	17

M 200 concrete is used for the tank portion in order to make it leak proof. *M* 150 concrete can be used for staging in the case of elevated tanks and also sometimes for roofs of closed tanks.

Steel. In direct tension σ_{st} = 1000 kg/cm²

Water face when thickness of the wall is > 225 mm, t = 1000 kg/cm²

Away from water face when the thickness of the wall > 225 mm = 1250 kg/cm²

In shear, wall thickness < 225 mm = 1000 kg/cm²

For wall thickness > 225 mm = 1250 kg/cm²

In compression columns etc = 1250/cm²

1·3. Minimum Reinforcement

A minimum reinforcement of 0·3% of gross area of concrete shall be provided in order to prevent shrinkage cracks.

Expansion and Contraction Joints. At every 30 m intervals expansion joints shall be provided. Contraction joints in roof slabs, walls shall be provided at intervals of 7·5 m.

Characteristic strengths. If σ_{cb} = 70 kg/cm²,

$\sigma_{st}=1000$ kg/cm^2 and $m=\dfrac{2800}{3\times 70}=13{\cdot}33=13$

$$\frac{\sigma_{cb}}{n}=\frac{\sigma_{st}}{m(d-n)},\quad \frac{70}{n}=\frac{1000}{13(d-n)}$$

$$70\times 13\ d-70\times 13\ n=1000\ n$$

$$910\ d\quad 910\ n=1000\ n$$

$$910\ d=1910\ n$$

$$n=\frac{910\ d}{1910}=0{\cdot}476\ d$$

$$jd=\text{lever arm}\left(d-\frac{n}{3}\right)=\left(d-\frac{0{\cdot}476}{3}d\right)=0{\cdot}84\ d$$

$$MR=b\times n\times\frac{\sigma_{cb}}{2}\times\left(d-\frac{n}{3}\right)=b\times 0{\cdot}476\ d\times\frac{70}{2}\times 0{\cdot}84\ d$$

$$=14\ bd^2$$

For $\sigma_{cb}=70$ kg/cm^2 and $\sigma_{st}=1250$ kg/cm^2,

$$n=\frac{910}{2160}=0{\cdot}42\ d$$

$$jd=\left(d-\frac{n}{3}\right)=\left(d-\frac{0{\cdot}42\ d}{3}\right)=0{\cdot}86\ d$$

$$MR=b\times n\times\frac{\sigma_{cb}}{2}\times\left(d-\frac{n}{3}\right)=b\times 0{\cdot}42\ d\times\frac{70}{2}\times 0{\cdot}86\ d$$

$$=12{\cdot}61\ bd^2$$

1·4. Circular Tanks—Free Base General

These tanks are used for larger capacities. The provision of sliding joint at base relieves the fixing moment and the tank walls are designed for hoop tension. These tanks are economical since for a given capacity, they have the least perimeter, and therefore the material requirement is the least. The formwork is, however, expensive.

1·5. Sidewall

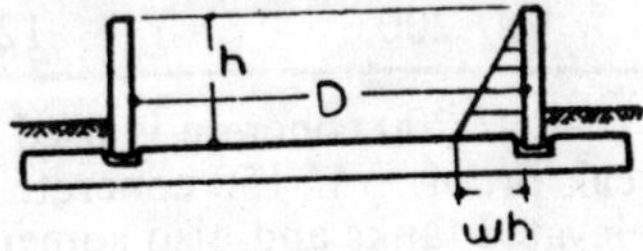

Let w = density of water,
h = depth of water,
D = diameter of the tank

1·5·1. Hoop Tension in the Wall

Maximum pressure at the base level $=wh$ kg/m

Maximum hoop tension in the wall

$$T=\frac{whD}{2}\text{ kg}$$

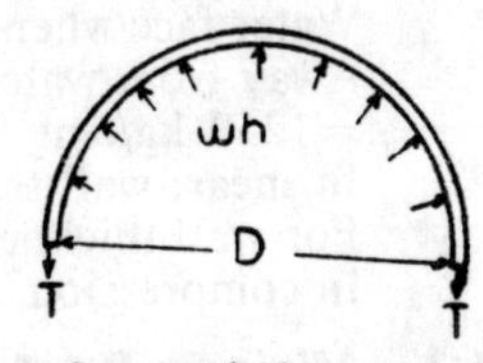

1·5·2. Amount of Reinforcement

Amount of reinforcement required to resist the hoop tension

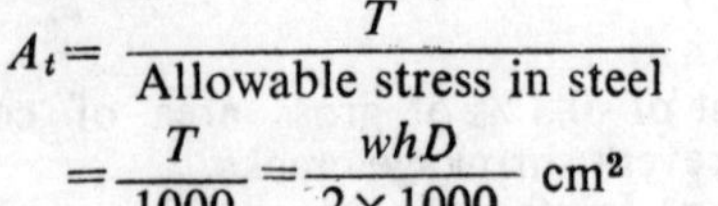

$$A_t=\frac{T}{\text{Allowable stress in steel}}$$

$$=\frac{T}{1000}=\frac{whD}{2\times 1000}\text{ cm}^2$$

Tensile stress in concrete $=\sigma_{ct}=\dfrac{T}{A_t+(m-1)A_t}$ kg/cm^2

If thickness of the wall is t_w

$$\sigma_{ct}=\frac{whD}{2[100\times t_w+(m-1)\,A_t]},\quad \text{where } m=\text{modular ratio.}$$

1·5·3. Secondary Reinforcement

A minimum reinforcement of 0·3% of concrete area shall be provided to resist shrinkage and temperature stresses

1·6. Minimum Thickness of Wall

Wall thickness is the greater of

(*i*) $t_w \nless 15$ cm (*ii*) 3 cm per meter depth of water+5 cm
(*iii*) as required by the concrete tensile stress σ_{ct}

1·7. Bottom Slab

Since the bottom slab rests on the ground there is no bending moment in the slab. However, when the tank is empty the slab is subjected to upward soil pressure developed due to the weight of the side walls and roof if any.

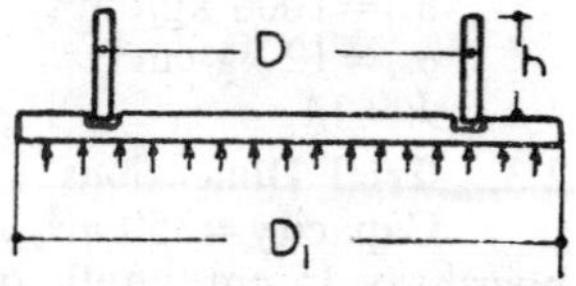

weight of side walls

$$=\pi(D+t_w)\cdot t_w\times h\times 2400 \text{ kg}$$

upward pressure at the bottom of the slab

$$=\frac{\pi(D+t_w)\,t_w\times h\times 2400}{\dfrac{\pi D_1^2}{4}}\text{ kg/m}^2$$

$$=q \text{ kg/m}^2$$

1·7·1. Design of Base Slab

$$\text{Maximum bending mement}=M=\frac{3}{16}q\left(\frac{D_1}{2}\right)^2$$

$$\text{depth of slab required}=d=\sqrt{\frac{M}{R\times b}}=\sqrt{\frac{M}{14\times b}}$$

1·7·2. Minimum Thickness of Base

The thickness of base slab shall not be less than 15 to 20 cm.

1·7·3. Minimum Reinforcement

Minimum reinforcement shall not be less than ·3% of concrete area and is distributed in both the directions and on both faces.

Circular Tank on Ground with Sliding Base 2
Capacity 350,000 Litres

2·1. Data

Capacity = 350 m^3

Tank rests on ground. Bearing capacity of soil = 10 t/m²

A sliding joint is provided between the base and the side walls.

Materials available ; M 200 grade concrete, Steel grade · 1.

2·2. Characteristic Strengths

$\sigma_{cb}=70$ kg/cm², M = 13

$\sigma_{ct}=1000$ kg/cm², $jd=\cdot 84\ d$, $\sigma_{st}=1250$ kg/cm²

$\sigma_{ct}=12$ kg/cm² $\quad R=12\cdot 6$

$R=14$

2·3. Trial Dimensions

Capacity = 350 m^3, Assume depth of water $h=3\cdot 5$ m and Free board say 15 cm, depth of the tank $h_1=3\cdot 65$ m

Let D be the diameter of the tank then

$$\frac{\pi D^2}{4}\times h = \text{Volume of the tank}$$

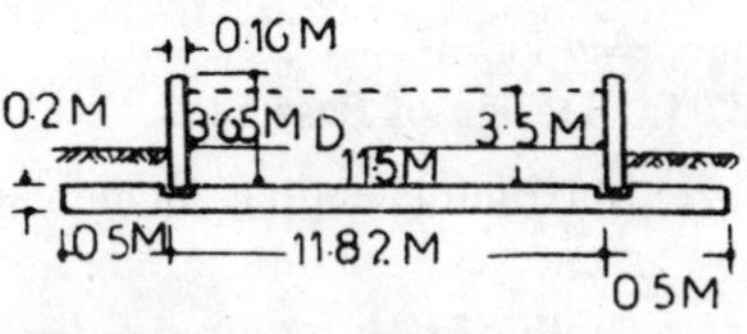

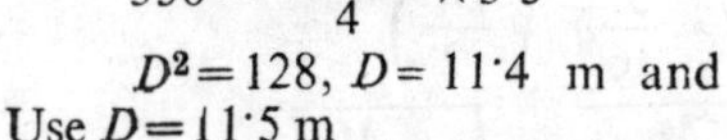

$$350=\frac{\pi\times D^2}{4}\times 3\cdot 5$$

$D^2=128$, $D=11\cdot 4$ m and

Use $D=11\cdot 5$ m

A tank of height 3·65 m and internal diameter of 11·5 m will be sufficient.

$$\text{Capacity provided}=\frac{\pi\times(11\cdot 5)^2\times 3\cdot 5}{4}=362\cdot 5\ \text{m}^3>350\ \text{m}^3$$

2·4. Sidewall

Maximum hoop tension in the sidewall per meter width

$$T=\frac{wh_1\ D}{2}\ \text{kg}=\frac{1000\times 3\cdot 65}{2}\times 11\cdot 5\ \text{kg}=21000\ \text{kg}$$

Area of steel required to resist maximum hoop tension T

$$=\frac{T}{1000}\ \text{cm}^2=\frac{21000}{1000}=21\ \text{cm}^2.$$ Use the 16 mm ϕ at 9·5 cm c/c for bottom 1 metre height on one face or 16 mm ϕ at 11 cm c/c on both faces

A_t provided = 21·16 cm²

2·4·1. Thickness of the Sidewall

(*i*) on cracking stress consideration

$$\sigma_{ct}=\frac{T}{A_c+(m-1)A_t}$$

$$12=\frac{21000}{100\times t_w+12\times 21\cdot 6}$$

$1200\ t_w + 12 \times 21{\cdot}16 \times 12 = 21000$

$1200\ t_w + 3050 = 21000$

$1200\ t_w = 17950$

$t_w = \dfrac{17950}{1200} = 14{\cdot}9$ cm

(*ii*) $\nless$ 15 cm

(*iii*) $(3 \times 3{\cdot}65 + 5) = 15{\cdot}95$ cm

Use 16 cm thick walls.

2·4·2. Main Reinforcement at Different Depths

As the hoop tension decreases towards top of the side wall, spacing can be increased towards top.

Depth from top (h) *m*	*Hoop tension* *5750h*	A_t *cm²*	*Spacing if provided on one face* *cm*	*Spacing if provided on both faces* *cm*
0 - 1	5750	5·75	12—19	12—38
1 2	11500	11·5	16—17	16—34
2 -3	17250	17·25	16—11·5	16—23
3 3·65	21000	21·0	16— 9·5	16—19

2·4·3. Secondary Reinforcement

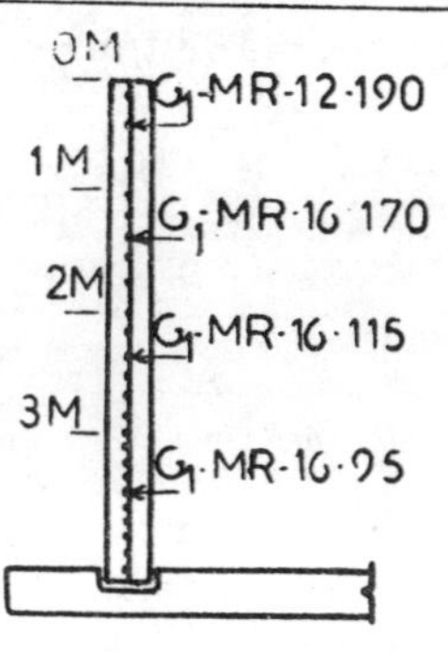

Minimum reinforcement

$= 0{\cdot}3\%$ of concrete area

$= \dfrac{0{\cdot}3 \times 100 \times 16}{100} = 4{\cdot}8\ \text{cm}^2.$

Use 10 mm - ϕ at 17 cm c/c if hoop reinforcement is provided on one face only and at 34 cm c/c if hoop reinforcement is provided on both faces of the side walls.

2·5. Base-Slab

Load from sidewalls when the tank is

empty $= \pi \times (11{\cdot}65) \times 3{\cdot}65 \times \dfrac{15}{100} \times 2400$

$= 48200$ kg

Intensity of soil pressure below the

base slab $= \dfrac{48200 \times 4}{\pi \times 12{\cdot}82} = 375\ \text{kg/m}^2$

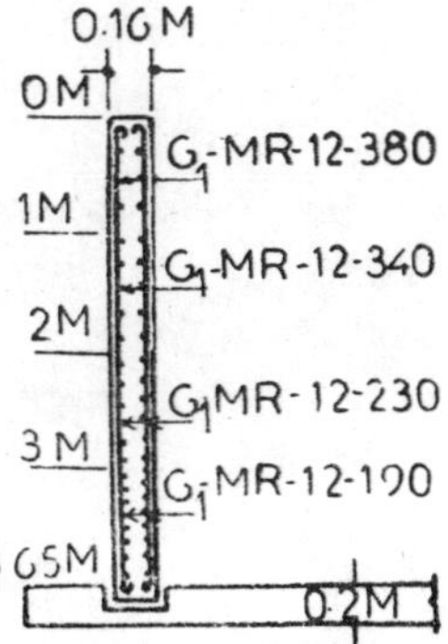

Maximum bending moment at centre of the span

$= \dfrac{3}{16}\, q \left(\dfrac{D_1}{2}\right)^2 = \dfrac{3}{16} \times 375 \times \left(\dfrac{12{\cdot}8}{2}\right)^2$

$= 3/16 \times 375 \times (6{\cdot}4)^2 = 28{,}80$ m kg

$= 2{,}880{,}00$ cm kg

.·5·1. Depth of the slab

$d = \sqrt{\dfrac{288000}{14 \times 100}} = \sqrt{206} = 14{\cdot}4$ cm

Use 20 cm thick slab with effective depth $= 15$ cm.

2·5·2 Main and Secondary Reinforcement

$$A_t = \text{required} = \frac{288000}{100 \times 0{\cdot}84 \times 15} = 22{\cdot}9 \text{ cm}^2$$

Use 12 mm at 9·5 cm c/c in both the directions

$A_t = \dfrac{0{\cdot}3 \times 20 \times 100}{100} = 6 \text{ cm}^2$. Use 8 mm ϕ, 16 cm c/c in both the directions at the bottom

2·5·3. Check on the Bearing Capacity of Soil

Weight of side walls $= 48{\cdot}2\ t$

Weight of bottom slab $= \dfrac{\pi D^2}{4} \times t_b \times 2400$

$$= \frac{\pi \times 12{\cdot}8^2}{4} \times \frac{20}{100} \times 2400 = 62{,}000 \text{ kg} = 62\ t$$

Weight of water when tank is full

$$= \frac{\pi \times 11{\cdot}5^2}{4} \times 3{\cdot}65 \times 1000 = 375\ t$$

Total load on the soil $= 48{\cdot}2 + 62 + 375 = 485{\cdot}2$ t

Maximum pressure on the soil $= \dfrac{485{\cdot}2 \times 4}{\pi(12{\cdot}8)^2}$

$= 3{\cdot}78 \text{ t/m}^2 < 10 \text{ t/m}^2$

Circular Tanks on Ground with Fixed Base 3

3·1. General

Circular tanks with rigid joint at the base is considered to be older type of construction. The side walls of these tanks are designed by one of the following methods.

(*i*) Reissner's method (*ii*) Carpenter's Simplified Reissner's method (*iii*) Approximate method

3·2. Reissner's Method—Side Walls

The water pressure produces hoop tension and cantilever bending moment in the sidewall. The side wall of the tank is assumed to be subjected partly to hoop action and partly to cantilever bending moment due to fixity at the base. The hoop tension in the wall and the fixing moment are seen to depend on the dimensions of tank.

Let h=Height of the tank,
t_w=thickness of the side wall
D=Diameter of the tank

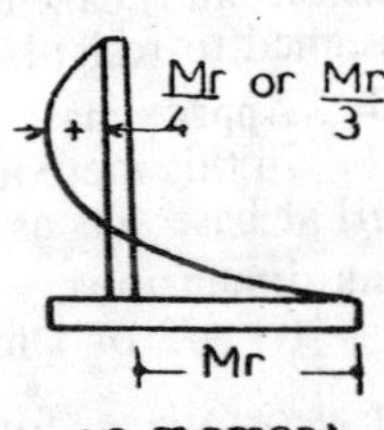

$$K = \frac{12\, h^4}{\left(\frac{D}{2}\right)^2 t_w^2}$$

For different values of K, coefficients for hoop tension and cantilever restraint moments are given

Coefficients for rectangular wall section

K	M_r Restraint moment	Maximum Tension T	Position of maximum tension from base
0	0·167 K_1	0	0
10	0·110 K_1	0·18 K_2	1·0 h
100	0·0582 K_1	0·27 K_2	1·0 h
1000	0·024 K_1	0·47 K_2	0·47 h
10000	0·0085 K_1	0·67 K_2	0·31 h
∝	0	1·0 K_2	0

$$K_1 = wh^3 \qquad K_2 = wh\,\frac{D}{2}$$

3·3. Design

Compute the value of K and determine the corresponding values of M_r and T from the table. The side wall is designed to resist both fixing moment and hoop tension.

3·4. Carpenter's Method

The magnitude and position of hoop tension in the side wall and the maximum cantilever moment coefficients K_3 and K_4 are given for different $\frac{h}{D}$ and $\frac{h}{t_w}$ ratios

3·5. Hoop Tension

(*i*) Maximum hoop tension in the wall

$$= w(h - K_3\,h)\frac{D}{2} = wh\,\frac{D}{2}\,(1 - K_3)$$

Position of maximum hoop tension $= K_3\,h$ metres above base

3·6. Maximum Cantilever Moment

(*ii*) Maximum cantilever moment $= K_4\,wh^3$

	Coefficients For K_4				K_3			
$\frac{h}{D}$ \ $\frac{h}{t_w}$	10	20	30	40	10	20	30	40
0·2	0·046	0·028	0·022	0·015		0·50	0·45	0·40
0·3	0·032	0·019	0·014	0·010	0·55	0·43	0·38	0·33
0·4	0·024	0·014	0·010	0·007	0·50	0·39	0·35	0·30
0·5	0·020	0·012	0·009	0·006	0·45	0·37	0·32	0·27
1·0	0·012	0·006	0·005	0·003	0·37	0·28	0·24	0·21
2·0	0·006	0·003	0·002	0·002	0·30	0·22	0·19	0·16
4·0	0·004	0·002	0·002	0·001	0·27	0·20	0·17	0·14

Using the coefficients for K_3, K_4 from the table, maximum hoop tension and cantilever moments are calculated. The side wall is designed to resist both hoop tension and cantilever moment.

3·7. Approximate Method

In this method, it is assumed that certain height of the tank side wall at base acts as a cantilever. This cantilever effect depends on the tank dimensions.

(*i*) $h/3$ or 1 m from the base, whichever is greater is taken to act as cantilever for tanks having $\frac{h^2}{D\,t_w}$ ratio between 6 and 12, and

(*ii*) For those tanks for which this ratio lies between 12 to 30. the cantilever portion, is either $h/4$ or 1 m whichever is greater.

Maximum hoop tension at the level D

$$= w(h - h_1)\,\frac{D}{2}$$

Maximum cantilever moment

$$= \tfrac{1}{2} \times wh \times h' \times \frac{h'}{3} = \frac{w(h_1)^2 h}{6}$$

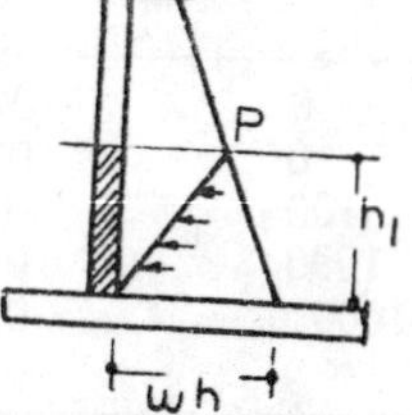

Hoop reinforcement $A_t = \frac{w(h - h_1)}{1000} \times \frac{D}{2}$ cm^2

This is provided on one face or at both faces at equal spacing upto the level D from the base. Above this level hoop reinforcement required may be calculated from the hoop tensions at different levels and the spacing of the bars may be increased towards top.

Vertical steel required to resist cantilever moment at the base

$$= \frac{wh(h_1)^2}{6 \times 1000 \times d_e}\ \text{cm}^2$$

This reinforcement should be placed vertically upto level D and extending into base slab horizontally.

Circular Tank on Ground with Fixed Base 4

Capacity 350,000 Litres by Reissner's Method

4·1. Data

Capacity = 350 m^3

Tank rests on ground and the side wall is fixed to the base slab

Bearing capacity of soil = 10 t/m^2

Materials available, Concrete – grade M 200, Steel – grade I

Design the tank using Reissner's method.

4·2 Characteristic Strengths

$\sigma_{cb} = 70$ kg/cm² $m = 13$

$\sigma_{st} = 1000$ kg/cm² $jd = \cdot 84d$

$\sigma_{ct} = 12$ kg/cm² $R = 14$

4·3. Trial Dimensions

Capacity = 350 m^3

depth (assumed) of tank $= h = 3.5$ m, free board (say) = 15 cm, height of the tank $h_1 = 3{\cdot}65$ m

Let D be the diameter of the tank, then

$$\frac{\pi D^2}{4}\, h = \text{volume} = \frac{\pi D^2}{4} \times 3{\cdot}5 = 350$$

$D^2 = 128$, $D = 11{\cdot}4$ m

Use $D = 11{\cdot}5$ m

A tank of height 3·65 m and internal diameter of 11·5 m will be sufficient.

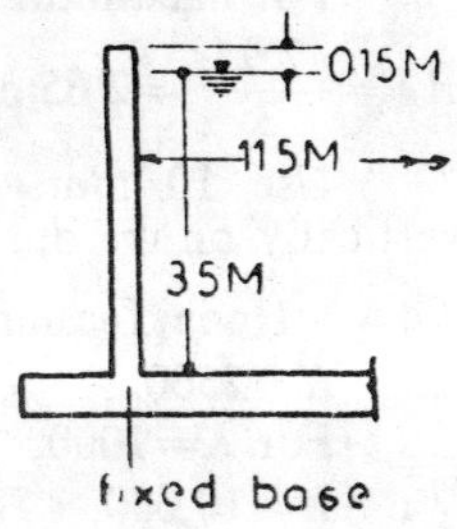

$$\text{Capacity provided} = \frac{\pi}{4} \times (11{\cdot}5)^2 \times 3.5$$

$= 362{\cdot}5$ $m^3 > 350$ m^3

4·3·1. Thickness of the Wall

Thickness of the wall t_w, (*i*) $\nless$ 15 cm (*ii*) $(3 \times 3{\cdot}65 + 5) = 16$ cm

Use $t_w = 16$ cm

4·4. Side Wall Design by Reissner's Method

$$K = \frac{12h^4}{\left(\frac{D}{2}\right)^2 t_w^2} = \frac{12 \times (3{\cdot}65)^4}{\left(\frac{11{\cdot}5}{2}\right)^2 \times (\cdot 16)^2} = 2500$$

4·4·0. Restraining Moment

K lies between 1000 and 10,000

for $K = 1000$, $M_r = 0{\cdot}024\, K_1$

$K = 10{,}000$. $M_r = 0{\cdot}0085\, K_1$

For $K=2500$, $M_r=C_1K_1$
by logarithmic interpolation

$$C_1=0{\cdot}0085+(.024-{\cdot}0085)\ \frac{\log 10{,}000-\log 2500}{\log 10{,}000-\log 1000}$$
$=0{\cdot}0085+(0{\cdot}024-0{\cdot}0085)\times 0{\cdot}6021=0{\cdot}0085+0{\cdot}0155\times 0{\cdot}6021$
$=0{\cdot}C025+0{\cdot}0094=0{\cdot}0179$

M_r=maximum − ve moment at the base
$=0{\cdot}0179\ K_1={\cdot}0179\ wh^3$
$={\cdot}0179\times 1000\times 3{\cdot}65^3$
$=870$ m kg.

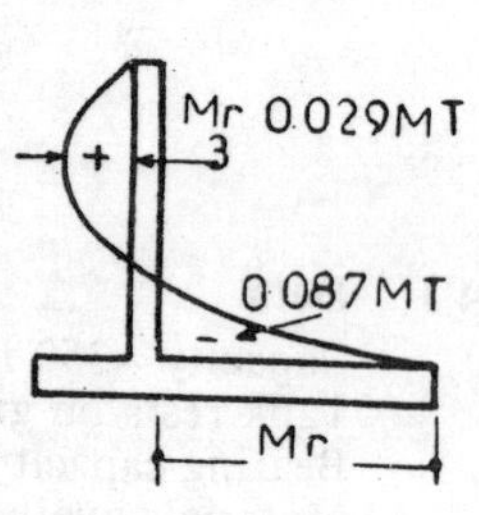

Max+Ve BM=Mr/4 or $\text{Mr}/3=\dfrac{870}{3}$

$=290$ m kg.

4·4·2. Thickness Required

$14\ bd^2=870\times 100$

$$t_w=d=\sqrt{\frac{87000}{100\times 14}}=\sqrt{62}\ =7{\cdot}85\text{ cm}$$

Thickness provided $=16$ cm
effective $=16-3$ (Cover) $=13$ cm

4·4·3. Main Reinforcement

$$A_t=\frac{87000}{1000\times 13\times {\cdot}84}=7{\cdot}95\text{ cm}^2.$$

Use 12 mm ϕ 14 cm c/c vertically on the inner face of the wall with a cover of 3 cm.

For maximum+Ve BM,

$$A_t=\frac{7{\cdot}95}{3}=2{\cdot}65\text{ cm}^2$$

Use 10 mm ϕ at 30 cm c/c vertically on the outer face

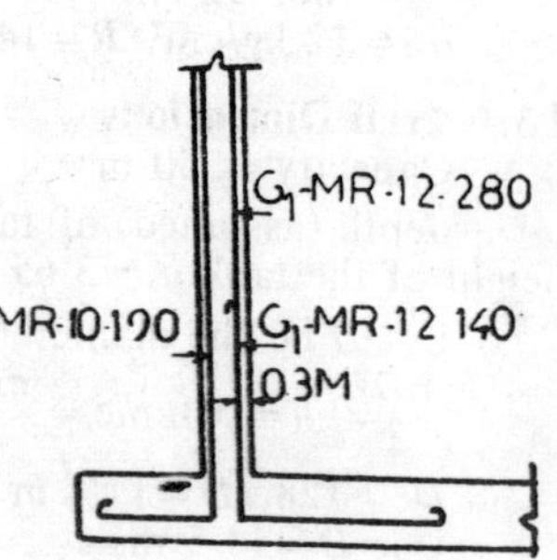

4·4·4. Hoop Tension in the side wall

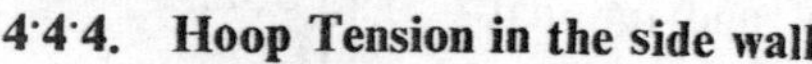

$K=2500$
For $K=1000$, $\quad T=0{\cdot}47\ K_2$
$K=10{,}000$, $\quad T=0{\cdot}67\ K_2$
$K=2500$, $\quad T=C_2\ K_2$
by logarithmic interpolation

$$C_2=0{\cdot}47+(0{\cdot}67-0{\cdot}47)\ \frac{\log 2500-\log 1000}{\log 10{,}000-\log 1000}$$
$=0{\cdot}47+0{\cdot}2\times{\cdot}3979=0{\cdot}47+{\cdot}07958=0{\cdot}54958=0{\cdot}5496$

Maximum hoop tension $={\cdot}5496\times wh\dfrac{D}{2}$

$=0{\cdot}5496\times 1000\times 3{\cdot}65\times\dfrac{11{\cdot}5}{2}=11500$ kg

Position of maximum hoop tension from the hase

$$=h\left[0{\cdot}47-(0{\cdot}47-0{\cdot}31)\frac{\log 2500-\log 1000}{\log 10{,}000-\log 1000}\right]$$

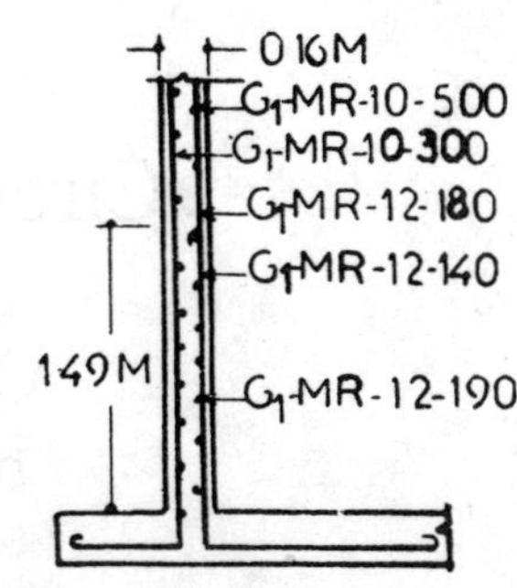

$=h(0{\cdot}47-0{\cdot}16\times0{\cdot}3979)$
$=(0{\cdot}47-.0636)\ h={\cdot}4064\ h$

height at which maximum hoop tension occurs from base $=0{\cdot}4064\times3{\cdot}65$
$=1{\cdot}49$ m

4·4·5. Hoop steel

Amount of steel required to resist hoop tension. $=\dfrac{11500}{100}=11{\cdot}5\ \text{cm}^2$.

Use 12mm ϕ at 9·5 c/c on one face or at 19 cm c/c on both faces. A_t provided $=11{\cdot}90\ \text{m}^2$

The main reinforcement for the cantilever moment can be curtailed above 1·49 m since there is no - ve bending moment. Therefore cut off alternate bars and use 12 mm ϕ at 28 cm c/c

4·4·6 On cracking stress consideration

$$\sigma_{ct}=\frac{11500}{100\times16+12\times11{\cdot}90}=\frac{11500}{1600\times142}=\frac{11500}{1742}=6{\cdot}6\ \text{kg/cm}^2$$

<12 kg/cm²

Secondary reinforcement$=\dfrac{{\cdot}3}{100}\times100\times16=4{\cdot}8\ \text{cm}^2$. Use 10 mm ϕ at 30 cm on both faces

4·4·7. Base Slab

Same as in the case of tank with sliding joint

Circular Tank on Ground with Fixed Base 5

Capacity 350 000 Litres by Carpenter's Method

5·1. Data

Capacity = 350 m³. Tank rests on ground and the side wall is fixed to the base slab.

Bearing capacity of soil = 10 t/m²
Materials available, Concrete—grade $M200$, Steel—grade—1
Design the tank using carpenter's method.

5·2. Characteristic Strengths

$\sigma_{cb} = 70$ kg/cm³ $\quad m = 13$
$\sigma_{st} = 1000$ kg/cm² $\quad jd = 0·84\, d$
$\sigma_{ct} = 12$ kg/cm². $\quad R = 14$

5·3. Trial Dimensions

Capacity = 350 m^3, Depth of tank (assumed) $h = 3·5$ m,
Free board = 15 cm,
Height of the tank $= h_t = 3·65$ m
Let D be the diameter of the tank, then

$$\frac{\pi D^2}{4} h = \text{volume} = \frac{\pi D^2}{4} \times 3·5 = 350$$

$D = 11·4$, Use $D = 11·5$ m

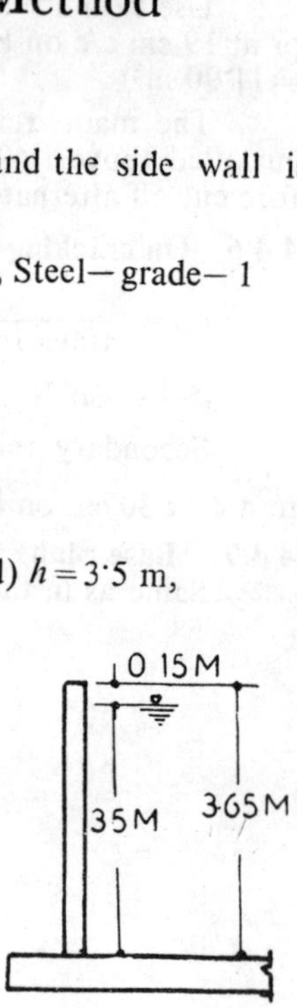

A tank of height 3·65 m and internal diameter of 11·5 m will be sufficient

Capacity provided $= \frac{\pi}{4}(11·5)^2 \times 3·5$

$= 362·5$ m³ > 350 m³

0 15 M
35 M
3·65 M

5·3·1. Thickness of the wall

Thickness of the wall t_w shall be, (*i*) ≮ 15 cm or
(*ii*) $(3 \times 3·65 + 5) = 16$ cm
Use $t_w = 16$ cm

5·3·2. Coefficients for Hoop Tension

$$\frac{h}{D} = \frac{3·65}{11·5} = 0·3175, \quad \frac{h}{t_w} = \frac{3·65}{0·16} = 22·8$$

From table :—For $\frac{h}{D} = 0·3$ and $\frac{h}{t_w} = 20$, $K_3 = 0·43$

For $\frac{h}{D} = 0·3$ and $\frac{h}{t_w} = 30$, $K_3 = 0·38$

By linear interpolation $h/D = 0·3$ and $h/t_w = 20$, $K_3 = 0·43$
$h/D = 0·3$ and $h/t_w = 22·8$,

$$K_3=0{\cdot}43-(0{\cdot}43-0{\cdot}38)\times\frac{22{\cdot}8-20}{30-20}=0{\cdot}43-\left(\cdot05\times\frac{2{\cdot}8}{10}\right)$$

$$=0{\cdot}43-0{\cdot}014\quad 0{\cdot}416$$

$h/D=0{\cdot}4$ and $h/t_w=20$, $K_3=0{\cdot}39$

$h/D=0{\cdot}4$ and $h/t_w=30$, $K_3=0{\cdot}35$

By linear interpolation

$h/D=0{\cdot}4$ and $h/t_w=22{\cdot}8$,

$$K_3=\cdot39-(\cdot39-\cdot35)\times\frac{22{\cdot}8-20}{30-20}=0{\cdot}39-\cdot04\times\frac{2{\cdot}8}{10}$$

$$=0{\cdot}39-0{\cdot}0112=\cdot3788$$

$hD/=0{\cdot}3$ and $h/t_w=22{\cdot}8$, $K_3=0{\cdot}4160$

$h/D=0{\cdot}4$ and $h/t_w=22{\cdot}8$, $K_3\quad 0{\cdot}3788$

For $h/D=0{\cdot}3175$, $h/t_w=22{\cdot}8$

by linear interpolation

$$K_3=0{\cdot}416-(\cdot4160-\cdot3788)\ \frac{0{\cdot}3175\quad 0{\cdot}3000}{0{\cdot}4000-0{\cdot}3000}$$

$$=0{\cdot}416-\left(0{\cdot}0372\times\frac{0{\cdot}0175}{0{\cdot}1}\right)=0{\cdot}416\cdot 0{\cdot}00651$$

$$=0{\cdot}40949=0{\cdot}4095$$

5·3·3. Hoop Tension

Maximum hoop tension at the level of K_3h

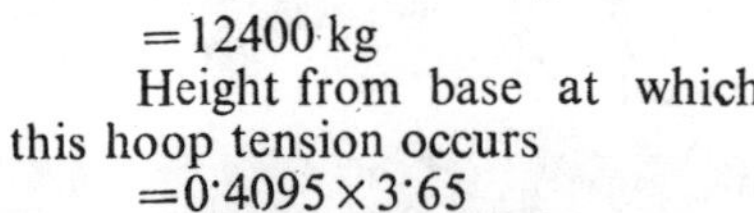

$$=w(h-K_3h)\ \frac{D}{2}$$

$$=wh\frac{D}{2}\ (1-K_3)$$

$$=1000\times3{\cdot}65\times\frac{11{\cdot}5}{2}\times0{\cdot}5905$$

$$=12400\ \text{kg}$$

Height from base at which this hoop tension occurs

$$=0{\cdot}4095\times3{\cdot}65$$

$$=1{\cdot}49\ \text{m}$$

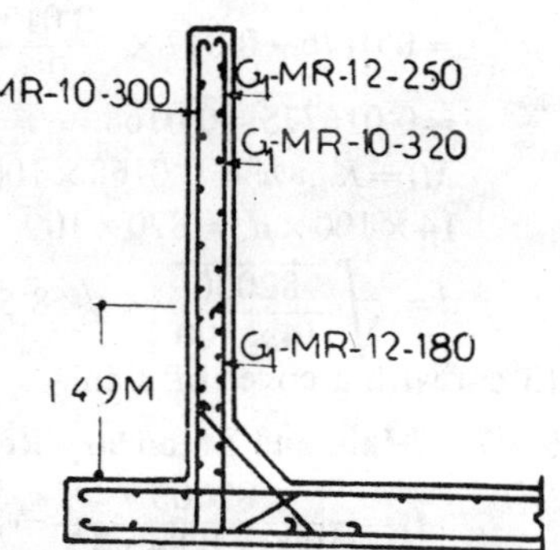

5·3·4. Hoop Steel Required

Hoop steel required to resist hoop tension

$$A_t=\frac{12400}{1200}=12{\cdot}4\ \text{cm}^2$$ Use 12 mm ϕ 18 cm c/c on both faces or at 9 cm c/c on one face.

A_t provided $=12{\cdot}56\ \text{cm}^2$.

5·3·5. Cracking Stress

Cracking stress in concrete,

$$\sigma_{ct}=\frac{12400}{100\times16+12\times12{\cdot}56}=\frac{12400}{1600+150}=7{\cdot}1\ \text{kg/cm}^2$$

$$<12\ \text{kg/cm}^2.$$

5·3·6. Maximum Cantilever Moment

$\frac{h}{D}=\frac{3·65}{11·5}=0·3175, \quad \frac{h}{t_w}=\frac{3·65}{0·16}=22·8$

$h/D=0·3$ and $h/t_w=20$, $K_4=·019$,

$h/D=0·3$ and $h/t_w=30$, $K_4=·014$, $h/D=0·3$ and $ht_w=22·8$

by linear interpolation $K_4=0·019-(·019-·014)\times 2·8/10$

$=0·019-0·005\times 2·8/10=0·019-0·0014=0·0176$

$h/D=0·4$ and $h/t_w=20$, $K_4=0·014$,

$\frac{h}{D}=0·4$ and $\frac{h}{t_w}=30$, $K_4=0·010$, $\frac{h}{D}=0·4$ and $\frac{h}{t_w}=22·8$,

by linear interpolation $K_4 \; 0·014-0·004\times\frac{2·8}{10}$

$=0·014-0·0112=0·01288=0·0129$

$\frac{h}{D}=0·3$ and $\frac{h}{t_w}=22·8$, $K_4=·0176$,

$\frac{h}{D}=0·4$ and $\frac{h}{t_w}=22·8$, $K_4=·0129$,

$\frac{h}{D}=·3175$ and $\frac{h}{t_w}=22·8$ by linear interpolation.

$K_4=·0176-(·0176-·0129)\times\frac{0·3175-0·3}{0·4-0·3}$

$=0·0176-0·047\times\frac{0·0175}{0·1}=0·0176-0·000825$

$=0·016775=0·0168.$

$M_r=K_4.wh^3=0·0168\times 1000\times 3·65^3=820$ m kg

$14\times 100\times d^2=820\times 100$

$d=\sqrt{\frac{82000}{14\times 100}}=\sqrt{58·5}=7·65$ cm. Use effective depth of 13 cm with a cover of 3 cm.

5·3·7. Main and Secondary Reinforcement

$A_t=\frac{82000}{1000\times 0·84\times 13}=7·5$ cm².

Use 12 mm ϕ, 14 cm c/c vertically on the inner face of the wall. Alternate bars can be curtailed above 1·49 m height from bottom and a spacing of 28 cm c/c shall be used.

Positive bending moment$=\frac{M_r}{3}$ or $\frac{M_r}{4}=\frac{820}{3}=273·3$ m kg.

Reinforcement required$=\frac{7·5}{3}=2·5$ cm². Use 10 mm bars at 30 cm c/c.

Secondary reinforcement$=\frac{0·3}{100}\times 100\times 16=4·8$ cm².

Use 10 mm ϕ at 16 mm c/c vertically on the outer face or at 32 cm c/c on both faces.

5·4. Base Slab. Same as in 2·5.

Circular Tank on Ground with Fixed Base 6

Capacity 350,000 Litres by Approximate Method

6·1. Data

Capacity = 350 m^3. Tank rests on ground and the side wall is fixed to the base slab.

Bearing capacity of soil = 10 t/m^2

Materials available concrete – grade *M* 200, steel – grade I.

Design the tank by approximate method

6·2. Characteristic Strengths

$\sigma_{cb} = 70$ kg/cm² $m = 13$

$\sigma_{st} = 1000$ kg/cm² $jd = \cdot 84d$

$\sigma_{ct} = 12$ kg/cm² $R = 14$

6·3 Trial Dimensions

Capacity = 350 m³, depth of tank (assumed) = $h = 3{\cdot}5$ m,

free board = 15 cm,

height of the tank = $h_1 = 3{\cdot}65$ m,

Let D be the diameter of the tank.

Then, $\pi D^2 h/4$ = volume = $\pi D^2/4 \times 3{\cdot}5 = 350$, $D = 11{\cdot}4$,

Use $D = 11{\cdot}5$ m

A tank of height 3·65 m and internal diameter of 11·5 m will be sufficient.

Capacity provided $= \frac{\pi}{4} (11{\cdot}5)^2 \times 3{\cdot}5 = 362{\cdot}5 \text{ m}^3 > 350\text{m}^3$

6·4. Thickness of the Wall

Thickness of the wall t_w shall be (*i*) $\not< 15$ cm

(*ii*) $(3 \times 3{\cdot}65 + 5) = 16$ cm

or (*iii*) thickness required to resist cracking stress σ_{ct}

6·5. Height of Sidewall which acts as Cantilever

$$K = \frac{h^2}{Dt_w} = \frac{3{\cdot}65 \times 3{\cdot}65}{11{\cdot}5 \times \cdot 16} = 7{\cdot}25$$

K lies between 6 and 12, h' or $h/3$ or 1m

Height over which wall acts as cantilever $= \frac{3{\cdot}65}{3} = 1{\cdot}22$ m

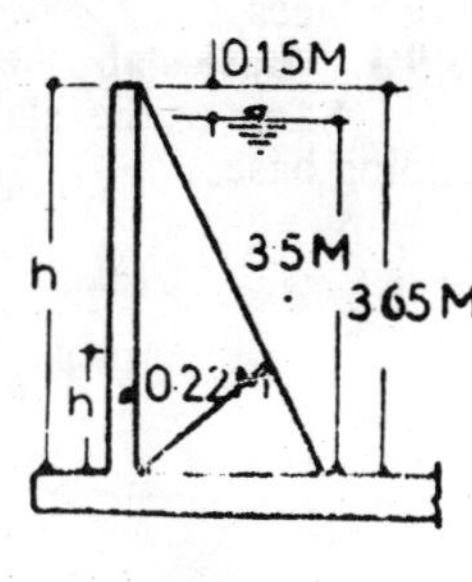

6·6. Maximum Hoop Tension

Maximum hoop tension

$$= w(h - h')\frac{D}{2} = 1000(3{\cdot}65 - 1{\cdot}22) \times \frac{11{\cdot}5}{2}$$

$$= 1000 \times 2{\cdot}43 \times \frac{11{\cdot}5}{2} = 14000 \text{ kg}$$

Amount of hoop Steel $A_t = \frac{14000}{1000} = 14 \text{ cm}^2$

Use 12 mm ϕ at 16 cm c/c on both faces

A_t provided $= 14{\cdot}14 \text{ cm}^2$

Cracking Stress in Concrete $\sigma_{ct} = \frac{14000}{100 \times 16 + 12 \times 14{\cdot}14}$

$= \frac{14000}{1600 + 170} = \frac{14000}{1770} = 7{\cdot}9 \text{ kg/cm}^2 < 12 \text{ kg/cm}^2$

Hoop Steel Required for Different Depths. The hoop tension at the top is zero and increases to a minimum at 2·43 m and therefore spacing of bars can be increased towards top

depth from top m	*Hoop tension T*	A_t *cm²*	*Spacing of 12 mm ϕ on both faces cm*
0 – 1	5750	5·75	39
1 – 2	11500	11·5	19·5
2 – 2·43	14000	14	16
2·43 – 3·65	14000	14	16

6·7. Cantilever Bending Momemt

Maximum cantilever moment $= \frac{wh(h')^2}{6} = \frac{1000 \times 3{\cdot}65 \times 1{\cdot}22^2}{6}$

$= 910$ m kg.

$d_e = \sqrt{\frac{910 \times 100}{14 \times 100}} = 8{\cdot}1$ cm

effective depth provided $= 16 - 3$ (cover) $= 13$ cm

6·8. Vertical Steel to Resist Cantilever Moment

$A_t = \frac{91000}{1000 \times {\cdot}84 \times 13} = 8{\cdot}3 \text{ cm}^2.$

Use 12 mm at 13·5 cm c/c at the inner face

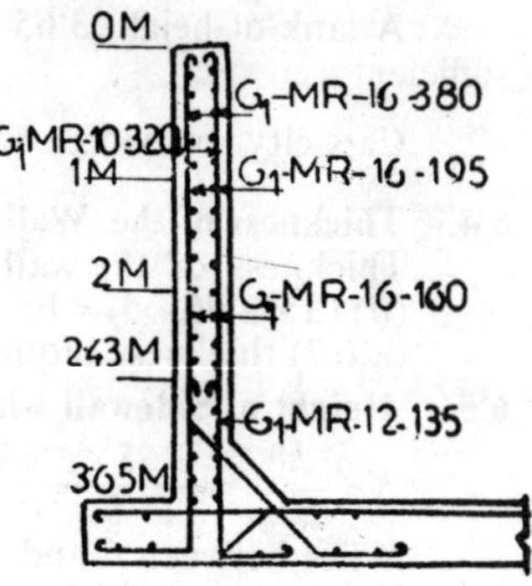

6·8·1. Secondary Steel

Amount of steel required

$= \frac{0{\cdot}3 \times 100 \times 16}{100} = 4{\cdot}8 \text{ cm}^2.$

Use 10 mm ϕ at 32 cm c/c on both faces

6·9·1. Base slab

Same as the slab for tank with sliding base.

Rectangular Tanks on Ground General Information 7

7·1. General

These tanks are used relatively for smaller capacities. Computation of exact moments in square or rectangular tanks is a difficult process. However, simplified methods which are sufficiently correct, for all practical purposes may be used.

Depending on the tank dimensions the tank walls may be subjected to either eantilever bending or horizontal bending or even bending in both the directions

7·1. Design of Tanks having L/B<2

The walls of these tanks are designed as a continuous frame subjected to a maximum pressure of $p = w(h \quad h\,)$ where h' is the height of the wall which acts as a cantilever and is taken as the larger of $h/4$ or 1 m. Moment distribution method is employed to determine the moments in the frame.

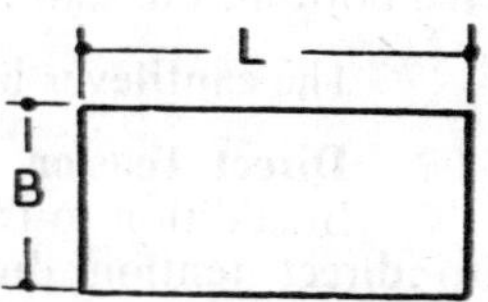

7·2. Approximate Moments For L/B=1

The moments are approximately taken as

at centre of the span $= \dfrac{whL^2}{16}$

at the end of the span $= \dfrac{whL^2}{12}$

7·3. Direct Tension in the Walls

In addition to the bending moments, the walls are subjected to direct tension due to water pressure.

Direct tension in the shorter walls $T_B = \dfrac{w(h-h')\,L}{2}$

and the direct tension in the longer walls $T_L = \dfrac{w(h-h')B}{2}$

7·4. Tanks having L/B>2

The side walls of rectangulas tanks with dimensions having a ratio of $L/B>2$ are designed to resist bending mom nt and direct tension.

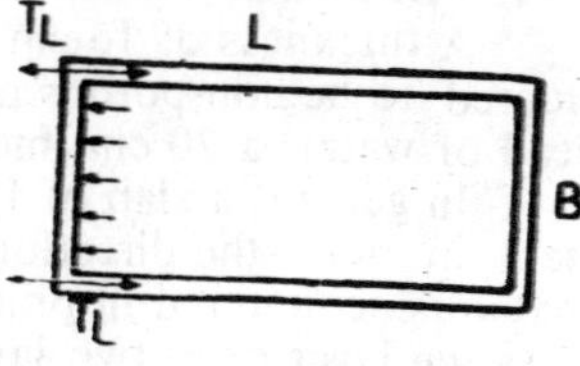

7·5. Longer Walls

Longer walls of these tanks are treated as cantilever slabs fixed to thc base slab and subjected to the water pressure.

The maximum moment at the base of the wall $M_e = \frac{1}{2}wh \times h \times h/3$

$$= \frac{wh^3}{6}$$

In addition to this bending moment these walls are subjected to direct tension due to water pressure at level *D* acting on the shorter wall over the whole length *B* of the wall

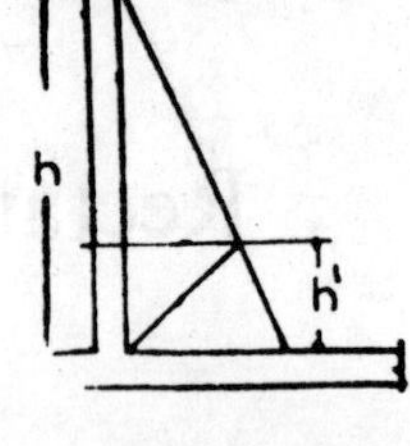

$$T_L=\frac{w(h\quad h')B}{2}$$

7·6. Short Walls Bending Moments

Short walls span horizontally between the longer walls.

Maximum positive bending moment at the centre of the span,

$$M_C=\frac{w(h-h')B^2}{16}$$

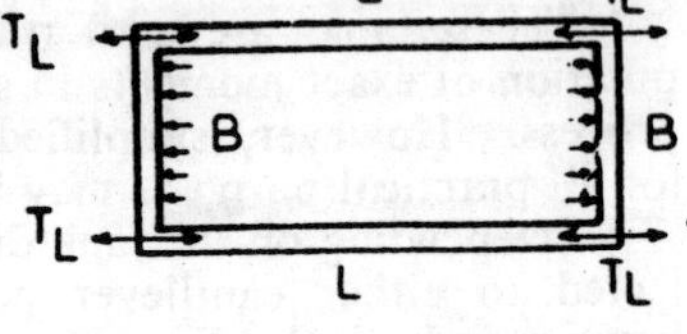

and the maximum negative bending moment at the end of the span or at support, $M_S=\frac{w(h-h')B^2}{12}$

In addition to these bending moments over the height of h' at the bottom, the slab acts as a cantilever.

The cantilever bending moment $M_C=\frac{1}{2}wh\times h'\times\frac{h'}{3}=\frac{wh(h')^2}{6}$

7·7. Direct Tension

In addition to the bending moments the short walls are subjected to direct tention due to water pressure acting on 1 metre width of long walls.

$T_B=w(h\quad h')\times 1=w(h-h')$

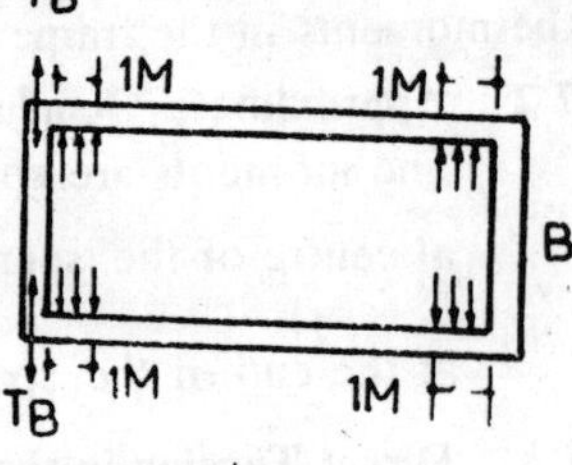

7·8. Floor Slab

tThe floor slab carries no effective strucural loading and is free from bending moments. However small moments are included due to the weight of the walls and roof, which causes tension in the top face of the slab. The slab is assumed freely supported at the edges for computing bending moments.

7·9. One Way Slab

If $L/B>2$. then the slab is designed as one way slab with shorter side as the span.

7·10. Two Way Slab

If $L/B<2$, then the slab is designed as a two way slab spaning in both the directions.

7·11. Minimum Thickness of the Base Slab

A thickness of 15 cm of *M* 200 concrete for floor slab is considered to be non-porous upto 3 to 3·5 m head of water. Upto 10 m head of water, a 20 cm thick base slab will be sufficient.

In general a slab of 15 cm thick *M* 200 concrete with 0·3% of steel in both the directions will be sufficient for all types of grounds that are encountered in practice. This amount of steel may be placed in single layer or in two layers one on each face.

Reservoirs on poor ground can be supported on sleeper walls. The floor slab may span in one direction or in two directions. The mass concrete sleeper beams form the concrete supports for the slab.

Open Rectangular Tank on Ground 8
Capacity 100,000 Litres L/B < 2

8·1. Data

Capacity = 100 m³, Soil bearing capacity = 20 t/m², Free board = 20 cm

Materials available concrete – grade M 200, steel – grade I

8·2. Characteristic Strengths

$\sigma_{cb} = 70$ kg/cm² $m = 13$

$\sigma_{st} = 1000$ kg/cm² $jd = 0{\cdot}84\ d$

$\sigma_{ct} = 12$ kg/cm² $R = 14$

8·3. Trial Dimensions

Required capacity = 100 m³, Depth of the tank (assumed) = h = 3 m, If A is the area of the tank in plan, $V = Ah$

$100 = A \times 3$, $A = 33{\cdot}3$ m²

If length $L = 7$ m, B = breadth = 5 m, Use 7 m × 5 m × 3·2 m

Actual capacity of the tank = 7 × 5 × 3 = 105m³ > 100 m³

$L/B = 7/5 = 1{\cdot}4 < 2$

The walls are treated as a continuous frame. Height of the tank from bottom, which acts as a cantilever = $h/4$ or 1 m = 3/4 m or 1 m

Take 1 m as the cantilever height. The wall for this height is designed as cantilever.

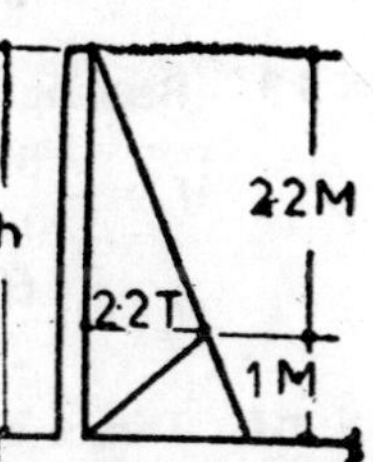

8.3.1. Bending Moment

By moment distribution, Maximum pressure at 2·2 m depth

$p = w(h - h') = 1000 \times 2{\cdot}2 = 2200$ kg

Fixed and moment

$= \frac{PL^2}{12}$ For span $AB = \frac{P \times 5 \times 5}{12} = \frac{25}{12} P$

For span $AC = \frac{P \times 7 \times 7}{12} = \frac{49}{12} P$

Moment at joint A

AC	AB	*Distrtbution facts*
$\frac{7}{12}$	$\frac{5}{12}$	$\frac{L_1}{L_1} : \frac{L_2}{L_2}$
$+\frac{25}{12} P$	$-\frac{49}{12}$	$\frac{1}{5AC} : \frac{1}{7AB}$
$+\frac{14}{12} P$	$+\frac{10}{12} P$	$\frac{1}{5} : \frac{1}{7}$
$+\frac{39}{12} P$	$-\frac{39}{12} P$	7 : 5
		$\frac{7}{12} : \frac{5}{12}$

Support moment

$=M_s=\frac{-39}{12}p$

Moment at the centre of the span shorter wall

$M_c=\frac{PL^2}{8}\quad\frac{-39}{12}P=\frac{25}{8}Px-\frac{39}{12}P=P\left(\frac{25}{8}-\frac{39}{12}\right)$

$=\frac{75-78}{24}=-\frac{3}{24}P$

Moment at the centre of the span—longer wal'

$M_c=\frac{PL^2}{8}-\frac{39}{12}p=\frac{49}{8}p-\frac{39}{12}P$

$=P\left(\frac{49}{8}-\frac{39}{12}\right)=\frac{147-78}{24}=\frac{69}{24}p$

8·3·2. Bending Moment in Shorter Walls

$M_s=\frac{-39}{12}\times2200=-7510$ mkg

$M_c=-\frac{1}{8}\times2200=-275$ mkg

8·3·3. Bending Moments in Longer Walls

$M_s=-\frac{39}{12}\times2200=-7510$ mkg

$M_c=+\frac{69}{24}\times2200=+6310$ mkg.

8·3·4. Direct Tension in Shorter Walls

$T_B=\frac{w(h-h')L}{2}=\frac{1000\times2{\cdot}2\times7}{2}=7700$ kg

8·3·5. Direct Tension in Longer Walls

$T_L=\frac{w(h-h')B}{2}=\frac{1000\times2{\cdot}2\times5}{2}=5500$ kg

8·3·6. Thickness of Walls

Maximum bending moment=+7510 mkg.

14 bd^2=7510

$d=\sqrt{\frac{7510\times100}{14\times100}}=\sqrt{540}=23{\cdot}25$ cm.

Use effective depth=25 cm and a overall depth=30 cm

8·3·7. Main Reinforcement : Shorter Walls

Net bending moment at the corners$=M_S-T_B\times x$

$=751000\quad 7700\times(25-15)=751000-77000=674{,}000$ cm kg.

Steel required to resist this bending moment

$A_{tb}=\frac{674000}{1000\times{\cdot}84\times25}=32{\cdot}1$ cm²

Direct tension in the wall $T_B=7700$ kg
Steel required to resist this tension

$$A_{tt}=\frac{7700}{1000}=7{\cdot}7 \text{ cm}^2$$

$A_t=A_{tb}+A_{tt}=39{\cdot}8$ cm².
Use 22 mm ϕ at 9·5 cm c/c, A_t provided 40·02 cm².

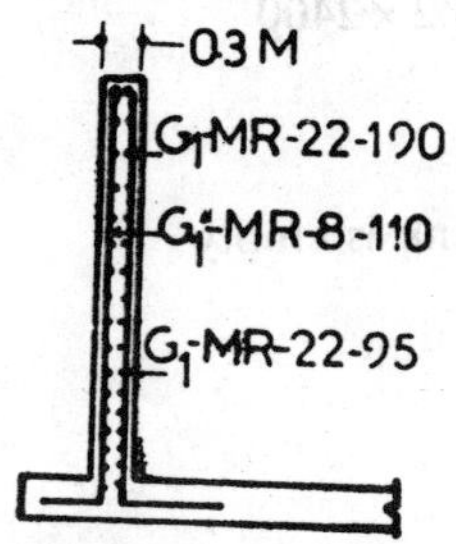

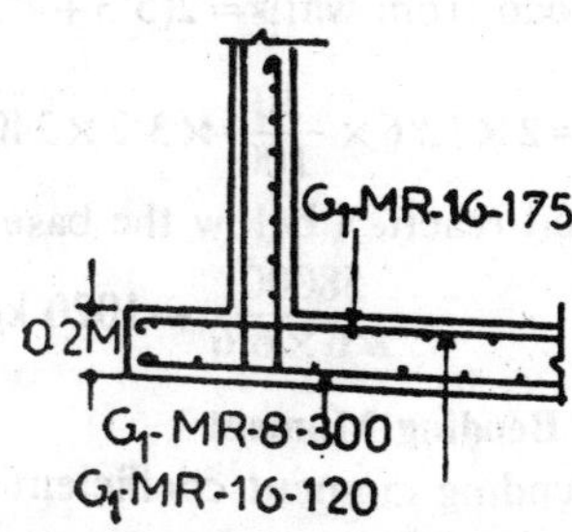

Reinforcement required at the centre of the span since BM is negative=27500 cm kg

$$A_t=\frac{27500}{1000\times{\cdot}84\times25}=1{\cdot}31 \text{ cm}^2$$

Therefore only nominal reinforcement is provided.

8·3·8. Secondary Steel Shorter Walls

$$0{\cdot}3\% \text{ of cross section of concrete}=\frac{0{\cdot}3}{100}\times30\times100=9 \text{ cm}^2$$

12 mm ϕ at 12·5 cm c/c, on one face or 8 mm ϕ at 11 cm c/c on both faces.

8·3·9. Main Reinforcement Longer Walls

At the ends of the wall provide same amount of steel as in the shorter wall. That is use 22 mm ϕ at 9·5 mm c/c.
Net bending moment at the centre of the wall

$$M_c=631000-T_L\times x=631000-5500(25-15)=631000-55000$$
$$=576{,}000 \text{ cm kg.}$$

This bending moment occurs away from water face and steel required to take this moment, $A_{tb}=\frac{576000}{1250\times{\cdot}86\times25}=21{\cdot}4$ cm²

Steel required to resist direct tension $A_{tt}=\frac{5500}{1000}=5{\cdot}5$ cm²

$A_t=26{\cdot}9$ cm²
Use 20 mm ϕ at 11·5 cm c/c and A_t provided=27·31 cm²

8·3·10. Secondary Steel

Same as in the short walls. That is 8 mm ϕ at 11 cm c/c on both faces.

8·3·11. Reinforcement for Cantilever Moment inside Walls

Cantilever moment on the bottom 1 m height of wall

$$=wh\times\frac{h'}{2}\times\frac{h'}{3}=wh\frac{h'^2}{6}=\frac{1000\times3{\cdot}2\times1}{6}=\frac{3200}{6}$$
$$=533{\cdot}33 \text{ m kg}=53330 \text{ cm kg.}$$

$A_t = \frac{53330}{1000 \times \cdot 84 \times 25} = 2{\cdot}55 \text{ cm}^2$. Use 8 mm ϕ at 19 cm c/c vertically.

8·4. Load on Base Slab

Bottom slab dimensions = 8·6 m × 6·6 m

Load from walls $= 2(5{\cdot}3 + 7{\cdot}3) \times \frac{30}{100} \times 3{\cdot}2 \times 2400$

$= 2 \times 12{\cdot}6 \times \frac{30}{100} \times 3{\cdot}2 \times 2400 = 58000$ kg

Soil reaction below the base when the tank is empty

$= \frac{58000}{8{\cdot}6 \times 6{\cdot}6} = 1020 \text{ kg/m}^2$

8·4·1. Bending Moment

Bending moment coefficient for

$\frac{L_x}{L_y} = \frac{7{\cdot}3}{5{\cdot}3} = 1{\cdot}38$, say 1·4

$Z'_x = \cdot 071, Z'_y = \cdot 05$

Maximum bending moment in the short span $= Z'_x w h_x^2$
$= \cdot 071 \times 1020 \times 5{\cdot}3^2 \times 100 = 204000$ cm kg

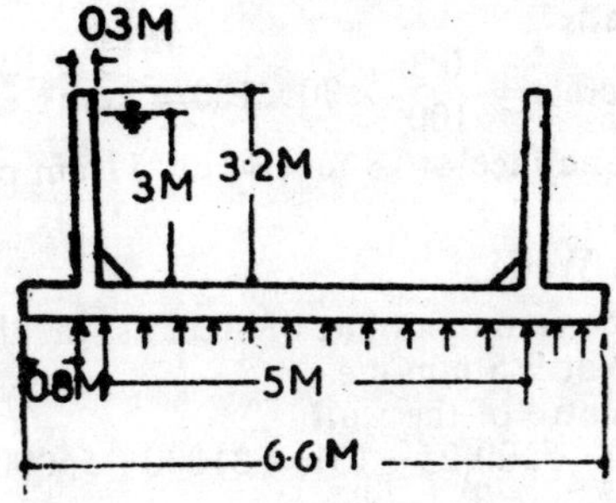

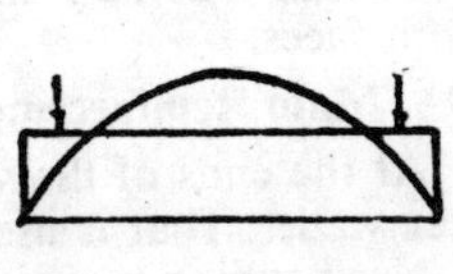

Maximum bending moment in the longer span
$= 0{\cdot}05 \times 1020 \times 5{\cdot}3^2 \times 100 = 144000$ cm kg

Depth of the slab $d = \sqrt{\frac{204000}{14 \times 100}} = \sqrt{146} = 12{\cdot}2$ cm

However a thickness of 20 cm shall be provided

8·4·2. Main and Secondary Reinforcement

A_t required $= \frac{204000}{1000 \times \cdot 84 \times 15} = 16{\cdot}2 \text{ cm}^2$

Use 16 mm ϕ at 12 cm c/c on the top face in the shorter direction

A_t required for longer span $= \frac{144{,}000}{100 \times \cdot 84 \times 15} = 11{\cdot}4 \text{ cm}^2$.

Use 16 mm ϕ at 17·5 cm c/c in the longer direction on the top face For the bottom face (secondary steel)

$$A_t = \frac{0\text{·}15}{100} \times 20 \times 100 = 3 \text{ cm}^2$$

Use 8 mm ϕ at 30 cm c/c in both the directions

8·4·3. Bearing Capacity

when the tank is full, water load $= 5 \times 7 \times 3\text{·}2 \times 1000$
$= 112{,}000$ kg

weight of walls $= 58{,}000$ kg

weight of base slab$= 8\text{·}6 \times 6\text{·}6 \times 2400 \times \frac{20}{100}$
$= 8\text{·}6 \times 6\text{·}6 \times 480 = 27200$ kg

Total load on the soil $= 197{,}200$ kg

Maximum pressure on the soil$= \frac{197{,}200}{8\text{·}6 \times 6\text{·}6 \times 1000}$
$= 3\text{·}48 \text{ t/m}^2 < 20 \text{ t/m}^2$.

Open Rectangular Tank on Ground 9
Capacity 100,000 Litres L/B > 2

9·1. Data

Capacity = 100 m³, Soil bearing capacity = 20 t/m²,
Free board = 15 cm.
Material available, concrete-grade M 200, steel-grade I.

9·2. Characteristic Strengths

$\sigma_{cb} = 70$ kg/cm² $\quad m = 13$
$\sigma_{st} = 1000$ kg/cm² $\quad jd = 0{\cdot}84\ d$
$R = 14$

$\sigma_{cb} = 70$ kg/cm² $\quad m = 13$
$\sigma_{st} = 1250$ kg/cm² $\quad jd = 0{\cdot}86\ d$
$R = 12{\cdot}6$

9·3. Trial Dimensions

Required capacity = 100 m³, Depth of the tank (assumed) = 3 m, Area of the tank = 33·3m². Let the length = L = 10m, breadth = B = 3·3m. Use 10 m × 3·5 m × 3 m tank.

Capacity of the tank provided = 10 × 3·5 × 2·85 = 100 cu m

$$\frac{L}{B} = \frac{10}{3{\cdot}5} = 2{\cdot}86 > 2$$

Therefore longer walls are designed as cantilevers and shorter walls as spanning horizontally between the longer walls

9·3·1. Longer Walls Bending Moment

$$\text{Maximum bending moment} = \frac{wh^3}{6} = \frac{1000 \times 3 \times 3 \times 3}{6}$$
$$= 4500 \text{ m kg}$$

9·3·2. Direct tension

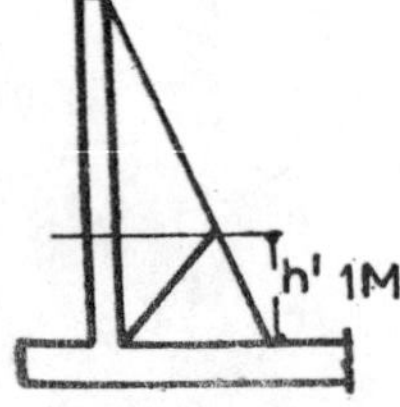

Height over which walls will be acting as cantilever $h' = h/4$ or 1 m $= \frac{3}{4}$ m < 1 m

Take $h' = 1$ m

Direct tension on the long wall

$$T_L = \frac{w(h-h')\,B}{2} = \frac{1000 \times 2}{2} \times 3{\cdot}5$$
$$= 3500 \text{ kg}$$

9·3·3. Thickness of the Wall

$$d = \sqrt{\frac{450000}{14 \times 100}} = \sqrt{320} = 17{\cdot}8 \text{ cm.}$$ Use 22 cm thick walls and effective depth = 18 cm.

Amount of reinforcement required for bottom 1 m height of wall, $A_t = \dfrac{450{,}000}{1000 \times {\cdot}84 \times 18} = 29{\cdot}6$ cm², Use 20 mm ϕ at 10·5 cm c/c.

9·3·4. Main and Secondary Reinforcement

Spacing for different depths

Depth *m*	*Bending moment* *cm kg*	A_t cm^2	*Spacing of 20 mm φ* *cm*
0–1	16,660	1·11	31·5
1–2	133,300	8·8	31·5
2–3	450,000	29·6	10·5

$A_t = 0{\cdot}3/100 \times 22 \times 100 = 6{\cdot}6$ cm²

A_t required per face $= 3{\cdot}3$ cm².

Use 8 mm φ at 15 cm c/c

9·3·5. Reinforcement for Direct tension

$$A_{tt} = \frac{3500}{1000} = 3{\cdot}5 \text{ cm}^2$$

Use 8 mm φ at 14 cm c/c

T_L 3·5M T_L

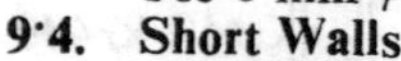

9·4. Short Walls

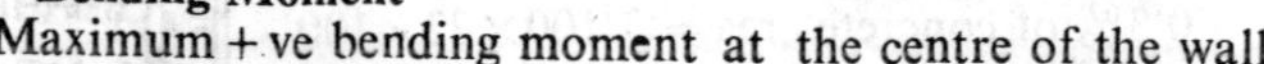

Short walls span horizontally between the longer walls

9·4·1. Bending Moment

Maximum +ve bending moment at the centre of the wall

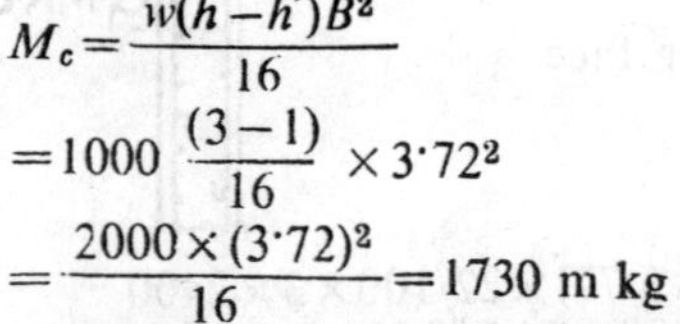

$$M_c = \frac{w(h-h')B^2}{16}$$

$$= 1000\frac{(3-1)}{16} \times 3{\cdot}72^2$$

$$= \frac{2000 \times (3{\cdot}72)^2}{16} = 1730 \text{ m kg}$$

0.22M 3.5M 3.72M

Maximum –ve moment at the ends of the wall $M_s = \dfrac{w(h-h')B^2}{12} = \dfrac{1000(3-1)(3{\cdot}72)^2}{12} = 2310$ m kg.

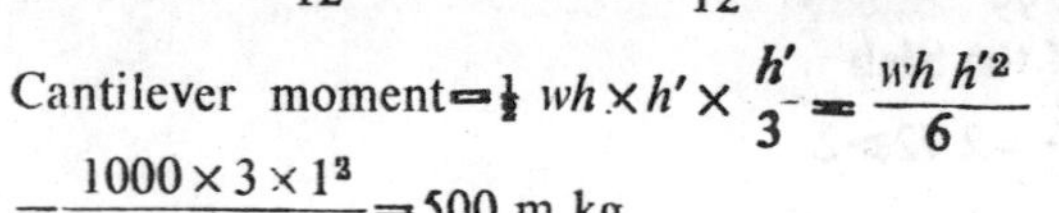

Cantilever moment $= \frac{1}{2}\, wh \times h' \times \dfrac{h'}{3} = \dfrac{wh\,h'^2}{6}$

$$= \frac{1000 \times 3 \times 1^2}{6} = 500 \text{ m kg.}$$

9·4·2. Direct Tension

$T_B = w(h-h') \times 1 = 1000 \times 2 \times 1 = 2000$ kg

9·4·3. Thickness of the Wall

$$d = \sqrt{\frac{2310 \times 100}{14 \times 100}} = \sqrt{165} = 12{\cdot}8 \text{ cm}$$

However use the same thickness as the longer wall—that is $d = 22$ cm and $d_e = 18$ cm

9·4·4. Main Reinforcement

Net bending moment on the shorter wall

$= 231000 - T \times x = 231000 - 2000\,(18-11)$

$= 231000 - 14{,}000 = 217{,}000$ cm kg

Steel required to resist bending

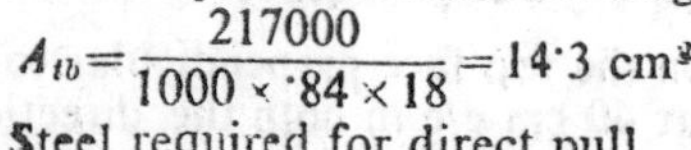

$$A_{tb} = \frac{217000}{1000 \times {\cdot}84 \times 18} = 14{\cdot}3 \text{ cm}^2$$

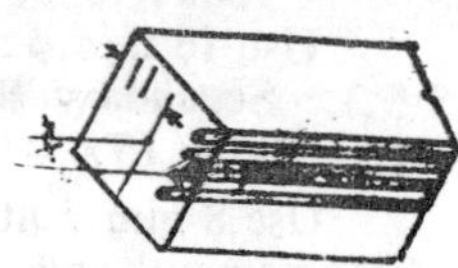

Steel required for direct pull

$A_{tt}= 2000/1000=2\ cm^2$. $A_{tb}+A_{tt}=A_t=16{\cdot}3\ cm^2$

Use 16 mm ϕ 12 cm c/c for bottom 1 m height. Vary the spacing towards top.

$$\text{Net+ve bending moment} = \frac{173000 - Tx}{173000-2000\times(18-11)}$$

$=159000$ cm kg.

A_{tb} required for +ve bending moment

$$A_{tb}=\frac{159000}{1250\times{\cdot}86\times18}=8{\cdot}2\ cm^2$$

Steel required for direct tension$=A_{tt}=2000/1250=1{\cdot}6\ cm^2$

$A_t=A_{tt}+A_{tb}=9{\cdot}8\ cm^2$, Use 12 mm ϕ 11·5 cm c/c

Reinforcement for cantilever moment

$$=\frac{50000}{1000\times{\cdot}84\times18}=3{\cdot}3\ cm^2.$$

Use 8 mm at 15 cm c/c vertically over bottom 1 metre height

G1MR-16-480

G1MR-16-160

G1-MR-8-150

9·4·5. Secondary Reinforcement

$A_t = 0{\cdot}3\%$ of concrete area$={\cdot}3/100\times22\times100$ $=6{\cdot}6\ cm^2$

Use 8 mm ϕ at 15 cm c/c per face

9·5. Load on Bottom Slab

Bottom slab dimensions are

$=11{\cdot}44\ m\times4{\cdot}99\ m$

Load from walls$=2(10{\cdot}22+3{\cdot}77)\times22/100\times3\times2400$

$=2\times13{\cdot}99\times22/100\times3\times2400=44{,}250$ kg

Soil pressure below the bottom slab when the tank is empty

$$=\frac{44{,}250}{11{\cdot}44\times4{\cdot}99}=770\ kg/m^2$$

9·5·1. Design of the Slab

$$\frac{L}{B}=\frac{10{\cdot}22}{3{\cdot}77}=2{\cdot}72>2$$

Slab spans in shorter direction only. Maximum bending moment

$$\text{approximately}=\frac{wB^2}{8}=\frac{770\times3{\cdot}77^2}{8}=1380\ m\ kg$$

$$\text{Depth of slab } d=\sqrt{\frac{1380\times100}{14\times100}}=\sqrt{985}=9{\cdot}9\ cm$$

However a minimum thickness of slab required to prevent leakage etc. is 15 cm

Use 15 cm thick slab

9·5·2. Main Reinforcement

$$A_t=\frac{135500}{1000\times{\cdot}84\times11}=14{\cdot}6\ cm^2$$

Use 16 mm ϕ at 13·5 cm c/c

9·5·3. Secondary Reinforcement

$A_t= 100\times15\times{\cdot}3/100=4{\cdot}5\ cm^2$.

Use 8 mm ϕ at 20 cm c/c on the top face perpendicular to main reinforcements and use 8 mm ϕ at 40 cm c/c in both the directions in the bottom face.

Open Square Tank on Ground 10
Capacity 100,000 Litres

10·1. Data

Capacity $=100\ m^3$, Bearing capacity of soil $=20\ t/m^2$. Shape required = Square

Materials available, Concrete-grade M 200, Steel-grade I, Free board = 15 cm.

10·2. Characteristic Strengths

$\sigma_{cb}=70\ kg/cm^2$ $\quad m=13$

$\sigma_{st}=1000\ kg/cm^2$ $\quad jd=0·84\ d,\quad R=14$

$\sigma_{cb}=70\ kg/cm^2$ $\quad m=13$

$\sigma_{st}=1250\ kg/cm^2$ $\quad jd=0·86\ d,\quad R=12·6$

10·3. Trial Dimensions

Required capacity $=100\ m^3$

Let the dimensions be $=6\ m\times 6\ m\times 3\ m$

depth of the tank $=3\ \ 0·15=2·85$ m

Capacity provided $=6\times6\times2·85=101·3$ cu m >100 cu m.

$\frac{L}{B}=\frac{6}{6}=1$, the walls are designed as a continuous frame.

The height over which walls will be acting cantilever $h'=h/4$ or 1 m.

$h/4=\frac{3}{4}=0·75$ m. Hence use $h'=1$ m.

Horizontal pressure at level $D=p=w(h-h')=1000\times(3-1)$ $=2000$ kg.

10·3·1. Bending Moments

Maximum negative bending moment at support

$$M_s=\frac{pL^2}{12}=\frac{2000\times6\times6}{12}=6000 \text{ m kg.}$$

Maximum positive bending moment at the centre of the wall

$$M_c=\frac{pL^2}{16}$$

It is the practice to use $\frac{pL^2}{16}$ although the exact value is $\frac{pL^2}{24}$

$$M_c=\frac{2000\times6\times6}{16}=4500 \text{ m kg.}$$

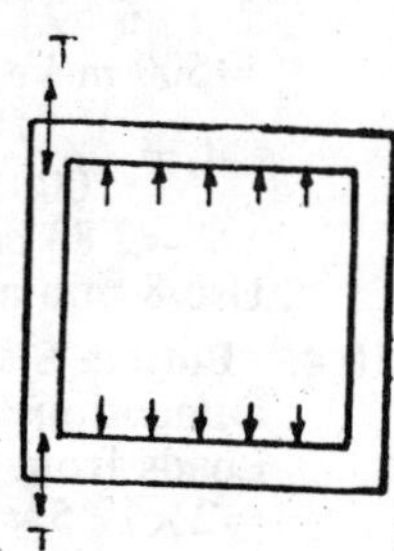

10·3·2. Direct Tension on the Wall

Direct tension on the wall due to water pressure on the perpendicular wall

$$T=\frac{w(h-h')\ L}{2}=\frac{1000\times(3-1)\times6}{2}$$

$$=1000\times\frac{2\times6}{6}=6000 \text{ kg}$$

10·3·3. Thickness of the Wall

$$d_e=\sqrt{\frac{6000\times100}{14\times100}}=\sqrt{430}=20{\cdot}75 \text{ cm}$$

Use $d=25$ cm and effective depth $d_e=21$ cm.

10·3·4. Main Reinforcement at Supports

Net bending moment at support

$=M_s-T_x=600000-6000\times(21-12{\cdot}5)$

$=600000\quad 6000\times8{\cdot}5=600{,}000-51{,}000$

$=549{,}000$ cm kg

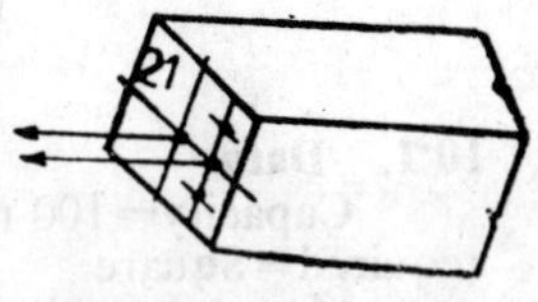

This bending moment occurs on the water face.

A_{tb} = Amount of steel required to resist bending

$$=\frac{549{,}000}{1000\times{\cdot}84\times21}=31 \text{ cm}^2$$

Steel required to resist direct tension

$A_{tt}=\frac{6000}{1000}=6$ cm², $A_t=37$ cm². Use 22 mm ϕ at 10 cm c/c for the bottom 2 m height and at 20 cm c/c for top 1 m.

10·3·5. Main Reinforcement at the Centre of the Wall

Net bending moment $=M_c-T_x$

$=450000-6000\times(21-12{\cdot}5)$

$=450000-6000\times8{\cdot}5=450{,}000-51000$

$=399{,}000$ cm kg.

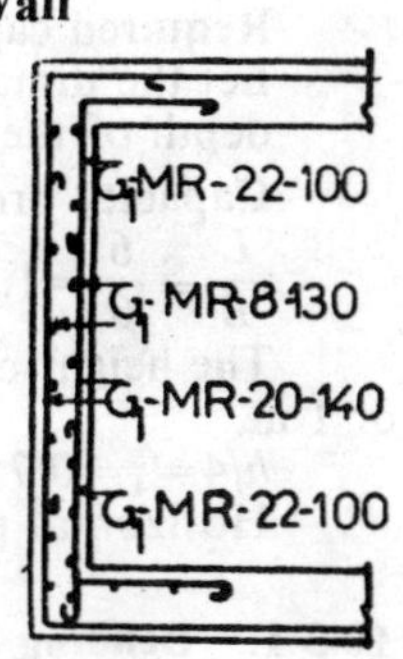

This bending moment occurs away from water face.

$$A_{tb}=\frac{399{,}000}{1250\times{\cdot}86\times21}=17{\cdot}5 \text{ cm}^2$$

$$A_{tt}=\frac{6000}{1250}=4{\cdot}78 \text{ cm}^2$$

$A_{tb}+A_{tt}=A_t=22{\cdot}28$ cm².

Use 20 mm at 14 cm c/c.

10·3·6. Secondary Reinforcement

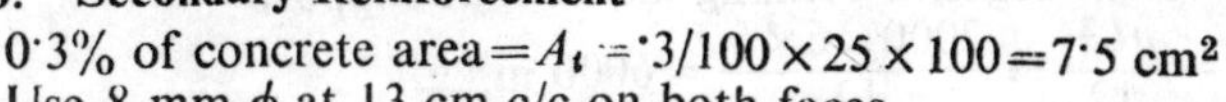

0·3% of concrete area $=A_t={\cdot}3/100\times25\times100=7{\cdot}5$ cm²

Use 8 mm ϕ at 13 cm c/c on both faces.

10·3·7. Main Reinforcement for Cantilever Moment at Bottom 1 m Height

$$\text{Cantilever moment}=wh\times\frac{h'}{2}\times\frac{h'}{3}$$

$$=\frac{wh\,h'^2}{6}=\frac{1000\times3\times1^2}{6}$$

$=500$ m kg.

$$A_t=\frac{50000}{1000\times{\cdot}84\times21}$$

$=2{\cdot}83$ cm².

Use 8 mm at 17 cm c/c vertically.

G1 MR-8-130

G1 MR-8 170

10·4. Bottom Slab

Dimensions of the bottom slab $=7{\cdot}5$ m $\times$ $7{\cdot}5$ m

Loads from walls $=2(6{\cdot}25+6{\cdot}25)\times25/100\times3\times2400$

$=2\times12{\cdot}5\times25/100\times3\times2400=45000$ kg.

Soil reaction below the base slab $= \frac{45000}{7{\cdot}5 \times 7{\cdot}5} = 800$ kg/m^2

$\frac{L}{B} = 1.$ Slab is designed spaning in two directions.

Approximate bending moment at centre when the tank is empty

$= 0{\cdot}5 \times \frac{wl^2}{8} = \frac{0{\cdot}5 \times 800 \times 6^2}{8} = 1800$ m kg.

Depth required $= \sqrt{\frac{1800 \times 100}{14 \times 100}} = \sqrt{128} = 11{\cdot}4$ cm.

Use 20 cm slab with effective depth $= 16$ cm

$A_t = \frac{1800 \times 100}{100 \times {\cdot}84 \times 16} = 13{\cdot}4$ cm^2

Use 16 mm ϕ at 15 cm c/c in both the directions, at top face

10·4·1. Secondary Reinforcement

$A_t = \frac{{\cdot}3}{100} \times 20 \times 100 = 6$ cm^2

Use 8 mm ϕ at 16·5 cm c/c in both directions at bottom face.

IV. CLOSED WATER TANKS ON GROUND

General Information 1

1·1. General

In this type of tanks the side walls are assumed to be supported at top. Depending on the dimensions of the tank, the walls are subjected to bending moments of different magnitudes.

The various components of the tank are

(*i*) Roof slab (*ii*) Side walls (*iii*) Base slab.

1·1·1. Roof Slab

If $L/B>2$, the roof slab is designed to span in shorter direction, that is as one way slab. If $L/B<2$ the roof slab is designed to span in both the directions, that is, as two way slab and the slab is assumed as freely supported at the edges

1·2. Side Walls

When the tank is closed, the side walls are assumed to be supported at top

Case (*i*) The ratio of horizontal span to vertical span of the side wall that is L/h or $B/h>2$, the walls are designed as cantilever walls, L and B are the length and breadth of the tank and h the depth.

1·3. Bending Moments

Maximum +ve bending moment in the side-wall $=\dfrac{wh^3}{33{\cdot}5}$

Maximum −ve bending moment in the side-wall $=wh^3/15$

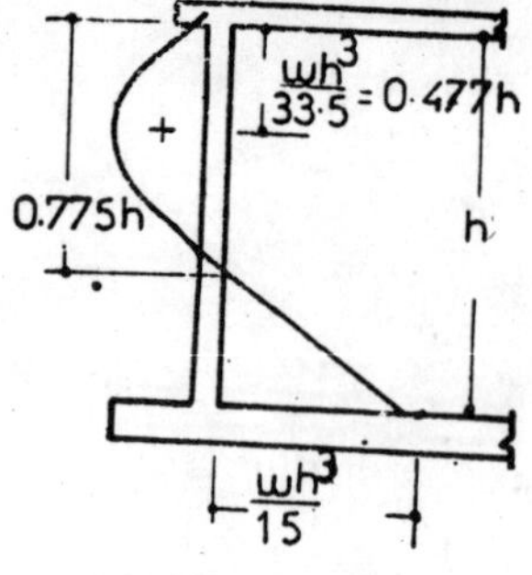

Case (*ii*) If L/h or B/h 0·5, the walls are assumed to span horizontally and the approximate moments are,

the maximum +ve bending moment at centre of the wall $=wh^3/16$

maximum −ve bending moment at support $=wh^3/12$

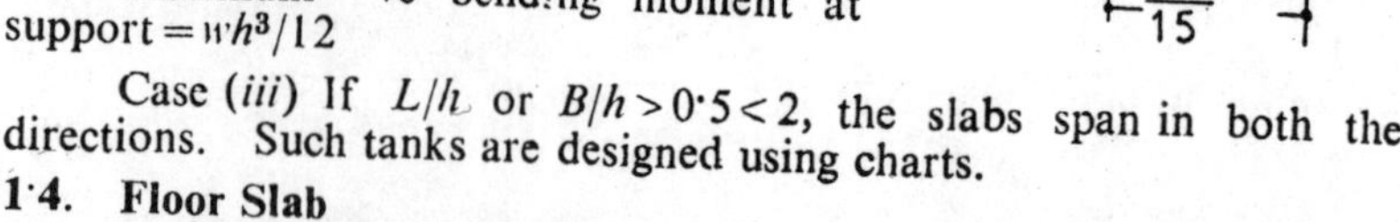

Case (*iii*) If L/h or $B/h>0{\cdot}5<2$, the slabs span in both the directions. Such tanks are designed using charts.

1·4. Floor Slab

The base or the floor slab can be designed as a slab spanning in one or two directions. For tanks resting on ground, the moments are produced due to the weight of the walls and roof and when the tank is empty, these moments cause tension in the top of the slab. It is assumed that the slab is freely supported for computation of these moments.

Closed Rectangular Tanks on Ground 2

Capacity 350,000 Litres

2·1. Data

Capacity – 350 m³, Free Board=20 cm, Shape desired – Rectangular and closed

Materials available : Concrete - grade *M* 200, Steel - grade – I

2·2. Characteristic Strengths

$\sigma_{cb}=70$ kg/cm² $\quad m=13$
$\sigma_{st}=1000$ kg/cm² $\quad jd=0{\cdot}84\ d,\ R=14$
$\sigma_{cb}=70$ kg/cm² $\quad m=13$
$\sigma_{st}=1250$ kg/cm² $\quad jd=0{\cdot}86\ d,\ R=12{\cdot}6$

2·3. Trial Dimensions

Required capacity=350 m³, Depth of tank (assume)=3·5 m, Area of the tank=100 m². Using 15·5 m×7 m×3·3 m

Actual capacity=358 m³>350 m³

∴ Use 15·5 m×7 m×3·5 m tank

$$\frac{L}{B}=\frac{15{\cdot}5}{7}>2,\ \frac{L}{h}=\frac{15{\cdot}5}{3{\cdot}5}>2,\ \frac{B}{h}=\frac{7}{3{\cdot}5}=2$$

The walls can be designed as propped cantilever.

2·4. Roof Slab Loads

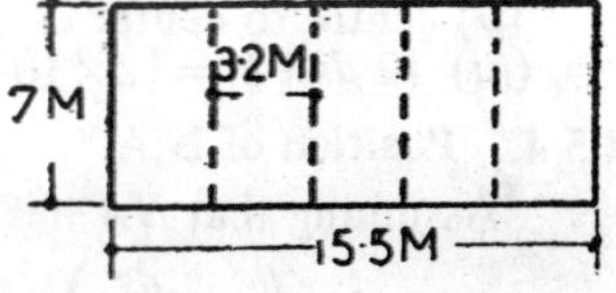

Use Slab and T-beam roof

Slab – span=3·2 m

Loads

Live load=150 kg/m²,

Screeding (5 cm)=110 kg/m²,

Self weight (1/35×320) at 10 cm thick=10×24=240 kg/m²

Total load=500 kg/m²

2·4·1. Bending Moment

Continuous slab – span=3·2 m

$$\text{Max. BM}=\frac{wl^2}{10}=\frac{500\times(3{\cdot}2)^2}{10}=512 \text{ m kg}$$

2·4·2. Depth of Slab

Using *M* 150 concrete for roof

$BM=8{\cdot}7\ bd^2$

$$d_e=\sqrt{\frac{512\times100}{8{\cdot}7\times100}}=\sqrt{59}=7{\cdot}7 \text{ cm.}$$

Use $d=10$ cm and $d_e=8{\cdot}5$ cm

2·4·3. Amount of Main and Secondary Reinforcement

$$A_t=\frac{51200}{1400\times\cdot87\times8\cdot5}=4\cdot95\ \text{cm}^2$$

Use 10 mm ϕ at 15·5 cm c/c

$$A_t=\frac{0\cdot15}{100}\times10\times100=1\cdot5\ \text{cm}^2$$

Use 6 mm ϕ at 16·5 cm c/c

2·5. T-Beam

Span = 7·25 m, assuming 25 cm bearing, centre to centre of beams = 3·2 m

Loads : From slab = 500 × 3·2 = 1600 kg/m

Self weight of rib (30 cm × 50 cm) $=\frac{30}{100}\times\frac{50}{100}\times1\times2400$

= 360 kg/m

Total load = (1600 + 360) = 1960 kg/m

2·5·1. Bending Moment

$$BM=\frac{wl^2}{8}=\frac{1960\times(7\cdot25)^2}{8}=12900\ \text{m kg}$$

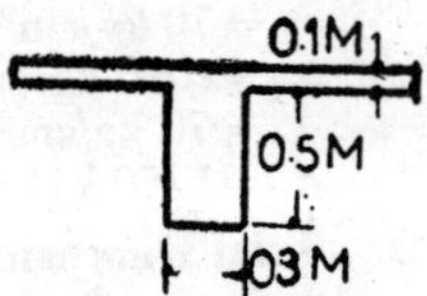

Approximate lever arm $=d_e-d_s/2=55-5$
$=50$ cm

2·5·2. Approximate Steel

Approximate steel required

$$=\frac{12900\times100}{1400\times50}=18\cdot4\ \text{cm}^2$$

Use 6—20 mm ϕ

A_t provided = 18·84 cm²

2·5·3. Effective width of Flange

The least of

(*i*) 1/3 of span $=\frac{7\cdot25}{3}=2\cdot42\text{m}=242$ cm,

(*ii*) centre to centre of beam = 3·2 m = 320 cm,

(*iii*) $12\ ds+b_r=12\times10+30=150$ cm

2·5·4. Position of N.A.

Assuming that *NA* lies below the flange

$$B\times ds\times\left(n-\frac{ds}{2}\right)=m\times A_t\,(d-n)$$

$$150\times10\times(n-5)=18\times18\cdot84\ (54-n)$$

$$1500\ n-7500=990-340\ n$$

$$1500\ n+340\ n=7500+18400$$

$$1840\ n=25900$$

$$n=14\cdot1\ \text{cm}$$

$$\bar{y}=\frac{3n-2ds}{2n-ds}\times\frac{ds}{3}=\frac{3\times14\cdot1-2\times10}{2\times14\cdot1-10}\times\frac{10}{3}$$

$$=\frac{42\cdot3-20}{28\cdot2-10}\times\frac{10}{3}=\frac{22\cdot3}{18\cdot2}\times\frac{10}{3}\quad4\cdot08\ \text{cm}$$

Actual lever arm $=d_e-\bar{y}=j_d=54-4\cdot08=49\cdot92$ cm

2·5·5. Stresses

Stress in steel $=\frac{1290000}{18\cdot84\times49\cdot92}=1370$ kg/cm² < 1400 kg/cm²

Stress in concrete $\frac{\sigma_{cb}}{n}=\frac{\sigma_{st}}{m(d-n)}$, $\sigma_{cb}=\frac{1370\times14\cdot1}{18\times39\cdot9}$

$=42\cdot7$ kg/cm² < 50 kg/cm²

2·5·6. Shear Stress Nominal Stirrups

Maximum shear at the support edge

$=\frac{1960\times7}{2}=6850$ kg

Shear stress $q_s=\frac{s}{jd+b_r}=\frac{6850}{49\cdot92\times30}=4\cdot6$ kg/cm² < 5 kg/cm²

Provide 8 mm ϕ two-legged at 30 cm c/c

2·6. Side Walls L/h and B/h > 2

Both shorter and longer walls are designed as propped cantilevers
Maximum water pressure at the base of the wall $p=wh$
$=1000\times3\cdot5=3500$ kg

2·6·1. Bending Moment

Maximum positive Bending Moment

$=\frac{wh^3}{33\cdot5}=\frac{1000\times(3\cdot5)^3}{33\cdot5}=1281$ m kg

Maximum negative bending moment

$=\frac{wh^3}{15}=\frac{1000\times(3\cdot5)^3}{15}=2860$ m kg

2·6·2. Thickness of the Walls

$d_e=\sqrt{\frac{2860\times100}{100\times14}}=\sqrt{205}=14\cdot3$ cm

Use $d=20$ cm thickwalls and $d_e=16$ cm

2·6·3. Main Reinforcement

Amount of steel required for −ve bending moment (water face)

$A_t=\frac{286000}{1000\times\cdot84\times16}=21\cdot3$ cm².

Use 20 mm ϕ at 14·5 cm c/c

Amount of steel required for +ve bending moment (away from water face)

$=\frac{128100}{1250\times\cdot86\times16}=7\cdot45$ cm². Use 12 mm ϕ at 15 cm c/c

2·6·4. Secondary Reinforcement

0·3% of the concrete area $=A_t=\frac{0\cdot3}{100}\times20\times100=6$ cm²

Use 8 mm ϕ at 15 cm c/c on both faces

2·7. Base Slab

Base slab dimensions $=16\cdot9$ m $\times$ 8·4 m

Loads : Roof slab $=7\cdot40\times15\cdot9\times500=58600$ kg

Beam ribs (4 Nos) $=7\cdot40\times\frac{30}{100}\times\frac{50}{100}\times4\times2400=10700$ kg

Sidewalls : weight of the side walls

$$=2(7{\cdot}2+15{\cdot}7)\times\frac{20}{100}\times 2400\times 3{\cdot}5$$

$$=2\times 22{\cdot}9\times\frac{20}{100}\times 2400\times 3{\cdot}5=77{,}100 \text{ kg}$$

Total load on the slab $58600+10700+77100=146{,}400$ kg

Soil reaction below the base

$$=\frac{146400}{16{\cdot}9\times 8{\cdot}4}=1035 \text{ kg/m}^2$$

2·7·1. Bending Moment

Approximate bending moment

$$=\frac{wl^2}{8}=\frac{1035\times(7{\cdot}2)^2}{8}=6700 \text{ m kg}$$

Actual moment. Total load $=1035\times 8{\cdot}4=8700$ kg

Reaction $=4350$ kg

Bending moment at the centre

$$=4350\times 3{\cdot}6-1035\times\frac{8{\cdot}4}{2}\times\frac{8{\cdot}4}{4}$$

$$=15400 \quad 9100 \quad 6300 \text{ mkg}$$

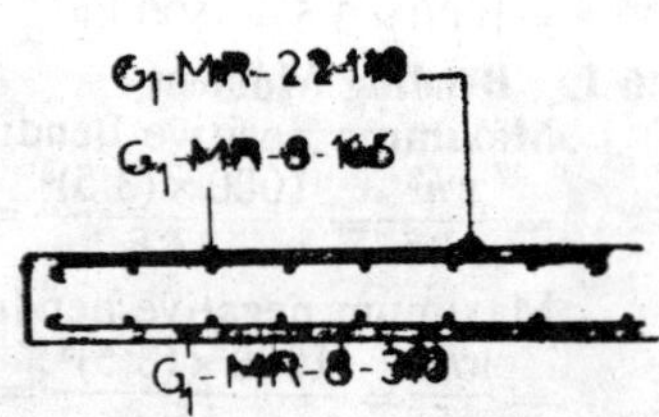

2·7·2. Depth of Slab Required

$$d_e=\sqrt{\frac{6300\times 100}{14\times 100}} \quad \sqrt{448}$$

$$=21{\cdot}2 \text{ cm}$$

Use $d=26$ cm and $d_e=22$ cm

2·7·3. Main and Secondary Reinforcement

$$A_t=\frac{630000}{1000\times{\cdot}84\times 22}=34 \text{ cm}^2$$ Use 22 mm ϕ 11 cm c/c at the top face

0·3% concrete area, $A_t=\frac{0{\cdot}3}{100}\times 20\times 100=6 \text{ cm}^2$

Use 8 mm ϕ at 16·5 cm c/c on the top face and 31 cm c/c in both directions at bottom face.

Closed Square Tanks on Ground General Information 3

3·1. General

In the case of square tanks with roofs, the wall slabs usually span in two directions.

$\frac{L}{h}$ that is ratio of horizontal span to vertical span lie between 0·5 and 2·5. These walls are assumed to span in both the directions.

Usual edge conditions assumed are the bottom edge continuous and the top edge simply supported.

3·1·1. Bending Moment Vertical Directions

Bending moment in the vertical span

$$M_v = 0.083\, wh^3 \left(\frac{K}{K+1}\right)$$

3·2. Horizontal Direction

$$M_h = \frac{0{\cdot}067\, w\, L^2\, h}{(1+K)}$$

It is possible to reduce the negative moments by 15% if equal moment is added to the positive moment.

The values of K, for different edge conditions are given in Fig.

Direct tension in the shorter walls

$$T_B = \left(\frac{1}{K+1}\right) wh \times 1 = p\left(\frac{1}{K+1}\right)$$

Closed Square Tank
Capacity 100,000 Litres

4

4·1. Data

Capacity = 100 m³, Free board = 15 cm, Bearing capacity = 20 t/m², Shape desired - square with roof.

Materials available : Concrete = M 200 grade for tank and M 150 grade for roof, Steel-grade – I.

4·2. Characteristic Strengths

M 150

$\sigma_{cb} = 50$ kg/cm²	$m = 18$	$R = 8.7$
$\sigma_{st} = 1000$ kg/cm²	$jd = \cdot 87\, d$	

M 200

$\sigma_{cb} = 70$ kg/cm²	$m = 13$	$R = 14{\cdot}0$
$\sigma_{st} = 1000$ kg/m²	$jd = \cdot 84\, d$	
$\sigma_{cb} = 70$ kg/cm²	$m = 13$	$R = 12{\cdot}6$
$\sigma_{st} = 1250$ kg/cm²	$jd = \cdot 86\, d$	

4·3. Trial Dimensions

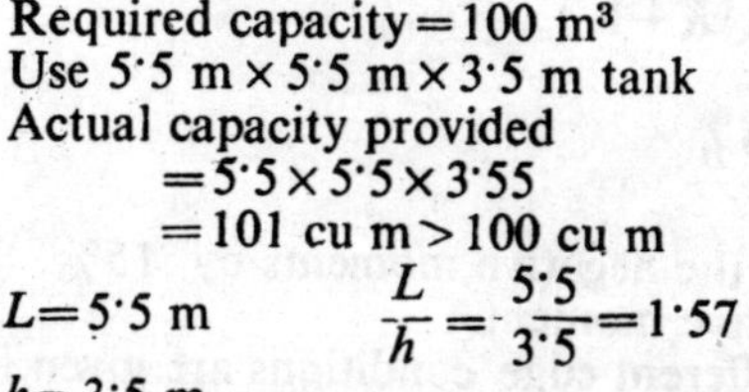

Required capacity = 100 m³

Use 5·5 m × 5·5 m × 3·5 m tank

Actual capacity provided

$= 5{\cdot}5 \times 5{\cdot}5 \times 3{\cdot}55$

$= 101$ cu m > 100 cu m

$L = 5{\cdot}5$ m $\qquad \frac{L}{h} = \frac{5{\cdot}5}{3{\cdot}5} = 1{\cdot}57$

$h = 3{\cdot}5$ m

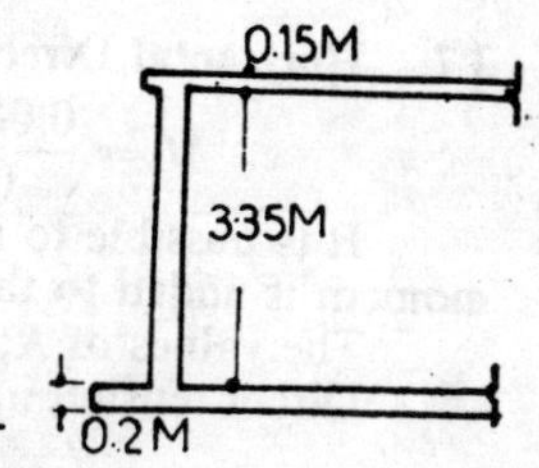

The side walls span in both the dimensions.

4·3·1. Roof Slab (Two Way Slab)

Span = 5·25 m, assuming 25 cm thick walls

Loads : Live load = 150 kg/m², Screeding = 100 kg/m²

Self weight (15 cm) = $\frac{15}{100} \times 1 \times 1 \times 2400 = 360$ kg/m²

Total load = 610 kg/m²

4·3·2. Bending Moment

Bending moment = $0{\cdot}05\, wL^2 = \cdot 05 \times 610 \times (5{\cdot}25)^2 = 840$ m kg

4·3·3. Depth of the Slab

Using M 150 concrete for roof

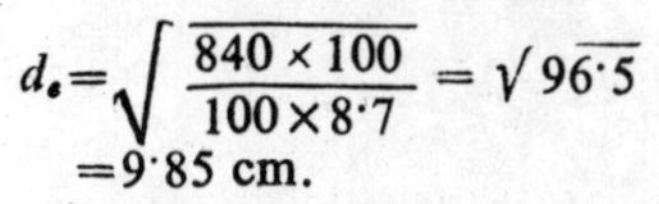

$$d_e = \sqrt{\frac{840 \times 100}{100 \times 8{\cdot}7}} = \sqrt{96{\cdot}5}$$

$= 9{\cdot}85$ cm.

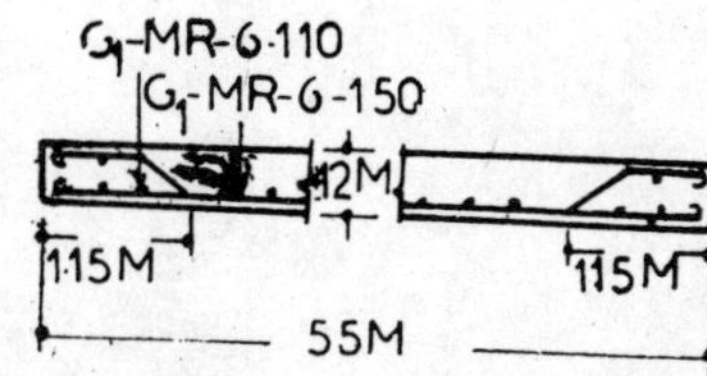

Use $d = 12$ cm and $d_e = 10$ cm

4·3·4. Amount of Reinforcement

$$A_t = \frac{84000}{1400 \times \cdot 87 \times 10} = 6 \cdot 9 \text{ cm}^2.$$

Use 10 mm ϕ at 11 cm c/c in both directions

4·4. Side Walls Vertical Span

$$\frac{L}{h} = \frac{5 \cdot 5}{3 \cdot 5} = 1 \cdot 57$$

Side wall is to be designed to span in both directions. K from graph for the case where top edge is simply supported and bottom edge is continuous. $K = 2 \cdot 5$.

4·4·1. Bending Moment

Free bending moment in the vertical span

$$M = \cdot 083 \times 1000 \times (3 \cdot 5)^3 \frac{2 \cdot 5}{(2 \cdot 5 + 1)} = \cdot 083 \times 1000 \times 3 \cdot 5^3 \times \frac{2 \cdot 5}{3 \cdot 5}$$

$= 2550$ m kg

with 15% adjustment in the moments. Maximum positive bending moment at the centre $= 0 \cdot 65\ M_v = 0 \cdot 65 \times 2250 = 1460$ m kg.

Maximum negative bending moment at the base $= 0 \cdot 89\ M_v$
$= 0 \cdot 89 \times 2250 = 2000$ m kg.

4·4·2. Horizontal Span

Free horizontal moment $M_h = \dfrac{\cdot 067 \times w\, L^2\, h}{(K+1)}$

$$= \frac{\cdot 067 \times 1000 \times 5 \cdot 5 \times 3 \cdot 5}{(2 \cdot 5 + 1)} = 2030 \text{ m kg}$$

4·4·3. Bending Moments

Negative moments at the corners
$= 0 \cdot 57\ M_h = \cdot 57 \times 2030$
$= 1160$ m kg

Positive moment at the mid span
$= 0 \cdot 43 \times M_h = 0 \cdot 43 \times 2030$
$= 875$ m kg.

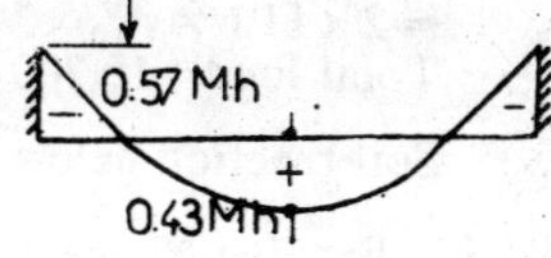

4·4·4. Maximum Moment

Maximum moment $= 2000$ m kg.

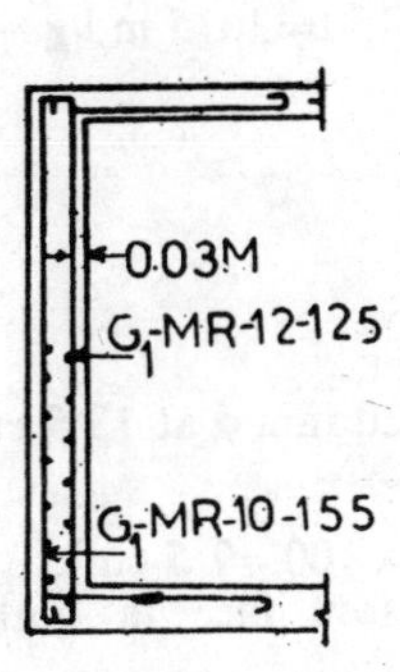

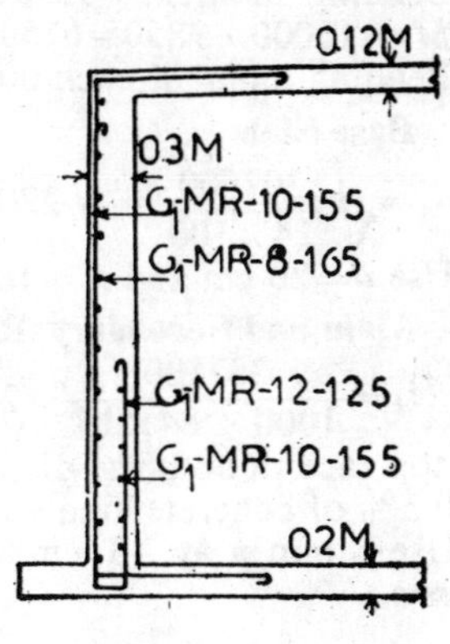

4·4·5. Thickness of Sidewalls

$$d_e=\sqrt{\frac{2000\times 100}{14\times 100}}=\sqrt{715}=26{\cdot}6 \text{ cm.}$$

Use $d=30$ cm and $d_e=27$ cm.

4·4·6. Main Steel Vertical Direction

Vertical steel required on the water face

$$A_t=\frac{200000}{1000\times{\cdot}84\times 27}=8{\cdot}8 \text{ cm}^2.$$ Use 12 mm ϕ at 12·5 cm c/c

Vertical steel required on the outer face (away from water).

$$A_t=\frac{146000}{1250\times{\cdot}86\times 27}=5{\cdot}01 \text{ cm}^2.$$ Use 10 mm ϕ at 15·5 cm c/c

4·4·7. Horizontal Steel

Steel required at corners (water face)

$$A_t=\frac{116000}{1000\times{\cdot}84\times 27}=5{\cdot}06 \text{ cm}^2.$$ Use 10 mm ϕ at 15·5 cm c/c

Steel required at mid span (away from water face)

$$A_t=\frac{87500}{1250\times{\cdot}86\times 27}=3{\cdot}02 \text{ cm}^2.$$ Use 8 mm ϕ at 16·5 cm c/c

4·5. Base Slab

5.5M
0.5M
5.5M
0.3M

Base slab dimensions$=7{\cdot}1$ m$\times 7{\cdot}1$ m when the tank is empty.

Loads : Roof slab self-weight

$=6{\cdot}1\times 6{\cdot}1\times\frac{12}{100}\times 2400=10{,}700$ kg

Screeding$=6{\cdot}1\times 6{\cdot}1\times 100=3720$ kg

Live load$=6{\cdot}1\times 6{\cdot}1\times 150=5{,}600$ kg

Sidewalls : self weight

$=2(5{\cdot}8+5{\cdot}8)\times\frac{30}{100}\times 2400\times 3{\cdot}5$

$=2\times 11{\cdot}6\times\frac{30}{100}\times 3{\cdot}5\times 2400=58{,}500$kg

Total load$=10{,}700+3720+5600+58{,}500=78{,}520$ kg.

Soil reaction below the base$=\frac{78{,}500}{7{\cdot}1\times 7{\cdot}1}=1560$ kg/m²

4·5·1. Bending Moment at Centre (Two Way Slab)

Total load$=7{\cdot}1\times 1560=11{,}100$ kg

Reaction $=\frac{11000}{2}=5{,}550$ kg

Bending moment$=5550\times 5{\cdot}8/2-1560\times 7{\cdot}1/2\times 7{\cdot}1/4$

$M_c=16000-9850=6150$ m kg

Bending moment each direction$=0{\cdot}5\times 6150=3075$ m kg

4·5·2. Base Slab

$$d_e=\sqrt{\frac{307500}{14\times 100}}=\sqrt{220}=14{\cdot}9 \text{ cm.}$$

Use $d=20$ cm and $d_e=16$ cm.

4·5·3. Main and Secondary Reinforcement

$$A_t=\frac{307500}{1000\times{\cdot}84\times 16}=22{\cdot}9 \text{ cm}^2.$$ Use 20 mm ϕ at 13·5 cm c/c

at the top face in both the directions.

0·3% of concrete area$=A_t={\cdot}3/100\times 25\times 100=7{\cdot}5$ cm².

Use 8 mm ϕ at 13 cm c/c at the bottom face in both the directions.

VI. UNDERGROUND WATER TANKS

General Information 1

1·1. General

These types of tanks are used in water purification works. The components of the tank, such as side walls and the base slab are not only subjected to water pressure but also to the earth pressure. The earth pressure may be due to dry soil or due to submerged soil in water logged areas.

1·1·1. Conditions of Loading

The following situations may arise in the design :

(*i*) Tank empty and dry soil earth pressure outside (*ii*) Tank empty and submerged soil outside (*iii*) Tank full and dry soil earth pressure outside (*iv*) Tank full and submerged soil outside.

1·2. Critical Conditions of Loading

The worst loading may be due to :

(*i*) Independent effect of water pressure during leak testing when the soil was not filled around the tank (*ii*) Independent effect of earth pressure either due to submerged or dry soil as the case may be (*iii*) Combined effect of earth and water.

1·3. Earth Pressure Dry Soils

Rankine's theory of earth pressure is used to compute the active earth pressure

$$P_a = w_e h\left(\frac{1-\sin\phi}{1+\sin\phi}\right)$$ where ϕ=angle of internal friction for dry soil and w_e=dry density of the soil

Total pressure $P = w_e h \times \frac{h}{2}\left(\frac{1-\sin\phi}{1+\sin\phi}\right)$

and acts at a height of $h/3$ above base

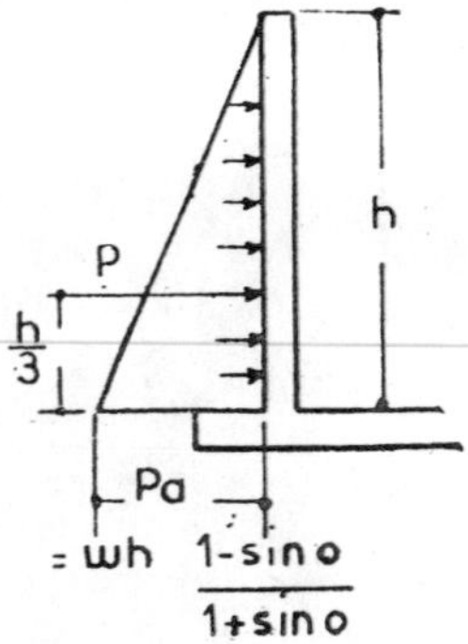

1·4. Earth Pressure Due to Submerged Soil

To estimate the earth pressure due to submerged soil angle of repose is assumed to be 6° and the weight of the earth as 1800 kg/cu m.

$$P_a = 1800 \times \frac{1-\sin 6°}{1+\sin 6°} \times h = 1800\, h \times \left(\frac{1-0·1045}{1+0·1045}\right)$$

$= 0·81 \times 1800\, h = 1450\, h$, which is sometimes called Grey's Rule. This can be taken as 1500 h.

1·5. Uplift of the Tank When Empty

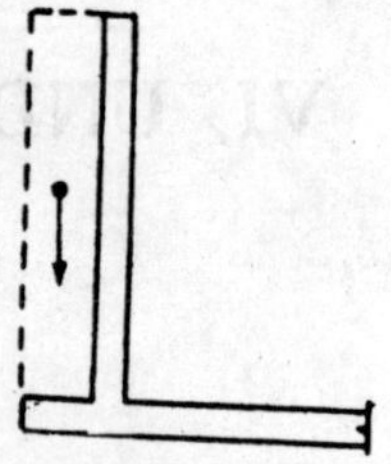

When the tank is empty and is surrounded by submerged soil outside, it is possible that the tank may float. If such a situation arises proper anchoring through screwpiles shall be provided to anchor the base slab to the foundations.

Secondly, the base slab may be projected so that the additional weight of the earth acting on the projecting slab will oppose the uplift forces and thus prevent floating.

1·6. Connection at the Base

The base slab and the side walls are usually rigidly connected.

1·7. Stability Requirements

Tank must not float when it is empty. To prevent uplift screw piles can be used. These piles consist of a 7·5 cm to 20 cm solid wrought iron or steel shaft with a cast iron or cast steel helicoid having a diameter 3 to six times that of the shaft. The pitch is about 1/3 of the diameter. The helicoid is normally of one turn.

Open Underground Tank in Dry Soil 2

Capacity 150,000 Litres

2·1. Data

Capacity = 150 m³, Shape required — Rectanglar-open

Weight of the soil = 150 kg/m³

Bearing capacity of soil = 20 t/m², Free board = 20 cm

Angle of internal friction of the soil, $\phi = 30°$

Ground water table is always much below the bottom of the tank, that is surrounding soil remains dry in all seasons. Leak test shall be carried out when there is no soil around.

Materials available Concrete – grade *M* 200 and steel – grade I.

2·2. Characteristic Strengths

$\sigma_{cb} = 70$ kg/cm² $\qquad m = 13$

$\sigma_{st} = 1000$ kg/cm² $\qquad jd = 0·84$ d $\qquad R = 14$

(Tension) $\sigma_{ct} = 12$ kg/cm² $\qquad \sigma_{ct} = 17$ kg/cm² (bending)

2·3. Trial Dimensions

Required capacity = 150 m³

Assume depth of water = 3·8 m

With free board allowance $h = 4$ m

Area of the tank in plan = 150/3·8 = 39·5 m²

Use 10 m × 4 m × 4 m size

Actual capacity provided = 10 × 4 × 3·8 = 152 m³

$L/B = 10/4 = 2·5 > 2$

The longer walls are designed as cantilevers and shorter walls as spanning between the longer walls.

2·4. Conditions

(1) When tank is full at the time of leak test and no earth pressure

(2) when the tank is empty and earth pressure outside

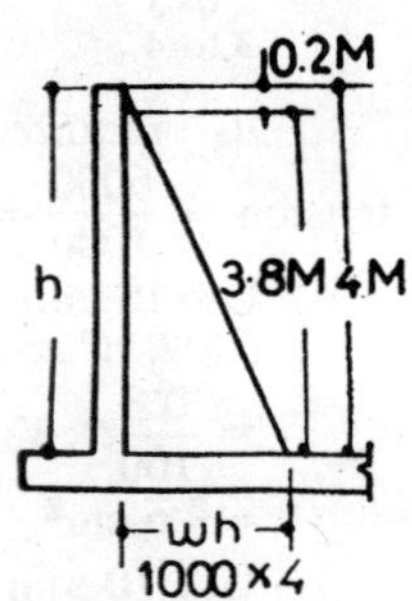

2·5. Longer Walls

Case (1) walls subjected to water pressure only

Maximum water pressure at base

$= wh = 1000 \times 4 = 4000$ kg

Maximum bending moment $= wh \times h/2 \times h/3$

$$= \frac{1000 \times 4 \times 4 \times 4}{6} = 10{,}667 \text{ m.kg}$$

Case (*ii*) when tank is empty and the wall is subjected to earth pressure only.

$$P = w_e h\left(\frac{1 - \sin\phi}{1 + \sin\phi}\right)$$

$$= 1500 \times 4 \times \left(\frac{1}{3}\right) = 2000 \text{ kg.}$$

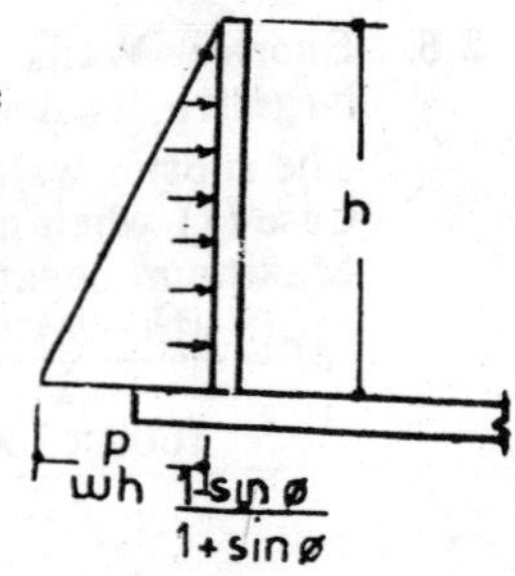

Maximum bending moment

$$= 2000 \times \frac{h}{2} \times \frac{h}{3} = 2000 \times \frac{4}{2} \times \frac{4}{3}$$

$=\frac{32000}{6}=5333{\cdot}3$ m kg.

2·5·1. Direct Tension in the Long Walls

Direct tension in the long walls is due to pressure acting on the shorter walls at a height equal to $h/4$ or 1 m

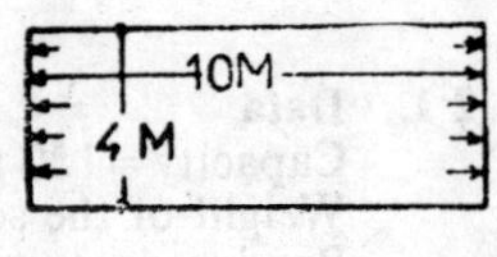

$$T_L=\frac{w\,(h-h')\,B}{2}=\frac{1000\quad 4\;)1-4}{2}$$

$=6000$ kg.

2·5·2. Thickness of the ᵛᵛalls

Maximum bending moment occurs in case (1) when only water pressure acts on the tank, $M=10{,}667$ m kg

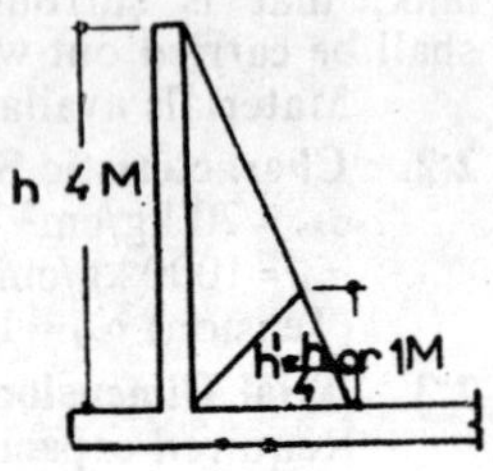

$$d_e=\sqrt{\frac{10{,}667\times100}{14\times100}}$$

$=\sqrt{762}=27{\cdot}6$ cm

Use $d=32$ cm, and $d_e=28$ cm

2·5·3. Main and Secondary Reinforcement

$$A_t \text{ required on the water face}=\frac{1066700}{1000\times{\cdot}84\times28}=45{\cdot}25\ \text{cm}^2$$

Use 22 mm ϕ at 8 cm c/c

depth *m*	*bending Momnnt* *m kg*	A_t *cm²*	*Spacing of* *20 mm* ϕ
0 to 1	166·7	0·71	38
1 to 2	1333·7	5-67	48
2 to 3	4500	19·2	16
3 to 4	10667	45·25	8

A_t required for direct tension $=\frac{6000}{1000}=6$ cm²

Use 10 mm ϕ at 13 mm c/c

0·3% of concrete area

$=\frac{0{\cdot}3}{100}\times32\times100$

$=9{\cdot}6$ cm²

Use 10 mm ϕ 16 mm c/c on each face

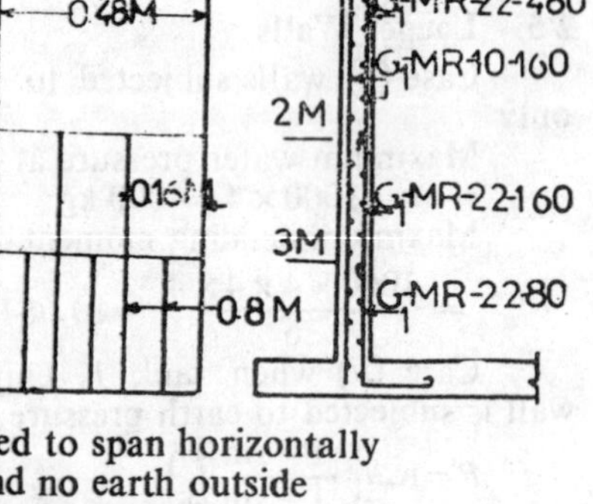

2·6. Shorter Walls Case (i) Negative Bending Moment

The shorter walls are designed to span horizontally Case (1) when tank is full and no earth outside Maximum negative moment at the corners

$$M=\frac{w(h-h')\times B^2}{12}=\frac{1000\ (4-1)\times4^2}{12}$$

$$=\frac{1000\times3\times4\times4}{12}=4000\ \text{m kg.}$$

2·6·1. Direct Tension

Direct tension $= w(h-h')\times 1 = 1000\ (4-1)\times 1 = 3000$ kg

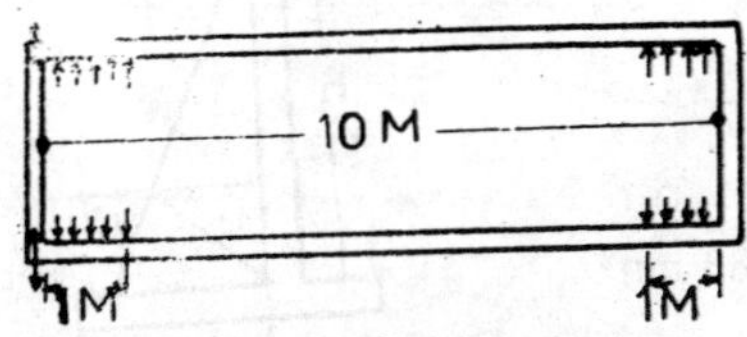

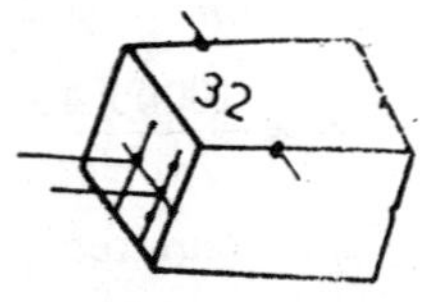

2·6·2. Net Bending Moment

$M_s = M - T_B\, x = 400000 - 3000\times(28-16)$

$= 400{,}000 - 3000\times(12) = 400{,}000 - 36{,}000 = 364{,}000$ cm kg

2·6·3 Main Reinforcement Water Face

Steel required to resist bending moment

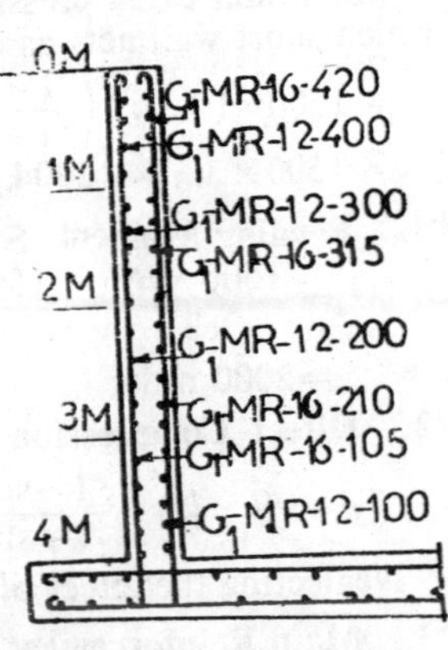

$$At_b = \frac{364000}{1000\times\cdot 84\times 28} = 15{\cdot}45\ \text{cm}^2$$

Steel required to resist direct tension

$$At_t = \frac{3000}{1000} = 3\ \text{cm}^2$$

Total At $15{\cdot}45 + 3 = 18{\cdot}45$ cm²

Use 16 mm ϕ at 10·5 cm c/c

From top to 1 m depth increased spacing can be used as the pressure reduces and therefore bending moment reduces.

2·6·4. Positive Bending Moment

Maximum positive bending moment at the centre of the wall

$$M = \frac{w(h-h')B^2}{16} = \frac{1000\times 3\times 4^2}{16} = 3000\ \text{m kg}$$

2·6·5. Direct Tension

$T_B = w(h-h')\times 1 = 1000\ (4-1)\times 1 = 3000$ kg

2·6·6. Net Bending Moment

Net bending moment $M_C = 300000 - 3000\times(28-16)$

$= 300{,}000 - 3000\times 12 = 300{,}000 \quad 36{,}000 = 264{,}000$ cm kg

2·6·7. Main and Secondary Steel Outerface Away From Water

Steel required to resist bending moment

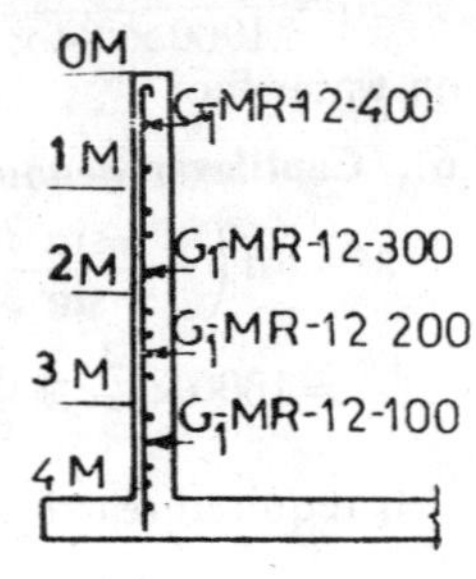

$$At_b = \frac{264{,}000}{1250\times\cdot 86\times 28} = 8{\cdot}75\ \text{cm}^2$$

Steel required to resist direct tension

$$At_t = \frac{3000}{1250} = 2{\cdot}4\ \text{cm}^2$$

Total $At = 8{\cdot}75 + 2{\cdot}4 = 11{\cdot}15$ cm²

Use 12 mm ϕ at 10 cm c/c at bottom 1 m height and reduce the spacing towards the top

0·3% of concrete area

$$A_t = \frac{0{\cdot}3}{100}\times 32\times 100 = 9{\cdot}6\ \text{cm}^2$$

Use 12 mm ϕ 23 cm c/c Vertically on both faces.

2·6·8. Cantilever Moment For Bottom 1 m Height

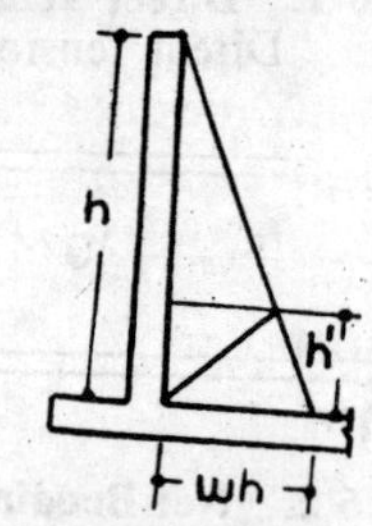

Cantilever moment

$$=wh\,\frac{h'}{2}\times\frac{h'}{3}=wh\,\frac{h'^2}{6}$$

$$=\frac{1000\times4\times1^2}{6}=666{\cdot}7 \text{ m kg}$$

$$A_t \text{ required}=\frac{66670}{1000\times{\cdot}84\times28}=2{\cdot}83 \text{ cm}^2$$

Use 8 mm ϕ at 17·5 cm c/c vertically.

2·7. Short Walls Case (ii)

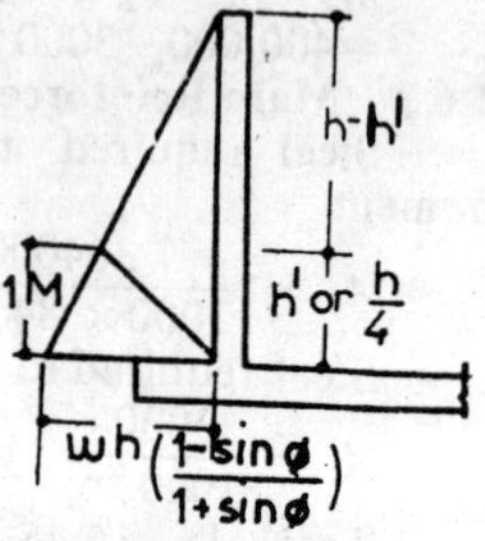

Tank empty and earth pressure outside.

Maximum earth pressure at the level up-to which short wall acts as cantilever

$$=1500\times(h-h')\left(\frac{1-\sin\phi}{1+\sin\phi}\right)$$

$$=1500\times3(\tfrac{1}{3})=1500 \text{ kg}$$

2·7·1. Bending Moments Support BM

$$M_s=\frac{1500\times B^2}{12}=\frac{1500\times4\times4}{12}$$

$$=2000 \text{ m kg}$$

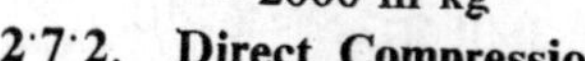

2·7·2. Direct Compression

$$T_B=w(h-h')\left(\frac{1-\sin\phi}{1+\sin\phi}\right)\times1=1500\times1=1500 \text{ kg}$$

Neglecting the effect of direct compression, $M_s=2000$ m kg

2·7·3. Main Reinforcement Earth Face

$$A_t=\frac{200000}{1250\times{\cdot}86\times28}=6{\cdot}7 \text{ cm}^2<11{\cdot}15 \text{ cm}^2$$ provided on earth face for case (i) and therefore no extra reinforcement is needed

2·7·4. Central Moment

$$M_c=\frac{1500\times B^2}{16}=\frac{1500\times4\times4}{16}=1500 \text{ m kg}$$

2·7·5. Main Reinforcement Water Face

$$A_t=\frac{150000}{1000\times{\cdot}84\times28}=6{\cdot}35 \text{ cm}^2<18{\cdot}45 \text{ cm}^2$$ provided for case (*i*) on water face

2·7·6. Cantilever Bending Moment

$$M=wh\left(\frac{1-\sin\phi}{1+\sin\phi}\right)\times\frac{h'}{2}\times\frac{h'}{3}=1000\times4\times1/3\times\frac{h'^2}{6}$$

$$=1000\times\frac{4}{3}\times\frac{1}{6}=\frac{4000}{18}=222 \text{ m kg}$$

$$A_t \text{ required (earth face)}=\frac{22200}{1250\times{\cdot}86\times28}={\cdot}74 \text{ cm}^2$$

As the quantity required is small, the secondary reinforcement provided will resist this cantilever moment.

2·8. Base Slab Case (i)

When tank is full and no earth pressure outside, the base slab is subjected to upward soil reaction due to self wt. of walls only, since the water pressure and ground soil pressure counteract each other.

Base slab dimensions $= 11{\cdot}6 \text{ m} \times 5{\cdot}6 \text{ m}$

Weight of walls

$= 2(4{\cdot}32 + 10{\cdot}32) \times \frac{32}{100} \times 4 \times 2400$

$= 2 \times 14{\cdot}64 \times \frac{32}{100} \times 4 \times 2400 = 90{,}500$ kg

$$\text{Upward soil reaction} = \frac{90500}{11{\cdot}6 \times 5{\cdot}6} = 1350 \text{ kg/m}^2$$

Maximum moment at centre

$$= \frac{1350}{2} \times 5{\cdot}6 \times \frac{4{\cdot}32}{2} - \frac{1350 \times 5{\cdot}6}{2} \times \frac{5{\cdot}6}{4}$$

$= 8200 - 5300 = 2900$ m kg

Bending moment at support

$= 1350 \times \frac{{\cdot}64 \times {\cdot}64}{2} = 257$ m kg. Produces tension on water face

2·8·1. Case (ii)

Tank empty and soil pressure acting outside.

Maximum bending moment at the base due to earth pressure $= 10{,}667$ m kg produces tension on earth face due to self weight of walls $= 257$ m kg

Net support moment $= 257 + 10667$ $= 10{,}924$ m kg tension on the earth side

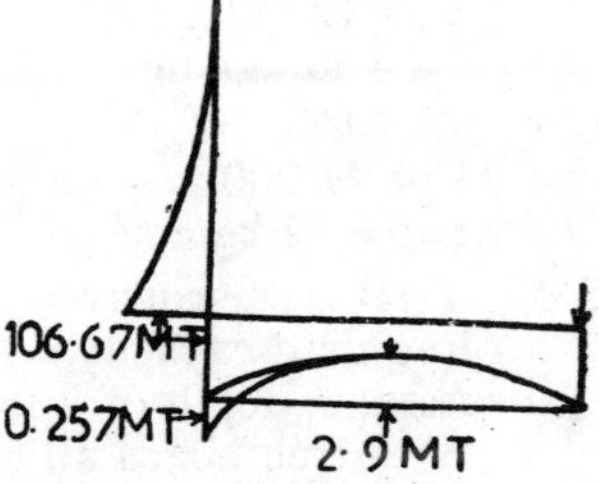

2·8·2. Depth of Base Slab

$$d_e = \sqrt{\frac{1092400}{14 \times 100}} = \sqrt{780} = 27{\cdot}8 \text{ cm}$$

Use $d = 32$ cm and $d_e = 28$ cm

2·8·3. Main Steel Away From Water Face

$$A_t = \frac{1092400}{1250 \times {\cdot}86 \times 28}$$

$= 36{\cdot}3 \text{ cm}^2$

Use 22 mm ϕ at 10 cm c/c at bottom face

2·8·4. Secondary Steel

$A_t = 0{\cdot}3\%$ of concrete

$= \frac{0{\cdot}3}{100} \times 32 \times 100 = 9{\cdot}6 \text{ cm}^2$

Use 10 mm ϕ 16 mm c/c both faces

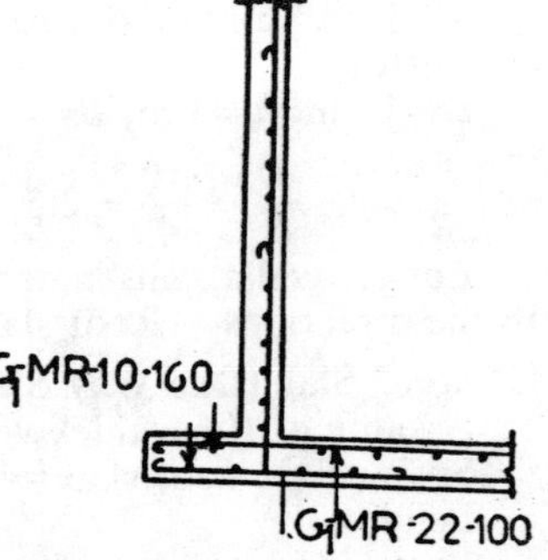

2·8·5. Main Reinforcement at Top Face When Tank Full No Earth Pressure

$$A_t = \frac{290000}{1000 \times {\cdot}84 \times 28} = 12{\cdot}3 \text{ cm}^2$$

Use 16 mm ϕ 16 cm c/c at top face

Underground Rectangular Tank in Dry Soil 3

Capacity 150,000 Litres

3·1. Data

Capacity = 150 m³, Shape = Rectangular-closed
Weight of the soil = 1500 kg/m³, Angle of internal friction $= \phi = 30°$, Bearing capacity of the soil = 20 t/m², Free board = 20 cm
Surrounding soil remains dry in all seasons.
Materials available : Concrete M 200 for walls and base and M 150 for roof slab, Steel – grade I.

3·2. Characteristic Strengths

$\sigma_{cb} = 70$ kg/cm²	$m = 13$	
$\sigma_{st} = 1000$ kg/cm²	$jd = \cdot 84\ d$	$R = 14$
$\sigma_{cb} = 70$ kg/m²	$m = 13$	
$\sigma_{st} = 1250$ kg/cm²	$jd = \cdot 86\ d$	$R = 12{\cdot}6$
$\sigma_{cb} = 50$ kg/cm²	$m = 18$	
$\sigma_{st} = 1400$ kg/cm²	$jd = 0{\cdot}87\ d$	$R = 8{\cdot}7$

For M 200
$\sigma_{ct_t} = 12$ kg/cm² $\qquad \sigma_{ct_b} = 17$ kg/cm²

3·3. Trial Dimensions

Required capacity = 150 m³
Assume depth of water = 3·8 m
With free board allowance $h = 4$ m

$$\text{Area of the tank in plan} = \frac{150}{3{\cdot}8} = 39{\cdot}5 \text{ m}^2$$

Use 10 m × 4 m × 4 m size
Actual capacity provided = 10 × 4 × 3·8 = 152 m³

3·4. Conditions of Loading

(*i*) Tank full and dry earth outside, (*ii*) Tank empty and dry earth outside

$L = 10$ m, $h = 4$ m, $B = 4$ m

$$\frac{L}{B} > 2, \quad \frac{L}{h} = \frac{10}{4} = 2{\cdot}5, \quad \frac{B}{h} = \frac{4}{4} = 1$$

Longer walls span in one direction and the shorter walls span in both the directions. Roof slab spans in one direction only.

3·5. Roof Slab (One Way Slab)

Assuming 30 cm thick walls
Span of the roof slab = 4·3 m
Loads :
Live load = 150 kg/m², Screeding = etc. = 100 kg/m²,
Self weight (15 cm) $= \frac{15}{100} \times 1 \times 1 \times 2400 = 360$ kg/m²
Total load = 610 kg/m²

3·5·1. Bending Moment

$$\text{Bending moment} = \frac{wl^2}{8} = \frac{610}{8} \times (4{\cdot}3)^2 = 1410 \text{ m kg}$$

3·5·2. Depth of Slab

Using M 150 concrete, $d_e=\sqrt{\dfrac{1410\times 100}{8{\cdot}7\times 100}}=\sqrt{152}=12{\cdot}35$ cm

Use $d_e=12{\cdot}5$ cm, $d=15$ cm

3·5·3. Amount of Reinforcement

$$A_t=\frac{1410\times 100}{1400\times{\cdot}87\times 12{\cdot}5}=9{\cdot}25\ \text{cm}^2$$

Use 12 mm ϕ at 12 cm c/c

3·5·4. Secondary Reinforcement

$$A_t=\frac{0{\cdot}15}{100}\times 15\times 100=2{\cdot}25\ \text{cm}^2$$

Use 6 mm ϕ at 12 cm c/c

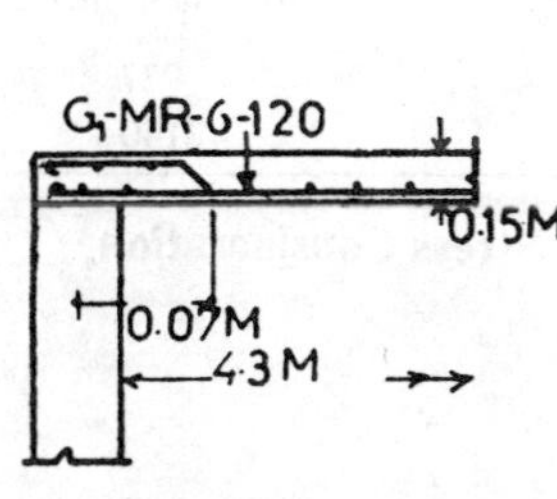

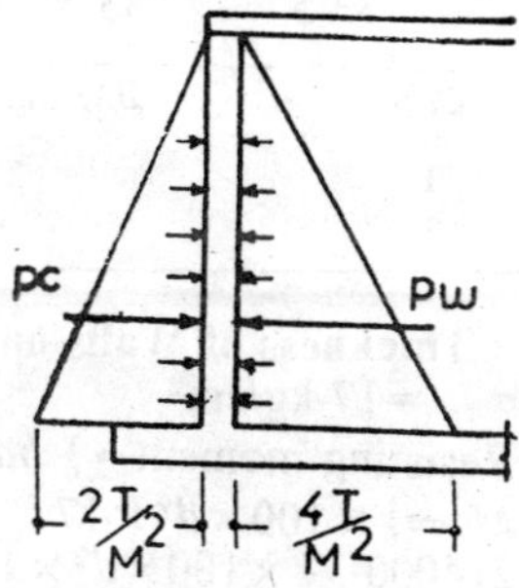

3·6. Longer Walls Case (i)

$$\frac{L}{h}=\frac{10}{4}=2{\cdot}5$$

Walls span in vertical direction only. Walls are assumed fixed at base and supported at top.

Tank full dry earth outside

Maximm water pressure $=w_w h=1000\times 4=4000$ kg/m²

Maximum earth pressure $=w_e h\left(\dfrac{1-\sin\phi}{1+\sin\phi}\right)=1500\times 4\left(\dfrac{1}{3}\right)$

$=2000$ kg/m²

3·6·1. Net Pressure Bending Moment

Net pressure on the wall

$p=4000-2000=2000$ kg/m²

Maximum negative bending moment (water face)

$$M_w=\frac{ph^2}{15}=\frac{2000\times 4\times 4}{15}$$

$=2140$ m kg

Maximum positive bending moment (earth face)

$$M_e=\frac{ph^2}{33{\cdot}5}=\frac{2000\times 4\times 4}{33{\cdot}5}$$

$=927$ m kg

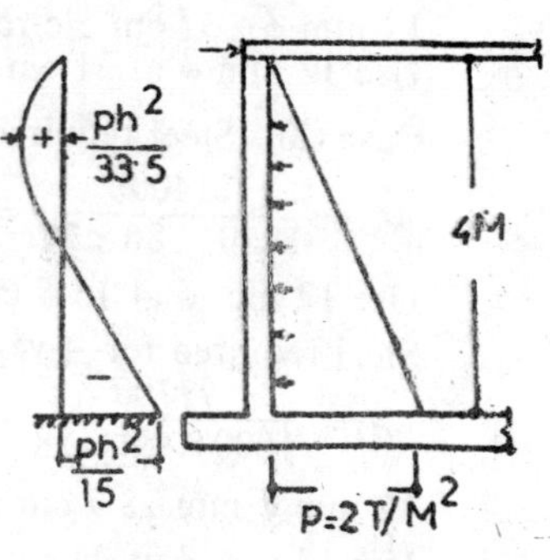

3·6·2. Case (2) Pressures

Tank empty—dry earth pressure outside

$$\text{Maximum earth pressure} = w_e h\left(\frac{1-\sin\phi}{1+\sin\phi}\right) = p = 1500 \times 4 \times \frac{1}{3}$$

$$= 2000 \text{ kg/m}^2$$

3·6·3. Bending Moments

Maximum negative bending moment (earth face)

$$M_e = \frac{ph^2}{15}, \quad M_e = \frac{2000 \times 4 \times 4}{15} = 2140 \text{ m kg}$$

Maximum positive bending moment (water face)

$$M_w = \frac{ph^2}{33{\cdot}5} = \frac{2000 \times 4^2}{33{\cdot}5} = 927 \text{ m kg}$$

Case	*BM water face m kg*	*BM earth face m kg*
1	2140	927
2	927	2140

3·6·4. Thickness of Walls on Cracking Stress Consideration

$\sigma_{ct_b} = 17$ kg/cm²

Resisting moment $= \frac{1}{6} bd^2 \sigma_{ct_b}$

$M_r = \frac{1}{6} \times 100 \times d^2 \times 17$

$214000 \quad \frac{1}{6} \times 100 \times d^2 \times 17$

$$d^2 = \frac{214000 \times 6}{100 \times 17}$$

$d = \sqrt{751} = 27{\cdot}5$ m. Use $d = 32$ cm $d_e = 28$ cm

3·6·5. Main Reinforcement Case (i)

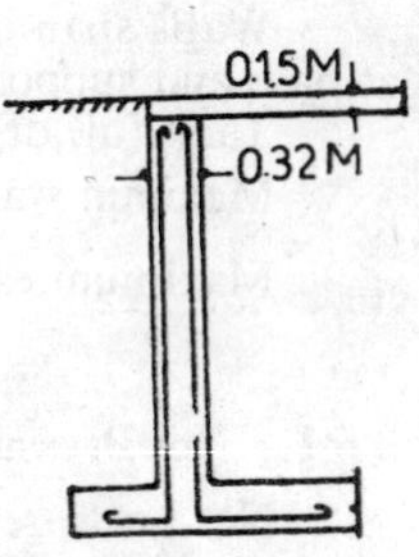

A_t required for $-$ve bending moment (water face)$= \frac{214000}{1000 \times {\cdot}84 \times 28} = 9{\cdot}1$ cm²

Use 12 mm ϕ – 12 cm c/c vertically

Steel required for +ve bending moment (earth face)

$$A_t = \frac{92700}{1250 \times {\cdot}86 \times 28} = 3{\cdot}07 \text{ cm}^2$$

12 mm ϕ – 37 cm c/c required

Use 12 mm ϕ at 31 cm c/ c

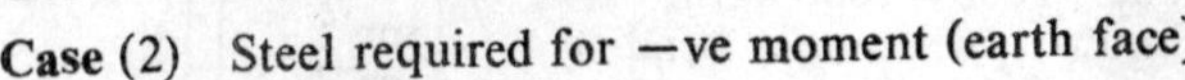

Case (2) Steel required for —ve moment (earth face)

$$A_t = \frac{214000}{1250 \times 86 \times 28} = 7{\cdot}1 \text{ cm}^2$$

Use 12 mm ϕ at 15·5 cm c/c

Steel required for +ve moment (water face)

$$A_t = \frac{92700}{1000 \times {\cdot}84 \times 28} = 3{\cdot}94 \text{ cm}^2$$

12 mm ϕ mm 28·7 cm c/c required

Use 12 mm ϕ at 24 cm c/c

3·6·6. Secondary Reinforcement

$$A_t = \frac{0{\cdot}3}{100} \times 32 \times 100 = 9{\cdot}6 \text{ cm}^2$$

Use 8 mm ϕ at 10 cm c/c both faces

3·7. Shorter Walls

$\frac{B}{h} = \frac{4}{4} = 1$, Walls span in both directions.

3·7·1. Bending Moment Vertical Direction Case (1)

Bending moment in the vertical direction

$$M_v = 0{\cdot}083\ wh^3\left(\frac{K}{K+1}\right)$$

K from graph 0·375

$$M_v = {\cdot}083 \times ph^2\left(\frac{K}{K+1}\right)$$

$$= {\cdot}083 \times 2000 \times 4^2 \times \frac{0{\cdot}375}{1{\cdot}375}$$

$= 750$ m kg

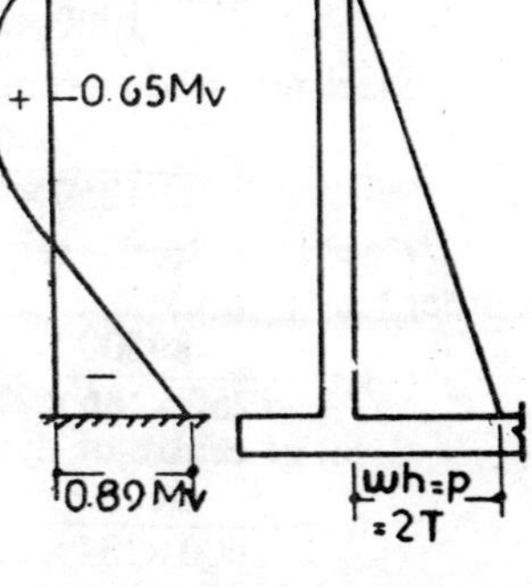

Maximum −ve moment (water face) $= 0{\cdot}89 \times 750 = 670$ m kg

Maximum +ve moment (earth face) $= {\cdot}65 \times 750 = 490$ m kg

Case (2) Maximum −ve moment (earth face) = 670 m kg

Maximum +ve moment (water face) = 490 m kg

3·7·2. Horizontal Span Case (1)

Free horizontal moment

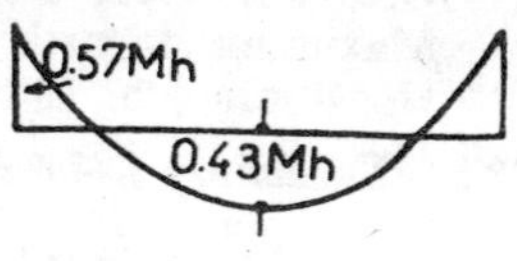

$$M_h = \frac{0{\cdot}067 \times wh\ L^2}{(K+1)}$$

$$= \frac{0{\cdot}067 \times 2000 \times 4^2}{({\cdot}375+1)}$$

$= 1570$ m kg

Negative moment at the corners (water face)
$= 0{\cdot}57\ M_h = 1570 \times 0{\cdot}57 = 895$ m kg

Positive moment at the centre (earth face)
$= 0{\cdot}43 \times M_h = 0{\cdot}43 \times 1570 = 680$ m kg

Case (2) Since $p = 2000$ kg/m² as in case (1) but acts outside.

Negative moment corner (earth face) = 895 m kg
Positive moment at the centre (water face) = 680 m kg

3·7·3. Direct Tension in Short Walls

$p = 2000$ kg/m²

$$T_B = \frac{2000 \times 1}{2} = 1000 \text{ kg}$$

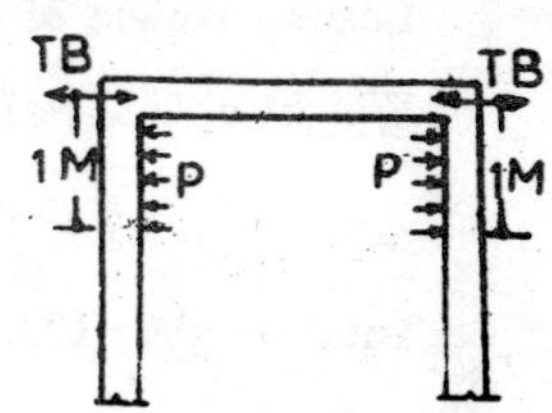

3·7·4. Maximum Moment in the Wall

Maximum moment occurs in the horizontal span = 875 m kg

3·7·5. Thickness of the Side Wall

$\frac{1}{6} bd^2 \sigma_{ct_b} = 87500$

$\frac{1}{6} \times 100 \times d^2 \times 17 = 87500$

$$d^2 = \frac{87500 \times 6}{100 \times 17}$$

$d = \sqrt{310} = 17·6$ cm. However use the same thickness as longer walls that is $d = 32$ cm, $d_e = 28$ cm

3·7·6. Main Reinforcement Case (1)

Steel required for maximum bending moment at corners (water face)

$$A_t = \frac{89500}{1000 \times ·84 \times 28} = 3·76 \text{ cm}^2$$

Steel required at centre (earth face)

$$A_t = \frac{68000}{1250 \times ·86 \times 28} = 2·27 \text{ cm}^2$$

Case (2) Steel required for maximum bending moment at corner (earth face)

$$A_t = \frac{89500}{1250 \times ·86 \times 28} = 2·75 \text{ cm}^2$$

Steel at centre of the span (water face)

$$A_t = \frac{68000}{1000 \times ·84 \times 28} = 2·9 \text{ cm}^2$$

Steel for direct tension

$$= \frac{1000}{1000} = 1 \text{ cm}^2$$

Maximum A_t (water face) $= 4·76$ cm²
Use 10 mm × 16·5 cm c/c
Maximum A_t earth face $= 2·75$ cm²
Use 8 mm ϕ at 18 cm c/c

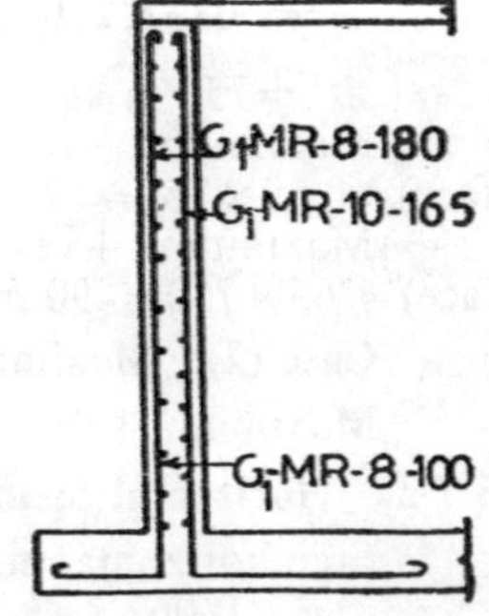

3·7·7. Secondary Reinforcement

$$A_t = \frac{0·3}{100} \times 32 \times 100 = 9·6 \text{ cm}^2 \text{ and per face } 4·8 \text{ cm}^2.$$

Use 8 mm ϕ at 10 cm c/c

3·8. Base Slab Case (1) Loads

$$\frac{L}{B} = \frac{10}{4} = 2·5 > 2 \text{ spans in one direction only.}$$

Tank full and dry soil outside

In this case the water pressure on base slab will be counteracted by the soil pressure below it.

Loads : Weight of roof $= 610 \times 4·64 \times 10·64 = 33{,}000$ kg

Weight of side walls $= 2(4·32 + 10·32) \times 4 \times \frac{32}{100} \times 2400$

$= 2 \times 14·64 \times 4 \times \frac{32}{100} \times 2400 = 90000$ kg

Total weight $= 123{,}000$ kg

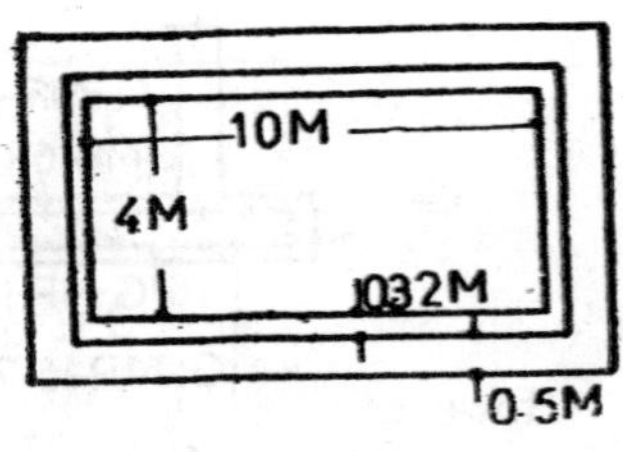

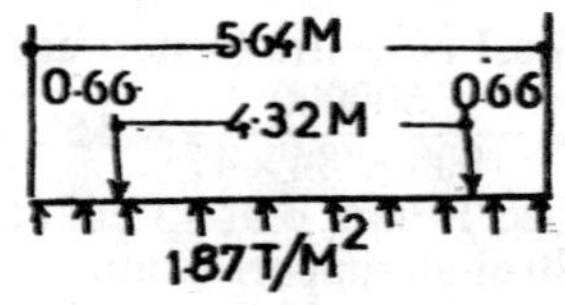

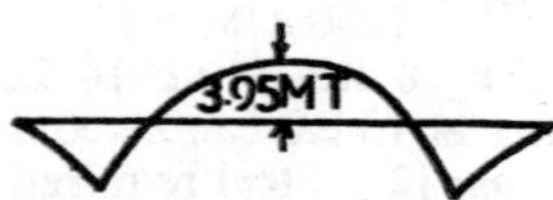

Base slab dimensions=5·64 m×11·64 m

Soil reaction below the base$=\dfrac{123000}{5{\cdot}64\times11{\cdot}64}=1870$ kg/m

3·8·1. Bending Moment

Bending moment at centre of the slab (water face)

$=1870\times\dfrac{5{\cdot}64}{2}\left(\dfrac{4{\cdot}32}{2}-\dfrac{5{\cdot}64}{4}\right)=1870\times\dfrac{5{\cdot}64}{2}\times{\cdot}75$

$==3950$ m kg

Bending moment at the support (earth face)

$=\dfrac{1890\times0{\cdot}66}{2}=406$ m kg

−ve Bending moment at the base due to loads on the side wall (water face)=2140 m kg

Net maximum bending moment at centre (providing tension on the water face)=3950+2140=6090 m kg

Net maximum bending moment at support (produces tension on the water face)=2140 - 406=1734 m kg

Case (2) Tank empty dry soil outside

Bending moment due to self weight of roof and side walls at centre (water face)=3950 m kg at support (earth face)=406 m kg

Bending moment at the base due to soil pressure outside (earth face)=2140 m kg

Net bending moment at centre of the slab (produces the tension on outer face)=3950−2140=1810 m kg

Net bending moment at support (producing tension on earth face)=2140+406=2546 m kg

Case	*Bending moment centre*	*Bending moment support*
1	6090	1734
2	1810	2546

3·8·2. Thickness of the Slab on Cracking Stress Consideration

$\frac{1}{6}\times b\times d^2\times\sigma_{ct_b}=609000$

$\frac{1}{6}\times100\times d^2\times17=609000$

$d^2=\dfrac{609000\times6}{100\times17}$

$d=\sqrt{2150}=46{\cdot}5$ cm. Use $d=52$ cm and $d_e=47$ cm

3·8·3. Main Reinforcement Case (1)

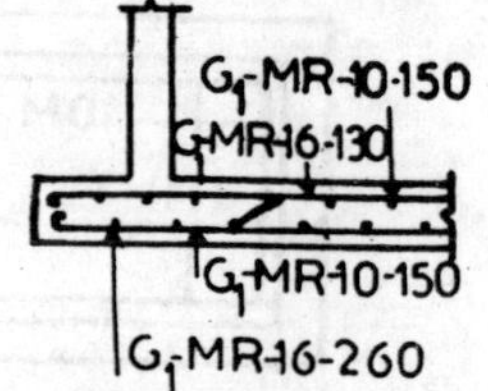

Steel at centre (water face)

$$=\frac{609000}{1000\times{\cdot}84\times47}=15{\cdot}2\ \text{cm}^2$$

Use 16 mm ϕ at 13 cm c/c

Steel at support (water face)

$$A_t=\frac{173400}{1250\times{\cdot}86\times47}=3{\cdot}43\ \text{cm}^2$$

Use 8 mm ϕ at 14·5 cm c/c. However 16 mm ϕ at 13 cm can be continued.

Case (2) Steel required at centre is less than that provided in case (1)

Steel requiring at support (earth face)

$$=\frac{254600}{1250\times{\cdot}86\times47}=5{\cdot}01\ \text{cm}^2.$$ Use 10 mm ϕ at 15·5 cm c/c

3·8·4. Secondary Reinforcement

$A_t=0{\cdot}2\%$ of concrete area ($d>45$ cm)

$$=\frac{0{\cdot}2}{100}\times52\times100=10{\cdot}4\ \text{cm}^2,$$ Steel per face$=5{\cdot}2$ cm². Use 10 mm ϕ at 15 cm c/c.

Open Underground Rectangular Tank in Wet Soil 4

Capacity 350,000 Litres

4·1. Data

Capacity = 350 m^3, Free Board = 20 cm,
Shape = Rectangular =(open), Weight of the soil = 1500 kg/m^3,
Angle of internal friction $\phi = 30°$,
Bearing capacity of soil = 20 t/m^2
Water table may rise to the ground level in rainy seasons
Material available. Concrete—*M* 200, Steel—grade *I*

4·2. Characteristic Strengths

$\sigma_{cb} = 70$ kg/cm^2 $\qquad m = 13$

$\sigma_{st} = 1000$ kg/cm^2 $\qquad j_d = 0{\cdot}84\ d,\ R = 14$

$\sigma_{ct_t} = 12$ kg/cm^2 $\qquad \sigma_{ct_b} = 17$ kg/cm^2

4·3. Trial Dimensions

Required capacity = 350 m^3
Assume depth of tank = 3·5 m
Area of the tank in plan

$$= \frac{350}{3{\cdot}5} = 100 \text{ m}^2$$

Use 20 m × 5 m × 3·7 m tank,
$L/B = 20/5 = 4 > 2$

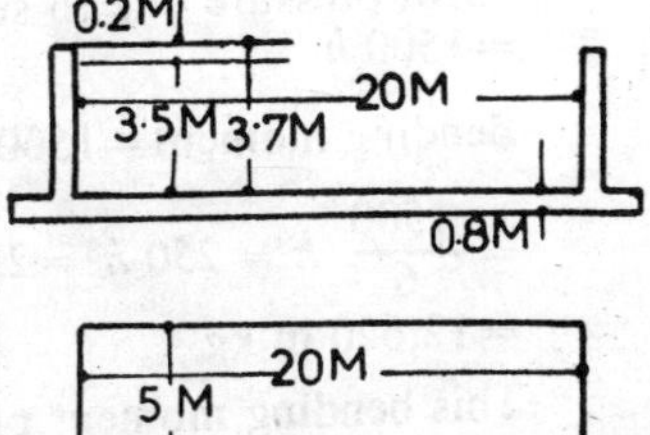

Longer walls span in vertical direction as cantilever and the shorter walls span horizontally between the longer walls.

4·4. General Conditions of Loading

(*i*) Tank full and dry soil outside, (*ii*) Tank full and submerged soil outside (*iii*) Tank empty dry soil outside (*iv*) Tank empty submerged soil outside.

4·4·1. Critical Loading Conditions

(1) Tank full and dry soil outside.
(2) Tank empty submerged soil outside.

4·5. Longer Walls case (1) Bending Moment

Walls are subjected to both water and earth pressure.

Bending moment due to water pressure

$$= \frac{wh^3}{6} = \frac{1000}{6} \times (3{\cdot}7)^3 = 8450 \text{ m kg}$$

Bending moment due to earth pressure

$$= wh\left(\frac{1-\sin\phi}{1+\sin\phi}\right) \times \frac{h}{2} \times \frac{h}{3}$$

$$= 1500 \times 3{\cdot}7 \times \tfrac{1}{3} \times \frac{3{\cdot}7}{2} \times \frac{3{\cdot}7}{3}$$

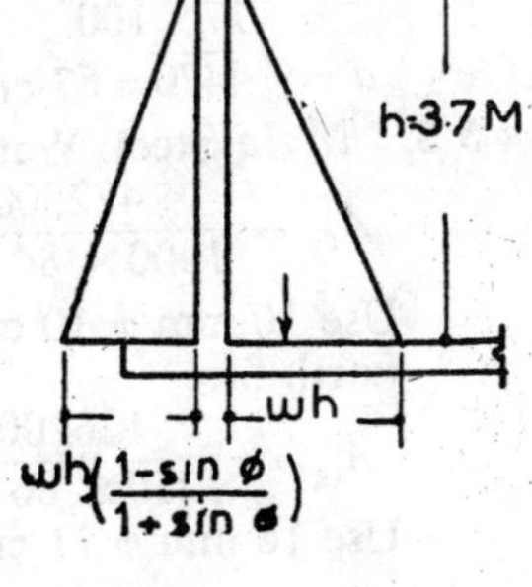

$= \frac{1500 \times 3.7^3}{18} = 4225$ m kg

Net bending moment $= 8450 - 4225 = 4225$ m kg

4·5·1. Direct Tension

This bending moment causes tension on water face. Direct tension due to pressure acting on shorter walls.

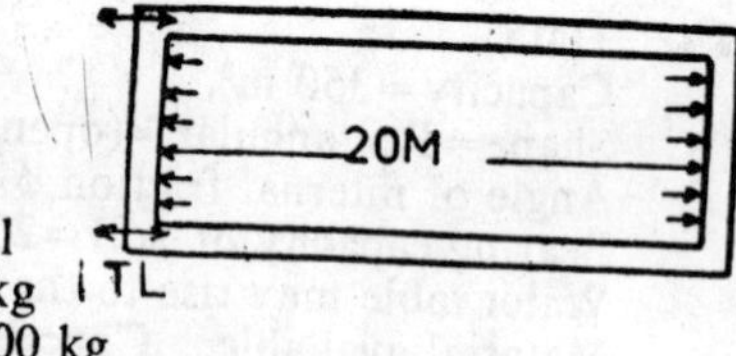

Earth pressure at the level upto which walls act as cantilever

$= w(h - h') \times \left(\frac{1 - \sin\phi}{1 + \sin\phi}\right)$

$= 1500 \times 2.7 \times \frac{1}{3} = 1550$ kg

Water pressure at the same level

$= w(h - h') = 1000 \times 2.7 = 2700$ kg

Net pressure $= 2700 - 1500 = 1200$ kg

T_l = direct tension in the long walls $= 1200 \times \frac{5}{2} = 3000$ kg

4·5·2. Longer Walls Case (2)

Tank empty submerged soil outside.

In order to estimate the earth pressure due to submerged soil Gray's rule may be employed.

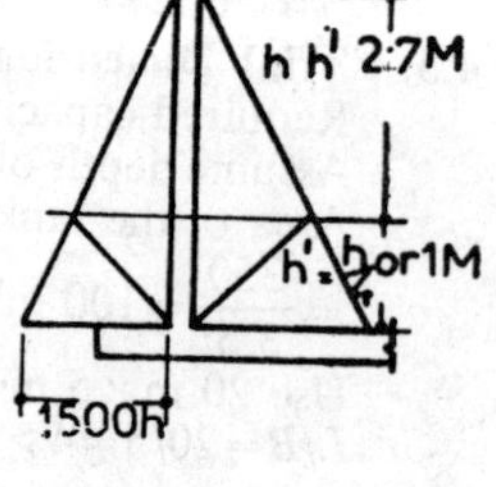

Earth pressure due to submerged soil $= 1500\,h$

Bending moment $= 1500\,h \times \frac{h}{2} \times \frac{h}{3}$

$= \frac{1500}{6} h^3 = 250\,h^3 = 250 \times (3.7)^3$

$= 12,620$ m kg

This bending moment produces tension on the earth face.

4·5·3. Thickness of the Walls Bending Moment Considerations

Maximum bending moment $= 12620$ m kg

$d_e = \sqrt{\frac{1262000}{14 \times 100}} = \sqrt{905} = 30.1$ cm

4·5·4. Thickness of the Wall on Cracking Stress Consideration

$1/6 \times 100 \times d^2 \times \sigma_{ctb} = M$

$1/6 \times 100 \times d^2 \times 17 = 1262000$

$d^2 = \frac{1262000 \times 6}{17 \times 100}$

$d = \sqrt{4470} = 67$ cm. Use $d = 70$ cm, $d_e = 66$ cm

4·5·5. Main Steel Water face

$A_{t_1} = \frac{422500}{1000 \times .84 \times 66} = 7.65$ m²

Use 10 mm ϕ 10 cm c/c (7·85 cm²)

Earth face

$A_{t_2} = \frac{1262000}{1250 \times .86 \times 66} = 17.9$ cm²

Use 16 mm ϕ 11 cm c/c (18·28 cm²)

4·5·6. Cracking Stresses

Position of N.A.

$$100\times n\times\frac{n}{2}+(m-1)\ A_{t_1}(n-4)$$

$$=\frac{(70-n)^2}{2}\times 100+(m-1)A_{t_2}(d-n-4)$$

$$100\frac{n^2}{2}+12\times 7{\cdot}85\ (n-4)$$

$$=\frac{100}{2}(70-n)^2+12\times 18{\cdot}28\times(66-12)$$

$50\ n^2+94{\cdot}2\ n-377=50(4900-140\ n+n^2)-220\ n+14500$
$=245{,}000-7000\ n+50\ n^2-220\ n+14500$
$94{\cdot}2\ n+7000\ n+220\ n=24500+14500+377$
$7314{\cdot}2\ n=259{,}879$

$$n=\frac{259877}{7314{\cdot}2}=35{\cdot}5 \text{ cm}$$

Total tension on the earth face

$$T=100\times 34{\cdot}5\times\frac{\sigma_{ct_b}}{2}+12\times 18{\cdot}28\times\frac{30{\cdot}5}{34{\cdot}5}\ \sigma_{ct_b}$$

$=1720\sigma_{ct_b}+210\ \sigma_{ct_b}=1930\ \sigma_{ct_b}$
$1930\ \sigma_{ct_b}\times{\cdot}82\times 66=1262000$

$$\sigma_{ct_b}=\frac{1262000}{1930\times{\cdot}82\times 66}=12{\cdot}1 \text{ kg/cm}^2<17 \text{ kg/cm}^2$$

Cracking stress at water face. Position of N.A.

$$100\times n\times\frac{n}{2}+18{\cdot}28\times 12\times(n-4)$$

$$=\frac{100}{2}(70-n)^2+7{\cdot}85\times 12\ (66-n)$$

$50\ n^2+220\ n\quad 880=245000-7000\ n+50\ n^2-94{\cdot}2\ n+6220$
$7000\ n+220\ n+94{\cdot}2\ n=245000+6220+880$

$$7314{\cdot}2\ n=252100,\ n=\frac{252100}{7314{\cdot}2}=34{\cdot}5 \text{ m}$$

Total tension on the water faec

$$T=100\times 35{\cdot}5\times\frac{\sigma_{ct_b}}{2}+12\times 7{\cdot}85\times\frac{31{\cdot}5}{35{\cdot}5}\ \sigma_{ct_b}$$

$=1780\ \sigma_{ct_b}+83{\cdot}5\ \sigma_{ct_b}=1863{\cdot}5\ \sigma_{ct_b}$
$1863{\cdot}5\ \sigma_{ct_b}\times{\cdot}82\times 66=422500$

$$\sigma_{ct_b}=\frac{422500}{1863{\cdot}5\times 66\times{\cdot}82}=7{\cdot}82 \text{ kg/cm}^2<17 \text{ kg/cm}^2$$

4·5·7. Main Reinforcement Curtailment

Towards the top, bending moment reduces and therefore the vertical bars can be curtailed.

Water Face Case (1).

Depth from top	*B.M. cm kg*	At_2	*Spacing of 10 mm* ϕ
m	$\frac{500}{6} \times h^3 \times 100$	cm^2	*cm*
0–1	8330	·15	40
1–2	66640	1·2	20
2–3	2225000	4·05	10
3–3·7	422500	7·65	10

Earth Face Case (2).

Similarly on the earth face bars can be curtailed

Bottom to 2 m level, 16 mm ϕ 11 cm c/c
2 to 1 m level, 16 mm ϕ 22 cm c/c
1 m to top of the wall, 16 mm ϕ 44 cm c/c

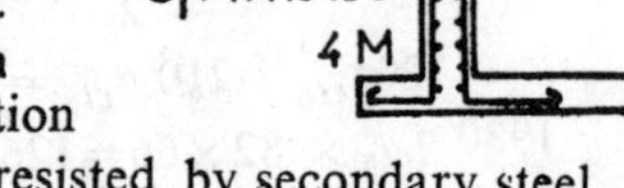

4·5·8. Secondary Reinforcement

0·2% of concrete area (thickness > 45 cm)
$= \cdot 2/100 \times 100 \times 70 = 14 \text{ cm}^2$
Use 10 mm ϕ 11 cm c/c both faces.

4·5·9. Reinforcement for Direct Tension

$A_t = 3000/1000 = 3 \text{ cm}^2 <$ distribution steel provided and the direct tension is resisted by secondary steel.

4·6. Shorter Walls Case (1)

Tankfull dry earth outside.

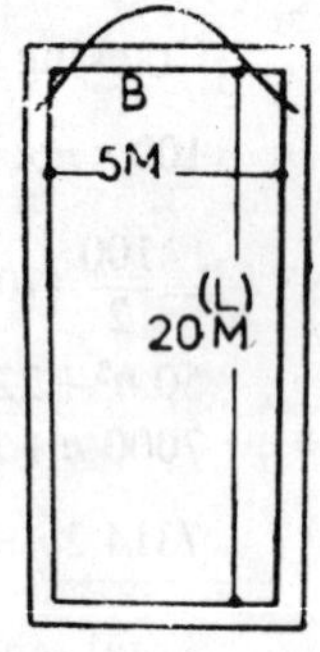

Negative bending moment due to earth pressure

$$= w_e(h-h')\left(\frac{1-\sin\phi}{1+\sin\phi}\right)\times\frac{B^2}{12}$$

$$= 1500 \times 2{\cdot}7 \times \tfrac{1}{3} \times \frac{5^2}{12} = 2812{\cdot}5 \text{ m kg.}$$

Negative bending moment due to water pressure

$$= w(h-h')\frac{B^2}{12} = 1000 \times 2{\cdot}7 \times \frac{5^2}{12} = 5625 \text{ m kg.}$$

Net negative bending moment at support

$$= 500 \times 2{\cdot}7 \times \frac{5^2}{12} = 2810 \text{ m kg}$$

This bending moment produces tension on the water face.

Positive bending moment at centre of the wall

$$= 500 \times \frac{2{\cdot}7 \times 5^2}{16} = 2110 \text{ m kg}$$

This moment produces tension on the earth-face

4·6·1. Direct Tension

$$T_B = w(h-h') \times 1 - w_e \times (h \times h')\left(\frac{1-\sin\phi}{1+\sin\phi}\right) \times 1$$

$$= 2{\cdot}7 \times 1000 - 1500 \times 2{\cdot}7 \times \tfrac{1}{3} \times 1$$

$$= 100 \times 2{\cdot}7 - 500 \times 2{\cdot}7$$

$$= 500 \times 2{\cdot}7 = 1350 \text{ kg}$$

4·6·2. Case (2) Bending Moment

Tank empty submerged soil outside. Pressure at the level upto which the wall acts as cantilever

$$=p_a=1500\ h\times\left(\frac{h-h'}{h}\right)=1500\times2{\cdot}7=4050\ \text{kg}$$

Maximum negative bending moment $=\dfrac{4050\times B^2}{12}$

$$=\frac{4050\times5^2}{12}=8420\ \text{m kg}$$

Maximum positive bending moment $=\dfrac{4050\times5^2}{16}=6300$ m kg

4·6·3. Thickness of Walls

Direct compression may be omitted.
Use same thickness as longer walls.

4·6·4. Main Reinforcement

Case (*i*) A_t required at suports (water face)

$$=\frac{281000}{1000\times{\cdot}84\times66}=5{\cdot}06\ \text{cm}^2$$

A_t required at centre (earth face)

$$=\frac{211000}{1250\times{\cdot}86\times66}=2{\cdot}83\ \text{cm}^2$$

A_t for direct tension

$$=\frac{1350}{1000}=1{\cdot}35\ \text{cm}^2$$

Total reinforcement water face
$=5{\cdot}06\times1{\cdot}35=6{\cdot}41$ cm²
Use 10 mm ϕ at 12 cm c/c
On the earth face, use 8 mm ϕ at 17·5 cm c/c

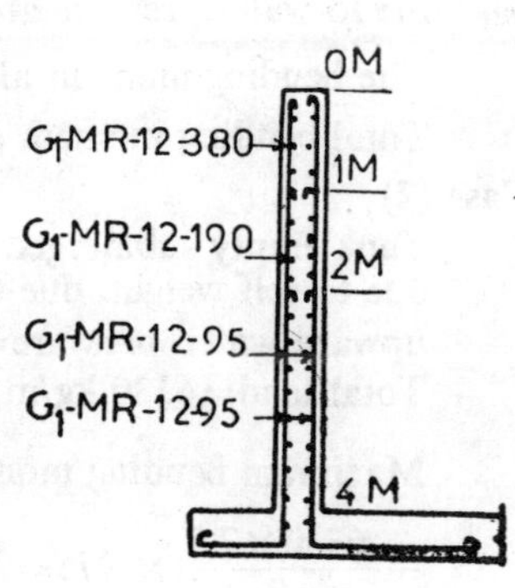

Case (2) A_t required at earth face

$$\text{at supports}=\frac{842000}{1250\times{\cdot}86\times66}=11{\cdot}9\ \text{cm}^2$$

Use 12 mm ϕ at 9·5 cm c/c
A_t required at water face at the

$$\text{centre of the wall}=\frac{630000}{1000\times{\cdot}84\times66}$$

$=11{\cdot}4$ cm² Use 12 mm ϕ at 9·5 cm c/c

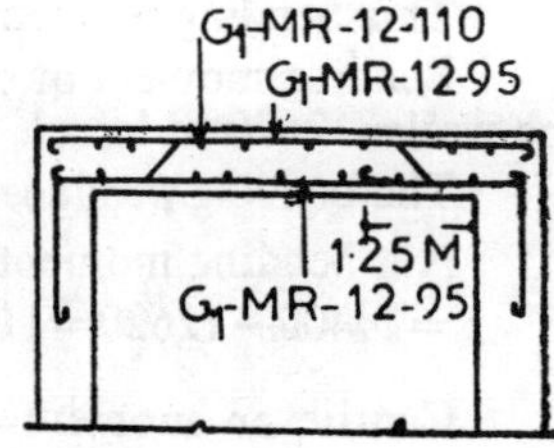

4·6·5. Secondary Reinforcement

0·2% of concrete area ($t>45$ cm)

$$=\frac{0{\cdot}2}{100}\times100\times70=14\ \text{cm}^2$$

Use 10 mm ϕ 11 cm c/c both faces.

4·7. Base Slab Case (i)

Tank full dry soil outside
Assume base slab dimensions
$=7{\cdot}5$ m $\times$ 22·5 m
Weight of walls

$$=2(5{\cdot}7+20{\cdot}7)\times3{\cdot}7\times\frac{70}{100}\times2400$$

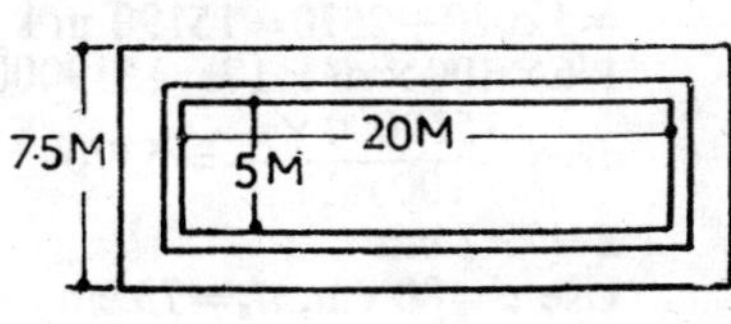

$$=2\times 26{\cdot}4\times 3{\cdot}7\times \frac{70}{100}\times 2400=326000 \text{ kg}$$

Water pressure and equivalent soil pressure will balance each other.

Upward soil reaction due to self weight of walls

$$=\frac{326000}{7{\cdot}5\times 22{\cdot}5}=1930 \text{ kg/m}^2$$

Maximum bending moment at centre

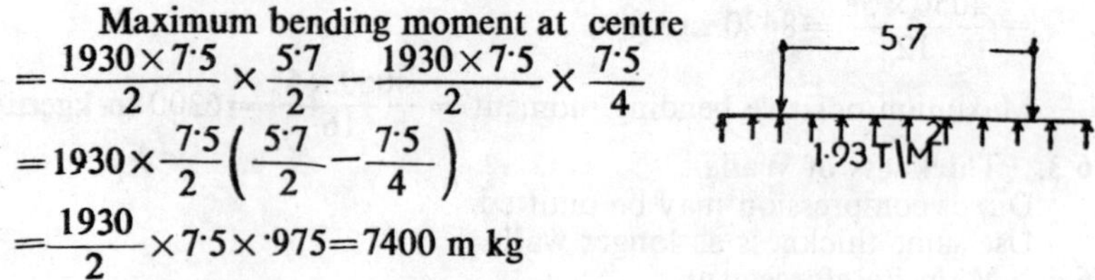

$$=\frac{1930\times 7{\cdot}5}{2}\times\frac{5{\cdot}7}{2}-\frac{1930\times 7{\cdot}5}{2}\times\frac{7{\cdot}5}{4}$$

$$=1930\times\frac{7{\cdot}5}{2}\left(\frac{5{\cdot}7}{2}-\frac{7{\cdot}5}{4}\right)$$

$$=\frac{1930}{2}\times 7{\cdot}5\times{\cdot}975=7400 \text{ m kg}$$

This bending moment produces tension on the water face.

Net cantilever bending moment in the longer walls acting at the base due to water pressure and earth pressure $=4225$ m kg

The bending moment also produces tension on water face.

Total bending moment at the centre $=7400+4225=11{,}625$ m kg

Case (2)

Tank empty submerged soil outside, soil reaction below the base due to self weight, due to reaction of soil $=1930$ kg/m
upward water pressure $=4400$ kg/m
Total load $=6330$ kg/m

Maximum bending moment at centre $=\dfrac{6330\times 7{\cdot}5}{2}\left(\dfrac{5{\cdot}7}{2}-\dfrac{7{\cdot}5}{4}\right)$

$$=\frac{6330\times 7{\cdot}5}{2}\times{\cdot}975=24{,}400 \text{ m kg}$$

This bending moment produces tension at the water face
Bending moment at centre due to submerged earth pressure on the wall $=12620$ m kg
The bending produces tension on earth face
Net bending moment producing tension on the water face
$=24400-12620=11780$ m kg

Cantilever moment $=\dfrac{6330\times 0{\cdot}9^2}{2}=2570$ m kg

4·7·1. Thickness of Wall on Cracking Stress Consideration

Maximum bending moment at
$=12620+2570=15190$ m kg
$1/6\times 100\times d^2\times 17=1519000$

$$d^2=\frac{1519000\times 6}{100\times 17}=5370$$

$d=73{\cdot}2$ cm
Use $d=80$ cm, $d_e=75$ cm

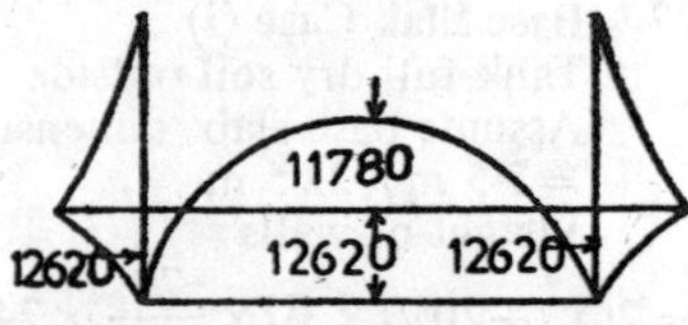

4·7·2. Main Reinforcement

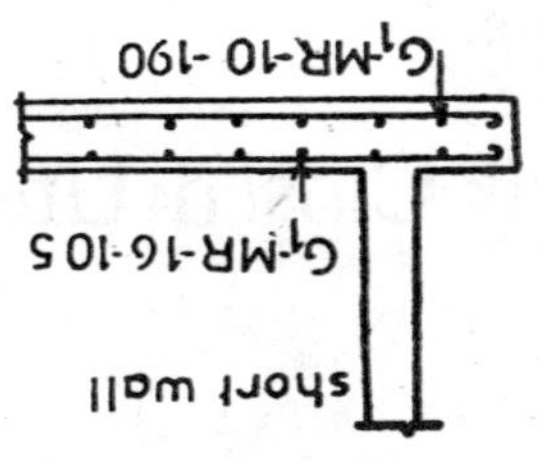

A_t at the ends (earth face)

$$A_t=\frac{1519000}{1250\times0{\cdot}86\times75}=18{\cdot}7 \text{ cm}^2$$

Use 16 mm ϕ 10·5 cm c/c

A_t at the centre (water face)

$$A_t=\frac{1178000}{1000\times0{\cdot}84\times75}=18{\cdot}5 \text{ cm}^2$$

Use 16 mm ϕ at 10·5 cm c/c

4·7·3. Secondary Reinforcement

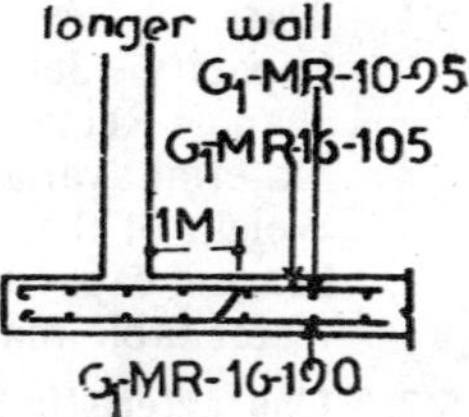

0·2% of concrete area (>45 cm thick)

$$=\frac{0{\cdot}2}{100}\times86\times100=16 \text{ cm}^2$$

10 mm at 9·5 cm c/c water face and 10 mm at 19 cm c/c in both directions on the earth face.

4·7·4. Check for Floatation or Uplift

Factor of safety against floatation required=1·1

Weight of walls and slab acting down resisting floatation, weight of walls=326,000 kg

$$\text{Weight of base slab}=\frac{80}{100}\times7{\cdot}5\times22{\cdot}5\times2400=325000 \text{ kg}$$

Total load=615,000 kg

Uplift due to water pressure=4·5×1000×7·5×22·5=760,000 kg

$$\text{Factor of safety}=\frac{651000}{760000}=0{\cdot}88<1{\cdot}1$$

Weight of earth on the projecting base slab

$$=2(7{\cdot}25+22{\cdot}25)\times\frac{50}{100}\times3{\cdot}7\times(1500-1000)$$

$$=2\times29{\cdot}5\times\frac{50}{100}\times3{\cdot}7\times500=54700 \text{ kg}$$

Total weight including earth acting down
=651000+54700=705,700 kg

Downward force required=1·1×760,000=835,000 kg

Available force=705700 kg

Force to be provided by screw piles=835,003−705,700
=129300 kg

4·8. Design of Screw Piles

Using two piles,

$$\frac{\pi d^2}{4}\times1400=\frac{129300}{2}$$

$$d^2=\frac{129300\times4}{2\times1400\times\pi}$$

$$d=\sqrt{59}=7{\cdot}7 \text{ cm}$$

Use 80 mm diameter steel shaft with a cast steel helicoid with a diameter=4×8=32 cm at the centre of base slab.

Closed Underground Rectangular Tank in Wet Soil 5

Capacity 350,000 Litres

5·1. Data

Capacity – 350 m³, Free board—20 cm
Shape – Rectangular – closed,
Material available Concrete – grade – M 200. Steel · grade – I
Weight of the soil = 1500 kg/m³, Angle of Internal friction ϕ=3(
Bearing capacity of the soil = 20 t/m²
Water table may rise to the ground level in rainy season.

5·2. Characteristic Strengths

$\sigma_{cb}=70$ kg/cm²	$m=13$
$\sigma_{st}=1000$ kg/cm²	$j_d=0{\cdot}84\ d,\ R=14$
$\sigma_{cb}=70$ kg/cm²	$m=13$
$\sigma_{st}=1250$ kg/cm²	$j_d=0{\cdot}86\ d,\ R=12{\cdot}6$
$\sigma_{cb}=50$ kg/cm²	$m=18$
$\sigma_{st}=1400$ kg/cm²	$j_d=0{\cdot}87\ d,\ R=8{\cdot}7$

5·3. Trial Dimensions

Required capacity = 350 m³, Depth of tank (assume) = 3·5 m
Area of tank = 100 m², Using 15·5 m × 7 m × 3·3 m
Actual capacity provided = 358 m³ > 350 m³
Use 15·5 m × 7 m × 3·5 m, allowing for a free board of 20 cm

$$L/B=\frac{15{\cdot}5}{7}>2,\ L/h=\frac{15{\cdot}5}{3{\cdot}5}>2$$

$$B/h=\frac{7}{35}=2$$

The side walls can be designed as propped cantilevers.

5·4. Roof Slab Loads

Slab and T – Beam roof

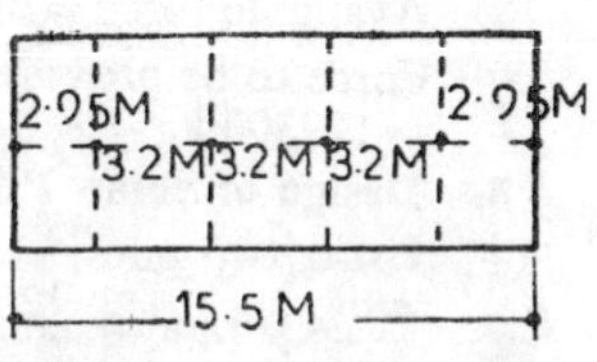

Slab – Assume 25 cm bearing on the side walls span = 3·2 m. Continuous over four T beam ribs.

Live load = 150 kg/m²,
Screeding (5 cm) = 110 kg/m²
Self weight (10 cm)

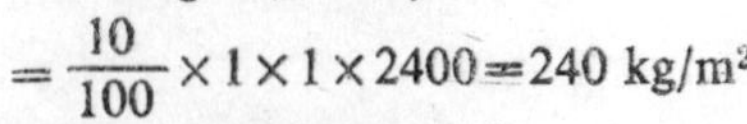

$$=\frac{10}{100}\times1\times1\times2400=240\ \text{kg/m}^2$$

Total load = 500 kg/m²

5·4·1. Bending Moments

Continuous slab span = 3·2 m

$$\text{Maximum bending moment}=\frac{wl^2}{10}=\frac{500\times(3{\cdot}2)^2}{10}=512\ \text{m kg}$$

5·5. Depth of Slab

Using M 150 concrete for roof

$M_r = 8{\cdot}7\ bd^2$

$$d_e = \sqrt{\frac{51200}{100 \times 8{\cdot}7}} = 7{\cdot}7 \text{ cm}$$

Use $d = 10$ cm and $d_e = 8{\cdot}5$ cm

5·5·1. Amount of Reinforcement

$$A_t = \frac{51200}{1400 \times {\cdot}87 \times 8{\cdot}5} = 4{\cdot}95 \text{ cm}^2$$

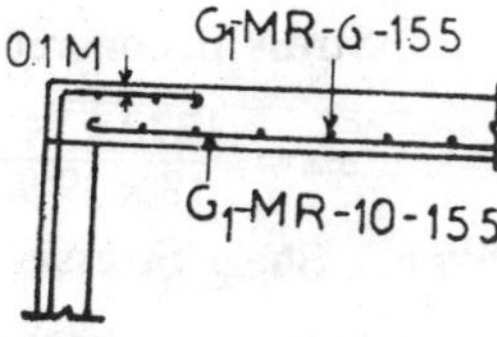

Use 10 mm ϕ at 15·5 cm c/c

5·5·2. Secondary Reinforcement

$$A_t = \frac{0{\cdot}15}{100} \times 10 \times 100 = 1{\cdot}5 \text{ cm}^2$$

Use 6 mm ϕ at 16·5 cm c/c

5·6. T-Beam Loads

Span = 7·25 m assuming 25 cm bearing centre to centre of beams = 3·2 m

From slab = $500 \times 3{\cdot}2 = 1600$ kg/m²

Self weight of rib (assume 30 cm × 50 cm)

$$= \frac{30}{100} \times \frac{50}{100} \times 2400 \times 1 = 360 \text{ kg/m}^2$$

Total load = $1600 + 360$ kg/m² = 1960 kg/m²

5·6·1. Bending Moment

Maximum bending moment

$$= \frac{wl^2}{8} = \frac{1960}{8} \times (7{\cdot}25)^2 = 12{,}900 \text{ m kg}$$

Approximate lever arm $= d_e - \dfrac{ds}{2} = 55 - 5 = 50$ cm

5·6·2. Approximate Steel

A_t = Approximate steel required

$$= \frac{129000 \times 100}{1400 \times 50} = 18{\cdot}4 \text{ cm}^2$$

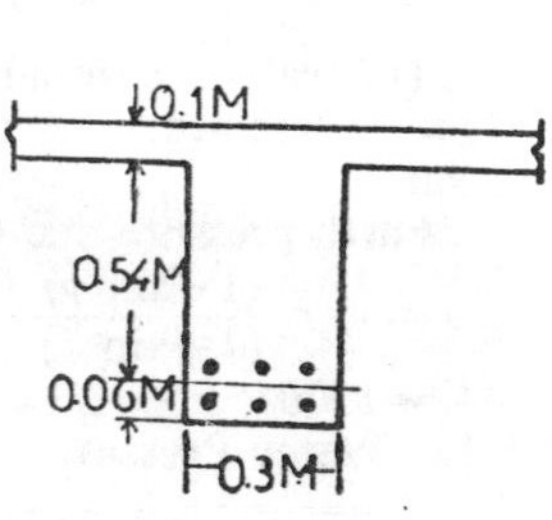

Use 6 – 20 mm ϕ

A_t provided 18·84 cm²

5·6·3. Effective Width of Flange

The least of : (*i*) 1/3 of span

$$= \frac{7{\cdot}25}{3} = 2{\cdot}42 \text{ m} = 242 \text{ cm}$$

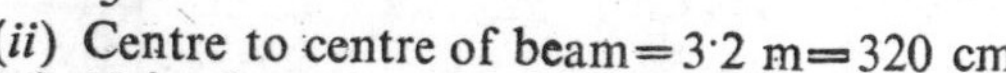

(*ii*) Centre to centre of beam = 3·2 m = 320 cm

(*iii*) $12ds + b_r = 12 \times 10 + 30 = 150$ cm

5·6·4. Position of N.A.

Assuming N.A. lies below the flange

$$B \times ds \times \left(n - \frac{ds}{2}\right) = m \times A_t\ (d - n)$$

$$150 \times 10 \times (n - 5) = 18 \times 18{\cdot}84\ (54 - n)$$

$$1500\ n - 7500 = 990 - 340\ n$$

$$1840\ n = 25900, \quad n = 14{\cdot}1 \text{ cm}$$

$$\bar{y}=\frac{3n-2ds}{2n-ds}\times\frac{ds}{3}=\frac{3\times14{\cdot}1-2\times10}{2\times14{\cdot}1-10}\times\frac{10}{3}$$

$$=\frac{22{\cdot}3}{18{\cdot}2}\times\frac{10}{3}=4{\cdot}08 \text{ cm}$$

Actual lever arm $jd=d-\bar{y}=54-4{\cdot}08=49{\cdot}92$ cm

5·6·5. Stresses

$$\text{Stress in steel}=\frac{1290000}{18{\cdot}84\times49{\cdot}92}=1370 \text{ kg/cm}^2 < 1400 \text{ kg/cm}^2$$

$$\text{Stress in concrete } \frac{\sigma_{cb}}{n}=\frac{\sigma_{st}}{m\,(d-n)}$$

$$\sigma_{cb}=\frac{1370\times14{\cdot}1}{18\times39{\cdot}9}=42{\cdot}7 \text{ kg/cm}^2 < 50 \text{ kg/cm}^2$$

5·6·6. Shear Stresses

$$\text{Maximum shear at the support edge}=\frac{1960\times7}{2}=6850 \text{ kg}$$

$$\text{Shear stress } q=\frac{Q}{j_d\times b_r}=\frac{6850}{49{\cdot}92\times30}=4{\cdot}6 \text{ kg/cm}^2 < 5 \text{ kg/cm}^2$$

5·6·7. Nominal Stirrups

Provide 8 mm ϕ two legged at 30 cm c/c in addition, the top row of bars may be bent up 1 m from the edge.

5·7. General Condition of Loading

(*i*) Tank full and dry soil outside, (*ii*) Tank full and submerged soil outside. (*iii*) Tank empty dry soil outside. (*iv*) Tank empty submerged soil outside.

5·7·1. Critical Loading Conditions

(1) Tank full and dry soil outside.

(2) Tank empty submerged or water logged soil outside.

5·8. Longer Walls case (1) Earth Pressure

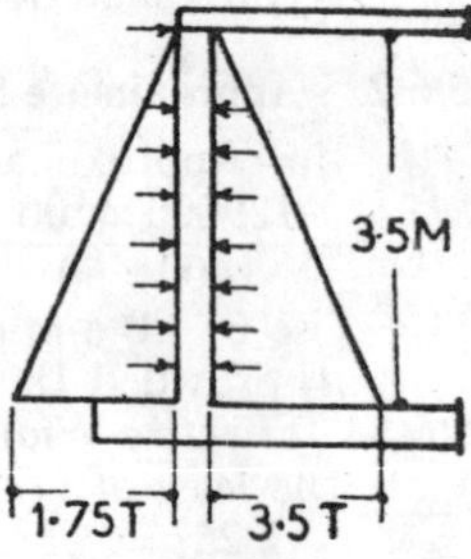

$$L/h=\frac{15{\cdot}5}{3{\cdot}5}>2$$

The walls are designed as propped cantilevers. These are subjected to earth and water pressures.

Earth pressure due to dry soil

$$=w_e\,h\,\frac{(1-\sin\phi)}{(1+\sin\phi)}$$

$$=1500\times3{\cdot}5\times1/3=1750 \text{ kg/m}^2$$

5·8·1. Water Pressure

Water pressure $=wh=3{\cdot}5\times1000$

$=3500$ kg/m^2

Net pressure at the base of the wall

$p=3500-1750=1750$ kg/m^2

5·8·2. Bending Moment

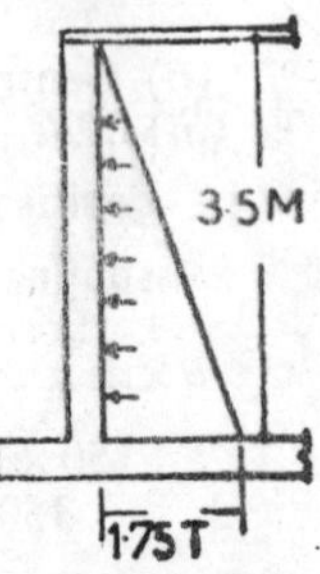

Maximum negative bending moment (water face)

$$M_w=\frac{ph^2}{15}=\frac{1750\times3{\cdot}5\times3{\cdot}5}{15}=1430 \text{ m kg}$$

Maximum positive bending moment (earth face)

$$M_e=\frac{ph^2}{33{\cdot}5}\times\frac{1750\times 3{\cdot}5^2}{33{\cdot}5}=640 \text{ m kg}$$

5·8·3. Case (2) Earth Pressure Due To Wet Soil

Tank empty and submerged soil outside.

Earth pressure due to submerged soil (Gray's rule) $=1500\ h=1500\times 3{\cdot}5=5250$ kg/m²

5·8·4. Bending Moment

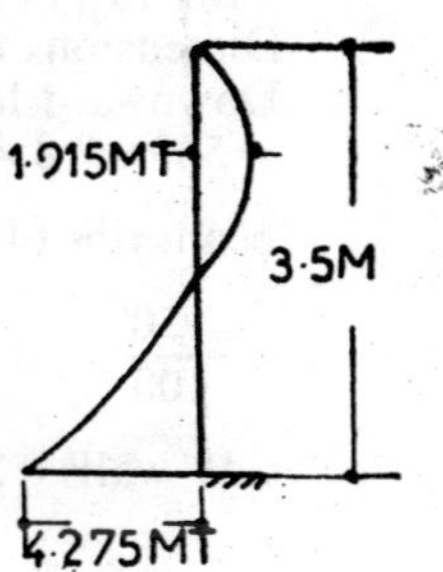

Maximum negative bending moment (earth face)

$$=\frac{ph^2}{15}=\frac{5250\times 3{\cdot}5\times 3{\cdot}5}{15}=4275 \text{ m kg.}$$

Maximum positive bending moment (water face)

$$\frac{ph^2}{33{\cdot}5}=\frac{5250\times 3{\cdot}5\times 3{\cdot}5}{33{\cdot}5}=1915 \text{ m kg.}$$

5·8·5. Thickness of Walls on Cracking Stress Consideration

$$\tfrac{1}{6}\times bd^2\ \sigma_{ct_b}=427500=\tfrac{1}{6}\times 100\times d^2\times 17=427500$$

$$d^2=\frac{427500\times 6}{100\times 17},\ d=39 \text{ cm},\quad \text{Use } d=40 \text{ cm and } d_e=36 \text{ cm}$$

5·8·6. Main Reinforcement Case (1)

Steel required for —ve bending moment (water face)

$$=\frac{1430\times 100}{1000\times{\cdot}84\times 36},\ A_t=4{\cdot}72 \text{ cm}^2$$

Use 10 mm ϕ 16 cm c/c

Steel required for +ve bending moment (earth face)

$$A_t=\frac{64000}{1250\times{\cdot}86\times 36}=1{\cdot}65 \text{ cm}^2$$

12 mm ϕ 60 cm c/c

Case (2)

Steel required for —ve bending moment (earth face)

$$A_t=\frac{427500}{1250\times{\cdot}86\times 36}=11{\cdot}1 \text{ cm}^2$$

12 mm ϕ 10 cm c/c

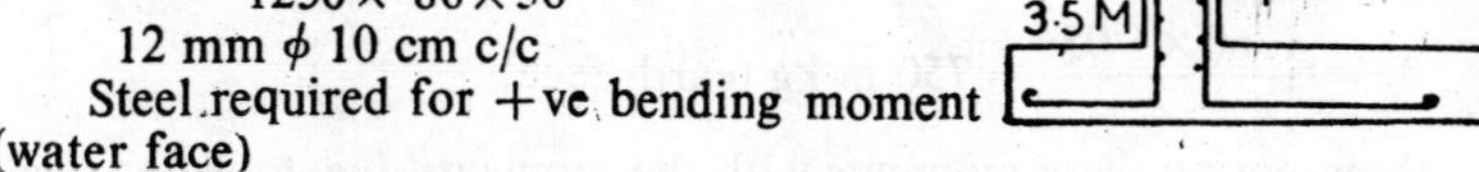

Steel required for +ve bending moment (water face)

$$A_t=\frac{191500}{1000\times{\cdot}84\times 36}=8{\cdot}35 \text{ cm}^2$$

10 mm ϕ 8 cm c/c.

5·8·7. Secondary Reinforcement

$$A_t=\frac{0{\cdot}3}{100}\times 40\times 100=12 \text{ cm}^2$$

A_t per face $=6$ cm²

Use 10 mm ϕ 13 cm c/c on both taces

5·9. Short Walls

$$\frac{B}{h}=\frac{7}{3{\cdot}5}=2$$

Walls span in one direction only and are designed as propped cantilevers. Therefore use the details as per longer walls.

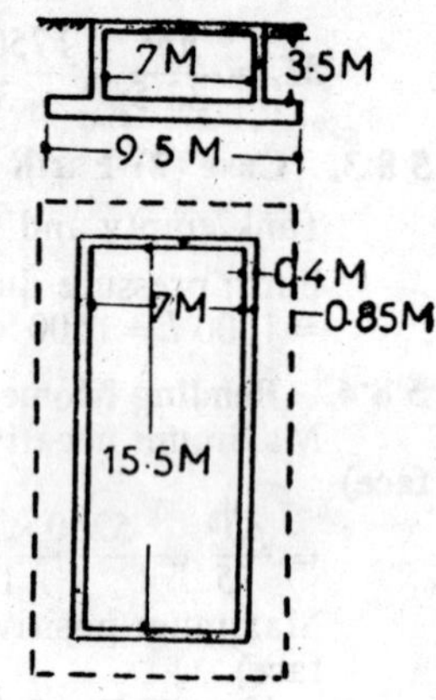

5·10. Bottom Slab Case (1)

Tank full dry soil outside

Dimensions of base $=9{\cdot}5$ m $\times$ 18 m

Downward loads due to roof slab

$=500\times7{\cdot}8\times16{\cdot}3=63{,}500$ kg

beam ribs (4 Nos) $=4\times7{\cdot}8\times\frac{30}{100}\times\frac{50}{100}\times2400=11{,}600$

side walls $=2(7{\cdot}4+15{\cdot}9)\times\frac{40}{100}\times3{\cdot}5\times2400$

$=2(23{\cdot}3)\times\frac{40}{100}\times3{\cdot}5\times2400=157{,}000$ kg

Total load $=232{,}100$ kg;

Water pressure and the soil pressure due to water in the tank will counteract.

Soil pressure below the base due to direct load

$$=\frac{232{,}100}{7{\cdot}5\times18}=1360 \text{ kg/m}^2$$

5·10·1. Bending Moments

Bending moment at centre (water face)

$$M=1360\times\frac{9{\cdot}5}{2}\left(\frac{7{\cdot}4}{2}-\frac{9{\cdot}5}{4}\right)$$

$$=1360\times\frac{9{\cdot}5}{2}(3{\cdot}7-2{\cdot}375)$$

$$=1360\times\frac{9{\cdot}5}{2}\times1{\cdot}325$$

$$=\frac{16900}{2}\text{ m kg}=8450\text{ m kg}$$

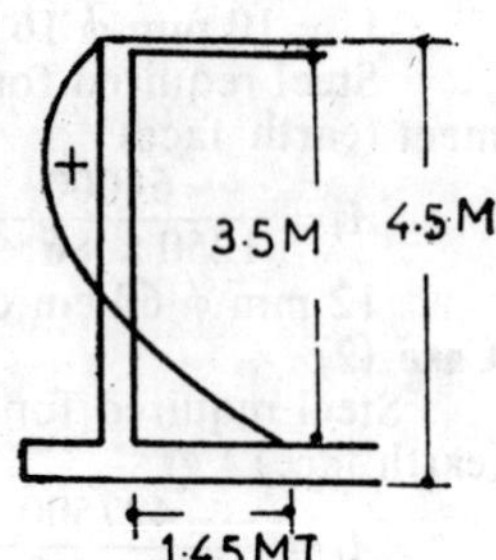

Bending moment at support

$$M=\frac{1360\times1{\cdot}05^2}{2}=750\text{ m kg (earth face)}$$

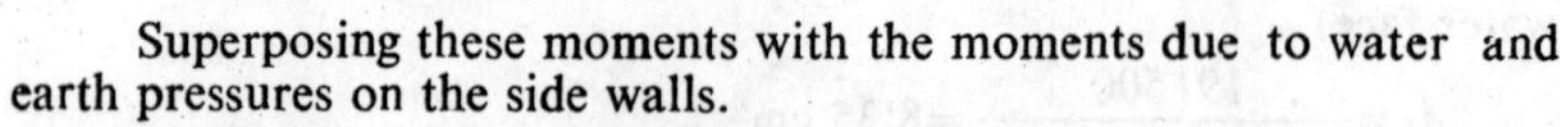

Superposing these moments with the moments due to water and earth pressures on the side walls.

Maximum moment at centre (water face)

$M=8450+1450=9900$ m kg

Maximum moment at support (water face)

$M_s=1450-750=700$ m kg.

5·10·2. Case (2)

Bending moment : Tank empty—submerged soil outside

Soil reaction due to self weight $=1360$ kg/m^2

Soil reaction due to water below in the soil assuming 1 m thick slab $=1000 \times 4\ 5 = 4500$ kg/m²

Total upward reaction below the base
$=4500+1360=5860$ kg/m²

Bending moment at centre of the slab (water face)

$$=5860 \times \frac{9{\cdot}5}{2}\left(\frac{7{\cdot}4}{2}-\frac{9{\cdot}5}{4}\right)$$

$$=5860 \times \frac{9{\cdot}5}{2} \times 1{\cdot}325 = 37500 \text{ m kg}$$

Bending moment at support (earth face)

$$M=\frac{5860 \times 1{\cdot}05^2}{2} = 3220 \text{ m kg}$$

Superposing these moments on the moments produced in the side wall in case (2),

Maximum moment at centre (water face)
$M_c = 37500 - 4275 = 33225$ m kg

Maximum moment at support (earth face)
$M_s = 4275 + 3220 = 7495$ m kg.

5·10·3. Thickness of Base Slab on Cracking Stress Consideration

$\frac{1}{6} b \times d^2 \times \sigma_{ctb} = 3222500$

$\frac{1}{6} \times 100 \times d^2 \times 17 = 3322500$

$$d^2 = \frac{3322500 \times 6}{17 \times 100}$$

$d = \sqrt{11750}$, $d = 108{\cdot}5$ cm

Use $d = 110$ and $d_e = 100$ cm

5·10·4. Main Reinforcement

Steel required on the water face at centre

$$A_t = \frac{3322500}{1000 \times {\cdot}84 \times 100} = 39{\cdot}75 \text{ cm}^2$$

Use 25 mm ϕ at 12 cm c/c at top

steel required on the earth face at support

$$A_t = \frac{749500}{1250 \times {\cdot}86 \times 100} = 6{\cdot}95 \text{ cm}^2$$

Use 12 mm ϕ at 16 cm c/c

5·10·5. Secondary Reinforcement

$A_t = 0{\cdot}2\%$ of concrete area
$= {\cdot}2/100 \times 100 \times 110 = 22$ cm²

per face $A_t = 11$ cm²

Use 12 mm ϕ at 10 cm c/c at both faces

5·10·6. Factor of Safety Against Floatation

Downward loads :

$$\text{Slab} = \frac{10}{100} \times 7{\cdot}8 \times 16{\cdot}3 \times 2400$$

$= 30{,}500$ kg

Screeding $= 110 \times 7{\cdot}8 \times 16{\cdot}3 = 14{,}000$ kg

$$\text{Ribs} = 4 \times 7{\cdot}8 \times \frac{30}{100} \times \frac{50}{100} \times 2400$$

$= 11{,}600$ kg

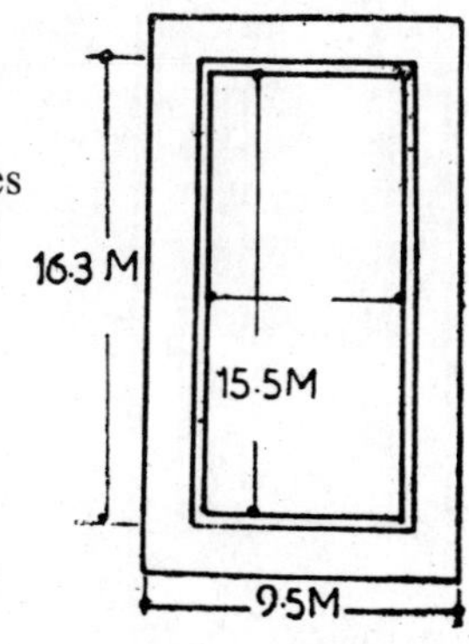

Sidewalls $=2(7.4+15.9)\ \frac{40}{100}\times 3.5\times 2400=157{,}000$ kg

Base slab $=9.5\times 18\times\frac{110}{100}\times 2400=452{,}000$ kg

Earth on the projecting slab

$=1500\times 2(9.5+16.3)\ \frac{85}{100}\times 3.5$

$=1500\times 2\times 25.8\times\frac{85}{100}\times 3.5$

$=230{,}000$ kg

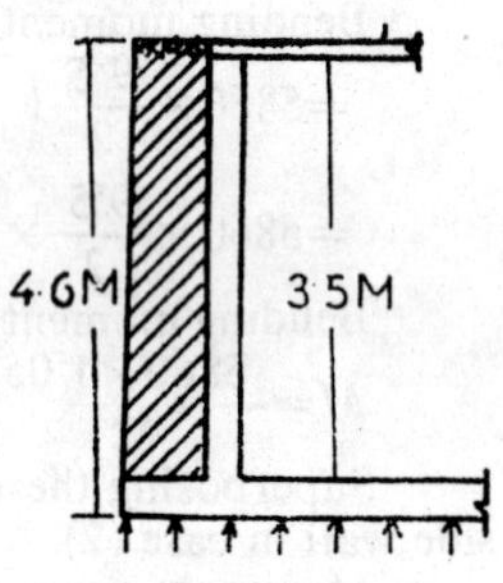

Total load $=738{,}100$ kg

Upward reaction or uplift

$=4.6\times 1000\times 9.5\times 18=785{,}000$ kg

Factor of safety $=\frac{738{,}100}{785000}=0.94<1.1$

Downward force required $=865{,}000$ kg

Available force $=738{,}100$ kg

Additional force required $=126{,}900$ kg.

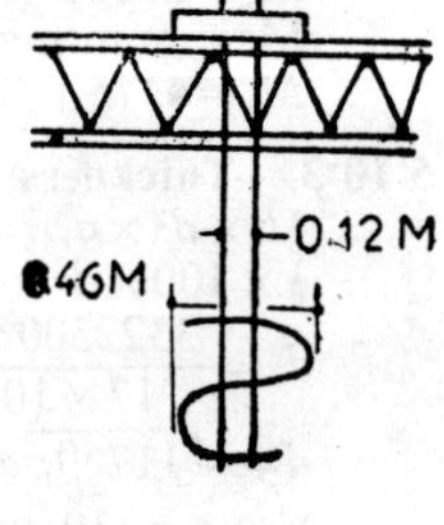

5.11. Diameter of Screw Pile

$\frac{\pi d^2}{4}\times 1400=126{,}900$

$d^2=\frac{126900\times 4}{1400\times\pi},\quad d=\sqrt{116}=10.8$ cm

Use 12 cm diameter steel shaft with 48 cm helicoid.

Closed Underground Circular Tank with Wet Soil 6

Capacity 350,000 Litres

6·1. Data

Capacity = 350 m^3, Free board = 15 cm, weight of soil = 1500 kg/m^3, Angle of internal friction ϕ = 30°, Bearing capacity of soil = 20 t/m^2

Water table may reach the ground level during rainy season.

Materials available : Concrete – M200, Steel – grade I

σ_{cb} = 70 km/cm^2 $\quad m = 13$

σ_{st} = 1000 kg/cm^2 $\quad jd = \cdot 84\ d,\ R = 14.$

6·2. Trial Dimensions

Required capacity = 350 m^3,

depth of tank = 3·5 m,

depth of water = 3·35 m

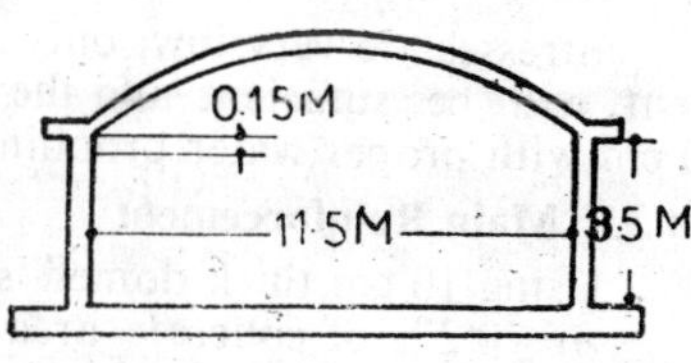

$$\frac{\pi D^2}{4} \times 3{\cdot}35 = 350$$

D = 11·5 m.

6·3. Roof Dome

Loads : Thickness of the dome = 15 cm

$$\text{self weight} = \frac{15}{100} \times 1 \times 1 \times 2400 = 360 \text{ kg/m}^2$$

Live and accidental loads = 300 kg/m^2

Water proofing and finish = 90kg/m^2

Total load = 750 kg/m^2

6·4. Geometry of the Dome

$$\text{Rise} = \frac{D}{6} = \frac{11{\cdot}5}{6} = 1{\cdot}9 \text{ m}$$

Use = r = 2 m

$$\text{Radius } R = \frac{l^2 + r^2}{2r} = \frac{5{\cdot}75^2 + 2^2}{2 \times 2}$$

$= 9{\cdot}25$ m

$$\sin \phi = \frac{l}{R} = \frac{5{\cdot}75}{9{\cdot}25} = 0{\cdot}623$$

$\phi = 38° - 30' < 51° 48'$

There is, therefore no tension in the dome

6·4·1. Force Components

Total load W = surface area × $w = 2\pi \times R \times r \times w$

$= 2\pi \times 9{\cdot}25 \times 2 \times 750 = 87000$ kg = 87 t

$$V_1 = \frac{W}{2\pi l} = \frac{87}{2 \times \pi \times 5{\cdot}75} = 2{\cdot}4 \text{ t/m}$$

$$\text{Cot } \phi = \frac{H_1}{V_1}$$

$$H_1 = V_1 \cot \phi = 2{\cdot}4 \times \frac{7{\cdot}25}{5{\cdot}75} = 3{\cdot}05 \text{ t}$$

$$T_1 = \frac{V_1}{\sin \phi} = 3{\cdot}05 \times \frac{9{\cdot}25}{5{\cdot}75} = 4{\cdot}9 \text{ t}$$

6·4·2. Stresses in the Dome

$$\text{Shear stresses} = \frac{V_1}{\text{area}} = \frac{2400}{100 \times 18} = 1{\cdot}6 \text{ kg/cm}^2$$

$$\text{Maximum hoop stress } (\phi = 0) = s = \frac{WR}{2t} = \frac{750 \times 9{\cdot}25}{2 \times 0{\cdot}15 \times 100 \times 100}$$

$$= 2{\cdot}3 \text{ kg/cm}^2$$

$$\text{Meridional stress} = c = \frac{WR}{t(1+\cos\phi)}$$

$$= \frac{750 \times 9{\cdot}25}{0{\cdot}15(1 + {\cdot}7826) \times 100 \times 100}$$

$$= \frac{0{\cdot}56 \times 750 \times 9{\cdot}25}{0{\cdot}15 \times 100 \times 100} = 3{\cdot}82 \text{ kg/cm}^2$$

Stresses are very low, only temperature and shrinkage reinforcement will be sufficient and the thickness of dome can be reduced to 10 cm with proper water proofiing.

6·4·3. Main Reinforcement

Using 10 cm thick domed slab

$A_t = 0{\cdot}2\%$ of concrete area

$$= \frac{0{\cdot}2}{100} \times 10 \times 100 = 2 \text{ cm}^2$$

Use 6 mm ϕ at 14 cm c/c in both directions.

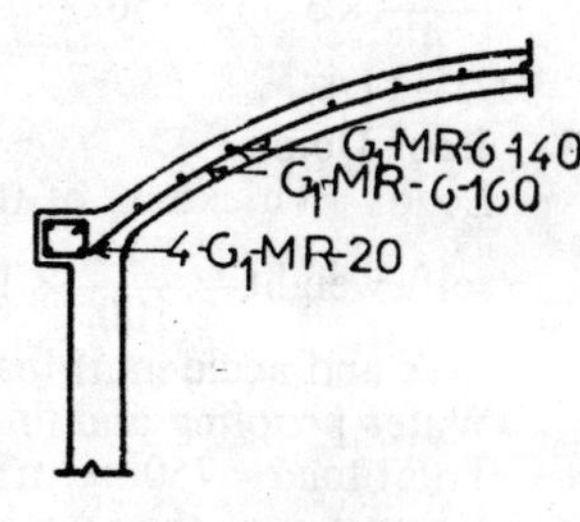

Steel required for shear at the edge

$$= \frac{V_1}{1400} = \frac{2400}{1400} = 1{\cdot}72 \text{ cm}^2$$

Use 6 mm at 16 cm c/c

6·4·4. Rib Design

$$\text{Maximum hoop tension} = T = H_1 \times \frac{D}{2}$$

$$= 3{\cdot}05 \times \frac{11{\cdot}5}{2} = 17{\cdot}5 \text{ t}$$

$$= 17500 \text{ kg}$$

$$A_t = \frac{17500}{1400} = 12{\cdot}55 \text{ cm}^2$$

Use 20 mm ϕ – 4 Nos,

A_t provided $= 12{\cdot}57$ cm²

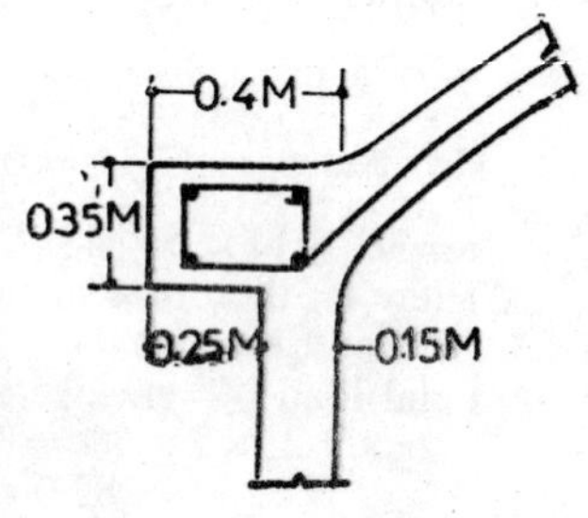

Tensile strength of concrete in direct tension $\sigma_{ct_t} = 12$ kg/cm²

$$\sigma_{ct_t} = \frac{T}{A_c + (m-1)A_t}$$

$$12 = \frac{17500}{A_c + 12 \times 12{\cdot}57}$$

$12\ A_c+12\times 12\times 12{\cdot}57=17500$

$12\ A_c+1810=17500$

$A_c=\dfrac{15690}{12}=1310$. Use 40×35 cm

6·5. Side Walls

General loading conditions (*i*) Tankfull—dry soil outside, (*ii*) Tankfull—submerged soil outside, (*iii*) Tank empty—dry soil outside, (*iv*) Tank outside—submerged soil outside, (*v*) Tankfull and no soil outside (when being tested for leakage, before filling in, the earth around).

6·5·1. Critical Loading Conditions Case (1)

(*i*) Tank full no soil outside (during leak test)

(*ii*) Tank empty and submerged soil outside

Tank full—no soil outside

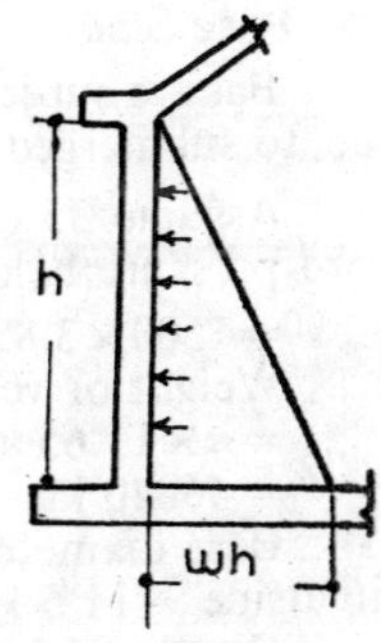

Hoop tension $=wh\dfrac{D}{2}$

$=1000\times 3{\cdot}5\times\dfrac{11{\cdot}5}{2}=20100$ kg

Area of steel $=\dfrac{20{,}100}{1000}=20{\cdot}1\ \text{cm}^2$

6·5·2. Thickness of the wall

$$\sigma_{ct_t}=\frac{T}{100\times t+(m-1)A_t}$$

$$12=\frac{20{,}100}{100\times t+12\times 20{\cdot}11}$$

$1200\ t+12\times 12\times 20{\cdot}11=20{,}100$

$1200\ t+2900=20{,}100$

$t=\dfrac{17200}{1200}=14{\cdot}3$ cm

Use 15 cm thick walls.

6·5·3. Case (ii)

Tank empty—submerged soil outside

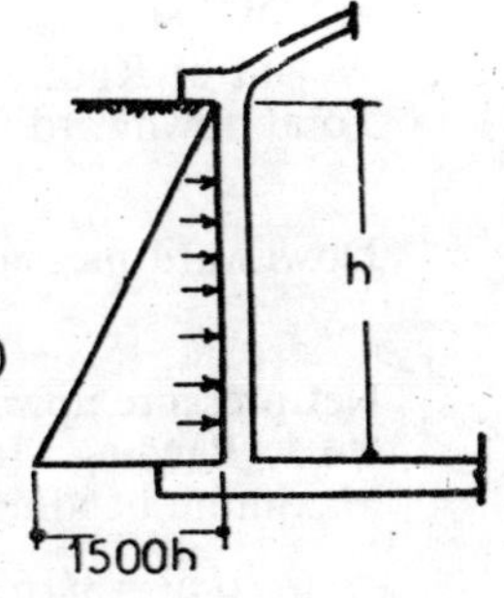

Maximum hoop compression (Gray's rule)

$=1500\times h\times\dfrac{D}{2}=1500\times 3{\cdot}5\times\dfrac{11{\cdot}5}{2}$

$=30250$ kg

Area of steel required $=\dfrac{30250}{1000}$

$=30{\cdot}25\ \text{cm}^2$

Use 20 mm ϕ at 10 cm c/c

Since the wall is subjected to hoop compression 15 cm provided will be sufficient.

6.5.4. Curtailment of Reinforcement

depth from top *m*	*Hoop compression* *kg*	A_t *cm²*	*Spacing 20 mm ϕ required*
0–1	8630	8·63	30
1–2	17260	17·26	18
2–3	25890	25·89	12
3–3·5	30250	30·25	10

6·5·5 Secondary Reinforcement

$A_t = 0.3\ \%$ of cross – section

$= \frac{\cdot 3 \times 15 \times 1000}{100} = 4.5\ \text{cm}^2$

Use 10 mm ϕ at 17 cm c/c

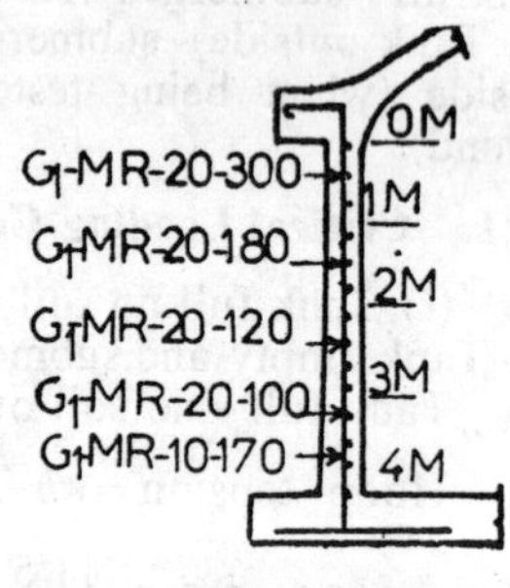

6·6. Base Slab

Base is subjected to upward pressure due to submerged soil.

Assume 35 cm thick base slab, upward pressure below the base = 1500 h

$= 1500 \times 3.85 = 5800\ \text{kg/m}^2$

Weight of vertical walls

$= \pi \times 11{\cdot}65 \times 0{\cdot}15 \times 3{\cdot}5 \times 2400$

$= 46040$ kg

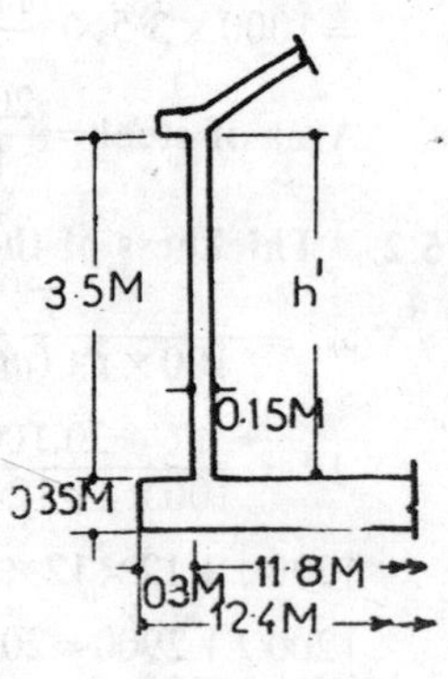

Base diameter with 30 cm projection on eitherside $= 11{\cdot}5 + {\cdot}15 + {\cdot}15 + {\cdot}30 + {\cdot}3 = 12{\cdot}4$ m

Weight of base slab

$= \frac{\pi \times 12{\cdot}4^2}{4} \times 0{\cdot}35 \times 2400 = 101{,}200$ kg

Weight of earth on the projection

$= \frac{\pi(12{\cdot}4^2 - 11{\cdot}8)^2 \times 3{\cdot}5 \times 1500}{4}$

$= 57500$ kg

Weight of Roof dome = 87000 kg

Total downward load = 46,040 + 101,200 + 57,000 + 87000

= 291,240 kg

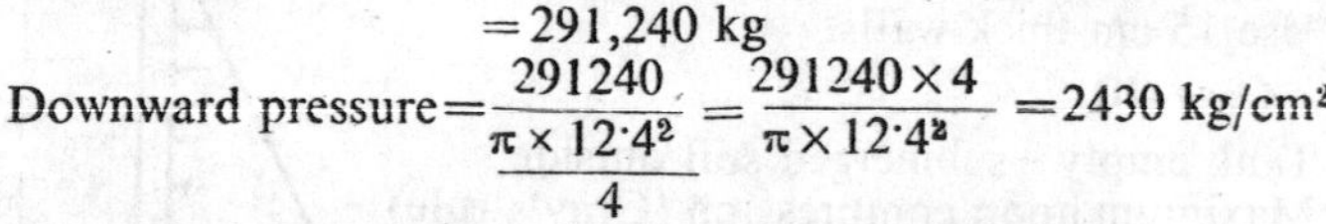

Downward pressure $= \frac{291240}{\frac{\pi \times 12{\cdot}4^2}{4}} = \frac{291240 \times 4}{\pi \times 12{\cdot}4^2} = 2430\ \text{kg/cm}^2$

Net pressure upwards $q = 5800 - 2430 = 3370\ \text{kg/m}^2$

6·6·1. Bending Moment

Maximum bending moment in the circular base slab

$= \frac{3}{16}\, q\, a^2 = 3/16 \times 3370 \times \left(\frac{12{\cdot}4}{2}\right)^2 = 3/16 \times 3370 \times (6{\cdot}2)^2$

$= 24400$ m kg $= 24{,}400{,}00$ cm kg

For $\sigma_{st} = 1000\ \text{kg/cm}^2$ and $\sigma_{cb} = 70\ \text{kg/cm}^2$, $R = 14{\cdot}01$

depth required $= \sqrt{\frac{2{,}440{,}000}{14{\cdot}01 \times 100}} = \sqrt{1730} = 41{\cdot}5$ cm > 35 cm

ıssumed

Use $d=50$ cm, $de=42{\cdot}5$ cm

Increase in the weight of base slab

$$=\frac{\pi\times(12{\cdot}4)^2}{4}\times 0{\cdot}15\times 2400=43{,}200 \text{ kg}$$

Downward pressure due to additional weight

$$=\frac{43200}{\frac{\pi(12{\cdot}4)^2}{4}}=\frac{43200\times 4}{\pi(12{\cdot}4)^2}=360 \text{ kg/m}^2$$

Revised upward pressure$=3370-360=3010$ kg/m²

$$\text{Revised } BM=\frac{3}{16}\times 3010\times(6{\cdot}2)^2\times 100=2{,}170{,}000 \text{ cm kg}$$

6·6·2. Main and Secondary Steel

$$A_t=\frac{2170000}{1000\times 0{\cdot}84\times 42{\cdot}5}=61 \text{ cm}^2$$

Use 22 mm ϕ at 12 cm c/c both ways at top (water face)

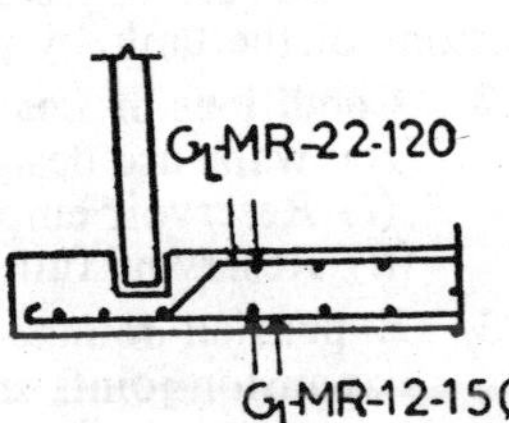

$$A_t=\frac{0{\cdot}3}{100}\times 50\times 100=15 \text{ cm}^2$$

Use 12 mm ϕ at 15 cm c/c bothways at bottom

6·6·3. Check for Uplift

Factor of cafety against floatation required 1·1

weight of tank$=291{\cdot}24+243{\cdot}2=334{\cdot}44$ t

Upward force due to pressure below the base when water logged

$$=\frac{\pi\times 12{\cdot}4^2}{4}\times 1500\times 3{\cdot}5=635{,}000 \text{ kg}=635 \text{ t}$$

$$\text{Factor of safety }=\frac{334{\cdot}44}{635}=0{\cdot}525<1{\cdot}1$$

Provide screw piles

6·7. Design of Screw Piles

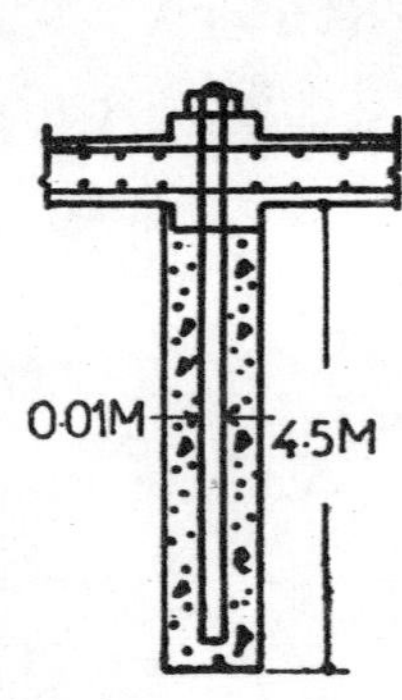

Downward force required for stability

$=1{\cdot}1\times 635=700$ t

Available downward force $=700-334{\cdot}44$

$=365{\cdot}56$ t

Force to be resisted by screw piles

$=365{\cdot}56$ t

Using 6 piles force per pile

$$=\frac{365{\cdot}56}{4}=91{\cdot}39 \text{ t}$$

Diameter the pile $=d$

$$\frac{\pi d^2}{4}\times 1400=91390,\quad d^2=\frac{91390\times 4}{\pi\times 1400}=83$$

$d=9{\cdot}3$ cm. Use 10 cm dia piles

Use 10 cm diameter steel shaft with steel helical coil with 40 cm helical plate.

Permissible bond stress$=8$ kg/cm²

$$\text{Length required}=nd=\frac{1400}{4\times 8}d=44\,d=440 \text{ cm}$$

Use 4·5 m length of pile

VII. OPEN UNDERGROUND RESERVOIRS

General Information 1

7·1. General Side Walls

The side walls of reservoirs are designed as retaining walls.

In the case of reservoirs, the subsoil water level is kept below the bottom of the tank, by providing proper drainage around the tank.

7·2. Conditions of Loading

The walls are designed for the following conditions.

(*i*) Reservoir empty-earth fill outside.

(*ii*) Reservoir full-with fill removed.

7·3. Expansion Joints

Expansion joints shall be provided at 6 to 9 m intervals.

7·4. Floors

Floors shall be laid at a slope 1 in 100 to make it easy for cleaning.

Water Bars

A stout galvanised steel or copper jointing strip is used in the floor joints

Open Underground Reservoir in a Dry Soil 2

2·1. Data

Capacity 5000 cu m. Free board 50 cm.

Bearing capacity of soil$=20$ t/m^2

Water table is much below the level of base and soil around the tank remains dry in all seasons.

Angle of internal fraction$=30°$, Weight of soil$=1500$ kg/m^3

Materials available :—Steel—grade I, Concrete - M 200.

2·2. Characteristic Strength

$\sigma_{cb}=70$ kg/cm^2 $\quad m=13$

$\sigma_{st}=1000$ kg/cm^2 $\quad J_d=\cdot84\ d,\ R$;4

$\sigma_{cb}=70$ kg/cm^2 $\quad m=13$

$\sigma_{st}=1250$ kg/cm^2 $\quad j_d=0\cdot86\ d,\ R=12\cdot6$

2·3. Trial Dimensions of The Tank

Required capacity of the reservoir$=5000$ cu m

Let, $h=4$ m

$$\text{Area of the tank}=\frac{5000}{4}=1250 \text{ m}^2$$

A reservoir of 50 m$\times$25 m$\times$4 m size will provide the required capacity.

2·4. Critical Loading Conditions

(*i*) Tank full—no earth fill outside.

(*ii*) Tank empty, dry earth fill outside.

The side walls are designed as cantilever retaining walls for the above conditions of loading.

2·5. Stability of the Trial Section Case (1)

Tank full—no earth fill outside. Vertical loads

Loads due to	*Magnitude* t	*Point of application from 0* m	*Moment* mt
W_1=stem (rectangular portion)	$\frac{20}{100}\times4\cdot5\times1\times\frac{2400}{1000}$ $=2\cdot16$	2·9	6·28
W_2=stem (triangular portion)	$\frac{1}{2}\times\frac{20}{100}\times4\cdot5\times1\times\frac{2400}{1000}$ $=1\cdot08$	2·73	2·95
W_3=Bas slab	$4\cdot5\times1\times\frac{40}{100}\times\frac{2400}{1000}$ $=4\cdot32$	2·25	9·75
W_4=weight of water above base	$2\cdot6\times\frac{1000}{1000}\times4$ $=10\cdot4$	1·30	13·50
W_5=weight of water (triangular wedge)	$\frac{20}{100}\times\frac{4}{4\cdot5}\times\frac{1\times\frac{1}{2}\times4\cdot5\times1000}{1000}$ $=0\cdot4$	2·66	1·06
	$\Sigma W=18\cdot36$		$\Sigma M=33\cdot54$

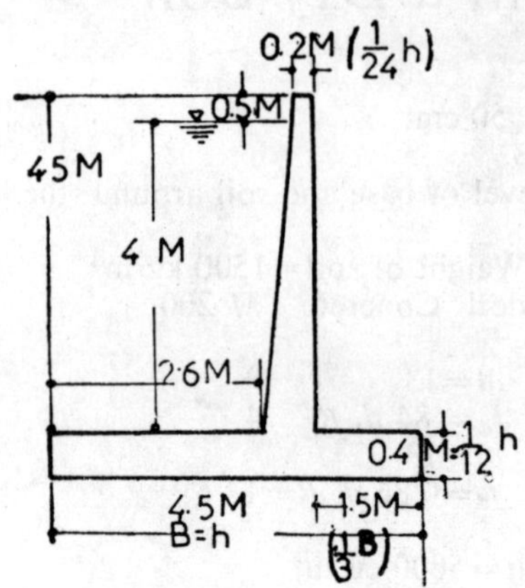

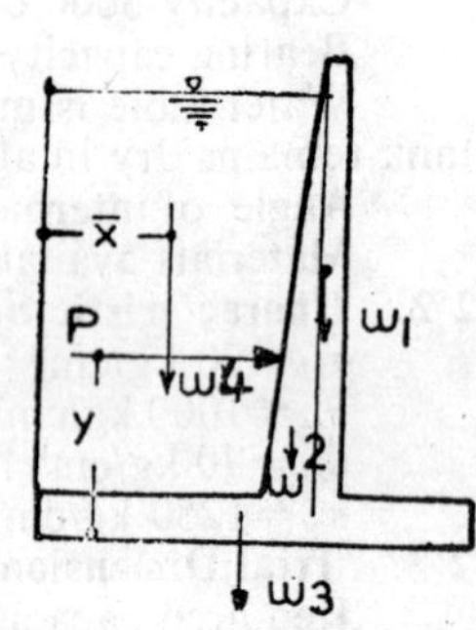

2·5·1. Point of Application of the Resultant Vertical Load

$$\bar{x}=\frac{33{\cdot}54}{18{\cdot}36}=1{\cdot}85 \text{ m}$$

Horizontal loads

Water pressure $=\frac{1}{2}\,wh\times h$

$$P=\tfrac{1}{2}\times1000\times4\times4=8000 \text{ kg}=8 \text{ t}$$

Point of application of horizontal force

$$\bar{y}=\frac{4}{3}+\frac{40}{100}=1{\cdot}33+{\cdot}4=1{\cdot}73 \text{ m}$$

2·5·2. Point of Application of The Resultant

Taking moments about o'; $W(\bar{x}_r-\bar{x})=P\bar{y}$

$$18{\cdot}36\,\bar{x}_r-33{\cdot}54=8\times1{\cdot}73$$

$$\bar{x}_r=\frac{48{\cdot}34}{18{\cdot}34}=2{\cdot}63 \text{ m}$$

2·5·3. Eccentricity

$$e=\bar{x}_r-\frac{b}{2}=2{\cdot}63-2{\cdot}25=0{\cdot}38 \text{ m}<0{\cdot}75$$

The resultant falls within the middle third.

2·5·4. Pressure Distribution Below the Bsea

$$P_{max}=\frac{W}{B}\left(1+\frac{6e}{b}\right)$$

$$=\frac{18{\cdot}34}{4{\cdot}5}\left(1+6\times\frac{0{\cdot}38}{4{\cdot}5}\right)$$

$$=6{\cdot}15 \text{ t/m}^2$$

$$P_{min}=\frac{18{\cdot}34}{4{\cdot}5}\left(1-\frac{6\times0{\cdot}38}{4{\cdot}5}\right)$$

$$=2{\cdot}02 \text{ t/m}^2$$

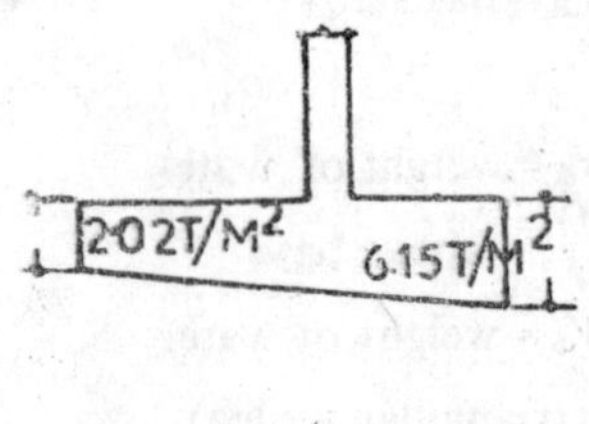

Trial section is safe for case (1)

2·6. Case (2)
Tank empty—dry soil outside, Vertical loads

Load due to	*Magnitude*	*Point of application from 0*	*Moment*
	t	*m*	*mt.*
W_1=stem (rectangular)	$\frac{20}{100}\times 4{\cdot}5\times 1\times\frac{2400}{1000}$ $=2{\cdot}16$	2·9	6·28
W_2=stem (triangular)	$1\times\frac{20}{1000}\times 4{\cdot}5\times 1\times\frac{2400}{1000}$ $=1{\cdot}08$	2·73	2·95
W_3=base slab	$4{\cdot}5\times 1\times\frac{40}{100}\times\frac{2400}{1000}$ $=4{\cdot}32$	2·25	9.75
W_4=weight of earth above base	$\frac{1500}{1000}\times 1{\cdot}5\times 1\times 4{\cdot}5$ $=10{\cdot}5$	3·75	38·00
	$\Sigma W=18{\cdot}06$		$\Sigma M=53{\cdot}98$

2·6·1. Resultant of the Vertical Loads

$$\bar{x}=\frac{56{\cdot}98}{18{\cdot}06}=3{\cdot}15 \text{ m}$$

2·6·2. Horizontal Earth Pressure

$$P=w_e\, h\left(\frac{1-\sin\phi}{1+\sin\phi}\right)\times\tfrac{1}{2}\times h$$

$$=\frac{1500}{1000}\times\frac{1}{3}\times\frac{4{\cdot}5}{2}\times 4{\cdot}5=5{\cdot}05 \text{ t}$$

2·6·3. Point of Application of the Resultant of All Loads

Taking moments about 0′

$$W(\bar{x}-\bar{x}_r)=P\bar{y}$$

$$18{\cdot}06(\bar{x}-\bar{x}_r)=5{\cdot}05\times 1{\cdot}9$$

$$56{\cdot}98-18{\cdot}06\,\bar{x}_r=9{\cdot}6$$

$$\bar{x}_r=\frac{47{\cdot}38}{18{\cdot}06}=2{\cdot}61 \text{ m}$$

2·6·4. Eccentricity

$$e=\bar{x}_r-\frac{b}{2}=2{\cdot}61-2{\cdot}25=0{\cdot}36 \text{ m} < 0{\cdot}75 \text{ m}$$

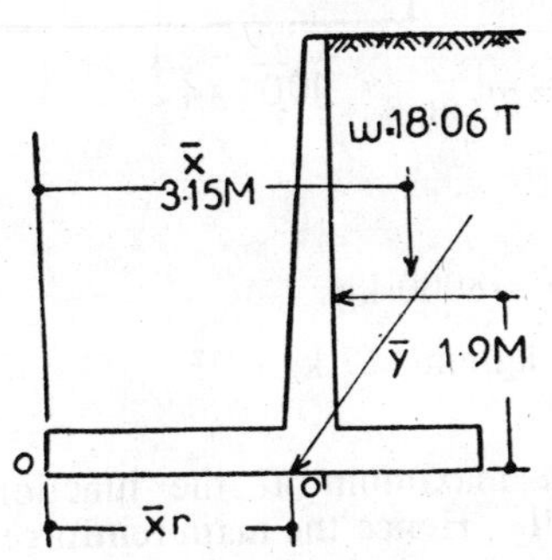

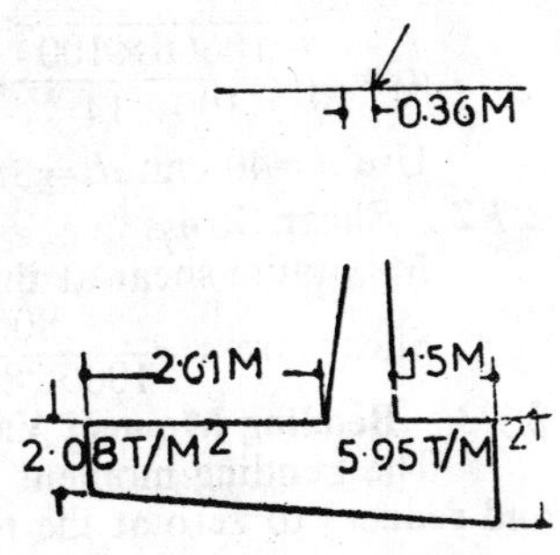

2·6·5. Pressure on the Base

$$P_{max}=\frac{W}{B}\left(1+\frac{6e}{8}\right)=\frac{18\cdot06}{4\cdot5}\left(1+\frac{6\times0\cdot36}{4\cdot5}\right)=5\cdot95 \text{ t/m}^2$$

$$P_{min}=\frac{18\cdot06}{4\cdot5}\left(1-\frac{6\times0\cdot36}{4\cdot5}\right)=2\cdot08 \text{ t/m}^2$$

The trial section is safe for case (2) loading and therefore trial section is suitable.

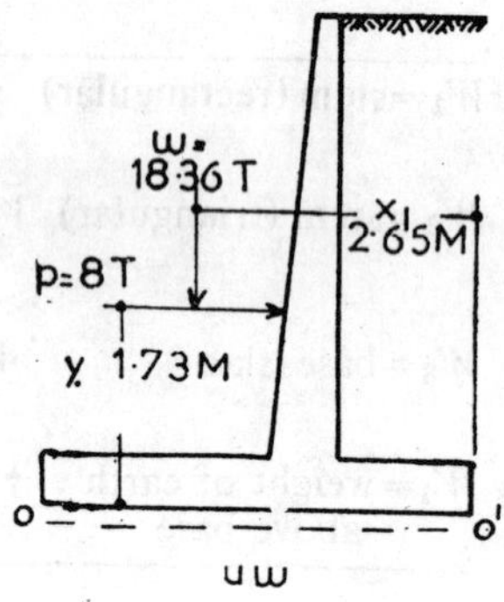

2·6·6. Stability Checks

Factor of safety against overtuning 0,

$$=\frac{\text{stabilizing moment}}{\text{overturning moment}}$$

$$=\frac{W\bar{x}_1}{P\bar{y}_1}=\frac{18\cdot36\times2\cdot65}{8\times1\cdot73}$$

$=3\cdot5 > 1\cdot5$

Factor of safety against sliding

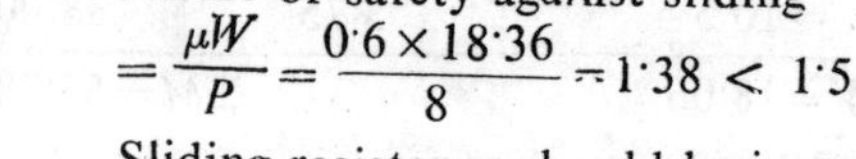

$$=\frac{\mu W}{P}=\frac{0\cdot6\times18\cdot36}{8}=1\cdot38 < 1\cdot5$$

Sliding resistance should be improved by providing proper joints between the base slab and bottom tank slab.

Case (*ii*)

Factor of safety against overturning about 0

$$=\frac{18\cdot06\times3\cdot15}{5\cdot05\times1\cdot9}=5\cdot95>1\cdot5$$

Factor of safety against sliding

$$=\frac{\mu W}{P}=\frac{0\cdot6\times18\cdot06}{5\cdot05}=2\cdot15>1\cdot5$$

Design of Stem Case (1)

Case (*i*) Tank full – no earth outside

Maximum water pressure on the wall

$P=\frac{1}{2}\times4000\times4=8000$ kg

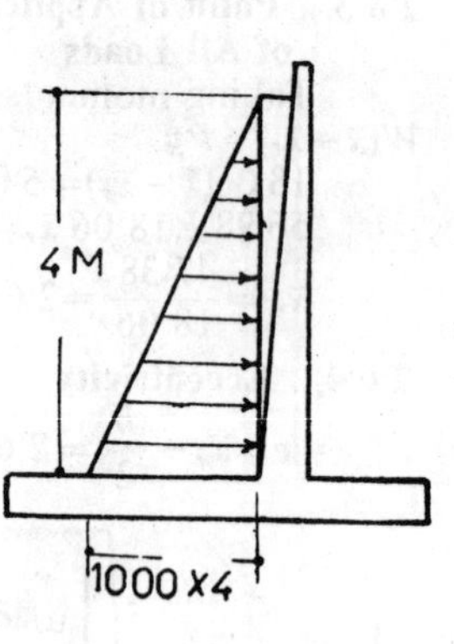

2·7. Bending Moment

Maximum bending moment at the base

$$P\times\frac{h}{3}=8000\times\frac{4}{3}=\frac{32000}{3}$$

$=10666$ m kg

2·7·1. Depth of Stem at the Junction

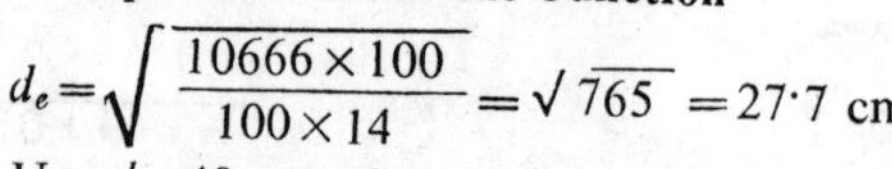

$$d_e=\sqrt{\frac{10666\times100}{100\times14}}=\sqrt{765}=27\cdot7 \text{ cm}$$

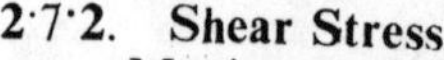

Use $d=40$ cm, $d_e=35$ cm

2·7·2. Shear Stress

Maximum shear at the junction $=P=8000$ kg

$$\text{Shear stress}=\frac{8000}{100\times\cdot84\times35}=2\cdot71 \text{ kg/cm} < 7 \text{ kg/cm}^2$$

2·7·3. Bending Moment Variation

The bending moment in the stem is maximum at the junction and reduces to zero at the top of the wall. Hence the main reinforcement can be suitably reduced towards top and the wall can be tapered reducing the thickness of stem towards the top to achieve economy.

Sample Calculations

Maximum bending moment at any level h from top = $\frac{wh_1^3}{6}$

Calculation for $h_1=1$ m

Bending moment= $\frac{1000\times 1^3}{6}=166{\cdot}6$ m kg

Effective depth needed $d_1=\sqrt{\frac{16660}{100\times 14}}=\sqrt{11{\cdot}9}=3{\cdot}45$ cm

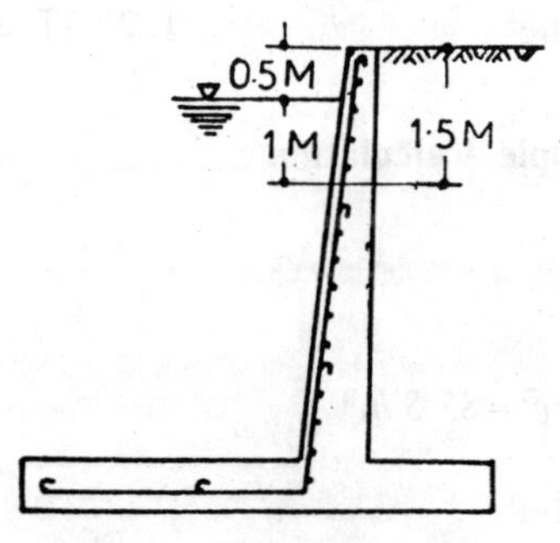

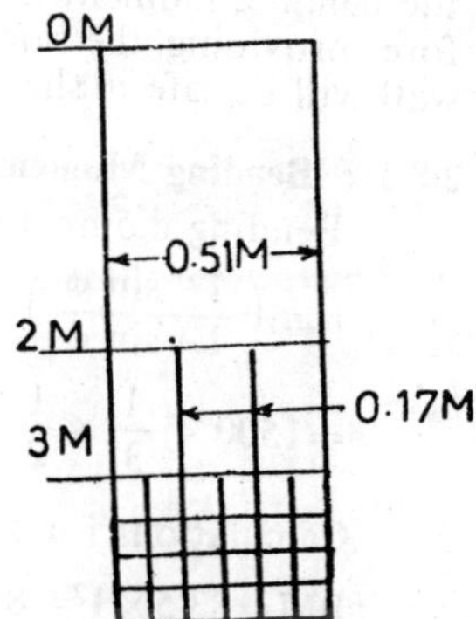

Overall depth provided$=20+\frac{20}{450}\times 150=26{\cdot}67$ cm

Effective depth provided$=26{\cdot}67-5$ (cover) $=21{\cdot}67$ cm

Amount of steel required$=\frac{16660}{1000\times{\cdot}84\times 21{\cdot}67}=0{\cdot}915$ cm^2

2·7·4. Curtailment of Reinforcement

Similarly these computations for different depths at an interval of 1 m are carried out as given in the table.

h_1	BM	effective depth needed	overall depth provided	effective depth provided	area of steel	spacing 20 cm φ	spacing adopted
m	m kg	cm	cm	cm	cm²	cm	
1	166·6	3·45	26·67	21·67	0·915	342	51
2	1333·0	9·75	31·10	26·1	6·1	51·5	51
3	4500·0	18·00	35·5	30·5	17·6	17·8	17
4	10666	27·70	40	35·00	36·2	8·65	8·5

Case (2)

Tank empty – dry earth outside

2·8. Stem Design

Maximum earth pressure on the stem at base $= w_e h \left(\frac{1-\sin\phi}{1+\sin\phi}\right)$

$= 1500 \times 4{\cdot}5 \times \frac{1}{3} = 2250$ kg/m²

$P = \frac{1}{2} \times 2250 \times 4{\cdot}5 = 5050$ kg

Maximum bending moment at base (earth face) $= 5050 \times 1{\cdot}5 = 7600$ m kg

This bending moment is less than the bending moment in case one. Therefore, providing the same thickness of the wall will be safe in shear.

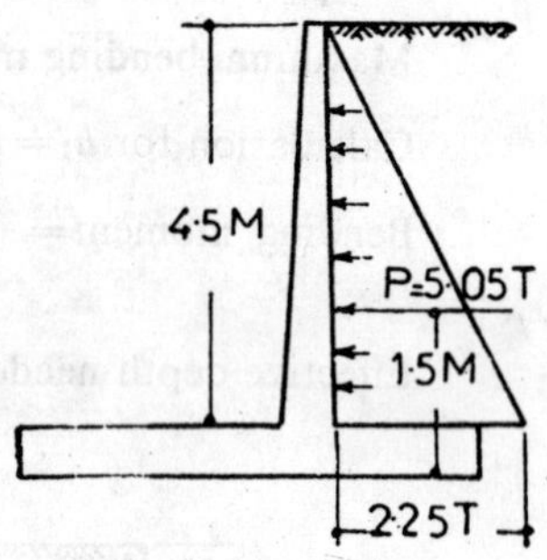

2·8·1. Bending Moment Variation Sample Calculation

Bending moment at any depth h_1

$w_e h_1 \left(\frac{1 \quad \sin\phi}{1+\sin\phi}\right) \times \frac{1}{2} \times h_1 \times \frac{h_1}{3}$

$= 1500 \times \frac{1}{3} \times \frac{1}{2} \times \frac{h_1^3}{3} = \frac{1500}{18} h_1^3 = 83{\cdot}5\, h_1^3$

Calculations for 1 m depth from top

$BM = 83{\cdot}5 \times 1^3 = 83{\cdot}5$ m kg

Effective depth needed $= \sqrt{\frac{8350}{12{\cdot}6 \times 100}} = 2{\cdot}48$ cm

Overall depth provided $= 20 + 20 \times \frac{100}{450} = 24{\cdot}45$ cm

Effective depth provided $= 24{\cdot}45 - 5$ (cover) $= 19{\cdot}45$ cm

Amount of steel required $= \frac{8350}{1250 \times {\cdot}86 \times 19{\cdot}45} = 0{\cdot}385$ cm²

Similarly these computations for different depths at an interval of 1 m are carried out as given in the table.

2·8·2. Curtailment of Main Reinforcement

h_1	BM	effective depth needed	overall depth provided	effective depth provided	area of steel	spacing 20 mm	spacing provided
m	*m kg*	*cm*	*cm*	*cm*	*cm²*	*cm*	*cm*
1	83·5	2·48	24·45	19·45	0·385	82·0	60
2	670·0	7·30	28·9	23·9	2·6	121·0	60
3	2250·0	13·40	33·35	28·35	7·4	42·5	30
4	5325·0	20·60	37·80	32·80	15·2	20·4	15
4·5	7600·0	24·60	40·00	35·00	20·00	15·7	15

2·8·3. Secondary Reinforcement

$A^2 = 0{\cdot}3\%$ of concrete area

$$= \frac{\cdot 3}{100} \times \left(\frac{20+40}{2} \right) \times 100$$

$= 0{\cdot}3 \times 30 = 9$ cm²

A_t per face $= 4{\cdot}5$ cm²

Use 10 mm ϕ 17 cm c/c on each face

0·075M
G₁MR-10-600
2M
G₁MR-10-300
3M
G₁MR-10-150
G₁MR-10-170

2·9. Base Slab Design Case (1) Heel Slab

Tank full—no earth pressure outside

Forces acting on the heel slab

Pressure ordinate at the edge of the wall

$$= 2{\cdot}02 + (6{\cdot}15 - 2{\cdot}02) \times \frac{2{\cdot}6}{4{\cdot}5} = 2{\cdot}02 + 4{\cdot}13 \times \frac{2{\cdot}6}{4{\cdot}5}$$

$= 2{\cdot}02 + 2{\cdot}38 = 4{\cdot}4$ t/m²

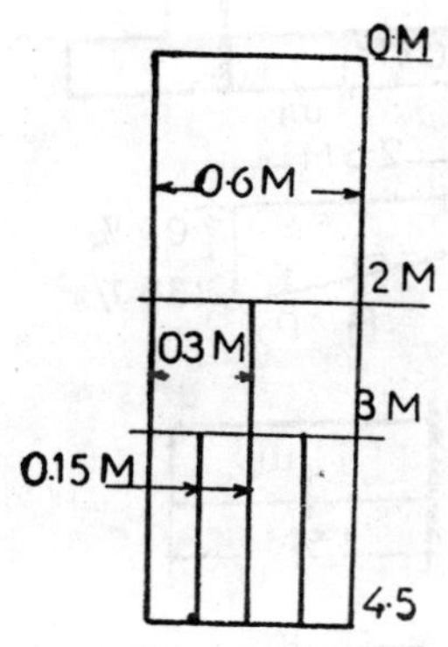

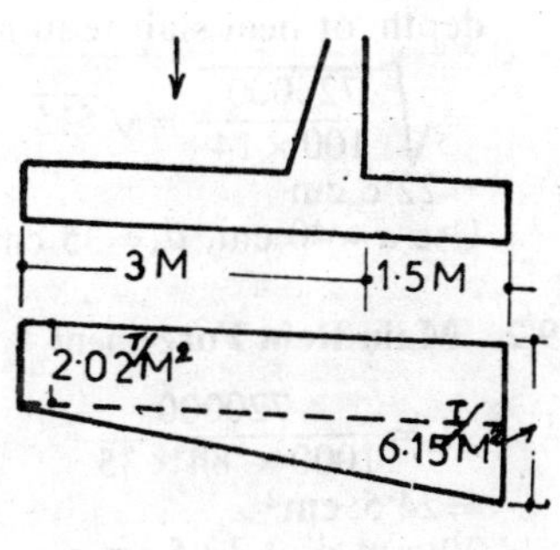

2·9·1. Bending Moments

Bending moment at the junction of the stem (*i*) due to upward forces

Force due to	*Magnitude*	*Point of application*	*Moment*
	t	*m*	*mt*
Rectangular Pressure P_1	$2{\cdot}02 \times 2{\cdot}6 = 5{\cdot}25$	1·3	6·82
Triangular pressure P_2	$\frac{1}{2} \times 2{\cdot}38 \times 2{\cdot}6 = 3{\cdot}1$	0·87	2·70
		Total	9·50

(*ii*) due to downward forces

Force due to	*Magnitude*	*Point of application*	*Moment*
	t	*m*	*mt*
Self weight of heel W_1	$\frac{\cdot 4\times 2\cdot 6\times 1\times 2400}{1000}=2\cdot 5$	1·3	3·20
Water above heel W_2	$\frac{2\cdot 6\times 1000\times 4\times 1}{1000}=10\cdot 4$	1·3	13·50
		Total	16·70

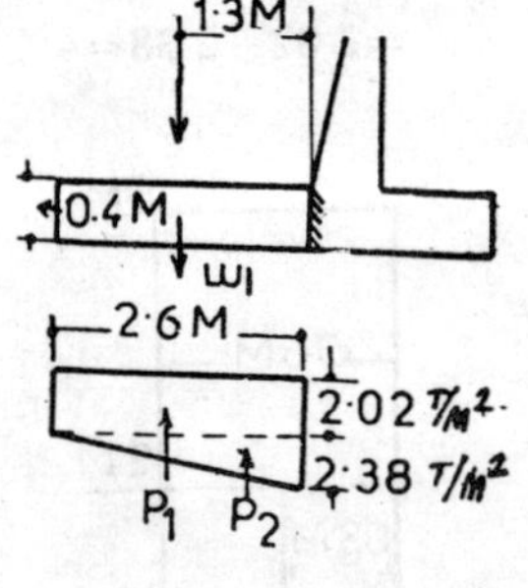

Net bending moment downward (water face)$=16\cdot 70-9\cdot 50=7\cdot 20$ mt
$=720000$ cm kg
depth of heel slab required

$=\sqrt{\frac{720000}{100\times 14}}=\sqrt{515}$

$=22\cdot 6$ cm

Use $d=40$ cm, $d_e=35$ cm

2·9·2. Main Rein Forcement

$A_t=\frac{720000}{1000\times \cdot 84\times 35}$

$=24\cdot 5$ cm²

20 mm ϕ at 12·5 cm c/c

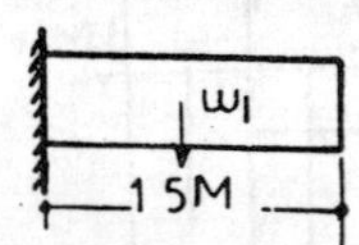

2·10. Te Design

Pressure ordinate at the junction$=2\cdot 02+4\cdot 13\times\frac{3}{4\cdot 5}$

$=2\cdot 02+2\cdot 76=4\cdot 78$ t/m²

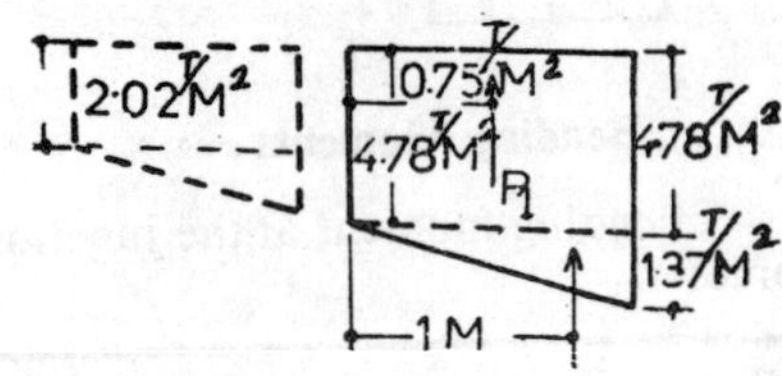

Bending moment upwards due to soil pressure

$=4\cdot 78\times 1\cdot 5\times\frac{1\cdot 5}{2}+\frac{1}{2}\times 1\cdot 37\times 1\cdot 5\times\frac{1\cdot 5\times 2}{3}$

$=5\cdot 375+1\cdot 03=5\cdot 405$ mt

Bending moment downwards due to soil weight

$=\frac{40}{100}\times 1\times 1\cdot 5\times\frac{2400}{1000}\times 0\cdot 75=1\cdot 08$ mt

Net bending moment upwards (tension at bottom face)

$=5.405-1\cdot 08=4\cdot 325$ mt $=432500$ cm kg

depth requird $=\sqrt{\frac{432500}{12\cdot 6\times 100}}=\sqrt{342}=18\cdot 55$ cm

2·11. Base Design Case (2) Heel Design

Use $d=40$ cm and d_e $=35$ cm

$$A_t=\frac{432500}{1250\times\cdot 86\times 35}$$

$=14{\cdot}4\ \text{cm}^2$

16 mm ϕ 13·5 cm c/c at the bottom face.

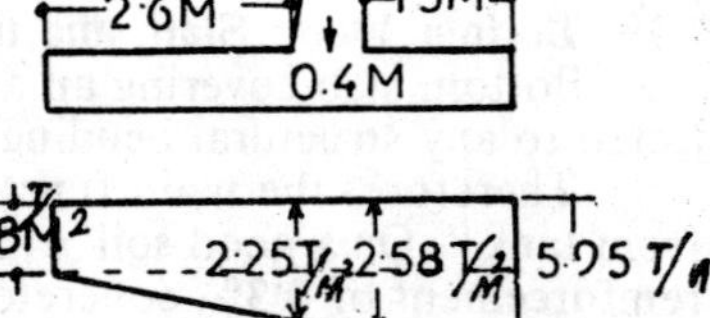

Tank empty – dry earth outside

(*i*) Bending moment due to upward pressure

$$BM=4{\cdot}66\times 1{\cdot}5\times\frac{1{\cdot}5}{2}+\tfrac{1}{2}\times 1{\cdot}29\times 1{\cdot}5\times 1{\cdot}5\times\tfrac{2}{3}=5{\cdot}24+0{\cdot}965$$

$=6{\cdot}205$ mt

(*ii*) Bending moment due to downward loads

$$BM=1{\cdot}5\times\frac{40}{100}\times 1\times\frac{2400}{1000}\times\frac{1{\cdot}5}{2}+4{\cdot}5\times 1\times 1{\cdot}5\times\frac{1500}{1000}\times\frac{1{\cdot}5}{2}$$

$=1{\cdot}08+7{\cdot}6=8{\cdot}68$ mt

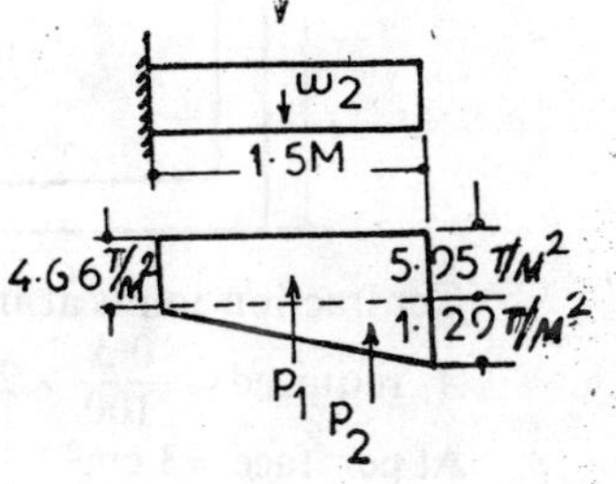

Net bending moment downwards

$=8{\cdot}68-6{\cdot}205=2{\cdot}475$ mt

$=247500$ cm kg

Depth of heel slab required

$$\sqrt{\frac{247500}{12{\cdot}6\times 100}}=\sqrt{197}=14{\cdot}1\ \text{cm}$$

However $d=40$ cm, $d_e=35$ cm has already been provided in case (*i*) therefore use the same dimensions

$$A_t\ \text{required}=\frac{247500}{1250\times\cdot 56\times 35}=6{\cdot}6\ \text{cm}^2$$

Use 10 mm ϕ 11·5 cm c/c

2.12. Toe Design : Bending Moment

Bending moment due to downward force

$$=2{\cdot}6\times 1\times\frac{40}{100}\times\frac{2400}{1000}\times\frac{2{\cdot}6}{2}=3{\cdot}25\ \text{mt}$$

Bending moment due to upward pressure

$$=2{\cdot}08\times 2{\cdot}6\times\frac{2{\cdot}6}{2}+\tfrac{1}{2}\times 2{\cdot}25\times 2{\cdot}6\times\frac{2{\cdot}6}{3}=7{\cdot}01+2{\cdot}54=9{\cdot}64\ \text{mt}$$

Net bending moment upwards (tension on the earth face)

$=9{\cdot}64-3{\cdot}25=6{\cdot}39$ mt

$=639000$ cm kg

Use $d=40$ cm $d_e=35$ cm

A^2 required

$$=\frac{639000}{1250\times\cdot 86\times 35}=16{\cdot}8\ \text{cm}^2$$

Use 16 mm ϕ 11·5 cm c/c at bottom face

G1-MR-10-130

G1-MR-20125

G1-MR-16-115

G1-MR-10-115

$A_t=0{\cdot}3\%$ of concrete area

$$=\frac{0{\cdot}3}{100}\times 40\times 100=12\ \text{cm}^2$$

2·12·1. Secondary Reinforcement

A_t per face=6 cm²

Use 10 mm ϕ 13 cm c/c, both faces

2·13. Bottom Floor Slab and the Base Slabs of Cantilever Side Walls

Bottom slab covering an area of 44·8 m × 19·8 m is not subjected to any structural bending.

Therefore, the main function is to prevent leakage of water from tank. On a good soil a 20 cm thick slab with a minimum reinforcement of 0·3% concrete area distributed in both directions at both the faces will prevent shrinkage and temperature cracks.

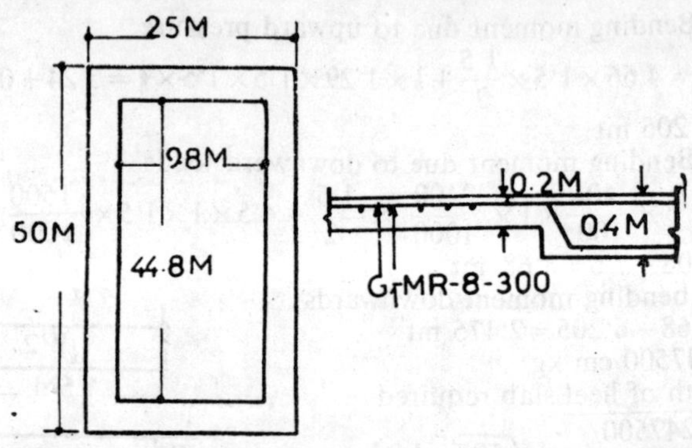

Contraction joints at intervals of 7·5 m shall be provided

$$A_t \text{ required} = \frac{0{\cdot}3}{100} \times 20 \times 100 = 6 \text{ cm}^2$$

At per face = 3 cm²

Use 8 mm ϕ 30 cm c/c at both faces and in both the directions.

VIII. ELEVATED WATER TANKS

General Information 1

1·1. General

(*i*) to act as service reservoirs
(*ii*) to serve as balancing tanks in water supply schemes
and (*iii*) replenishing the tanks of locomotives.

1·2. Advantages

Advantages of concrete water towers are :
concrete (*i*) is not affected by climatic changes
(*ii*) is leak proof
(*iii*) provides greater rigidity
(*iv*) is adoptable for all shapes
(*v*) provides wide choice in design

1.3 Choice of Shape

The cost of the tank is the combined cost of the concrete, reinforcement and shuttering and it might be thought that, if the unit cost of these three items were known, it would be possible to calculate which shape would be cheapest.

However not only do the relative costs vary with economic conditions, but they vary with actual volume of the tank.

1·4. Components of Water Tower

(*a*) Tank portion consists of :
(*i*) Roof and roof beams, (if any)
(*ii*) Sidewalls
(*iii*) Floor or bottom slab, and
(*iv*) Floor beams including circular girders, (if any)
(*b*) Staging portion consists of :
(*v*) Columns
(*vi*) Bracings and
(*vii*) Foundations

1·5. Types

(*i*) Square-open or closed.
(*ii*) Rectangular-open or closed.
(*iii*) Circular-open or closed.

Square and rectangular types are normally used for smaller capacities. The circular types are preferred for larger capacities. The most common type is known as Intze tank.

Circular tanks may have flat floors or domical floors and these are supported on circular girders.

1·6. Design of Components of Tank Portion

The components such as roof slabs, roof beams, side walls are designed on the same principles used in the design of Tanks resting on ground.

Floor slab and beams are designed as per roof slabs and beams for water load and other roof and dead loads.

1·7. Design of Staging or Tower Components General

Diagonal members are seldom used in panels of tower constructed with reinforced concrete. If, however, the diagonals are omitted from towers, stability under lateral loads, can only be attained by filling the panels with R.C. slabs or by designing the towers as rigid frames, the corners being stiffened to resist the negative moments at the junctions of the braces and the columns.

The three dimensional frame work formed by a series of columns arranged on the circumference of one or more circles and braced with ring beams at one, two or more levels is a common type of support for an elevated water tank.

The exact determination of stresses in the frame due to dead loads and wind pressure is too complicated for use in practice.

1·8. Wind Stresses

Wind loads on towers and tanks are taken as per I.S. 875—1964. It is usual to allow a reduction of the pressure intensity and the reduction factors depend on the shape of the tank and tower.

1·9. Approximate Analysis Assumptions

(*i*) The rigidity of a member subjected to bending is directly proportional to EI/L.

(*ii*) The braces are assumed to be stiff enough to hold the columns effectively.

(*iii*) The point of contraflexure is at the mid point of the panel in both columns and braces, when one end of the member is hinged there is of course, no point of contraflexure inthat member.

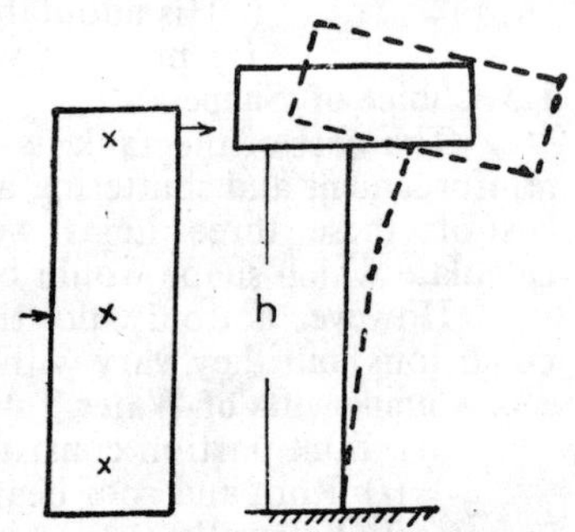

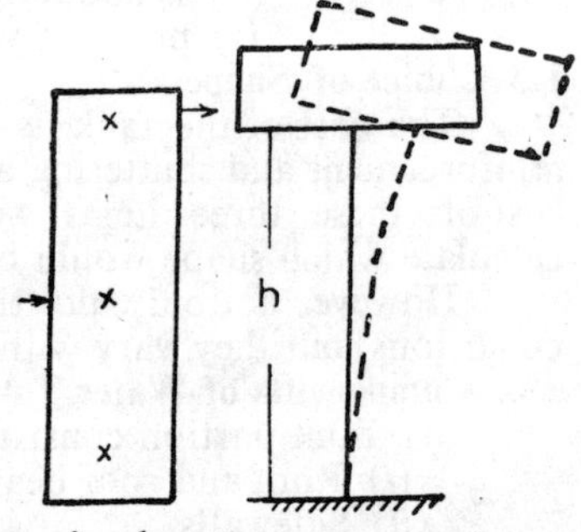

Inner columns take twice as much shear as outer columns, owing to the stiffening effect of double bracing.

1·10. Different Cases

1. Single row of columns

Let P = total wind pressure

n = total number of columns resisting wind pressure

h = height of the column

$$\text{Shear in each column} = \frac{P}{n}$$

$$BM = \frac{P}{n} \times h$$

1·10·1. Case (2)

2. Two rows of columns

(*a*) Hinged at base

$$\text{Shear in each column} = \frac{P}{n}$$

$$BM \text{ at the top of the column} = \frac{P}{n} \times h_1$$

Vertical reaction in the column due to wind load P

Taking moments about B

$$V\times l\ \frac{-2ph}{n}=0$$

$$V=\frac{2P}{n}\ \frac{h}{l}$$

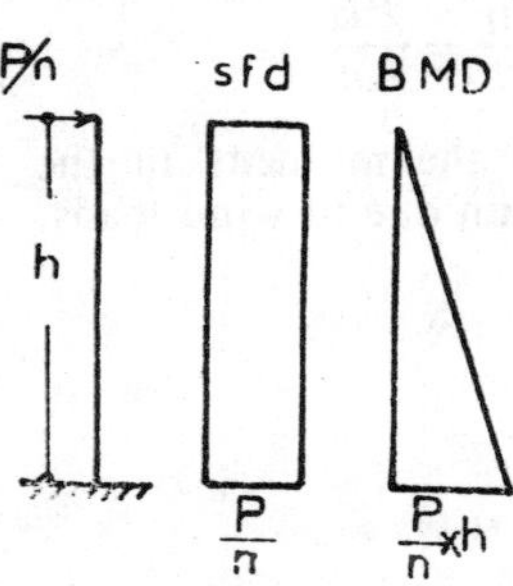

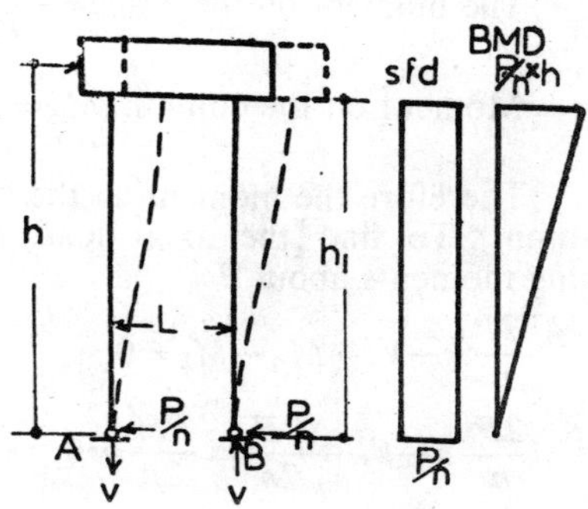

(*b*) Fixed Base.

When the columns are fixed at the base the point of contraflexure in the beam is assumed at mid height of the column.

Let M_A and M_B be the fixing moments at the base of the columns.

$$M_A=M_B=\frac{P}{n}\times\frac{h_1}{2}$$

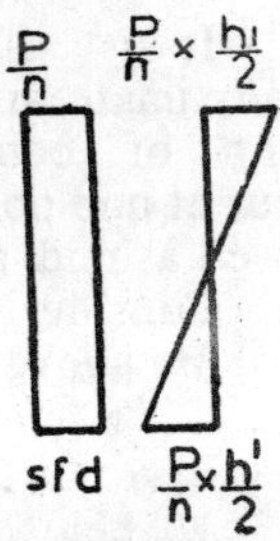

Taking moments about B, the vertical reaction in the column V can be found

$$-Vl+\frac{2P}{n}\times h-M_A-M_B=0$$

$$Vl=\frac{2Ph}{n}-\frac{Ph_1}{2n}-\frac{Ph_1}{2n}$$

$$V=\frac{2P}{nl}\left(h-\frac{h_1}{2}\right)$$

1·11. Design of Columns

If M=bending moment in the column

Q=total axial load in the column due to dead live and wind loads.

A=area of the concrete section of the column.

I=moment of inertia of column including steel area.

Y= extreme fibre distance.

Stress in bending $\sigma_{cb}=M/I\times y$

Stress in axial compression $\sigma_c=Q/A$

The combined stress$=\sigma_c\pm\sigma_{cb}=\dfrac{Q}{A}\pm\dfrac{My}{I}$

should be less than the allowable stresses in compression and tension.

1·12. Braces At Regular Intervals

It is usual practice to place the braces at regular intervals, that is $h_1 = h_2 = h_3 \ldots = H$

Substituting $V_{dc} = \frac{P}{nl} \times 2\,H$

The moment on the brace $= \frac{2PH}{nl} \times \frac{l}{2} = \frac{PH}{n}$

Moment on the column $M_A = M_B = \frac{P}{n} \times \frac{h_1}{2} = \frac{PH}{2n}$

Therefore the moment in the brace is twice the moment in the column. To find the axial load in the column due to wind loads, taking moments about B

$$\frac{2P}{n}h - Vl - M_A - M_B = 0$$

$$\frac{2P}{n}h - Vl - \frac{PH}{2n} - \frac{PH}{2n} = 0$$

$$Vl = \frac{2Ph}{n} - \frac{2PH}{2n}$$

$$= \frac{2P}{nl}\left(h - \frac{H}{2}\right)$$

1·13. Towers With Horizontal Braces Irregular Intervals

Fixed at the base-Two rows of columns.

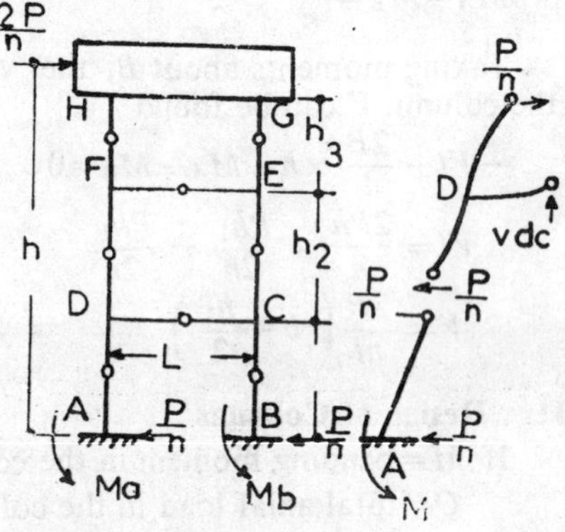

It is assumed in the approximate analysis that the point of contraflexure will occur at mid point of columns as well as mid point of braces.

Consider joint D

M_d above brace

$$= \frac{P_2}{n} \times \frac{h_2}{2}$$

M_d below brace

$$= \frac{P}{n} \times \frac{h_1}{2}$$

For equilibrium at joint D

$\Sigma M = 0$

Let V_{de} = shear in the brace

$$-V_{de} \times \frac{l}{2} + \frac{P}{n} \times \frac{h_2}{2} + \frac{P}{n}\,\frac{h_1}{2} = 0$$

$$\text{or } \frac{P}{2n} - (h_1 + h_2) = V_{de} \times \frac{l}{2}$$

$$V_{de} = \frac{2P}{2n}\,\frac{(h_1 + h_2)}{l} = \frac{P}{nl}(h_1 + h_2)$$

The brace DC has to be designed for a shear $= V_{de}$ and for a moment $= \frac{PH}{n}$

The column has to be designed for a direct load of V and a moment $PH/2n$.

Three or more Rows of columns.

In this case the interior columns being braced on both sides are held more stiffly than the exterior columns and may be assumed to resist double the amount of shear.

Let, n = number of interior columns

$2n$ = number of exterior columns

P = total wind pressure on the tank

S = shear taken up by the exterior column

$2S$ = shear taken up by the interior column.

$$S \times 2n + 2S \times n = P$$

$$4\,Sn = P$$

$$S = \frac{P}{4n}$$

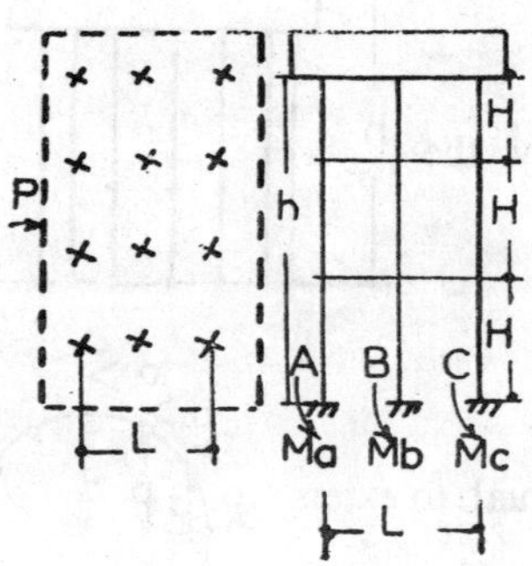

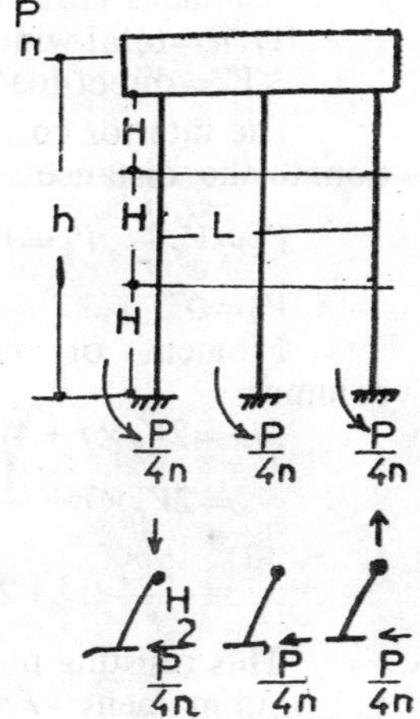

Shear in exterior column $= \frac{P}{4n}$

Shear in interior column $= \frac{P}{2n}$

For one horizontal row of columns wind load

$$= \frac{P}{4n} + \frac{P}{4n} + \frac{P}{2n} = \frac{P}{n}$$

$$M_A = M_C = \frac{P}{4n} \times \frac{H}{2} = \frac{PH}{8n}$$

$$M_B = \frac{P}{2n} \times \frac{H}{2} = \frac{PH}{4n}$$

Direct column load due to wind is zero on column B.

To find V in the exterior columns A_c, take moments about c

$$-Vl + \frac{P}{n} h - M_A - M_B - M_C = 0$$

$$-Vl+\frac{P}{n}h-\frac{PH}{8n}-\frac{PH}{4n}-\frac{PH}{8n}=0$$

$$-Vl+\frac{Ph}{n}-\frac{1}{2}\frac{PH}{n}=0$$

$$Vl=\frac{Ph}{n}-\frac{1}{2}\frac{PH}{n}=\frac{P}{n}\left(h-\frac{H}{2}\right)$$

$$V=\frac{P}{nl}\left(h-\frac{H}{2}\right)$$

1·14. Circular Group of Columns

Types (*i*) columns hinged at the footing
(*ii*) columns fixed at the footing
(*iii*) columns braced horizontally

1·15. Columns Hinged at the Base

(*i*) P=total wind pressure on the tank
V_r=direct load in the outermost column

The interior columns have to carry direct load in direct proportion to the distance

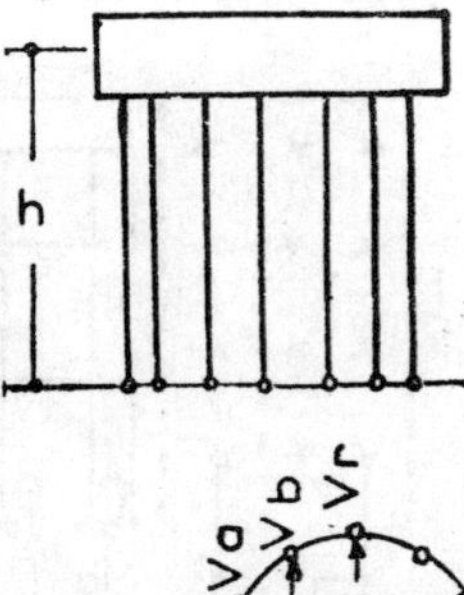

$$V_b=V_r\frac{b}{r},\ V_a=V_r\frac{a}{r}$$

$$V_0=0$$

Moment of resistance offered by the columns

$$M_R=2V_r\times r+4V_b\times b+4V_a a$$

$$=2V_r\times r+\frac{4V_r b^2}{r}+\frac{4V_r a^2}{r}$$

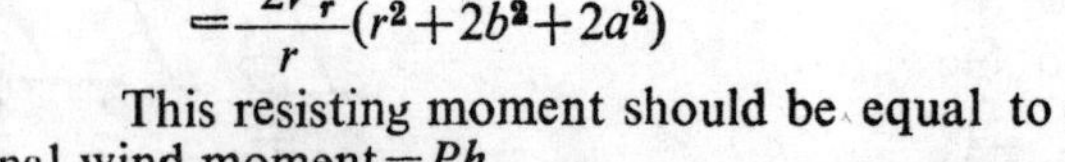

$$=\frac{2V_r}{r}(r^2+2b^2+2a^2)$$

This resisting moment should be equal to external wind moment$=Ph$

$$Ph=\frac{2V_r}{r}(r^2+2b^2+2a^2)$$

Vr=direct load in the column due to wind load can be obtained

$$V_r=\frac{Phr}{2\Sigma r^2}$$

1·16. Column Fixed at Base

Shear per column$=\frac{P}{n}$

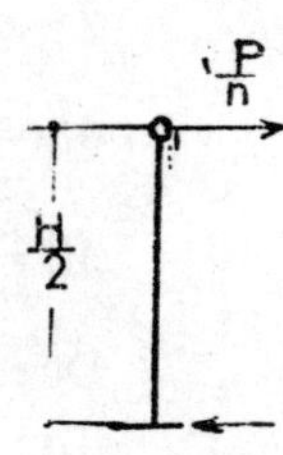

Moment at the base$=\frac{P}{n}\times\frac{H}{2}$

Total moment of resistance

$$=\frac{P}{n}\times\frac{H}{2}\times n=\frac{PH}{2}$$

Due to the wind load

$$=2\frac{V_r}{r}(r^2+2b^2+2a^2)=2\frac{V_r}{r}\Sigma r^2$$

$$\frac{PH}{2}+2\frac{V_r}{r}\Sigma r^2=Ph$$

From which V_r can be determined.

1·17. Base Slab Rectangular Tanks Bending Moments

Slabs spanning in one directions $L/B>2$

Bending moment at the ends of the slab. $M_e=\frac{wh^3}{6}$ and is maximum when the tank is full

Direct pull $T=\frac{wh^2}{2}$

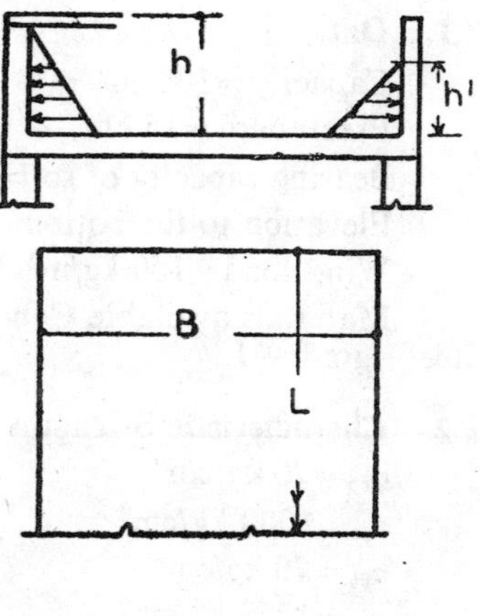

Bending moment at centre or mid span for any given depth of water h_1

$$M_c=\frac{wh_1\,B^2}{8}-\frac{wh_1^3}{6}+w_d\frac{B^2}{8}$$

Where w_d=self weight of the base slab

The bending moment is maximum when

$$\frac{\partial M}{\partial h}=0$$

$$\frac{wB^2}{8}-\frac{3wh_1^2}{6}=0$$

$h_1^2=B^2/4$ $\qquad h_1=B/2$

Substituting, maximum bending moment

$$M_{c\ max}=\frac{wB^3}{24}+\frac{w_dB^2}{8} \text{ and } T_{max}=\frac{wB^2}{8}$$

1·18. Base Slab Square Tanks

Base slabs spanning in two directions $L/B=1$

Bending moment at the centre

$$M_e=\frac{wh_1B^2}{16}-\frac{wh_1^3}{6}+\frac{w_dB^2}{16}$$

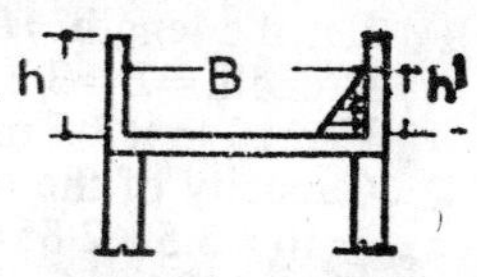

This moment is maximum when $\frac{\partial M}{\partial h_1}=0$

$$\frac{wB^2}{16}-\frac{wh_1^2}{2}=0$$

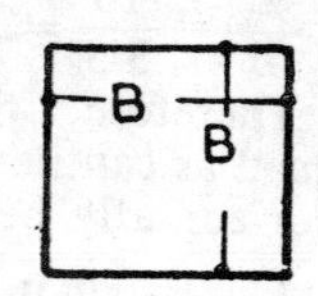

$$h_1^2=\frac{B^2}{8}$$

$$h_1=\frac{B}{2\sqrt{2}}$$

$$M_{e\ max}=\frac{wB^3}{48\sqrt{2}}+\frac{w_dB^2}{16}$$

$$T_{e\ max}=\frac{wh_1^2}{2}=\frac{wB^2}{16}$$

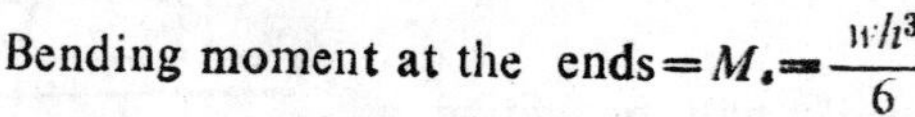

Bending moment at the ends$=M_e=\frac{wh^3}{6}$

Pull $=T_e=\frac{wh^2}{2}=\frac{wB^2}{16}$.

Elevated Open Rectangular Tank Capacity 1,00,000 Litres 2

2·1. Data

Capacity $=100\ m^3$
Free board $=15$ cm
Bearing capacity of soil $=20\ t/m^2$
Elevation to the bottom of the tank above ground level $=15$ m
Wind load $=150\ kg/m^2$
Materials available Concrete—grade M 150 and M 200, Steel—grade—I.

2·2. Characteristic Strengths

$\sigma_{cb}=70\ kg/cm^2$ $\quad m=13$
$\sigma_{st}=1000\ kg/cm^2$ $\quad jd=0{\cdot}84\ d, R=14$
$\sigma_{cb}=70\ kg/cm^2$ $\quad m=13$
$\sigma_{st}=1250\ kg/cm^2$ $\quad jd={\cdot}86\ d, R=12{\cdot}6.$

2·3. Trial Dimensions

Required capacity $=100\ m^3$
Depth of the tank (assumed) $=3$ m
Area of the tank $=33{\cdot}3\ m^2$
Let the length $=L=10$ m
Breadth $=B=3{\cdot}3$ m
Use 10 m × 3·5 m × 3 m tank
Capacity of the tank provided
$=10\times3{\cdot}5\times2{\cdot}85=100$ cu m.

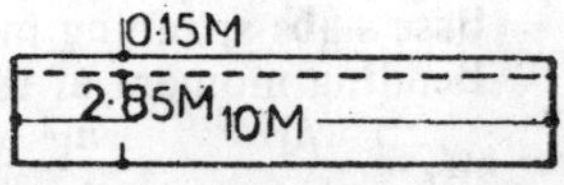

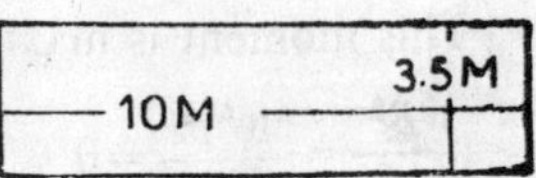

$$\frac{L}{B}=\frac{10}{3{\cdot}5}=2{\cdot}86>2$$

Therefore, the longer walls are designed as cantilevers and shorter walls as spanning horizontally between the longer walls.

2·4. Longer Walls Bending Moment

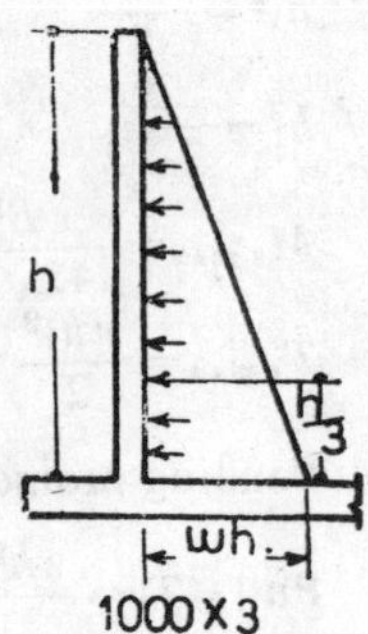

Maximum bending moment

$$=\frac{wh}{6}=\frac{1000\times3\times3\times3}{6}=4500 \text{ m kg}$$

$=450{,}000$ cm kg

2·4·1. Direct Tension

Height over which walls will be acting as cantilever $h'=h/4$ or 1 m

$=\frac{3}{4}$ m <1 m

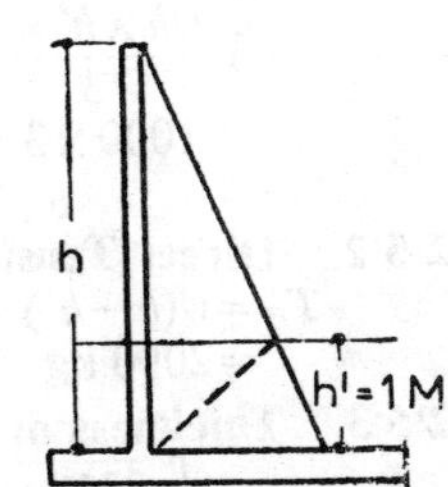

Take $h'=1$ m

Direct tension on the long wall

$$T_L=\frac{w(h-h')B}{2}$$

$$=\frac{1000\times 2\times 3\cdot5}{2}=3500 \text{ kg.}$$

2·4·2. Thickness of the Wall

$$d=\sqrt{\frac{450000}{14\times 1000}}=\sqrt{320}=17\cdot8 \text{ cm}$$

Use 22 cm thick walls and $d_e=18$ cm

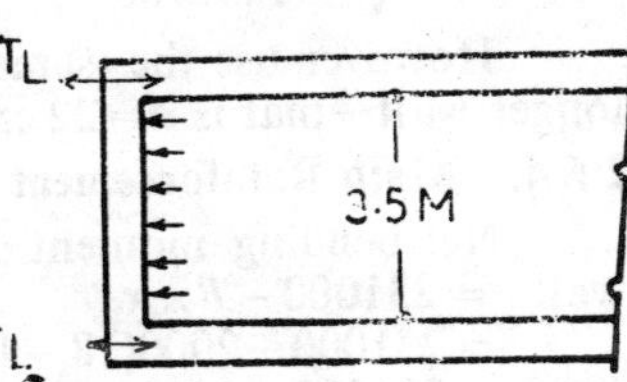

2·4·3. Main Reinforcement

Amount of reinforcement required for bottom 1 m height of wall

$$A_t=\frac{450000}{1000\times\cdot84\times 18}=29\cdot6 \text{ cm}^2$$

Use 20 mm ϕ at 10·5 cm c/c

2·4·4. Curtailment of Main Reinforcement at Different Depths

Depth m	*Bending moment cm kg*	*A_t cm²*	*Spacing of 2o mm ϕ cm*
0—1	16,660	1·11	63
1—2	133,300	8·8	31·5
2—3	450,000	29·6	10·5

2·4·5. Secondary Reinforcement

$$A_t=\frac{\cdot3}{100}\times 22\times 100=6\cdot6 \text{ cm}^2$$

Steel per face$=3\cdot3$ cm²

Use 8 mm ϕ at 15 cm c/c

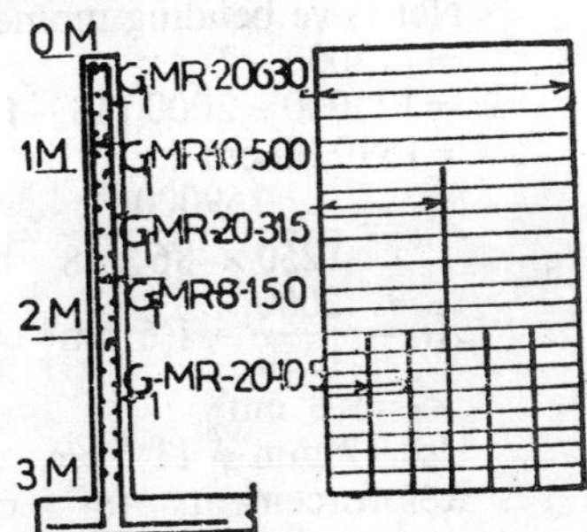

2·4·6. Reinforcement for direct tension

$$A_{tt}=\frac{3500}{1000}=3\cdot5 \text{ cm}^2$$

Use 8 mm ϕ at 14 cm c/c

2·5. Short Walls

Shorter walls span horizontally between the longer walls.

2·5·1. Bending Moments

Maximum +ve bending moment at the centre of the wall

$$M_c=\frac{w(h-h')B^2}{16}=\frac{1000(3\cdot1)3\cdot72^2}{16}$$

$$=\frac{2000\times 3\cdot72^2}{16}=1730 \text{ m kg.}$$

Maximum −ve moment at the ends of the wall

$$M_e=\frac{w(h-h')B^2}{12}=\frac{1000(3-1)3\cdot72^2}{12}=2310 \text{ m kg.}$$

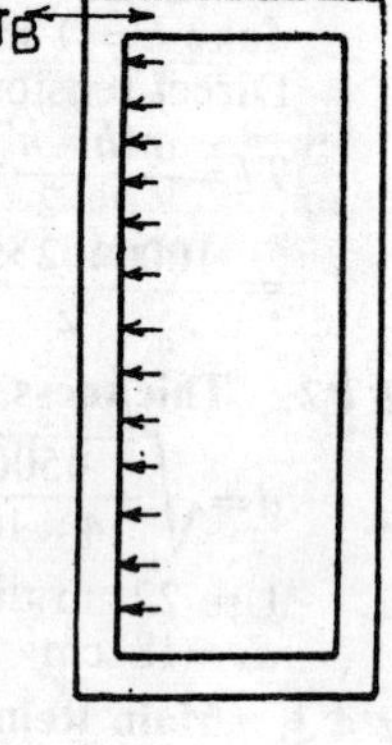

Cantilever moment

$$=\frac{1}{2}\,\frac{wh\times h'\times h'}{3}=\frac{whh_1^2}{6}$$

$$=\frac{1000\times 3\times 1^2}{6}=500 \text{ m kg.}$$

2·5·2. Direct Tension

$$T_B=w(h-h')\times 1=1000(3-1)\ 1$$
$$=2000 \text{ kg}$$

2·5·3. Thickness of the Wall

$$d=\sqrt{\frac{2310\times 100}{14\times 100}}=\sqrt{165}=12{\cdot}8 \text{ cm}$$

However use the same thickness as the longer wall—that is $d=22$ cm, $d_e=18$ cm.

2·5·4. Main Reinforcement

Net bending moment on the shorter wall $=231000-T\times x$

$=231000-2000(18-11)$

$=231000-14000$

$=217{,}000$ cm kg

Steel required to resist bending

$$A_{tb}=\frac{217000}{1000\times{\cdot}84\times 18}=14{\cdot}3 \text{ cm}^2$$

Steel required for direct pull

$$A_{tt}=\frac{2000}{1000}=2 \text{ cm}^2$$

$$A_{tb}+A_{tt}=A_t=16{\cdot}3 \text{ cm}^2$$

Use 16 mm ϕ 12 cm c/c for bottom 1 m height. The spacing of these bars is increased towards top.

Net +ve bending moment

$=173000-T\times x$

$=173000-2000\ (18-11)$

$=1590{,}00$ cm kg.

$$A_{tb}=\frac{159000}{1250\times{\cdot}86\times 18}=8{\cdot}2 \text{ cm}^2$$

$$A_{tt}=\frac{2000}{1250}=1{\cdot}6 \text{ cm}^2$$

$$A_t=9{\cdot}8 \text{ cm}^2$$

Use 12 mm ϕ 11·5 c/c

Reinforcement for cantilever moment

$$=\frac{50000}{1000\times{\cdot}84\times 18}=3{\cdot}3 \text{ cm}^2$$

Use 8 mm ϕ 15 cm c/c vertically over bottom 1 metre length

0 M
G-MR-8-150
G₁MR-16-230
G₁-MR-16-240
1M
G₁-MR-8-150
G₁MR-16-115
G₁MR-16-140
2 M
G₁-MR-16-120
G₁MR-16-115
G₁MR-16-120

2·5·5. Secondary Reinforcement

$$A_t={\cdot}3\% \text{ of concrete area}=\frac{{\cdot}3}{100}\times 22\times 100=6{\cdot}6 \text{ cm}^2$$

Use 8 mm ϕ at 15 cm c/c on each face.

2·6. Base Slab

$\frac{L}{B}=\frac{10{\cdot}22}{3{\cdot}77}>2$ slab spans in the shorter direction.

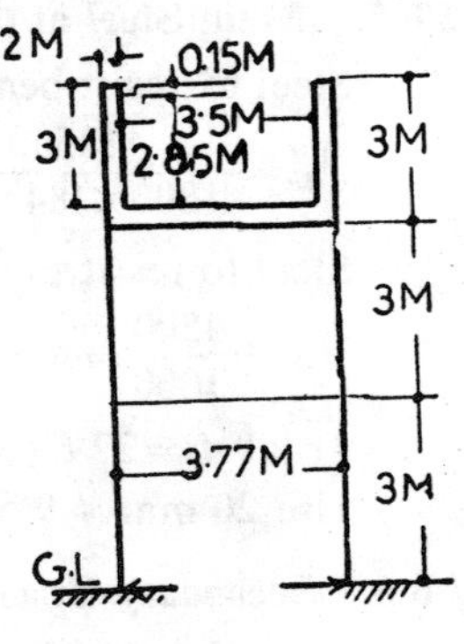

Central Moment

Maximum moment at the centre of the base slab due to water

$$=\frac{wB^3}{24}=\frac{1000\times 3{\cdot}77^3}{24}$$

$=2240$ m kg

Bending moment due to self weight of slab assuming (30 cm)

$$=\frac{30}{100}\times\frac{2400\times 3{\cdot}77^2}{8}=1285 \text{ m kg}$$

Total moment$=2240+1285=3525$ m kg$=352500$ cm kg

2·6·1. Depth of the Slab

$$d_e=\sqrt{\frac{352500}{14\times 100}}=\sqrt{252}=15{\cdot}9 \text{ cm}$$

Use $d=22$ cm, $\quad d_e=18$ cm

$$\text{Pull}=\frac{wB^2}{8}=\frac{1000\times 3{\cdot}77^2}{8}=1780 \text{ kg}$$

Net bending moment$=M-T\times x$

$=352500-1780\times(18-11)$

$=352500-1780\times 7=352500-12460$

$=340,040$ cm/kg

2·6·2. Main Steel at Centre

Steel for bending moment

$$A_{tb}=\frac{340,040}{1250\times{\cdot}86\times 18}=17{\cdot}6 \text{ cm}^2$$

Steel for Pull or direct tension

$$A_{tt}=\frac{1780}{1250}=1{\cdot}42 \text{ cm}^2$$

$$A_t=17{\cdot}6+1{\cdot}42=19{\cdot}02 \text{ cm}^2$$

Use 16 mm ϕ 10·5 cm c/c

2·6·3. Bending Moment at end Section

Maximum ($-$ve) moment at the ends

$$=\frac{wh^3}{6}=\frac{1000\times 3^3}{6}=4500 \text{ m kg}$$

$$\text{Pull}=\frac{wh^2}{2}=\frac{1000\times 3^2}{2}=4500 \text{ kg}$$

Net bending moment$=M-Tx=450000-4500(18-11)$

$=450000-31500=418,500$ cm kg.

2·6·4. Check for Depth

$$d_e=\sqrt{\frac{418,500}{14\times 100}}=\sqrt{298}=17{\cdot}3 \text{ cm} < 18 \text{ cm}$$

2·6·5. Main Steel at the End Section

Steel to resist bending moment (water face)

$$A_{tb}=\frac{418500}{1000\times{\cdot}84\times18}=27{\cdot}6\ \text{cm}^2$$

Steel to resist pull

$$A_{tt}=\frac{4500}{1000}=4{\cdot}5\ \text{cm}^2$$

Total $A_t=27{\cdot}6+4{\cdot}5=32{\cdot}1\ \text{cm}^2$

Use 20 mm ϕ 9·5 cm c/c

2·6·6. Secondary Steel

$$A_t=\frac{{\cdot}3}{100}\times22\times100=6{\cdot}6\ \text{cm}^2$$

Use 8 mm ϕ 15 cm c/c on both faces

2·7. Floor Beams Loads

Loads : weight of water = 100 t

Weight of side walls

$$=2(10{\cdot}22+3{\cdot}77)\times3\times\frac{2400}{1000}\times\frac{22}{100}$$

$$=2\times13{\cdot}99\times3\times2{\cdot}4\times{\cdot}22=44{\cdot}25\ \text{t}$$

Weight of base slab

$$=10{\cdot}44\times3{\cdot}99\times\frac{22}{100}\times\frac{2400}{1000}$$

$$=10{\cdot}44\times3{\cdot}99\times{\cdot}22\times2{\cdot}4=21{\cdot}8\ \text{t}$$

Total loads

$$=100+44{\cdot}25+21{\cdot}8=166{\cdot}05\ \text{t}$$

4 columns will be sufficient.

2·7·1. Beam B₁

Span = 10·22 m

Loads : From water, side walls, and base slab

$$=166{\cdot}05\times\frac{(6{\cdot}45+10{\cdot}22)}{2}\times\frac{3{\cdot}77}{2}\times\frac{1}{10{\cdot}22\times3{\cdot}77}$$

$$=166{\cdot}05\times\frac{16{\cdot}67}{2}\times\frac{3{\cdot}77}{2}\times\frac{1}{10{\cdot}22\times3{\cdot}77}=68\ \text{t}$$

Total self weight of the beam (assuming 80×40 cm)

$$=\frac{80}{100}\times\frac{40}{100}\times\frac{2400}{1000}\times10{\cdot}22=7{\cdot}8\ \text{t}$$

W = the ordinate of the trapizoidal loading

$$2\times\tfrac{1}{2}\times\frac{3{\cdot}77}{2}\times W+W\times6{\cdot}45=68$$

$$1{\cdot}885\,W+6{\cdot}45\,W=68$$

$$8{\cdot}335\,W=68$$

$$W=68/8{\cdot}335=8{\cdot}15\ \text{t}$$

2·7·2. Bending Moment

Reaction at $A = \frac{68}{2} + \frac{7·8}{2} = 34 + 3·9 = 37·9$ t

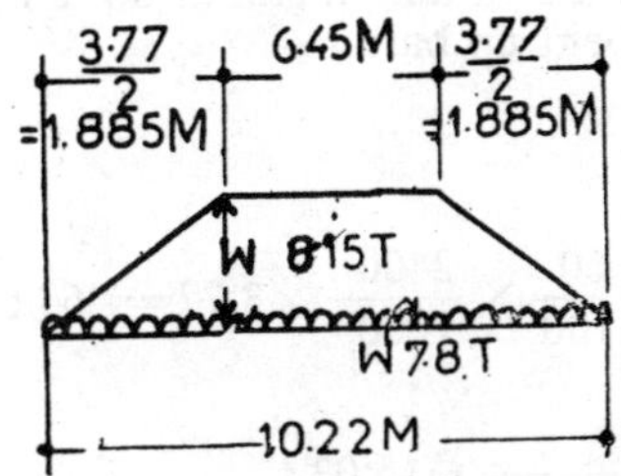

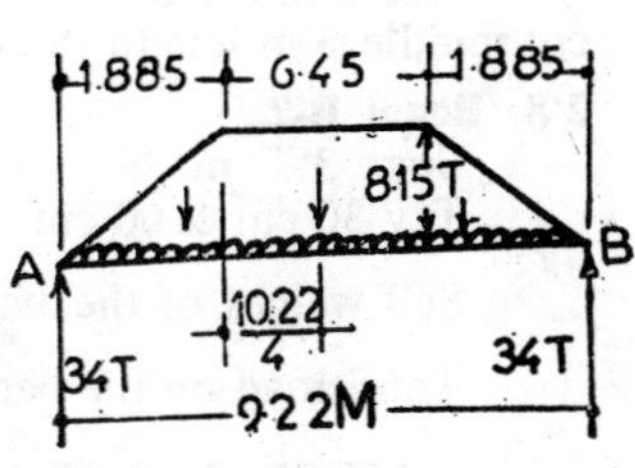

Maximum bending moment

$= 37·9 \times \frac{10·22}{2} - \frac{1}{2} \times 8·15 \times 1·885$

$\times \left(\frac{6·45}{2} + \frac{1}{3} \times 1·885\right) - \frac{6·45}{2} \times 8·15 \times \frac{6·45}{4} - \frac{7·8}{2} \times \frac{10·22}{4}$

$= 194 - 29·75 - 42·4 - 10 \quad 194 - 82·15 = 111·85$ mt

$= 11{,}185{,}000$ cm kg

Maximum shear force $= 37·9$ t

2·7·3. Depth of the Beam

Using $M200$, $d_e = \sqrt{\frac{11185000}{12·1 \times 40}} = \sqrt{23000}$

$= 151 > 80$ cm assumed

Assume 130 cm × 65 cm

Additional load $= \frac{25}{100} \times \frac{50}{100} \times \frac{2400}{1000} \times 10·22 = 3·52$ t

Additional $BM = \frac{3·52 \times 10·22}{8} = 4·95$ mt $= 495{,}000$ cm kg

Addition $SF = \frac{3·52}{2} = 1.76$ t

Maximum $SF = 37·9 + 1·76 = 39·66$ t

Maximum revised beam $= 11185000 + 495000 = 11680{,}000$ cm kg

$d_e = \sqrt{\frac{11680000}{12·1 \times 65}} = \sqrt{14800} = 122$ cm

Use $d = 130$ cm, $d_e = 122$ cm

2·7·4. Main Reinforcement

$A_t = \frac{11680000}{1400 \times ·87 \times 122}$

$= 79 \text{ cm}^2$

Use 10 Nos – 32 mm ϕ in two layers

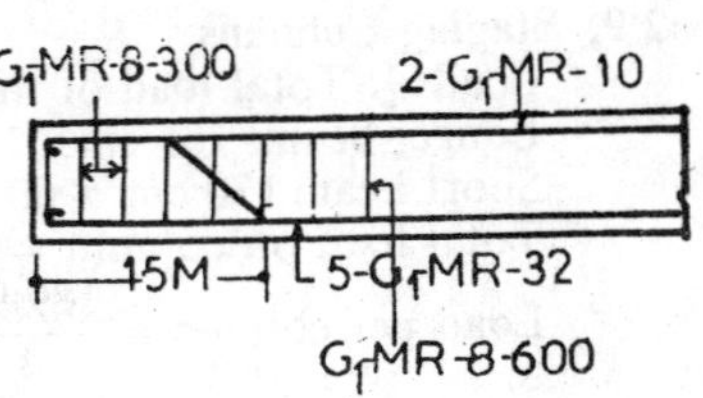

2·7·5. Shear Reinforcement

Maximum shear at the support $= 39·66$ t $= 39660$ kg

$$\text{Shear stress}=\frac{39660}{65\times\cdot87\times122}=5{\cdot}76\ \text{kg/cm}^2<7\ \text{kg/cm}^2$$

2·7·6. Nominal Stirrups

Use 8 mm ϕ at 30 cm c/c at the end 2.5 m length and at 60 cm c/c middle 5 m length in addition to the bent up bars.

2·8. Beam B-2

Span 3·77 m.

Try 30 cm × 60 cm

$$\text{Self weight of the beam}=\frac{30}{160}\times\frac{60}{100}\times\frac{2400}{1000}\times3{\cdot}77=1{\cdot}63\ \text{t}$$

Total load on the base

$$=166{\cdot}05\times\tfrac{1}{2}\times3{\cdot}77\times\frac{3{\cdot}77}{2}\times\frac{1}{10{\cdot}22\times3{\cdot}77}=15{\cdot}03\ \text{t}$$

Maximum *SF* Reaction R_A

$$=\frac{15{\cdot}03}{2}+\frac{1{\cdot}63}{2}=7{\cdot}515+{\cdot}815$$

$=8{\cdot}330$ t

Maximum bending moment

$$=8{\cdot}33\times\frac{3{\cdot}77}{2}-\frac{15{\cdot}03}{2}\times\tfrac{1}{3}\times\frac{3{\cdot}77}{2}$$

$$-\frac{1{\cdot}63}{2}\times\frac{1{\cdot}63}{4}$$

$=15{\cdot}8-4{\cdot}75-{\cdot}33=10{\cdot}72$ mt $=1072000$ cm kg

2·8·1. Depth of the Beam

$$d_e=\sqrt{\frac{1072000}{12{\cdot}1\times30}}=\sqrt{2950}=54{\cdot}1\ \text{cm}$$

Use $d=60$ cm, $d_e=55$ cm

2·8·2. Main Steel

$$A_t=\frac{1072000}{1400\times\cdot87\times55}=16\ \text{cm}^2$$

Use 8 Nos 16 mm ϕ

2·8·3. Shear Reinofrcement

$$\text{Shear stress}=\frac{8330}{30\times\cdot87\times55}$$

$=5{\cdot}8$ kg/cm^2 <7 kg/cm^2

Use nominal shear stirrups.

8 mm ϕ at 20 mm c/c for 1 m length at supports and at 40 cm c/c over the remaining length.

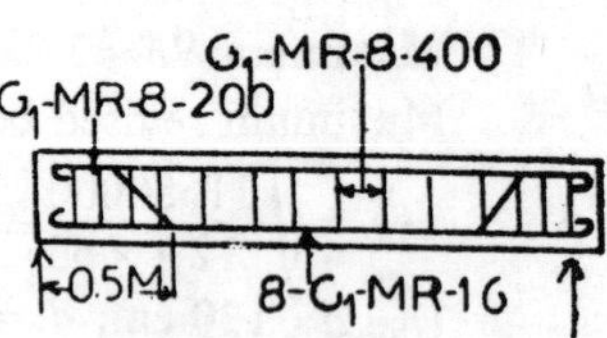

2·9. Staging Columns

Loads : Total load of the tank including water $=166{\cdot}05$ t

Longer beams (65 cm × 130 cm) 2 Nos $=10{\cdot}32\times2=20{\cdot}64$ t

Short beam (30 cm × 60 cm) 2 Nos $=1{\cdot}67\times2=3{\cdot}26$ t

Total load on column $=189{\cdot}95$ t

$$\text{Load per column}=\frac{189{\cdot}95}{4}=47{\cdot}49\ \text{t}$$

2·9·1. Wind Loads

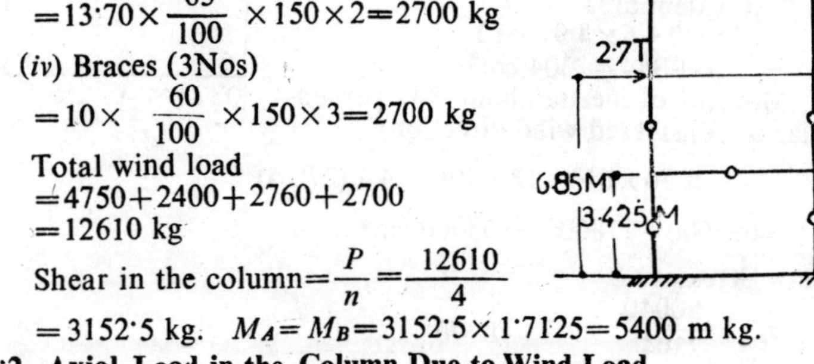

Wind loads on

(i) Tank portion $=150\times3\times10{\cdot}44$
$=4750$ kg

(ii) Base slab and longer beam $=10{\cdot}44\times1{\cdot}52\times150$
$=2400$ kg

(iii) Columns (2 Nos)

$$=13{\cdot}70\times\frac{65}{100}\times150\times2=2700 \text{ kg}$$

(iv) Braces (3Nos)

$$=10\times\frac{60}{100}\times150\times3=2700 \text{ kg}$$

Total wind load
$=4750+2400+2760+2700$
$=12610$ kg

$$\text{Shear in the column}=\frac{P}{n}=\frac{12610}{4}$$

$=3152{\cdot}5$ kg. $M_A=M_B=3152{\cdot}5\times1{\cdot}7125=5400$ m kg.

2·9·2. Axial Load in the Column Due to Wind Load

$$V\times l-M_A-M_B+\frac{4750}{2}\times16{\cdot}72+\frac{2400}{2}\times14{\cdot}46+\frac{5460}{2}\times6{\cdot}85=0$$

$$V\times3{\cdot}77-5400-5400+39600+17500+18700=65000$$

$$V=\frac{65000}{3{\cdot}77}=17{,}250 \text{ kg}=17{\cdot}25 \text{ t}$$

Maximum axial load in the leeward column due to dead and wind loads $=47{\cdot}49+17{\cdot}25=64{\cdot}74$ t

Self weight of the column (65 cm $\times$ 65 square)

$$=\frac{65}{100}\times\frac{65}{100}\times\frac{2400}{1000}\times15=15{\cdot}3 \text{ t}$$

Total axial load on the column $=64{\cdot}74+15{\cdot}3=80{\cdot}04$ t.

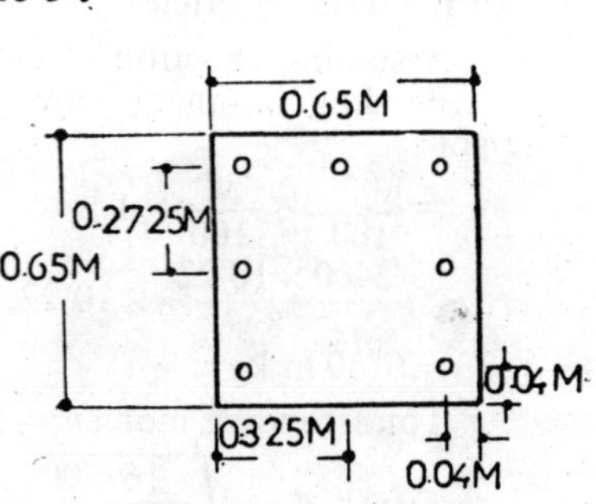

2·9·3.. Bending Moment

Bending moment at the base of the column $=5400$ m kg
$=540000$ cm kg

2·9·4. Trial Section

Try 65 cm$\times$65 cm column vith 8 – 25 mm ϕ

Equivalent area
$=65\times65+12\times8\times4{\cdot}91$
$=4225+472=4697$ cm^4

$$\text{Moment of interia}=\frac{1}{12}\,65\times65^3+12\times6\times4{\cdot}91(27{\cdot}25)^2$$

$=1500{,}000+268{,}000=178{,}000$ cm^4

$$\sigma_c'=\frac{P}{A}=\frac{80040}{4697}=17{\cdot}2 \text{ kg/cm}^2$$

$$\sigma_{cb}' = \frac{My}{I} = \frac{540000 \times 32{\cdot}5}{1768000} = 9{\cdot}91 \text{ kg/cm}^2$$

$$\frac{\sigma_c'}{\sigma_c} + \frac{\sigma_{cb}'}{\sigma_{cb}} = \frac{17{\cdot}2}{66{\cdot}7} + \frac{9{\cdot}91}{93{\cdot}3} = {\cdot}258 + {\cdot}103 = {\cdot}361 < 1$$

Section is too uneconomical

Try 65 cm × 30 cm rectangular section with 6 – 25 mm ϕ

Equivalent area

$= 65 \times 30 + 6 \times 4{\cdot}91 \times 12$

$= 1950 + 354 = 2304 \text{ cm}^2$

Moment of inertia about XX (perpendicular to considered wind direction)

$$= \frac{1}{12} \times 30 \times 65^3 + 12 \times 4{\cdot}91 \times 4 \times (27{\cdot}25)^2$$

$= 685000 + 178000 = 863000 \text{ cm}^4$

2·9.5. Stresses

$$\sigma_c' = \frac{80040}{2304} = 34{\cdot}6 \text{ kg/cm}^2$$

$$\sigma_b' = \frac{540{,}000 \times 32{\cdot}5}{863{,}000} = 20{\cdot}25 \text{ kg/cm}^2$$

$$\frac{\sigma_c'}{\sigma_c} + \frac{\sigma_{cb}'}{\sigma_{cb}} = \frac{34{\cdot}6}{66{\cdot}7} + \frac{20{\cdot}25}{93{\cdot}3} = {\cdot}520 + {\cdot}217 = 0{\cdot}737 < 1$$

Use 65 cm × 30 cm with 6 - 25 mm ϕ

Use 8 mm ϕ ties at 30 cm c/c

2·10. Braces

Maximum bending moment in the brace $= 3152{\cdot}5 \times 1{\cdot}7125 \times 2$

$= 5400 \times 2 = 10{,}800$ m kg

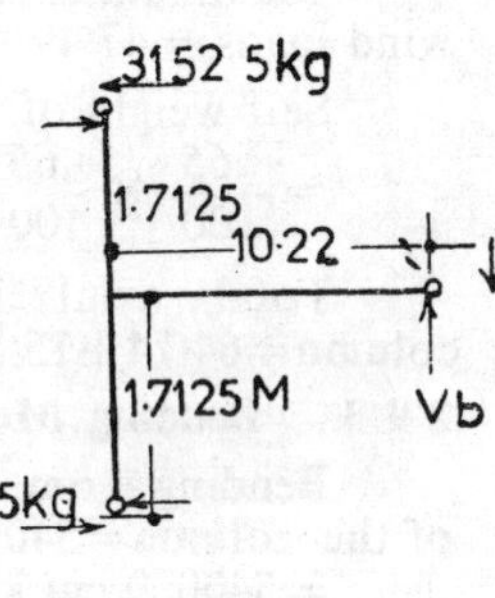

2·10·1. Trial Section

Assuming a section 65 cm × 30 cm

Bending moment due to self weight

$$= \frac{65}{100} \times \frac{30}{100} \times 1 \times \frac{2400 \times 10{\cdot}22}{8} \times 10{\cdot}22$$

$= 6150$ m kg

Total bending moment $= 10800 + 6150 = 16950$ m kg.

$$\text{Depth } d_e = \sqrt{\frac{1695000}{30 \times .121}}$$

$= \sqrt{4660} = 68$ cm

Use $d = 65$ cm. $d_e = 55$ cm.

$$A_t \text{ required} = \frac{1695000}{1400 \times 55}$$

$= 22 \text{ cm}^2$

Use 8 bars of 20 mm at botto m and top,

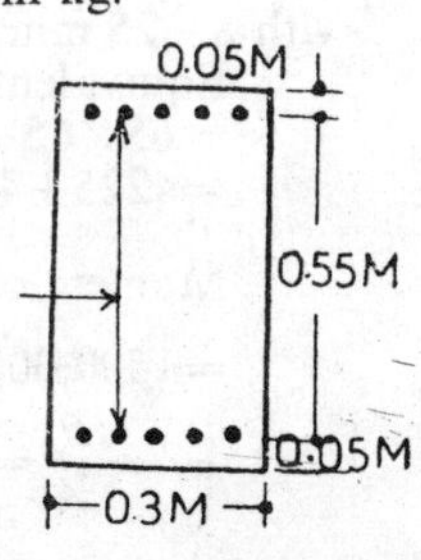

2·10·2. Shear Force

Shear force in the longer brace

$$V_b \times \frac{10{\cdot}22}{2} = 16950$$

$$V_b = \frac{16950 \times 2}{10{\cdot}22} = 2280 \text{ kg}$$

$$\text{Shear stress} = \frac{2280}{30 \times 55 \times {\cdot}87} = 1{\cdot}59 \text{ kg/cm}^2$$

Use nominal stirrups 8 mm ϕ at 30 cm c/c.

2·11. Shorter Brace

The same section can be used for shorter brace also.

Bending moment due to wind loads
$= 10800$ m kg

Self weight bending moment

$$= \frac{65}{100} \times \frac{30}{100} \times 1 \times 2400 \times \frac{3{\cdot}77 \times 3{\cdot}77}{8} = 832 \text{ m kg}$$

Total BM$=10800+832=11632$ m kg

$$A_t = \frac{1163200}{1400 \times 55} = 15{\cdot}2 \text{ cm}^2$$

5 Nos$-$20 mm ϕ at top and bottom

2·11·1. Shear Force

$$V_b \times \frac{3{\cdot}77}{2} = 11632$$

$$V_b = \frac{11632 \times 2}{3{\cdot}77} = 6200 \text{ kg}$$

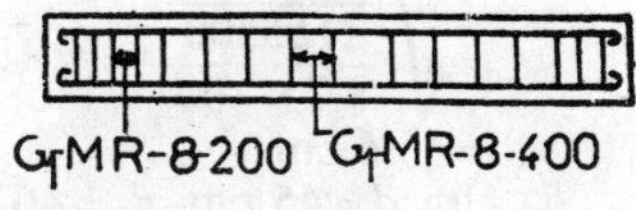

$$\text{Shear stress} = \frac{6200}{55 \times 30} = 3{\cdot}74 \text{ kg/cm}^2 < 7 \text{ kg/cm}^2$$

Use nominal stirrups 8 mm ϕ at 20 cm over 1 m length at supports and at 40 cm in the rest of the span.

2·12. Foundation Isolated Footings

Total load at the base of the column$=80{\cdot}04$ t

Add 10% for footings 8·0 t

Total$=88{\cdot}04$ t, Say 88 t

Moment at the base of the column$=5400$ m kg

$$\text{Eccentricity} = \frac{540000}{88000} = 6{\cdot}15 \text{ cm} = {\cdot}0615 \text{ m}$$

2·13. Area for the Footings

Approximate area for the footings

$$= \frac{88}{20} = 4{\cdot}4 \text{ m}^2$$

Use 3 m$\times$1·75 m size $A=5{\cdot}25$ m²

Foundation pressures

$$\text{Max. pressure} = \frac{P}{A}\left(1 + \frac{6e}{B}\right)$$

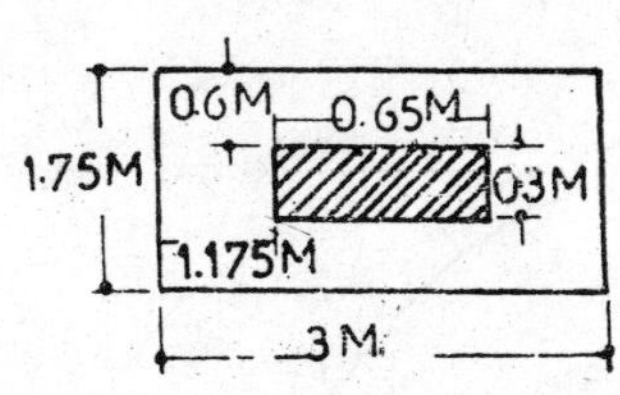

$$=\frac{88}{5{\cdot}25}\left(1\pm\frac{6\times0\,615}{3}\right)=18{\cdot}9\ \text{t/m}^2$$

$$\text{Min. pressure}=\frac{88}{5{\cdot}25}\times{\cdot}877=14{\cdot}7\ \text{t/m}^2$$

Soil pressure at the edge of the footing

$$=14{\cdot}7+(18{\cdot}9-14{\cdot}7^2\times\frac{1{\cdot}825}{3}=14{\cdot}7+4{\cdot}2\times\frac{1{\cdot}825}{3}$$

$$=14{\cdot}7+2{\cdot}55=17{\cdot}25\ \text{t/m}^2$$

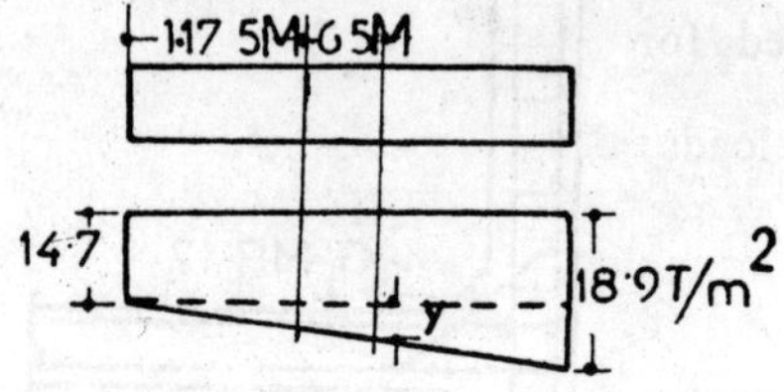

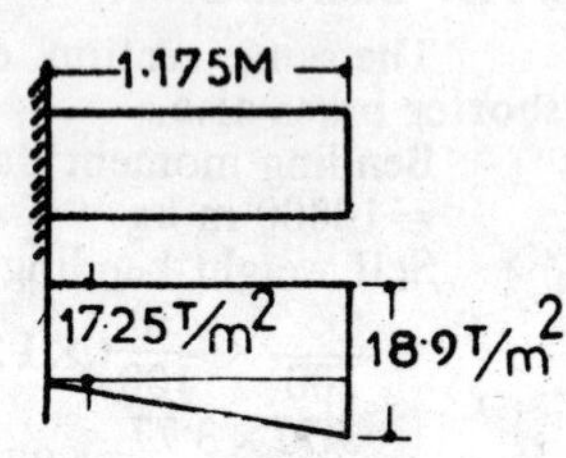

Bending moment at the edge of the footing

$$=(18{\cdot}9-17{\cdot}25^2\times\tfrac{1}{2}\times1{\cdot}75\times\tfrac{2}{3}\times1{\cdot}75+17{\cdot}25\times1{\cdot}175\times\frac{1{\cdot}175}{2}$$

$$={\cdot}755+0{\cdot}8=11{\cdot}555\ \text{mt}=1155500\ \text{cm kg}$$

Using M 150

Depth of footing

$$=\sqrt{\frac{1155500}{8{\cdot}7\times100}}=\sqrt{1320}$$

$=36{\cdot}6$ cm

Use $d=45$ cm, $d_e=40$ cm

$$A_t=\frac{1155500}{1400\times{\cdot}87\times40}$$

$=23{\cdot}6\ \text{cm}^2$

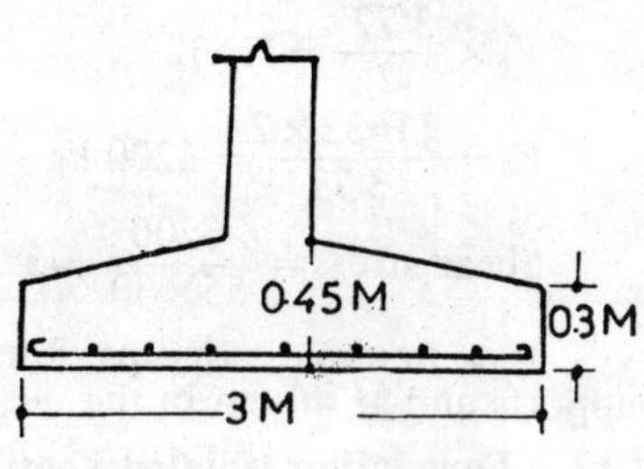

Use 20 mm ϕ at 13 cm c/c both ways.

Closed Rectangular Tanks 3
Capacity 350,000 Litres

3·1. Data

Capacity = 350 m^3
Free board = 20 cm
Bearing capacity of soil = 20 t/m^2
Elevation to the bottom of the floor beams of the tank above $GL = 12$ m.
Wind load = 150 kg/m^2
Materials available, Concrete – grade—M 150, M 200, Steel—grade—1.

3·2. Characteristic Strengths

$\sigma_{cb} = 70$ kg/cm^2 $M = 13$
$\sigma_{st} = 1000$ kg/cm^2 $jd = \cdot 84\ d$ $R = 14$
$\sigma_{cb} = 70$ kg/cm^2 $M = 13$
$\sigma_{st} = 1250$ kg/cm^2 $jd = 0{\cdot}86\ d$ $R = 12{\cdot}6$

3·3. Trial Dimensions

Required capacity = 350 m^3
Depth of tank (assumed) = 3·5 m
Area of the tank = 100 m^2
Using 15·5 m × 7 m × 3·3 m
Actual capacity provided = 358 $m^3 > 350$ m^3

$$\frac{L}{B} = \frac{15{\cdot}6}{7} > 2,\quad \frac{L}{h} = \frac{15{\cdot}5}{3{\cdot}5} > 2,\quad \frac{B}{h} = \frac{7}{2{\cdot}5} > 2$$

The walls can therefore be designed as propped cantilevers.

3·4. Roof Slab

Use slab and T-beam roof
Slab Span = 3·2 m

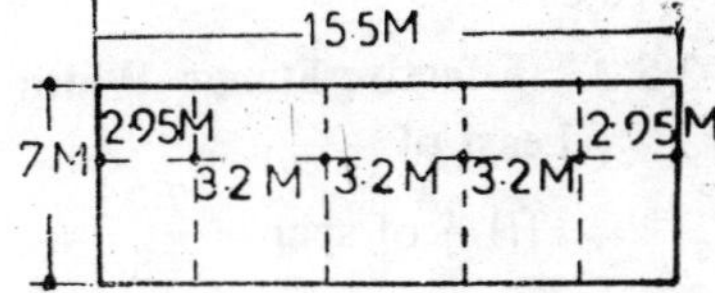

3·4·1. Loads

Loads : Live load = 150 kg/m^2
Screeding (5 cm) = 110 kg/m^2
Self weight (10 cm)

$$= \frac{10}{100} \times 2400 = 240 \text{ kg/m}^2$$

Total load = 500 kg/m^2.

3·4·2. Bending Moment

Slab is designed as a continuous slab

$$\text{Maximum bending moment} = \frac{wl^2}{10}$$

$$= \frac{500 \times 3{\cdot}2^2}{10} = 512 \text{ m kg}$$

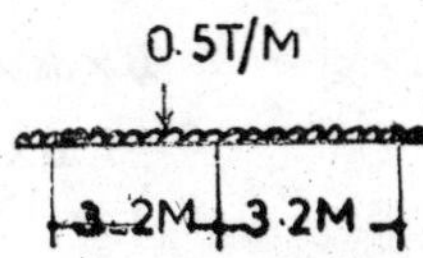

3·4·3. Depth of Slab

Using M 150 concrete for roof

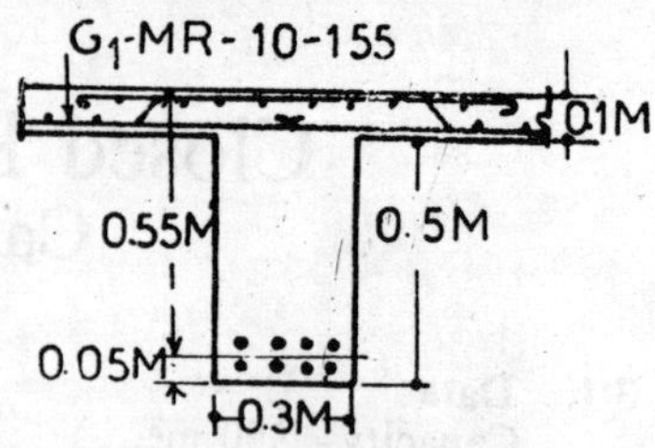

$$d_e=\sqrt{\frac{512\times100}{8{\cdot}7\times100}}$$

$$=\sqrt{59}=7{\cdot}7 \text{ cm}$$

Use $d=10$ cm, $d_e=8{\cdot}5$ cm

3·4·4. Main Steel

$$A_t=\frac{51200}{1400\times{\cdot}87\times8{\cdot}5}$$

$=4{\cdot}95$ cm². Use 10 mm ϕ at 15·5 cm c/c.

3·4·5. Secondary Reinforcement

$$A_t=\frac{{\cdot}15}{100}\times10\times100=1{\cdot}5 \text{ cm}^2$$

Use 6 mm ϕ at 16·5 cm c/c.

3·5. T-Beam

Span $=7{\cdot}25$ m (assuming 25 cm bearing

Centre to centre of beam ribs $=3{\cdot}2$ m

Loads

Floor slab $=500\times3{\cdot}2=1600$ kg/m

Self weight of the rlb (30 cm × 50 cm)

$$=\frac{30}{100}\times\frac{50}{100}\times1\times2400=360 \text{ kg/m}$$

Total load $=1960$ kg/m.

3·5·1. Bending Moment

$$M\ \frac{wl^2}{8}=\frac{1960}{8}(7{\cdot}25)^2=12{,}900 \text{ m kg}$$

$$\text{Approximate lever arm}=d_e-\frac{d_s}{2}=55-\frac{10}{2}=50 \text{ cm}$$

3·5. Approximate Steel

$$A_t=\frac{12900\times100}{1400\times50}=18{\cdot}4 \text{ cm}^3$$

Use 6 – 20 mm ϕ, A_t provided $=18{\cdot}84$ cm².

3·5·3. Effective Flange Width

Least of

(*i*) $\frac{1}{3}$ of span $=\frac{7{\cdot}25}{3}$

$=2{\cdot}42$ m $=242$ cm

(*ii*) Centre to centre of beam $=3{\cdot}2$ m $=320$ cm

(*iii*) $12\ ds+b_r=12\times10+30=150$ cm.

3·5·4. Position of N.A.

Assuming that N.A. lies below the flange

$$B\times ds\times\left(n-\frac{ds}{2}\right)=m\times A_t(d-n)$$

$$150\times10\times(n-5)=18\times18{\cdot}84\ (54\quad n)$$

$$1500\ n-7500=990-340\ n$$

$$1840\ n=25900$$

$$n=14{\cdot}1 \text{ cm}$$

$$\bar{y}=\frac{3n-2\,ds}{2n-ds}\times\frac{d_s}{3}=\frac{3\times14{\cdot}1-2\times10}{2\times14{\cdot}1-10}\times\frac{10}{3}$$

$$=\frac{42{\cdot}3-20}{28{\cdot}2-10}\times\frac{10}{3}=\frac{22{\cdot}3}{18{\cdot}2}\times\frac{10}{3}=4{\cdot}08\text{ cm}$$

3·5·5. Lever Arm

Actual lever arm $= d_e - \bar{y},\ j_d = 54-4{\cdot}08 = 49{\cdot}92$ cm

3·5·6. Stresses

$$\text{Stress in steel}=\frac{1{,}290{,}000}{18{\cdot}84\times49{\cdot}92}=1370\text{ kg/cm}^2 < 1400\text{ kg/cm}^2$$

$$\frac{\sigma_{cb}}{n}=\frac{\sigma_{st}}{m(d\quad n)}$$

$$\sigma_{cb}=\frac{1370\times14{\cdot}1}{18\times39{\cdot}9}=42{\cdot}7\text{ kg/cm}^2 < 50\text{ kg/cm}^2$$

3·5·7. Shear Stresses

$$\text{Maximum shear at the support edge}=\frac{1960\times7}{2}=6850\text{ kg}$$

$$\text{Shear stresses}=\frac{s}{j_d\times b_r}=\frac{6850}{49{\cdot}92\times30}=4{\cdot}6\text{ kg/cm}^2 < 5\text{ kg/cm}^2$$

3·5·8. Nominal Stirrups

Provide 8 mm ϕ two legged at 30 cm c/c

3·6. Side walls $\frac{L}{h}$ and $\frac{B}{h}>2$

Both shorter and longer walls are designed as propped cantilever

Maximum water pressure at the base of the wall

$p=wh=1000\times3{\cdot}5=3500$ kg

3·6·1. Bending Moments

Maximum positive bending moment

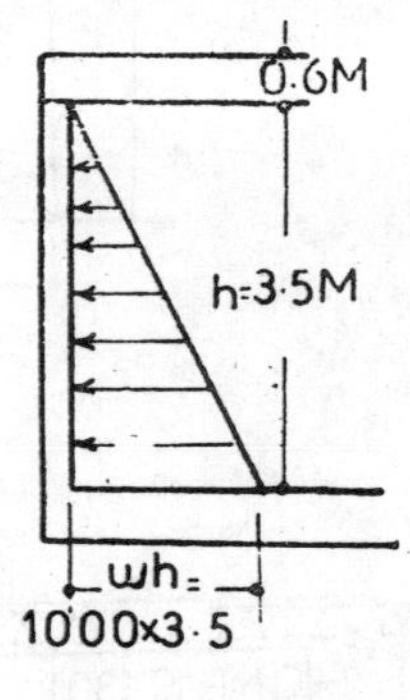

wh³/33.5
0.775h
wh³/15

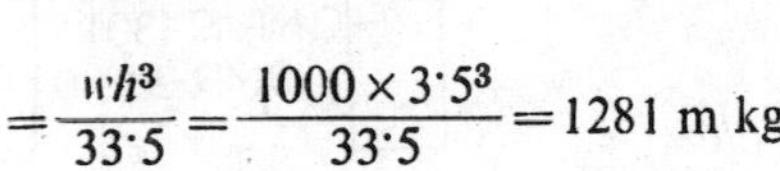

$$=\frac{wh^3}{33{\cdot}5}=\frac{1000\times3{\cdot}5^3}{33{\cdot}5}=1281\text{ m kg}$$

Maximum negative bending moment

$$=\frac{wh^3}{15}=\frac{1000\times3{\cdot}5^3}{15}=2860\text{ m kg}$$

3·6·2. Thickness of the walls

$$d_e=\sqrt{\frac{2860\times100}{100\times14}}=\sqrt{205}=14{\cdot}3 \text{ cm}$$

Use d=20 cm thick walls and d_e=16 cm

3·6·3. Main Reinforcement

Amount of steel reinforcement for −ve moment (water face)

$$A_t=\frac{286000}{1000\times.84\times16}$$

$=21{\cdot}3 \text{ cm}^2$

Use 20 mm ϕ at 14·5 cm c/c

Amount of steel required for +ve bending moment (away from water face)

$$A_t=\frac{128100}{1250\times{\cdot}86\times16}=7{\cdot}45 \text{ cm}^2$$

Use 12 mm ϕ at 15 cm c/c

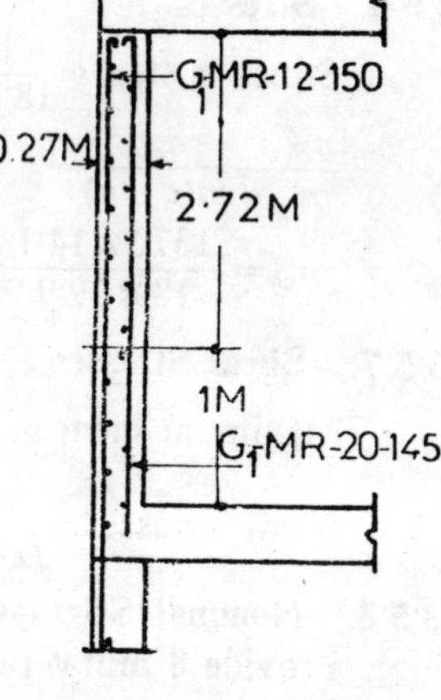

3·6·4. Secondary Reinforcement

A_t=·3% of the concrete area

$$=\frac{{\cdot}3}{100}\times20\times100=6 \text{ cm}^2$$

Use 8 mm ϕ at 15 cm c/c on both faces

3·7. Base Slab Loads

Slab is designed as a continuous over T-beam ribs.

Loads: Self weight of slab (25 cm thick)

$$=\frac{20}{100}\times2400\times1=480 \text{ kg/m}^2$$

Weight of water=1000×3·5=3500 kg/m²

Total load=3980 kg/m³

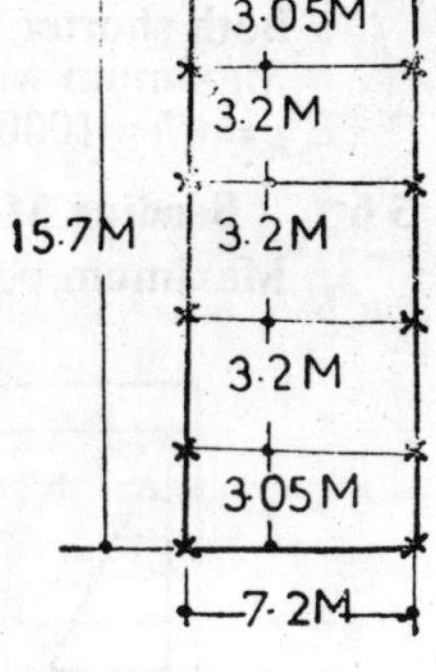

3·7·1. Bending moments

Span for the slab=3·2 m

Maximum bending moment

$$=\frac{wl^2}{10}=\frac{3980\times3{\cdot}2^2}{10}=4060 \text{ m kg}$$

3·7·2. Depth of the Slab

$14\,bd^2=406000$

$$d_e=\sqrt{\frac{406000}{14\times100}}=\sqrt{291}=17 \text{ cm}$$

Use d=20 cm, d_e=17 cm

3·7·3. Main Reinforcement

$$A_t=\frac{406000}{1000\times{\cdot}84\times17}$$

$=28{\cdot}5 \text{ cm}^2$

Use 20 mm ϕ at 11 cm c/c

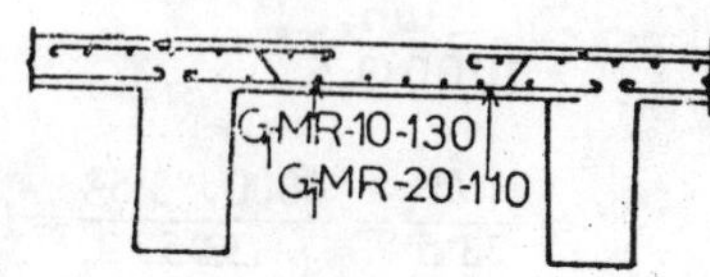

3·7·4. Secondary Steel

$$A_t=\frac{{\cdot}3}{100}\times20\times100=6 \text{ cm}^2$$

Use 10 mm, ϕ at 13 cm c/c

3·8. T-Beams

Loads : Assume 35×70 cm ribs

Load from slab $=3{\cdot}2\times3980=12700$ kg/m

Self weight of the slab $=\frac{35}{100}\times\frac{70}{100}\times1\times2400=590$ kg/m

Total load per metre length $=13290$ kg

3·8·1. Bending Moment

Span $=7{\cdot}2$ m

The beam at the ends is restrained due to monolithic connection between the side walls and the floor.

Maximum bending moment at centre section (+ve)

$$=\frac{wl^2}{100}=\frac{13290\times7{\cdot}2^2}{10}=69000\text{ m kg}=6900000\text{ cm kg}$$

3·8·2. Trial Mid-Section

Effective depth of T-beam $=90-6$
$=84$ cm

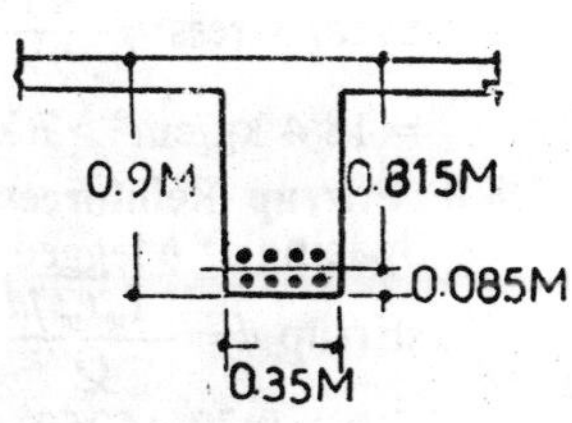

Lever arm (approximate) $=d-\frac{ds}{2}$

$=84-10=74$ cm

Steel required $=\frac{6900{,}000}{1250\times74}=75\text{ cm}^2$

Use 32 mm ϕ – 10 Nos – (80·42 cm²)

3·8·3. Effective Flange Width

The least of

(*i*) $\frac{1}{3}$ span $=\frac{1}{3}\times720=240$ cm

(*ii*) c/c of beam $=320=320$ cm

(*iii*) $12ds+b_r=12\times20+35=240+35=275$ cm

Assuming the *NA* is outside the flange

$$B\times ds\left(n-\frac{ds}{2}\right)=\text{m }A_t\,(d-n)$$

$$240\times20\,(n-10)=13\times80{\cdot}42\,(81{\cdot}5-n)$$

$$4800\,n-48000=85{,}000-1045\,n$$

$$5845\,n=55000+48000=133000$$

$$n=\frac{133000}{5845}=22{\cdot}7\text{ cm}$$

3·8·4. Lever Arm

$$\bar{y}=\frac{3n-2ds}{2n\quad ds}\times\frac{ds}{3}=\frac{3\times22{\cdot}7-2\times20}{2\times22{\cdot}7-20}\times\frac{20}{3}$$

$$=\frac{68{\cdot}1\quad40}{45{\cdot}4\quad20}\times\frac{20}{3}=\frac{28{\cdot}1}{25{\cdot}4}\times\frac{20}{3}=7{\cdot}4\text{ cm}$$

lever arm $j_d=d-\bar{y}=81{\cdot}5-7{\cdot}4\quad74{\cdot}1$ cm

3·8·5. Stresses

Stress in steel $=\sigma_{st}=\frac{6900{,}000}{80{\cdot}42\times74{\cdot}1}=1150\text{ kg/cm}^2$

Compressive stress in concrete

$$\frac{\sigma_{cb}}{n}=\frac{\sigma_{st}}{m(d-n)}$$

$$\sigma_{cb}=\frac{n\times\sigma_{st}}{m(d-n)}=\frac{22{\cdot}7\times1150}{13(81{\cdot}5-22{\cdot}7)}=\frac{22{\cdot}7\times1150}{13\times58{\cdot}8}=34{\cdot}3\text{ kg/cm}^2$$

3·8·6. Negative BM at support

—ve bending moment at support (water face)
$=6900{,}000$ cm kg

The end section is designed as doubly reinforced steel beam

$A_t \times (90-8{\cdot}5\times 2)\times 1000=6900{,}000$

$A_t \times 73\times 1000=6900{,}000$

$$A_t=\frac{6900{,}000}{73\times 1000}=94{\cdot}5 \text{ cm}^2$$

Use 32 mm ϕ 12 Nos at top over a length from end to the point of contraflexure that is 2 m from each end.

3·8·7. Shear Stresses

$$\text{Maximum shear force}=\frac{13290\times 7{\cdot}2}{2}=48{,}000 \text{ kg}$$

$$\text{Shear stress}=\frac{48{,}000}{35\times 74.1}$$

$=18{\cdot}4$ kg/cm^2 >7 kg/cm^2

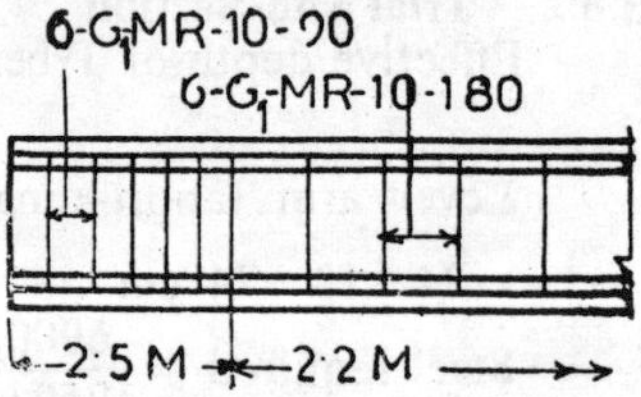

6-G$_1$MR-10-90

6-G$_1$MR-10-180

2·5 M — 2·2 M

3·8·8. Stirrup Reinforcement

Spacing of 6 legged 10 mm

$$\text{stirrup } \phi=\frac{A_w t_w jd}{Q}$$

$$=\frac{6\times 0{\cdot}79\times 1250\times 74{\cdot}1}{48000}=9{\cdot}2 \text{ cm}$$

Use 10 mm ϕ —6 legged stirrups at 9 cm c/c

3·9. Beam B-2

The beam on the longer side is designed to carry the weight of side walls and roof loads. The beam is supported on columns and is designed as a continuous beam over number of supports.

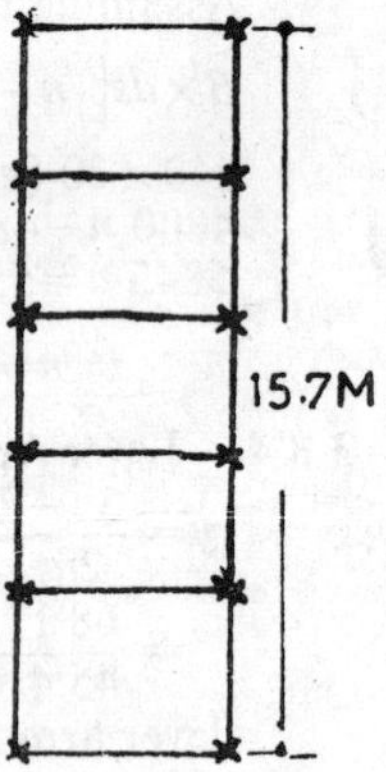

Loads : weight of side wall

$$=\frac{20}{100}\times 3{\cdot}5\times 1\times 2400=1680 \text{ kg/m}$$

$$\text{Roof slab}=\frac{7{\cdot}40}{2}\times 500\times 1=1850 \text{ kg/m}$$

$$\text{Ribs (4 Nos)}=\frac{7{\cdot}40}{2}\times 4\times\frac{30}{100}\times\frac{50}{100}\times\frac{2400}{15{\cdot}7}$$

$=340$ kg/m

15.7M

Total uniformly distributed load

$=w=1680+1850+340=3870$ kg/m

3·9·1. Bending moments

$$\text{Maximum bending moment}=\frac{wl^2}{10}=\frac{3870}{10}\times 3{\cdot}2^2=3980 \text{ mkg.}$$

3·9·2. Depth of the Beam

Assuming 20 cm wide beam

$$d_e=\sqrt{\frac{398000}{14\times 20}}=\sqrt{1420}=38 \text{ cm}$$

Use 20×45 cm, $d=45$ cm and $d_e=40$ cm

3·9·3. Main Steel

$$A_t = \frac{398000}{1250 \times \cdot 84 \times 40} = 9{\cdot}5 \text{ cm}^2$$

Use 5 Nos 16 mm ϕ

3·9·4. Shear Reinforcement

$$\text{Shear force} = \frac{3870 \times 2{\cdot}2}{2} = 6220 \text{ kg}$$

$$\text{Shear stress} = \frac{6220}{20 \times \cdot 86 \times 40} = 9{\cdot}05 \text{ kg}/m^2 > 7 \text{ kg/cm}^2$$

3·9·5. Stirrup Reinforcement

$$\text{Spacing of 10 mm } \phi \text{ two legged stirrups} = \frac{A_w t_w\, jd}{Q}$$

$$= \frac{2 \times 0{\cdot}79 \times 1250 \times \cdot 86 \times 40}{6220} = 10{\cdot}9 \text{ cm c/c}$$

Use 10 mm ϕ two legged stirrups at 10 cm c/c

3.10 Staging Columns

Loads form :

(*i*) Roof slab $= 7{\cdot}40 \times 15{\cdot}90 \times 500 = 58{,}600$ kg $= 58{\cdot}6\ t$

(*ii*) Roof Beam ribs (4 Nos)

$= 7{\cdot}40 \times \frac{30}{100} \times \frac{50}{100} \times 4 \times 2400$

$= 10{,}700$ kg $= 10{\cdot}7\ t$

(*iii*) Side walls

$= 2(7{\cdot}2 + 15{\cdot}7) \times \frac{20}{100} \times 2400 \times 3{\cdot}5$

$= 77{,}100$ kg $= 77{\cdot}1\ t$

(*iv*) Floor slab

$$= 7{\cdot}40 \times 15{\cdot}90 \times \frac{20}{100} \times 2400$$

$= 56{,}200$ kg $= 56{\cdot}2\ t$

(*v*) Floor beam

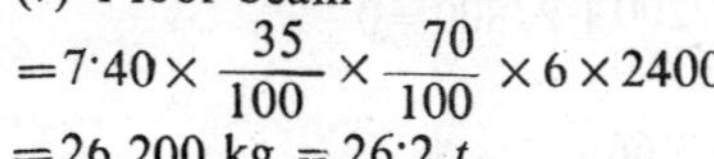

$$= 7{\cdot}40 \times \frac{35}{100} \times \frac{70}{100} \times 6 \times 2400$$

$= 26{,}200$ kg $= 26{\cdot}2\ t$

(*iv*) Beams under longer walls

$$= 15{\cdot}9 \times \frac{20}{100} \times \frac{45}{100} \times 2400 \times 2$$

$= 6850$ kg $= 6{\cdot}85\ t$

(*vii*) Weight of water $= 358\ t$

Total loads on the columns

$= 58{\cdot}6 + 10{\cdot}7 + 77{\cdot}1 + 56{\cdot}2 + 26{\cdot}2 + 6{\cdot}85 + 358$

$= 593{\cdot}65\ t$ Say $600\ t$

Axial load per column

$$= \frac{600}{12} = 50\ t$$

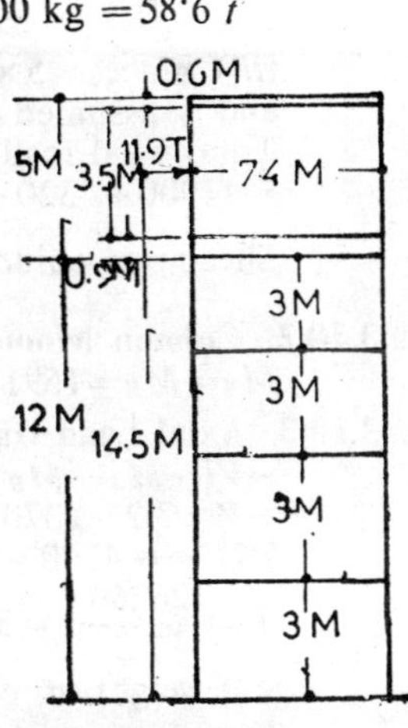

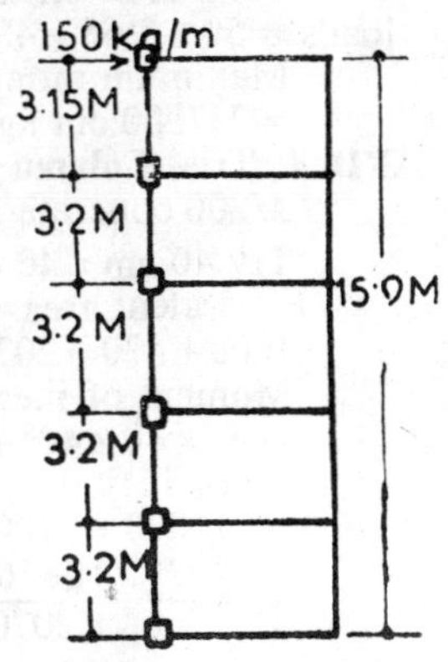

3·10·1. Wind Loads

Wind load on

(*i*) On the tank including roof slab beams and floor slab and beam

$= 5 \times 15{\cdot}9 \times 150 = 11{,}900$ kg and acts at 14·5 m above *GL*.

(*ii*) Columns (assuming 40 cm × 40 cm)
$= 6 \times \frac{40}{100} \times 12 \times 150 = 4320$ kg
and is assumed to act at 6 m above *G.L*

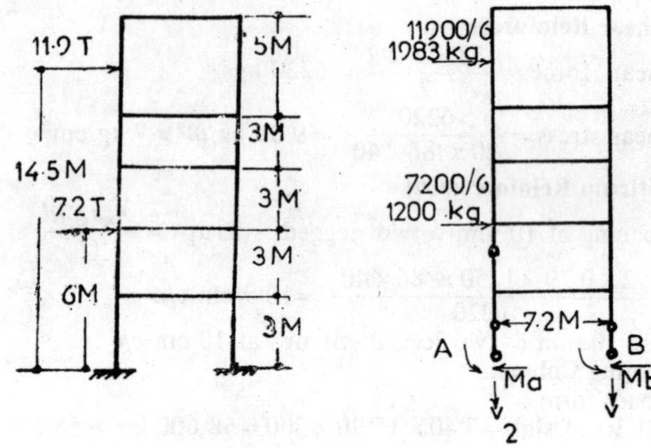

(*iii*) braces $= 5 \times 3 \times 3{\cdot}2 \times \frac{40}{100} \times 150 = 2880$ kg
and is assumed to act at 6 m above *G.L*
Total wind load on the tank and tower
$= 11900 + 4320 + 2880 = 19000$ kg

Shear per column $= \dfrac{19000}{12} = 1591$ kg

3.10·2. Column Moments

$M_A = M_B = 1591 \times 1{\cdot}5 = 2370$ m kg

3.10·3. Axial Load Due to Wind Moment

$-Vl - M_A - M_B + 1200 \times 6 + 1982 \times 14{\cdot}5 = 0$
$-V \times 7{\cdot}2 - 2370 - 2370 - 7200 + 23800 = 0$
$7{\cdot}2V = -4740 + 31000 = 26260$

$V = \dfrac{26260}{7{\cdot}2} = 3660 \text{ kg} = 3{\cdot}66\ t$

Self-weight of column $= \frac{40}{100} \times \frac{40}{100} \times 12 \times 2400 = 4600 \text{ kg} = 4{\cdot}6\ t$

Total load on the column due to dead and live loads $= 50 + 3{\cdot}65 + 4{\cdot}60 = 58{\cdot}26\ t$

Maximum moment on the column = 2370 m kg
= 237000 cm kg

3·10·4. Trial Column Section

*M*200 concrete

Try 40 cm × 40 cm with 8–25 mm ϕ (39·27 cm²)
Equivalent area $= 40 \times 40 + 12 \times 39{\cdot}27$
$1600 + 470 = 2070$ cm²
Moment of inertia
$= \frac{1}{12} \times 40 \times 40^3 + 12 \times 4{\cdot}91 \times 6 \times (14{\cdot}75)^2$
$= 214000 + 77000 = 391{,}000$ cm⁴

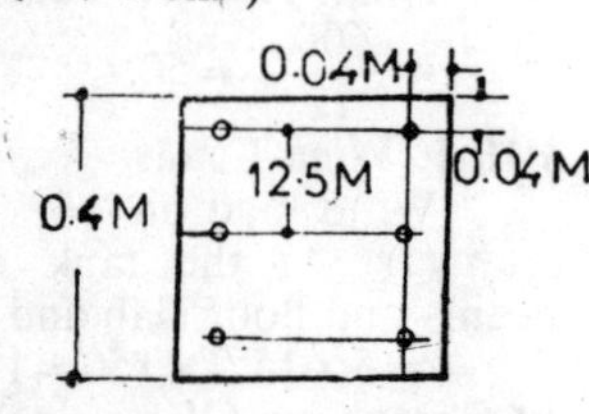

$\sigma_e' = \dfrac{P}{A} = \dfrac{58260}{2070} = 28{\cdot}2$ kg/cm²

$$\sigma_{cb}'=\frac{Mv}{I}=\frac{237000\times 20}{391000}=12{\cdot}1 \text{ kg/cm}^2$$

$$\frac{\sigma_c'}{\sigma_c}+\frac{\sigma_{cb}'}{\sigma_{cb}}=\frac{28{\cdot}2}{66{\cdot}7}+\frac{12{\cdot}1}{93{\cdot}3}={\cdot}425+{\cdot}13={\cdot}555<1$$

Trial section is not economical

Try $M-150$ concrete, 40 cm $\times$40 cm with 8$-$20 mm ϕ

Equivalent area$=40\times 40+12\times 3{\cdot}14\times 8=1600+302=1902$ cm^2

Moment of inertia$=\frac{1}{12}\times 40\times 40^3+12\times 3{\cdot}14\times 6(15)^2$

$=214000+50{,}100=264{,}100$ cm^4

$$\sigma_c'=\frac{58260}{1902}=30{\cdot}6 \text{ kg/cm}^2$$

$$\sigma_{cb}'=\frac{237000\times 20}{264100}=18 \text{ kg/cm}^2$$

$$\frac{\sigma_c'}{\sigma_c}+\frac{\sigma_{cb}'}{\sigma_{cb}}=\frac{30{\cdot}6}{53{\cdot}1}+\frac{18}{66{\cdot}7}=0{\cdot}578+0{\cdot}27={\cdot}848<1.$$

Or Try 35×35 cm with 8 bars of 20 mm ϕ M 200 concrete

Equivalent area$=35\times 35+12\times 3{\cdot}14\times 8$

$=1235+302=1537$ cm^2

$$\text{MI}=\frac{1}{12}\times 35\times 35^3+12\times 3{\cdot}14\times 6$$

$=181000+12\times 3{\cdot}14\times 6\times 10^2=181{,}300+21{,}600=202{,}600$ cm^4

$$\sigma_c'=\frac{58260}{1537}=38{\cdot}25 \text{ kg/cm}^2$$

$$\sigma_b'=\frac{237000\times 20}{202600}=23{\cdot}4 \text{ kg/cm}^2$$

$$\frac{\sigma_c'}{\sigma_c}+\frac{\sigma_{cb}'}{\sigma_{cb}}=\frac{38{\cdot}25}{66{\cdot}7}+\frac{23{\cdot}4}{93{\cdot}3}={\cdot}575+{\cdot}25={\cdot}825<1$$

Use either 40×40 cm, M 150 concrete or 35×35 cm M 200 concrete.

3·11. Brace

Maximum bending moment in the brace = 2$\times$moment in the column$=2\times 2370=4740$ m kg

Assume 40 cm $\times$ 40 cm section

Self weight of the brace

$$=\frac{40}{100}\times\frac{40}{100}\times 2400=384 \text{ kg/m}$$

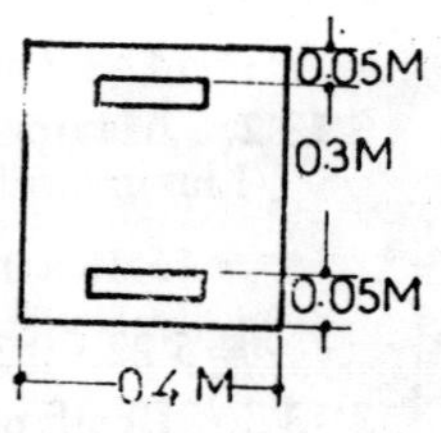

Bending moment (longer brace) due to

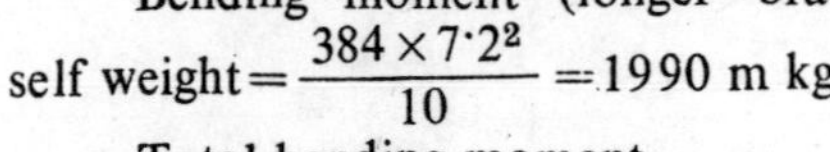

$$\text{self weight}=\frac{384\times 7{\cdot}2^2}{10}=1990 \text{ m kg}$$

Total bending moment

$=4740+1990=6730$ m kg.

$$A_t=\frac{673000}{30\times 1400}=16 \text{ cm}^2$$

Use 16 mm 8 Nos at top and bottom.

3·11·1. Shear Force

Shear force in shorter brace

$$V\times\frac{3{\cdot}2}{2}=4740$$

$V=\dfrac{4740\times 2}{3{\cdot}2}=2960$ kg

Shear stress $=\dfrac{2960}{40\times 30}$

$=2{\cdot}47$ kg/cm$^2<5$ kg/cm^2

Use nominal stirrups 8 mm ϕ 30 cm c/c.

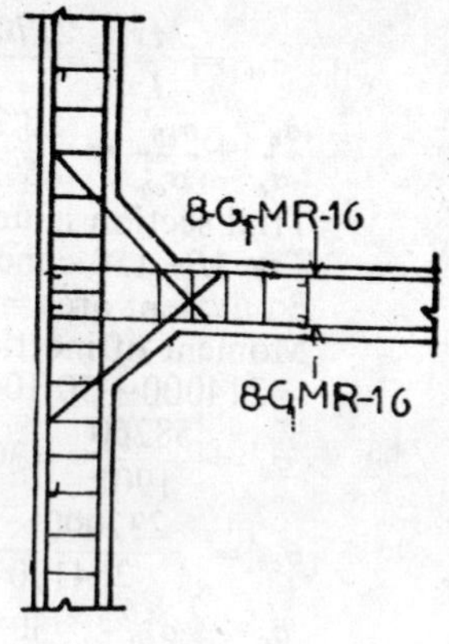

3·12. Foundation : Isolated Footing

Total load at the base of the column $=58.26$ t

Adding 10% for footing total load

$=58{\cdot}26+5{\cdot}82=64{\cdot}08$ t $=$ Say 64 t

Moment at the base of column

$=2370$ m kg

Eccentricity $=\dfrac{2370}{64}={\cdot}0372$ m

3·13. Area of Footing

Area $=\dfrac{64}{20}=3{\cdot}2$ m^2

Use 2 m $\times$ 2 m footing

3·13·1. Pressure

Maximum pressure

$=\dfrac{P}{A}\left(1+\dfrac{6e}{B}\right)$

$=\dfrac{64}{4}\left(1+\dfrac{6\times{\cdot}0372}{2}\right)$

$=16(1+{\cdot}112)=16\times 1{\cdot}112$

$=17{\cdot}8$ t/m$^2<20$ t/m^2

Minimum pressure $=16\times{\cdot}188=14{\cdot}27$ t/m^2

Pressure at the edge of the column

$=14{\cdot}2+(17{\cdot}8-14{\cdot}2)\times\dfrac{1{\cdot}2}{2}=14{\cdot}2+3{\cdot}6\times\dfrac{1{\cdot}2}{2}$

$=142+2{\cdot}16=16{\cdot}36$ t/m^2

3·13·2. Maximum Bendiug Moment

Maximum bending moment in the footing

$=16{\cdot}36\times 0{\cdot}8\times\dfrac{0{\cdot}8}{2}+\frac{1}{2}\times{\cdot}8\times 1{\cdot}44\times\dfrac{2}{3}\times 0{\cdot}8$

$=5{\cdot}22+0{\cdot}307=5{\cdot}527$ mt.

3·13·3. Depth of Footing

Maximum bending moment $=552700$ cm kg

$d_e=\sqrt{\dfrac{552700}{100\times 8{\cdot}7}}=\sqrt{640}=25{\cdot}25$ cm

$d=30$ cm, $d_e=25$ cm

Use 30 cm deep footing and at the edge 15 cm can be used

$A_t=\dfrac{552700}{1400\times{\cdot}87\times 25}=18{\cdot}3$ cm^2

Use 16 mm ϕ at 10·5 cm c/c in both directions.

Small Open overhead Circular Tank Capacity 25,000 Litres 4

4·1. Data

Capacity=25 m³
Height of the tank to the bottom above the ground level=5 m
Bearing capacity of the soil=20 t/m²
Wind load=150 kg/m²
Shape factor=0·7
Free board=20 cm
Materials available
Concrete – grade – *M* 200, *M* 150, Steel grade I
Relevant codes—IS – 456, IS – 875

4·2 Characteristic Strengths

σ_{cb} 70 kg/cm²	$m=13$
$\sigma_{st}=1000$ kg/cm²	$jd=\cdot 84\ d$
$\sigma_{ct}=12$ kg/cm²	$R=14$
$\sigma_{cb}=50$ kg/cm²	$m=18$
$\sigma_{st}=1400$ kg/cm²	$jd=\cdot 87\ d$
	$R=8\cdot 7$

4·3. Trial Dimensions

Required capacity=25 cu m
Diameter of the tank (assumed)=3 m

$$\text{Volume}=\frac{\pi D^2}{4}h=75$$

$$h=\frac{25\times 4}{\pi\times 3^2}=3\cdot 54 \text{ m}$$

Use $D=3$ m and $h=3\cdot 75$ m

4·4. Cylindrical Wall Hoop Tension

Maximum water pressure$=wh=1000\times 3\cdot 75=p=3750$ kg

$$\text{Hoop tension } T=\frac{PD}{2}=3750\times\frac{3}{2}=5625 \text{ kg}$$

$$A_t=\text{Area of steel required}=\frac{5625}{1000}=5\cdot 625 \text{ cm}^2$$

Use 10 mm at 13·5 cm c/c, Actual A_t used$=5\cdot 82$ cm²

4·4·1. Thickness of the Wall

The thickness of wall is determined on the basis of cracking tensile stress of concrete in tension.

$$\sigma_{ct}=\frac{T}{b\times t_w+(m-1)A_t}$$

$$12=\frac{5625}{100\times t_w+(13-1)\times 5\cdot 82}$$

$1200\ t_w + 12 \times 12 \times 5{\cdot}82 = 5625$

$1200\ t_w + 840 = 5625$

$1200\ t_w = 4785$

$t_w = \frac{4785}{1200} = 4 \text{ cm}$

Generally there is no advantage in using walls thinner than 12·5 cm since the extra cost of steel fixing and placing and tamping the concrete increases the cost of construction.

An error in displacement of steel is not necessarily harmful in a circular tank since the steel has to take only hoop tension.

4·4·2. Main Reinforcement

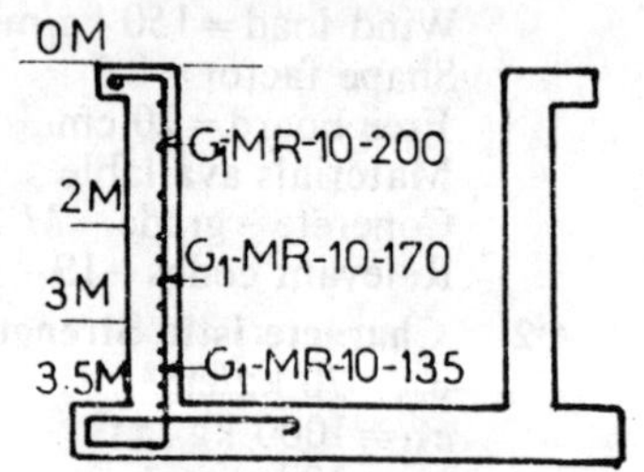

Smaller diameter of bars are preferred in tanks of small diameter on account of their increased flexibility and also because they give a more closely spaced arrangements of steel, inecrease the hond stress and prevent large shrinkage cracks in the concrete during the process of hardening. A minimum of 10 mm ϕ at 15 cm c/c reinforcement is desirable.

Depth from Dtop *m*	*Hoop tension* *kg*	A_t *cm*²	*Spacing required* *cm*	*Spacing used* *cm*
0 – 1	1500	1·5	52·5	20
1 – 2	3000	3	26·5	20
2 – 3	4500	4·5	17·0	17·0
3 – 3·75	5625	5·625	13·5	13·5

4·4·3. Secondary Reinforcement

·3% of concrete area

$A_t = \frac{\cdot 3}{100} \times 12{\cdot}5 \times 100 = 3{\cdot}75 \text{ cm}^2$

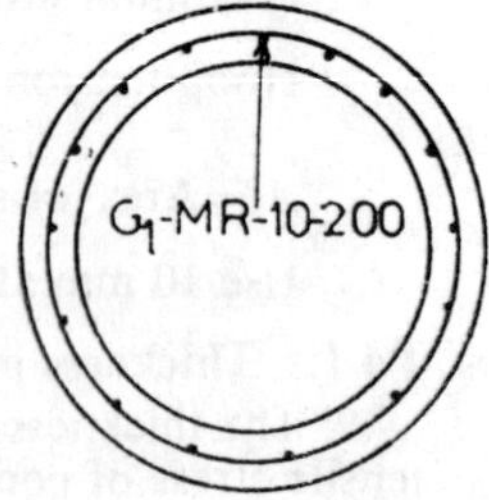

Use 10 mm at 20 cm c/c vertically inside the hoops. In this position they are most effective in resisting the cantilever moment and tendency to crack at the junction of the wall and the floor slab.

4·5. Floor Slab

Base slab is supported on 4 beams and is designed as a two way slab to take a moment equal to $= \frac{W}{26}$ m kg in each direction

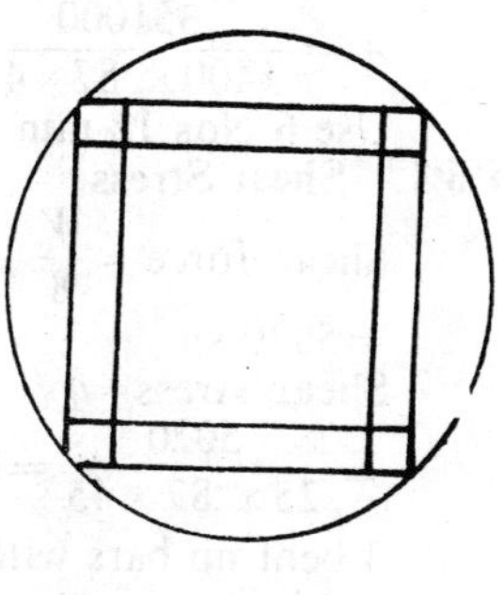

Loads on the floor slab

Weight of water

$= \frac{\pi \times 3^2}{4} \times 3{\cdot}75 \times 1000 \quad 26500$ kg

Weight of top rib

$= \pi \times 3{\cdot}30 \times \frac{15}{100} \times \frac{5}{100} \times 2400$

$= 186{\cdot}5$ kg

Weight of wall

$= \pi \quad 3{\cdot}125 \times \frac{12{\cdot}5}{100} \times 3{\cdot}75 \times 2000$

$= 10{,}300$ kg

Weight of bottom rib

$= \pi \times 3{\cdot}315 \times \frac{1}{2} \times \frac{10}{100} \times \frac{20}{100} \times 2400 = 250$ kg

Weight of the bottom slab (assuming 20 cm thick)

$= \pi \times 3{\cdot}45^2 \times \frac{20}{100} \times 2400 = 4475$ kg

Total load $= W = 41711{\cdot}5$ kg

Bending moment $= M = \frac{W}{26} = \frac{41711{\cdot}5}{26} = 1610$ m kg.

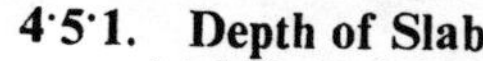

4·5·1. Depth of Slab

$14\,bd^2 = 161000$

$d = \sqrt{\frac{161000}{14 \times 100}} = \sqrt{115} = 10{\cdot}7$ cm

Use $d = 15$ cm, $d_e = 11$ cm

4·5·2. Main Reinforcement

$A_t = \frac{161000}{1000 \times {\cdot}84 \times 11} = 17{\cdot}1 \text{ cm}^2$

Use 16 mm ϕ at 11·5 cm c/c in both the directions.

4·6. Floor Beams

Each beam carries a triangular load.

Maximum bending moment per beam

$= \frac{WL}{24} = \frac{41711{\cdot}5 \times 2}{24} = 3360$ m kg

Self weight of the beam (assume 50 cm × 25 cm)

$\frac{50}{100} \times \frac{25}{100} \times 2 \times 2400 = 600$ kg

Bending moment

$= \frac{600 \times 2}{8} = 150$ m kg

Total bending moment $= 3510$ m kg

Using m 150 concrete

$87\,bd^2 = 351000$

$d = \sqrt{\frac{351000}{8{\cdot}7 \times 25}} = \sqrt{1610} = 40{\cdot}2$ cm

Use $d = 50$ cm $d_e = 45$ cm

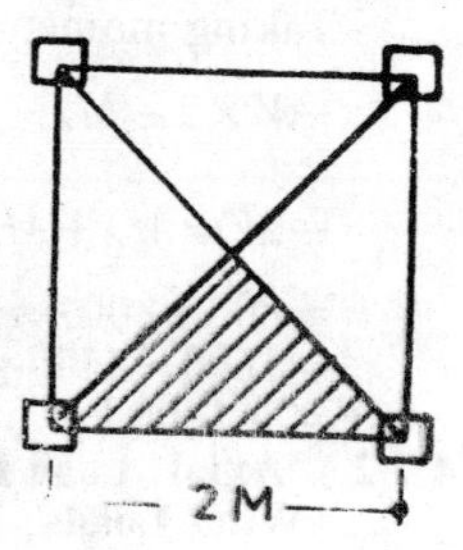

$$A_t = \frac{351000}{1400 \times \cdot 87 \times 45} = 6{\cdot}4 \text{ cm}^2$$

Use 6 Nos 12 mm ϕ

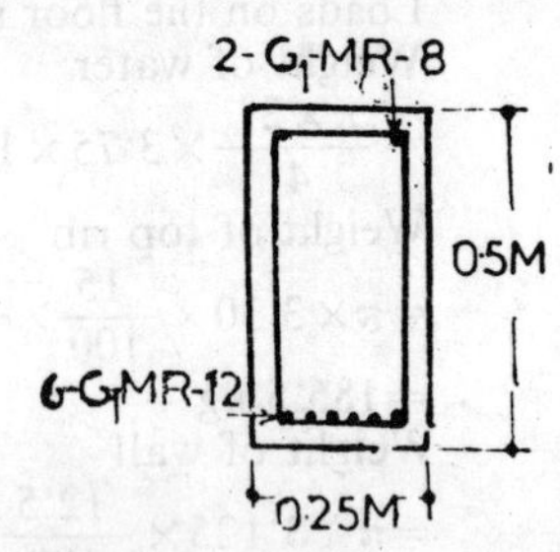

4·6·1. Shear Stress

$$\text{Shear force} = \frac{W}{8} = \frac{41711{\cdot}5}{8}$$

$= 5020$ kg

Shear stress $= q$

$$= \frac{5020}{25 \times \cdot 87 \times 45} = 5{\cdot}15 \text{ kg/cm}^2$$

3 bent up bars will take the excess shear stress.

Use 6 mm ϕ 15 cm c/c two-legged nominal stirrups

4·7. Staging Columns

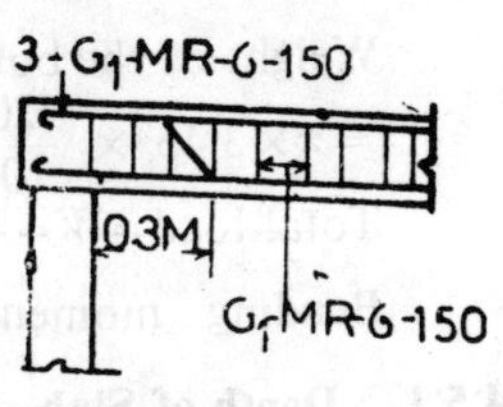

Loads: Total load of the tank with water $= 41711{\cdot}5$ kg

Self weight of beam $= 4 \times 600 = 2400$ kg

Total load $= 44111{\cdot}5$ kg

Load per column

$$= \frac{44111{\cdot}5}{4} = 11{,}030 \text{ kg}$$

Self weight of the column (assuming 35×35 cm)

$$= \frac{35}{100} \times \frac{35}{100} \times 5 \times 2400 = 1470 \text{ kg}$$

Total axial load $= 12{,}500$ kg

Wind loads $p = 150$ kg/m²

Wind load on the tank $= 0.7 \times 3{\cdot}25 \times 3{\cdot}75 \times 150 = 1280$ kg

$$\text{Shear per column} = \frac{1280}{4} = 320 \text{ kg}$$

$M_A = M_B = 320 \times 5/2 = 800$ m kg

4·7·1. Bending Moment in the columns

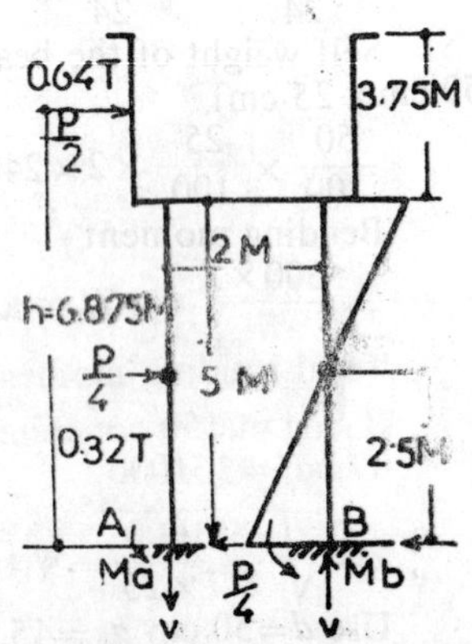

Taking moments about B

$$-V \times 2 - M_A - M_B + \frac{P}{2} \times h = 0$$

$$-2V = M_A + M_B - \frac{P}{2} h$$

$= 800 + 800 - 640 \times 6{\cdot}875$

$= 1600 - 4400 + 2V = +2800$

$V = +1400$ kg

4·7·2. Axial Load in the Column Due to Wind Loads. Use M 150

Maximum axial load in the column $= 12{\cdot}5 + 1{\cdot}4 = 13{\cdot}9$ t

Maximum moment at the base of the column $= 800$ m kg

4·7·3. Trial Section

Use $M-150$ concrete.

Try 30 cm × 30 cm with 4—20 mm ϕ

Equivalent area = 30 × 30 + 17 × 3·14 × 4 = 900 + 226 = 1126 cm²

Moment of inertia

$$I=\frac{1}{12}\times 30\times 30^3+4\times 3{\cdot}14\times 17(10)^2=67{,}5000+21250$$

$$=88{,}750 \text{ cm}^4$$

4·7·4. Stress in Column

$$\sigma_c{}'=\frac{P}{A}=\frac{13900}{1126}=12{\cdot}4 \text{ kg/cm}^2$$

$$\sigma_{cb}{}'=\frac{Mv}{I}=\frac{80000\times 15}{88750}=13{\cdot}6 \text{ kg/cm}^2$$

$$\frac{\sigma_c{}'}{\sigma_c}+\frac{\sigma_{cb}{}'}{\sigma_{cb}}=\frac{12{\cdot}4}{53{\cdot}3}+\frac{13{\cdot}6}{66{\cdot}7}={\cdot}236+{\cdot}205={\cdot}441<1$$

Section is uneconomical. A section 25 cm × 25 cm can be tried.

Lateral ties. Use 8 mm ϕ at 30 cm c/c

4·8. Foundations

Total axial load on the leeward column

= 125000 + 1400 = 13900 kg = 13·9 t

Adding 10% for the footing, load on the soil

= 13·9 + 1·39 = 15·29 t Say 16 t

Bending moment = 800 m kg

$$\text{Eecentricity, } e=\frac{80000}{15290}=5{\cdot}25 \text{ cm}={\cdot}0525 \text{ m}$$

$$\text{Area of footing}=\frac{16}{20}=0{\cdot}75 \text{ m}^2$$

Use 1 m × 1 m footing

4·8·1. Pressures Below The Footing

Maximum pressure

$$=\frac{P}{A}\left(1+\frac{6e}{B}\right)$$

$$=\frac{16}{1\times 1}\left(1+\frac{6\times {\cdot}0525}{1}\right)$$

$$=16(1+{\cdot}314)$$

$$=21 \text{ t/m}^2>20 \text{ t/m}^2$$

Increase the footing size

Use 1·25 m × 1·25 m

Maximum pressure

$$=\frac{16}{1{\cdot}25\times 1{\cdot}25}\left(\frac{1+6\times 0{\cdot}0525}{1{\cdot}25}\right)$$

$$=10{\cdot}25(1+{\cdot}25)=12{\cdot}8 \text{ t/m}^2<20 \text{ t/m}^2$$

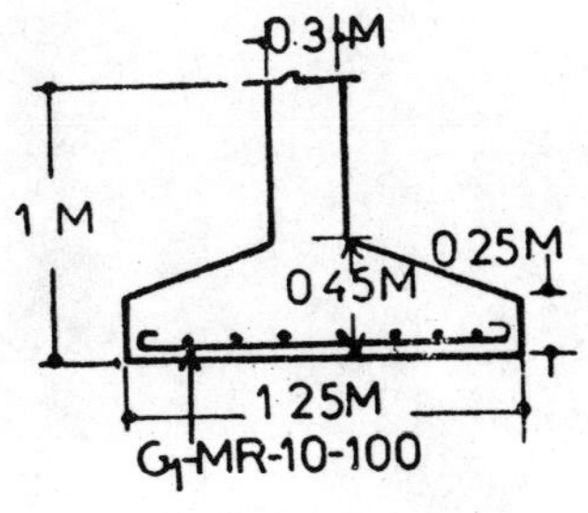

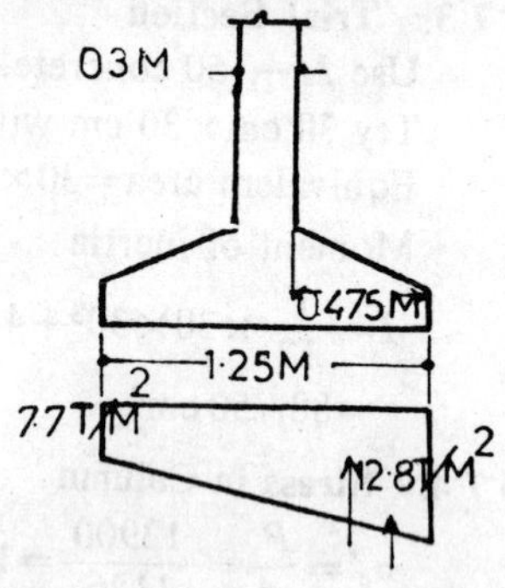

Minimum pressure

$$=\frac{16}{1{\cdot}25\times1{\cdot}25}\left(1-\frac{6\times0{\cdot}0525}{1{\cdot}25}\right)$$

$=10{\cdot}25+0{\cdot}75=7{\cdot}7\ t/m^2$

Maximum pressure at the junction of the column and the footing

$$=7{\cdot}7+\frac{(12{\cdot}8-7{\cdot}7)}{1{\cdot}25}\times{\cdot}795$$

$$=7{\cdot}7+\frac{5{\cdot}1\times{\cdot}795}{1{\cdot}25}$$

$=7{\cdot}70+3{\cdot}24=10{\cdot}94\ t/m^2$

4·8·2. Bending Moment

$$M=10{\cdot}94\times{\cdot}475\times\frac{{\cdot}475}{2}+\tfrac{1}{2}\times(12{\cdot}80-10{\cdot}94)\times{\cdot}475\times\tfrac{2}{3}({\cdot}475)$$

$=1{\cdot}24+\frac{1}{2}\times1{\cdot}96\times{\cdot}475\times\frac{2}{3}\times{\cdot}475=1{\cdot}24+{\cdot}147=1{\cdot}387$ mt

$=138700$ cm kg

4·8·3. Depth

$$d_e=\sqrt{\frac{138700}{8{\cdot}7\times100}}=\sqrt{158}=12{\cdot}60\text{ cm}$$

Use $d=20$ cm $d_e=15$ cm

4·8·4. Steel

Amount of steel $A_t=\dfrac{138700}{1400\times{\cdot}87\times15}=7{\cdot}65\text{ cm}^2$

Use 10 mm ϕ 10 cm c/c

Flat Bottom Elevated Tank 5

Capacity 114 cu. m. (25,000 gals)

5·1. Data

Elevation of the tank=15·2 m

Bearing capacity of soil=15 t/m^2 (assumed)

Bracing =3 levels

Materials and allowable stresses, Tank portion $M200$ concrete, grade - 1 steel

$\sigma_{cb}=70$ kg/cm² $\sigma_{st}=1000$kg/cm²

$m=13$ $jd=0{\cdot}84\ d$

$\sigma_{ct}=12$ kg/cm² $R=14$

Top dome, Columns, Braces etc.

$M150$ concrete, grade—1 steel

$\sigma_{cb}=50$ kg/cm² $\sigma_{st}=1400$ kg/cm²

$m=18$ $R=8{\cdot}7$

$jd=0{\cdot}87$

5·2. Trial Dimensions

Required capacity=114 cu m

Try D=diameter of the tank=6·6 m

h=depth of tank =3·5 m

Depth of water in the tank =3·50−·15 =3·35 m

Volume of the tank using flat bottom tank$=\dfrac{\pi\times D^2\times h}{4}$

$$=\frac{3{\cdot}14\times 6{\cdot}6^2\times 3{\cdot}35}{4}=114 \text{ cu m}$$

Use 6·6 m dia 3·5 m height cylindrical tank

5·3. Domed Roof

Loads : Assuming 10 cm thick dome

Self weight of the dome$=\dfrac{10}{100}\times 1\times 1\times 2400=240$ kg/m²

Wind and accidental loads=150 kg/m²

Water proofing etc. =110 kg/m²

Total load =500 kg/m²

5·3·1. Geometry of the Dome

Rise$=r=\dfrac{1}{5}$ to $\dfrac{1}{6}D=\dfrac{6{\cdot}6}{6}$ =1·1 m to 1·32 m

Use =1·2 m

Span $=2l=6{\cdot}6$ m

$l=3{\cdot}3$ m

R=Radius of the dome$=\dfrac{l^2+r^2}{2r}$

$$=\frac{3{\cdot}3^2+1{\cdot}2^2}{2\times 1{\cdot}2}=\frac{10{\cdot}9+1{\cdot}44}{2{\cdot}4}=5{\cdot}12 \text{ m}$$

$$\sin\phi=\frac{3{\cdot}3}{5{\cdot}12}=0{\cdot}645$$

$\phi=40°-12'<51°-48'$

Therefore the entire dome will be subjected to compressive stresses only.

5·3·2. Stresses in the Dome

At crown $\phi = 0$

Hoop stress $= s = \dfrac{wR}{2t} = \dfrac{500 \times 5{\cdot}12}{2 \times {\cdot}1} = 12800$ kg/m²

$= 1{\cdot}28$ kg/cm² < 10 kg/cm²

Meridional stress $c = \dfrac{wR}{t(1 + \cos\phi)} = \dfrac{500 \times 5{\cdot}12}{0{\cdot}1 \times (1 + {\cdot}7638)}$

$= 14420$ kg/cm²

$= 1{\cdot}442$ kg/cm² < 10 kg/cm²

These stresses are very low

5·3·3. Force Components

Surface area of dome $= 2\pi Rr$

$= 2 \times \pi \times 5{\cdot}12 \times 1{\cdot}2 = 38{\cdot}8$ sq. m

Total load $= W$ $= 38{\cdot}8 \times 500 = 19400$ kg $= 19{\cdot}4\ t$

V_1 = vertical load per metre of the circumference of the dome

$= \dfrac{W}{2 \times \pi \times l} = \dfrac{19{\cdot}4}{2 \times \pi \times 3{\cdot}3} = 0{\cdot}965\ t = 965$ kg

H_1 = horizontal thrust $= V_1 \cot\phi = 965 \times \left(\dfrac{R-r}{l}\right)$

$= 965 \times \left(\dfrac{5{\cdot}12 - 1{\cdot}2}{3{\cdot}3}\right) = \dfrac{965 \times 3{\cdot}92}{3{\cdot}3} = 1140$ kg

T_1 = thrust along the meridion $= \dfrac{V_1}{\text{sion}\ \phi} = V_1 \dfrac{R}{l}$

$= \dfrac{965 \times 5{\cdot}12}{3{\cdot}3} = 1490$ kg.

5·3·4. Shear Stress at the Edge

Shear stress along the perimeter of the dome

$= \dfrac{V_1}{\text{area}} = \dfrac{965}{10 \times 100} = 0{\cdot}965$ kg/cm² < 5 kg/cm²

Therefore 10 cm thick *M* 150 concrete slab will be sufficient

5·3·5. Main Reinforcement

Provide minimum reinforcement to allow for indeterminate stresses due to wind forces shrinkage and temperature etc.

$A_t = 0{\cdot}2\%$ of concrete area $= \dfrac{0{\cdot}2}{100} \times 10 \times 100 = 2$ cm²

Use 6 mm ϕ at 14 cm c/c both ways

[cr 10 mm ϕ at $\left(\dfrac{0{\cdot}79}{2{\cdot}0} \times 100\right) = 39{\cdot}5$ cm c/c

Maximum spacing $= 3 \times d = 3 \times 7{\cdot}5 = 22{\cdot}5$ cm]

Maximum hoop tension $= T = \dfrac{H_1 \times D}{2} = \dfrac{1140 \times 6{\cdot}6}{2} = 3760$ kg

Amount of steel required $= \dfrac{3760}{1400} = 2{\cdot}7$ cm²

Use 4–10 mm ϕ bars. A_t provided $= 3{\cdot}14$ cm²

Stirrup steel required $= \dfrac{V_1}{1400} = \dfrac{965}{1400} = 0{\cdot}69$ cm²

Steel provided in the dome will be sufficient

i.e 6 mm ϕ – at 14 cm c/c

5·3·6. Rib Dimensions

$\sigma_{ct}=11$ kg/cm^2

$$\sigma_{ct}=\frac{T}{A_c+(m-1)A_t}$$

$$11=\frac{3760}{A_c+17\times 3{\cdot}14}$$

$$11\,A_c+11\times 17\times 3{\cdot}14=3760$$

$$A_c=\frac{3175}{11}=290\text{ cm}^2$$

Use 20 cm × 15 cm

5·4. Cylindrical Side Wall

M 200 concrete $\quad \sigma_{cb}=70$ kg/cm^2 $\quad \sigma_{st}=1000$ kg/cm^2

Tank wall is assumed to be free at top and fixed at the base

Try t=thickness of wall=15 cm

$h=3{\cdot}5$ m, $D=6{\cdot}6$ m

$$\frac{h^2}{D\times t}=\frac{3{\cdot}5\times 3{\cdot}5}{6{\cdot}6}=12{\cdot}4$$

5·4·1. Hoop Tension

Maximum hoop tension$=0{\cdot}640\;wh\;D/2$

$$=0{\cdot}640\times 1000\times 3{\cdot}35\times\frac{6{\cdot}6}{2}=7050\text{ kg}$$

Maximum hoop tension occurs at $=0{\cdot}7=0{\cdot}7h\times 3{\cdot}35$

$=2{\cdot}35$ m from top

5·4·2. Bending Moments

Maximum negative BM at base$=0{\cdot}0104\times w\times h^3$

$=0{\cdot}0104\times 1000\times 3{\cdot}53^3=391$ m kg

Maximum +ve BM (occurs at $0{\cdot}8\;h$)$={\cdot}00254\times wh^3$

$={\cdot}00254\times 1000\times 3{\cdot}35^3=95{\cdot}5$ m kg

This occurs at a depth from top$=0{\cdot}8\times 3{\cdot}35=2{\cdot}775$ m

5·4·3. Shear Force

Maximum shear force at base

$V=0{\cdot}143\times wh^2=0{\cdot}143\times 1000\times 3{\cdot}35^2=1600$ kg

5·4·4. Main Steel

Steel required for hoop tension

$$=\frac{7050}{1000}=7{\cdot}05\text{ cm}^2$$

Use 10 mm $\phi=\left(\frac{0\;79\times 100}{7{\cdot}05}\right)=2{\cdot}25$ cm c/c on both faces

5·4·5. Secondary Steel

$A_t=0{\cdot}3\%$ of concrete area$=0{\cdot}3/100\times 15\times 100=4{\cdot}5$ cm^2

Use 10 mm ϕ at 34 cm c/c on both the faces

Steel required for −ve BM (with 4 cm cover)

$$=\frac{M}{t\times jd}=\frac{39100}{1000\times{\cdot}84\times 110}=4{\cdot}22\text{ cm}^2$$

Use 10 mm ϕ at 17 cm c/c, water face up to 1 m height

Steel required for +ve $BM=\frac{9550}{1000\times{\cdot}84\times 11}=1{\cdot}24$ cm^2

10 mm—at 34 cm c/c on the outerface provided as secondary steel will resist this BM.

5·5. Floor or Bottom Slab

Floor slab is designed as a circular slab with the edge fixed to the sidewall.

Loads: weight of water $=1000\times3{\cdot}35=3350$ kg/m²

Self weight of slab (assuming 25 cm thick) including water proofing $=\frac{25}{100}\times1\times1\times2400=600$ kg/m²

Total load/sq m $=w=3950$ kg

Circumferential moment $=\dfrac{wR_1^2}{16}=\dfrac{3950\times3{\cdot}3^2}{16}=2680$ m kg

Radial moments

(*i*) Positive moment $=\dfrac{wR_1^2}{16}=2680$ m kg

(*ii*) Negative moment $=\dfrac{2\times wR_1^2}{16}=2680\times2=5360$ m kg

Depth of slab required for the $-$ve BM

$=\sqrt{\dfrac{536000}{14\times100}}=\sqrt{382}=19{\cdot}5$ cm

Use $d=25$ cm, $d_e=21$ cm

5·5·1. Main Steel at the Edge

A_t ($-$ve steel-water face) $=\dfrac{536000}{{\cdot}84\times21\times1000}=30{\cdot}5$ cm²

Use 20 mm ϕ at 10 cm c/c for the length

$\dfrac{D}{5}=\dfrac{6{\cdot}6}{5}=1{\cdot}32$ m from the edge at top

A_t required for circumferential moment $=15{\cdot}25$ cm²

Use 16 mm $\phi\times13$ cm c/c both directions.

5·6. Ring Beam

Loads : Weight of water $=114$ t

Load from dome $=19{\cdot}4$ t

weight of rib $=\frac{15}{100}\times\frac{5}{100}\times2{,}400\times6{\cdot}95={\cdot}392$ t

weight of cylindrical wall $=\pi\times6{\cdot}75\times\frac{15}{100}\times3{\cdot}5\times2{\cdot}4=26{\cdot}7$ t

weight of base slab $=\dfrac{\pi\times6{\cdot}9^2}{4}\times{\cdot}25\times2{\cdot}4=22{\cdot}3$ t

weight of ring beam $=\dfrac{35}{100}\times\dfrac{40}{100}\times11\times6{\cdot}5\times2{\cdot}4=6{\cdot}85$ t

Total vertical load on the beam

$=114+19{\cdot}4+0{\cdot}392+26{\cdot}7+22{\cdot}3+6{\cdot}85$

$W=189{\cdot}642$t, Say $=190$ t

5·6·1. Bending Moments

Bending moment at support ($-$ve) $=.0148\ WR$

$={\cdot}0148\times190\times\dfrac{6{\cdot}5}{2}=9{\cdot}15$ mt

Bending moment at centre of supports

$={\cdot}0075\times190\times\dfrac{6{\cdot}5}{2}=4{\cdot}62$ mt

Torsional moment $=0{\cdot}0015\times190\times\dfrac{6{\cdot}5}{2}=0{\cdot}93$ mt

5·6·2. Shear Forces

$$\text{Shear force} = \frac{\text{Total load}}{2 \times \text{No. of columns}}$$

$$= \frac{W}{2 \times 6} = \frac{190}{12} = 15{\cdot}83 \text{ t}$$

Point of maximum torsion $\theta = 12{\cdot}75°$

Shear force at point of maximum torsion

$$= 15{\cdot}83 - \frac{15{\cdot}83 \times 12{\cdot}75 \times 2}{60} = 15{\cdot}83 - 6{\cdot}75 = 9{\cdot}08 \text{ t}$$

Shear stress at point of maximum torsion

$$q = \frac{9080}{40 \times {\cdot}87 \times 56} = 4{\cdot}7 \text{ kg/cm}^2$$

Shear stress due to torsional moment

$$q' = \frac{T_m(3+2\,b/d)}{b^2D} = \frac{93000(3+2 \times \frac{40}{60})}{40 \times 40 \times 60}$$

$$= \frac{93000 \times 4{\cdot}33}{40 \times 40 \times 60} = 4{\cdot}2 \text{ kg/cm}^2$$

Total shear stress $= 4{\cdot}7 + 4{\cdot}2 = 8{\cdot}9$ kg/cm²
> 5 kg < 20 kg/cm²

Therefore shear reinforcement shall be provided

$$d = \sqrt{\frac{915000}{14 \times 40}} = \sqrt{1640} = 40{\cdot}5 \text{ cm}$$

Use $d = 60$ cm (assumed value) $\therefore d_e = 54$ cm

5·6·3. Main Steel

A_t for $-$ve moment (water face)

$$= \frac{915000}{1000 \times {\cdot}84 \times 54} = 20{\cdot}2 \text{ cm}^2$$

Use 7 – 20 mm ϕ at top of support

A_t required for +ve moment

$$= \frac{462000}{1250 \times {\cdot}86 \times 54} = 7{\cdot}95 \text{ cm}^2$$

Use 7 Nos – 12 mm ϕ at bottom at centre

5·6·4. Shear Reinforcement

Shear stirrups at support using 12 mm ϕ 4 legged.

$$\text{Spacing } p = \frac{4 \times 1{\cdot}13 \times 1000 \times {\cdot}86 \times 54}{15830} = 13{\cdot}3 \text{ cm c/c}$$

Use 12 mm ϕ 4 legged at 13 cm c/c at support

5·6·5. Shear Stirrups at Point of Maximum Torsion

Area of steel required for shear force

$$Aw_1 = \frac{9080 \times 13}{1000 \times {\cdot}86 \times 54} = 2{\cdot}55 \text{ cm}^2$$

Using 13 cm c/c spacing area of steel required for torsional shear stress

$$Aw_2 = \frac{T_m \times p}{0{\cdot}8 \times t_w \times X \times Y} = \frac{93000 \times 13}{0{\cdot}8 \times 1000 \times 30 \times 50} = 1{\cdot}01 \text{ cm}^2$$

Total shear reinforcement $= 2{\cdot}55 + 1{\cdot}01 = 3{\cdot}56$ cm²

Using 12 mm ϕ

Area of each leg $= 1{\cdot}13$.

No. of legs reqd. $= \frac{3{\cdot}56}{1{\cdot}13} = 3{\cdot}2$

Use 12 mm ϕ 4—legged at 13 cm c/c

5·6·6. Longitudinal Reinforcement for Torsion

Volume of longitudinal reinforcement per unit length
= Volume of stirrups per unit length provided for torsion

$= \frac{1{\cdot}01 \times 50}{13} = 3{\cdot}9 \text{ cm}^2$

Use 4 – 12 mm ϕ two per face

5·7. Staging Columns

Loads: (*i*) Tank full direct axial load on each column
$= \frac{190}{6} = 31{\cdot}67$ t

Self weight of the columns (40 cm dia columns)
$= \pi/4 \times (0{\cdot}4)^2 \times 15{\cdot}2 \times 2{\cdot}4 = 4{\cdot}54$ t

weight of braces $= 3 \times \frac{35}{100} \times \frac{25}{100} \times 3{\cdot}25 \times 2{\cdot}4 = 2{\cdot}06$ t

Total axial load $= 31{\cdot}67 + 4{\cdot}54 + 2{\cdot}06 = 38{\cdot}27$ t

(*ii*) Tank empty, weight of water = 114 t
load on the columns when tank is empty = 190 – 114 = 76 t
load per column $= \frac{76}{6} = 12{\cdot}67$ t
self weight of column = 4·54 t
weight of braces = 2·06 t
Total load on each column $= 12{\cdot}67 + 4{\cdot}54 + 2{\cdot}06 = 19{\cdot}27$ t

5·7·1. Wind Loads

Wind pressure (central zone) = 150 kg/m²
reduction factor = 0·7

Wind loads on (*i*) domed roof and cylindrical portion including ring beam
$= 4{\cdot}75 \times 6{\cdot}9 \times 150 \times 0{\cdot}7 = 3450$ kg, acts at 17·575 m from G.L.

(*ii*) columns $= 4 \times 0{\cdot}4 \times 15{\cdot}2 \times 150 \times 0{\cdot}7 = 2550$ kg

acts at $\frac{15{\cdot}2}{2} = 7{\cdot}6$ m from G.L.

(*iii*) braces $= 3 \times 6{\cdot}9 \times {\cdot}35 \times 150 = 1090$kg, acts at 7·6 m from G.L.

5·7·2. Wind Moment

Wind moment on the tank
$= 3450 \times 17{\cdot}57 + 2550 \times 7{\cdot}6 + 109 \times 7{\cdot}6$
$= 60500 + 19400 + 8300 = 88200$ m kg

Maximum axial load in the farthest columns due to wind load

$= \frac{Mr}{\Sigma r^2}$

$r^2 = 4\left(\frac{6{\cdot}5}{2} \cos 30°\right)^2 = 4(3{\cdot}25 \times {\cdot}866)^2 = 4 \times 8 = 32$

axial load due to wind $= \frac{88200 \times 2{\cdot}82}{32 \times 1000} = 7{\cdot}8$ t

Maximum load on the leeward column due to all loads $= 38{\cdot}27 + 7{\cdot}8 = 46{\cdot}07$ t (comp)

Minimum load on the windward column
$= 19{\cdot}27 - 7{\cdot}8 = 11{\cdot}47$ t (comp)

Total horizontal loads due to wind on the columns
$= 3450 + 2550 + 1090 = 7090$ kg

Horizontal shear force per column

$=\frac{7090}{6}=1181{\cdot}67$ kg$=1182$ kg say

5·7·3. Bending Moment In the Column

Maximum moment in the column

$=1182\times\frac{3{\cdot}8}{2}=2260$ m kg

5·7·4. Column Trial Section

M 150 concrete. Try 40 cm dia column, 6—16 mm ϕ bars

Equivalent area of the column

$=\frac{\pi\times 40^2}{4}+17\times 2{\cdot}01\times 6=\frac{\pi\times 1600}{4}+17\times 12{\cdot}06=1250+205$

$=1455$ cm^2

Moment of intertia of the section

$=\frac{\pi\times 40^4}{64}+17\times 4\times 2{\cdot}01\times(16\times{\cdot}866)^2$

$I=\frac{3{\cdot}14\times 40^4}{64}+17\times 4\times 2{\cdot}01\times 115=126{,}000+15{,}800$

$=141{,}800$ cm^4

5·7·5. Stresses in Column

direct stress$=\sigma_c'=\frac{P}{A}=\frac{46070}{1455}=31{\cdot}8$ kg/cm^2

bending stress$=\sigma_{cb}'=\frac{My}{I}=\frac{226000\times 20}{141{,}800}=31{\cdot}5$ kg/cm^2

$\frac{\sigma_c'}{\sigma_c}=\frac{\sigma_{cb}'}{\sigma_{cb}}<1$

$\frac{31{\cdot}8}{1{\cdot}33\times 40}+\frac{31{\cdot}5}{1{\cdot}33\times 50}=0{\cdot}6+{\cdot}475=1{\cdot}075>1$

Try 40 cm dia column with 6—20 mm ϕ

$A=\frac{\pi\times 40\times 40}{4}+17\times 6\times 3{\cdot}14=1250+322=1572$ cm^2

$I=\frac{3{\cdot}14\times 40^4}{64}+17\times 4\times 3{\cdot}14\times 115=126{,}000+23{,}700$

$=149{,}700$ cm^4

$\sigma_c'=\frac{46070}{1572}=29{\cdot}5$ kg/cm^2

$\sigma_{cb}'=\frac{226000\times 20}{149700}=30{\cdot}25$ kg/cm^2

$\frac{\sigma_c'}{\sigma_c}+\frac{\sigma_{cb}'}{\sigma_{cb}}=\frac{29{\cdot}5}{1{\cdot}33\times 40}+\frac{30{\cdot}25}{1{\cdot}33\times 50}$

$={\cdot}551+{\cdot}457=1{\cdot}008\simeq 1$

40 cm dia column with 6—20 mm ϕ,

Use 6 mm at 25 cm c/c links

5·8. Brace

Max moment in the brace$=2\times$moment in the column

$=2\times 2260=4520$ m kg

Try a section 35 cm × 25 cm with equal tensile and compressive reinforcement

$A_t \times 28 \times \sigma_{st} = M,$

$A_t = \dfrac{4520 \times 100}{28 \times 1400} = 11{\cdot}6\ \text{cm}^2,$

Use – 6 16 mm ϕ in both flanges

5·8·1. Shear Stress

$\text{Shear force} = \dfrac{M}{\text{half span}} = \dfrac{4520}{3{\cdot}25 \sin 30°} = \dfrac{4520 \times 2}{3{\cdot}25} = 2800\ \text{kg}$

$\text{Shear stress} = \dfrac{2500}{0{\cdot}87 \times 31 \times 25} = 4{\cdot}15\ \text{kg/cm}^2,$

Provide nominal stirrups 6 mm ϕ—25 cm c/c

5·9. Foundation

Loads : total load on the foundation when the tank is full

$= 46{\cdot}07 \times 6 = 276{\cdot}42$ t

Add 10% for self weight of foundation = 27·60 t

Total load = 304·02 t

Area required for the raft (assuming 15 t/m² bearing capacity)

$= \dfrac{304{\cdot}02}{15} = 20{\cdot}1$ sq. m

Try 5 m inner diameter and 8 m outer diameter

$\text{Area available} = \dfrac{\pi}{4}(8^2 - 5^2) = \dfrac{\pi}{4}(64 - 25) = \dfrac{\pi}{4} \times 39 = 30{\cdot}5$ sq. m

Try 5·5 m inner and 7·5 m outer diameter

$\text{Area} = \dfrac{\pi}{4}(7{\cdot}5^2 - 5{\cdot}5^2) = \dfrac{\pi}{4}(56 - 30{\cdot}25) = \dfrac{\pi}{4} \times 25{\cdot}75$

$= 20{\cdot}25$ sq. m.

$\text{Upward pressure on the raft} = \dfrac{304{\cdot}02}{30{\cdot}5} = 9{\cdot}95\ \text{t/m}^2 < 15\ \text{t/m}^2$

5·10. Circular Beam. Bending Moments

Total load on the beam (including wind) = W = 276·42 t

Bending Moment at support (–ve) = ·0148 WR

$= {\cdot}0148 \times 276{\cdot}42 \times \dfrac{6{\cdot}5}{2} = 13{\cdot}3$ mt

$\text{BM at centre } (+\text{ve}) = {\cdot}0075 \times 276{\cdot}42 \times \dfrac{6{\cdot}5}{2} = 6{\cdot}75$ mt

$\text{Torsional moment} = {\cdot}0015 \times 276{\cdot}42 \times \dfrac{6{\cdot}5}{2} = 1{\cdot}35$ mt

5·10·1. Shear Forces

$\text{Shear force} = \dfrac{\text{total load}}{2 \times \text{No. of columns}} = \dfrac{W}{2 \times 6}$

$= \dfrac{276{\cdot}42}{12} = 22{\cdot}25$ t

Point of maximum torsion = 12·75°

Shear forces at point of maximum torsion

$= 22{\cdot}25 - \dfrac{22{\cdot}25 \times 12{\cdot}75 \times 2}{60} = 22{\cdot}25 - 9{\cdot}45 = 12{\cdot}80$ t

5·10·2. Shear Stresses

Shear stress at point of maximum torsion (assuming 40 cm × 70 cm) $q=\dfrac{12800}{40\times\cdot 87\times 6\cdot 6}=5\cdot 6\ \text{kg/cm}^2$

Shear stress due to torsional moment

$$q'=\frac{T_m\left(3\times 2\dfrac{b}{D}\right)}{b^2D}=\frac{135000\left(3+2\times\dfrac{40}{70}\right)}{40\times 40\times 70}$$

$$=\frac{135000\times 4\cdot 14}{40\times 40\times 70}=4\cdot 97\ \text{kg/cm}^2$$

Total shear stress$=5\cdot 6+4\cdot 97=10\cdot 57$ kg/cm² > 5 kg < 20 kg/cm²

Shear stirrups shall be provided

5·10·3. Depth of Beam

$$d=\sqrt{\frac{1330000}{8\cdot 7\times 40}}=\sqrt{3820}=62\ \text{cm}$$

Use 40 × 70 cm beam, $d_e=64$ cm

5·10.4. Main Steel

A_t for −ve moment (water force)

$$=\frac{1330000}{1000\times\cdot 84\times 54}=24\cdot 7\ \text{cm}^2.\quad \text{Use } 8-20\ \text{mm}\ \phi$$

A_t required for +ve moment $=\dfrac{675000}{1250\times\cdot 86\times 64}=9\cdot 75\ \text{cm}^2$

Use 5 Nos − 16 mm ϕ

5·10·5. Shear Reinforcement

Shear stirrups at support, Using 12 mm ϕ 6 legged

$$\text{Spacing}=p=\frac{6\times 1\cdot 13\times 1000\times\cdot 86\times 64}{22250}=16\cdot 7\ \text{cm c/c}$$

Use 12 mm ϕ 6−legged 16 cm c/c

5·10·6. Shear Stirrups at Point of Maximum Torsion

Area of steel required for shear force

$$A_{w_1}=\frac{72800\times 16}{1000\times\cdot 86\times 64}=3\cdot 72\ \text{cm}^2$$

Using 16 cm c/c spacing A_t required for torsional shear stress

$$A_{w_2}=\frac{T_m\times p}{0\cdot 8\times t_w\times X\times Y}=\frac{135000\times 16}{0\cdot 8\times 1000\times 30\times 60}=1\cdot 5\ \text{cm}^2$$

Total shear reinforcement$=3\cdot 72+1\cdot 5=5\cdot 22$ cm²

Using 12 mm ϕ, No. of legs required $=\dfrac{5\cdot 22}{1\cdot 13}=4\cdot 65$

Use 12 mm ϕ−6 legged stirrups at 16 cm c/c

5·10·7. Longitudinal Reinforcement

Volume of stirrups per unit length provided for torsion

$=\dfrac{1\cdot 5\times 60}{16}=5\cdot 62$ cm². Use 6−12 mm ϕ 3 per face

5·11. Raft Slab

Projection of the raft slab from the edge of beam = 55 cm

Bending moment $=9950\times\cdot 55\times\dfrac{0\cdot 55}{2}=1510$ m kg

$=151{,}000$ cm kg

Depth $d_e=\sqrt{\dfrac{151000}{8{\cdot}7\times100}}=\sqrt{174}=13{\cdot}2$ cm

Use 15 cm thick raft, $d_e=13$ cm

$A_t=\dfrac{151000}{{\cdot}87\times1400\times13}=9{\cdot}35\ \text{cm}^2$. Use 12 mm ϕ at 12 cm c/c

Secondary steel at ·15% $=\dfrac{{\cdot}15\times15\times100}{100}=2{\cdot}25\ \text{cm}^2$

Use 6 mm ϕ at 12·5 cm c/c

5·12. Balcony and Landings

75 cm wide. Loads : Live loads $=150$ kg/m²
Slab (10 cm) $=240$ kg/m²
dead load of railings etc $=10$ kg/m²
total load $=400$ kg/m²

$\text{BM}=400\times{\cdot}75\times\dfrac{{\cdot}75}{2}=112{\cdot}5$ m kg

$\text{depth}=\sqrt{\dfrac{11250}{8{\cdot}7\times100}}=\sqrt{13}=3{\cdot}6$ cm

Use 10 cm at the edge and 7·5 cm at the end

$A_t=\dfrac{11250}{1400\times{\cdot}87\times8}=1{\cdot}15\ \text{cm}^2$

Use min 10 mm at 15 cm c/c

Secondary steel $=\dfrac{{\cdot}15}{100}\times10\times100=1{\cdot}5\ \text{cm}^2$

Use 6 mm ϕ at 18·5 cm c/c

5·13. Stability

Total load on the Soil including weight of foundation $=304{\cdot}02$ t
Total weight when tank is empty $=W=304{\cdot}02-114=190{\cdot}02$ t
Total wind moment $=M_w=88200$ m kg $=88{\cdot}20$ mt
$P=$ Total horizontal force $=7{\cdot}09$ t

Factor of safty against overturning $=\dfrac{\text{Stabilizing moment}}{\text{Overturning moment}}$

$=\dfrac{W_x}{M_w}=\dfrac{190{\cdot}02\times6{\cdot}5}{2\times88{\cdot}200}=7{\cdot}01$

Factor of safety against sliding $=\dfrac{\mu W}{P}=\dfrac{0{\cdot}5\times190{\cdot}02}{7{\cdot}09}=13{\cdot}6$

Intze Type Elevated Water Tank General Information 6

6·1. General

This type of tank was originally designed to obtain balanced inward and outward thrusts on the ring beam at the summits of the columns.

This type of tank is an improvement on the simple domed floor and was devised to obtain equilibrium between the forces on the annular support due to water in the portions above the dome and above the core.

The equilibrium is obtained when

$$W_1 \cot \alpha = W_2 \cot \beta$$

where W_1 = Sum of weight of domed floor and water above it.

W_2 = Sum of weights of roof and wall loads, inclined portion of floor and wieght of water above the portion.

The floors for these tanks may be flat or domed. A central shaft is also sometimes provided above the domed floor.

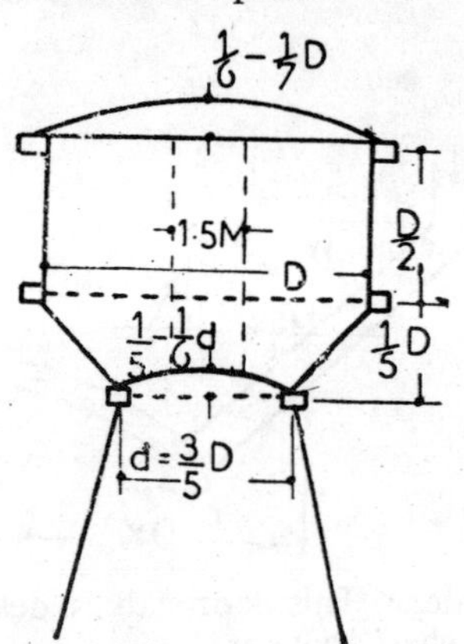

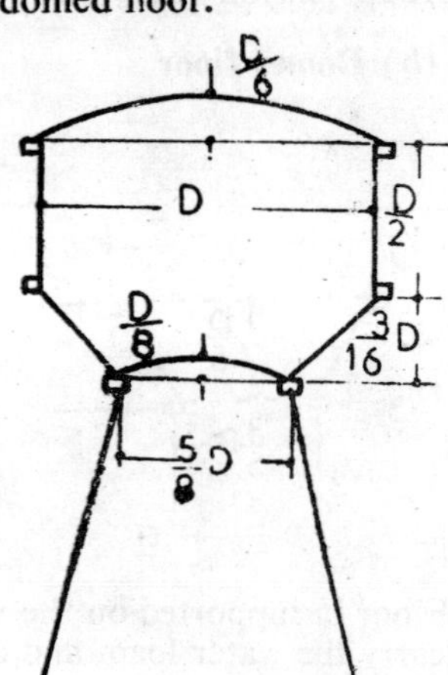

6·2. Components of the Tank

Tank portion : (*i*) domed roof, (*ii*) cylinder, (*iii*) conical or inclined slab (*iv*) domed floor

Tower portion : (*v*) circular ring girder (*vi*) columns (*vii*) braces (*viii*) Foundation.

6·3. Proportioning

(*i*) Gray's, (*ii*) Reynauld's

6·4. Analysis

(1) Domed roof : It is designed to take : (1) self weight of dome (2) weight of lantern (3) weight of water proof covering (4) insulation if any (5) snow loads.

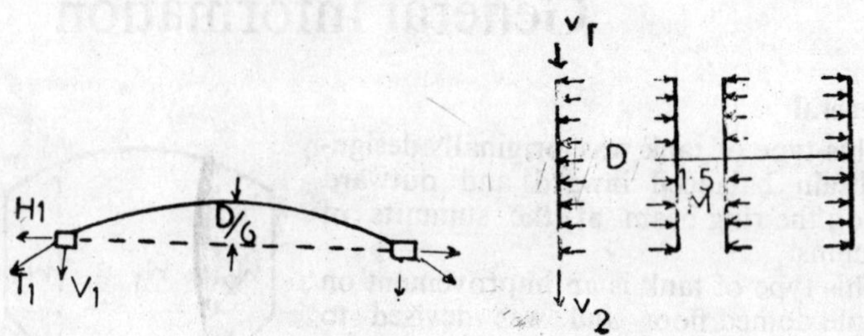

Usually 10 to 15 cm thick slab with minimum reinforcement of 0·2% will be ample. A ring beam is provided to take horizontal thrust H_1. The roof may be of spherical or conical dome.

(*ii*) The cylindrical portion of the tank is designed for hoop tension due to water pressure.

The central shaft is designed for hoop compression.

6·4·1. (*iii*) Conical or Inclined Slab

This slab is designed to span between the two ring beams and also for hoop tension due to water pressure as a conical dome

Horizontal thrust at any height $H_x = P \text{ cosec } \alpha + w \cot \alpha$

Hoop tension $= H_x \times \frac{D_x}{2}$

6·4·2. (*iv*) Domed floor

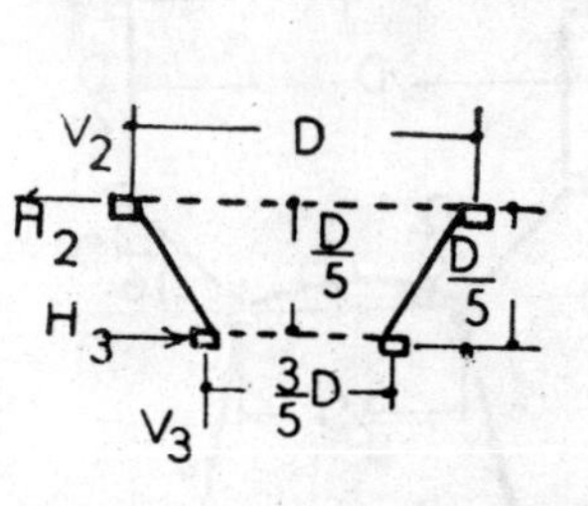

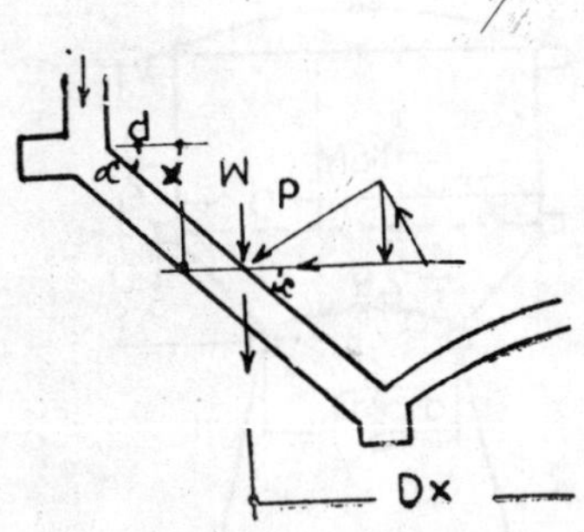

Floor is supported on the ring girder. This floor slab is designed to carry the water load, and central shaft above it.

6·4·3. Ring Girder

The design of this girder depends on the number of supports on which it rests. The ring grinder is subjected to bending and torsional moments. The moments are computed using the coefficients in the table.

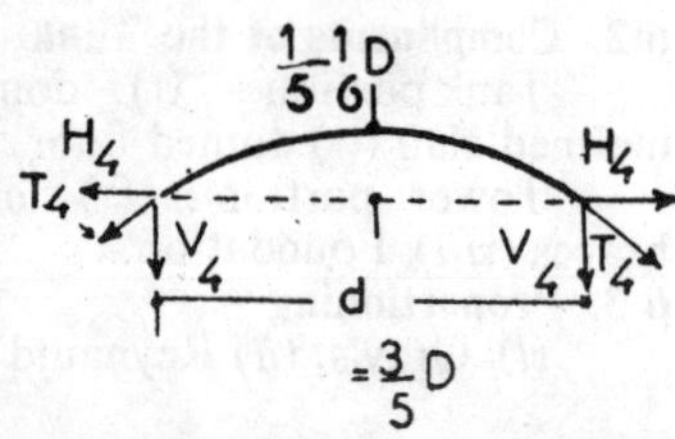

Number of supports	Bending moment at support	Bending moment at centre	Torsional moment	Point of maximum torsion
4	0·0342 *WR*	0·0176 *WR*	0·0053 *WR*	19°—12′
6	0·0148 *WR*	0·0075 *WR*	0.0015 *WR*	12°—44′
8	0·0083 *WR*	0·00416*WR*	0·0006 *WR*	9°—33′
10	0·0054 *WR*	0·00230*WR*	0·0003 *WR*	7°—30′
12	0·00375*WR*	0·00142*WR*	0·00017*WR*	6°—15′

where W=total vertical load on the girder, R=radius of the girder

6·4·4. Columns

Columns are designed for both axial loads and bending moment due to wind loads on the tank.

6·4·5. Braces

Braces are designed as a steel beam, that is equally reinforced in both tension compression zones, to resist +ve and −ve moments produced by the columns above and below it, due to wind loads. Moment in the brace is twice the moment in the column if the braces are placed at regular intervals.

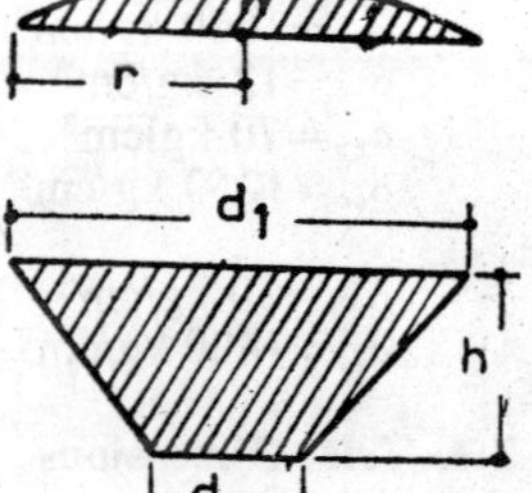

6·4·6. Foundations

The foundation normally consists of a circular ring beam on a circular slab.

6·5. Volume Computation

Volume of a segment of a sphere $V=\frac{\pi h}{6}(3r^2+h^2)$

Volume of frustum of a cone$=\frac{\pi h}{12}(d_1^2+d_2^2+d_1d_2)$

Intze Type Elevated Water Tank 7

Capacity 900,000 Litres

7.1. Data

Capacity=900 cu m. Elevation of the tank to the ring girder above the ground level=12 m. Bearing capacity of soil=$40 t/m^2$.

Materials available, Concrete–grade *M* 200 and *M* 150 Steel–grade–1.

7·2. Characteristic Strengths

$\sigma_{cb}=70$ kg/cm² $\quad m=13$
$\sigma_{st}=1000$ kg/cm² $\quad jd=\cdot 84d$
$\sigma_{ct}=12$ kg/cm² $\quad R=14$
$\sigma_{cb}=70$ kg)cm² $\quad m=13$
$\sigma_{st}=1250$ kg/cm² $\quad jd=\cdot 86d$
$\quad R=12\cdot 6$
$\sigma_{cb}=50$ kg/cm² $\quad m=18$
$\sigma_{tt}=1400$ kg/cm² $\quad jd=\cdot 87d$
$\quad R=8\cdot 7$

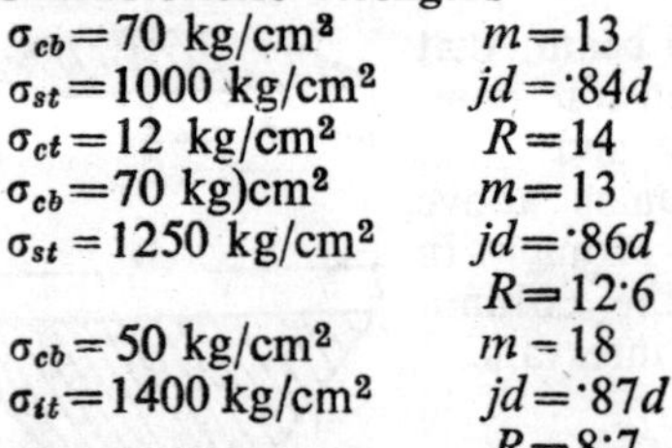

7·3. Trial Dimensions

Required capacity=900 cu m

Adopting Gray's proportions, let

$$D=12\cdot 5 \text{ m}$$

Volume of the cylindrical portion

$$V_1=\frac{\pi\times(12\cdot 5)^2}{4}\times 6\cdot 25=765 \text{ cu m}$$

Volume of the conical portion for $h=2\cdot 5$m

$$V_2=\frac{\pi h}{12}(D^2+d^2+dD)$$

$$=\frac{\pi\times 2\cdot 5}{12}(12\cdot 5^2+7\cdot 5^2+12\cdot 5\times 7\cdot 5)=199\cdot 5 \text{ cu m.}$$

V_3=Volume of the segment of a sphere

$$=\frac{\pi h_1}{6}(3\,r^2+h_1^2)=\frac{\pi\times 1\cdot 5}{6}\left[3\times\left(\frac{7\cdot 5}{2}\right)^2+1\cdot 5^2\right]=44\cdot 5 \text{ cu m}$$

V_4=Volume of the central shaft$=\frac{\pi}{4}\times(1\cdot 5)^2\times 7\cdot 25=12\cdot 8$ cu m

Actual volume provided$=V_1+V_2-(V_3+V_4)$

$=765+199\cdot 5-(44\cdot 5+12\cdot 8)=964\cdot 5-57\cdot 3$

$=907\cdot 2$ cu m $>$ 900 cu m

Therefore the assumed section gives the required capacity.

7·4. Domed Roof

Loads : Assuming —15 cm thick self weight of dome

$=\frac{15}{100}\times 1\times 1\times 2400=360$ kg/m²

Wind and accidental loads $=150$ kg/m²

Water proofing, insulation etc. $=90$ kg/m²

Total uniformly distributed load $w=600$ kg/m²

7·4·1. Geometry of the dome

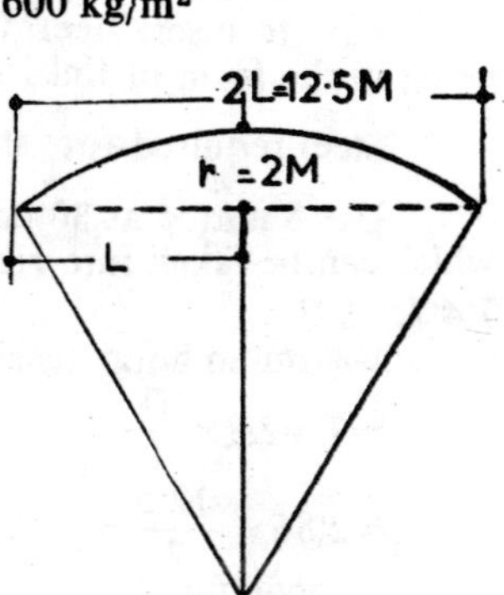

r=rise=2 m, $2l$ = span=12·5 m,

R=radius

$$R=\frac{l^2+r^2}{2r}=\frac{6{\cdot}25^2+2^2}{2\times 2}$$

$=10{\cdot}75$ m

$$\text{Sin }\phi=\frac{6{\cdot}25}{10{\cdot}75}=0{\cdot}585$$

$$\phi=35^\circ-48' < 51^\circ-48'$$

Therefore entire dome will be subjected to compressive stresses

7·4·2. Stresses in the Dome

Hoop stress at crown ($\phi=0$)

$$=\frac{w\,R}{2t}=\frac{600\times 10{\cdot}75}{2\times{\cdot}15}=21400\text{ kg/m}^2=2{\cdot}14\text{ kg/cm}^2$$

Meridional stress :

$$C=\frac{w\times R}{t(1+\cos\phi)}=\frac{600\times 10{\cdot}75}{{\cdot}15(1+{\cdot}8111)}=23{,}600\text{ gk/m}^2=2{\cdot}36\text{ kg/cm}^2$$

The stresses are however low.

7·4·3. Force Components

Surface area $=2\pi Rr=2\pi\times 10{\cdot}75\times 2$

$=134$ sq. m

Total load $=W=134\times 600$

$=80400$ kg $=80{\cdot}4$ t

$$V_1=\frac{W}{2\pi l}=\frac{80{\cdot}4}{2\pi\times 6{\cdot}25}=2{\cdot}05\text{ t}$$

$$H_1=V_1\cot\phi=2{\cdot}05\times\left(\frac{R-r}{l}\right)$$

$$=2{\cdot}05\times\frac{(10{\cdot}75-2)}{6{\cdot}25}=\frac{2{\cdot}05\times 8{\cdot}75}{6{\cdot}25}$$

$=2{\cdot}86$ t

$$T_1=\frac{V_1}{\sin\phi}=V_1\frac{R}{l}=\frac{2{\cdot}05\times 10{\cdot}75}{6{\cdot}25}=3{\cdot}53\text{ t}$$

7·4·4. Shear Stress at the Edge

Shear stress along the perimeter of the dome

$$=\frac{V_1}{\text{area}}=\frac{2050}{100\times 15}=1{\cdot}37\text{ kg/cm}^2 < 5\text{ kg/cm}^2$$

A 15 cm thick slab of *M* 150 concrete provides good protection against leakage etc, although a 10 cm thick roof would have been structurally safe.

7·4·5. Minimum Reinforcement

Since there are no tensile stresses anywhere in the dome, only a nominal reinforcement is required, to allow for such indeterminate stresses due to wind, distortion, shrinkage and temperature etc.

$$A_t = \cdot 2\% \text{ of concrete area} = \frac{\cdot 2}{100} \times 15 \times 100 = 3 \text{ cm}^2$$

Use 8 mm ϕ 16 cm c/c in both the directions

7·4·6. Steel for the Shear at the Junction of the Rib and Dome

Not withstanding the low shear stress along the perimeter, it is customary to insert steel to resist all the shear, the reinforcement being in the form of links continuous around main bars in the rib.

$$\text{Steel required for shear} = \frac{V_1}{1400} = \frac{2050}{1400} = 1{\cdot}46 \text{ cm}^2$$

Use 8 mm ϕ at 30 cm c/c in addition to the dome reinforcement which can be taken into rib round the rib reinforcement.

7·4·7. Rib

Maximum hoop tension

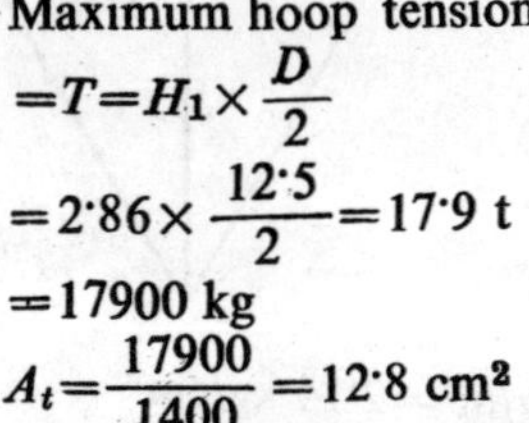

$$= T = H_1 \times \frac{D}{2}$$

$$= 2{\cdot}86 \times \frac{12{\cdot}5}{2} = 17{\cdot}9 \text{ t}$$

$$= 17900 \text{ kg}$$

$$A_t = \frac{17900}{1400} = 12{\cdot}8 \text{ cm}^2$$

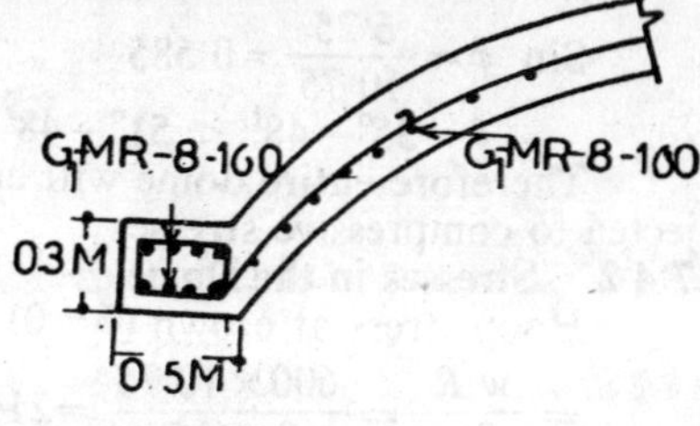

Use 16 mm ϕ 8 Nos. (A_t provided 16·08 cm²)

$\sigma_{ct} = 11$ kg/cm²

$$\sigma_{ct} = \frac{T}{A_c + (m-1)A_t}, \quad 11 = \frac{17900}{A_c + 17 \times 16{\cdot}08}$$

$$11\,A_c + 17 \times 16{\cdot}08 \times 11 = 17900$$

$$11\,A_c + 3000 = 17900$$

$$A_c = \frac{13900}{11} = 1270 \text{ cm}^2. \text{ Use } 50 \text{ cm} \times 30 \text{ cm}$$

7·5. Cylindrical Wall Hoop Tension

The cylindrical wall is designed to resist hoop tension.

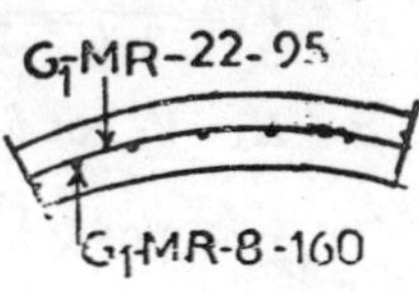

Maximum water pressure at the bottoom of the cylindrical wall, $p = wh = 1000 \times 6{\cdot}25$
$= 6250$ kg/m²

Maximum hoop tension in the wall

$$T = p \times \frac{D}{2} = 6250 \times \frac{12{\cdot}5}{2} = 39000 \text{ kg}$$

7·5·1. Main Steel

$$\text{Area of steel required} = A_t = \frac{39000}{1000}$$

$= 39$ cm²

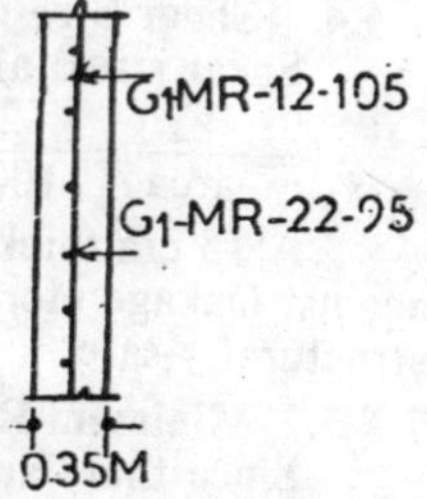

Use 22 mm ϕ at 9·5 cm c/c centre of the wall. A_t provided $= 40{\cdot}02$ cm²

Use 16 mm ϕ 10 cm c/c both faces (A_t provided 40·22 cm²)

7·5·2. Thickness of the wall

$$\sigma_{ct} = \frac{T}{A_c + (m-1)A_t}$$

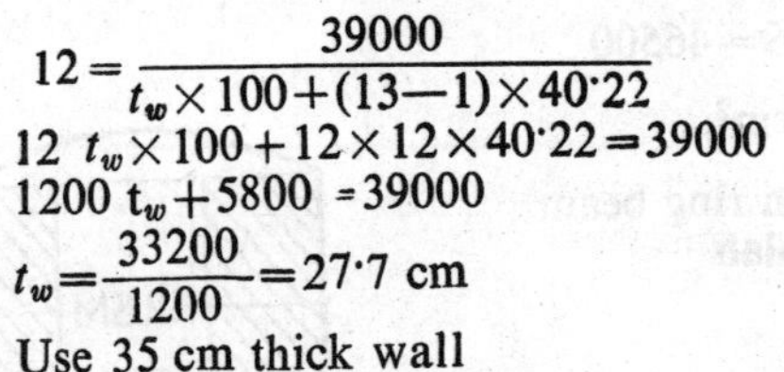

$$12 = \frac{39000}{t_w \times 100 + (13-1) \times 40{\cdot}22}$$

$$12\ t_w \times 100 + 12 \times 12 \times 40{\cdot}22 = 39000$$

$$1200\ t_w + 5800 = 39000$$

$$t_w = \frac{33200}{1200} = 27{\cdot}7 \text{ cm}$$

Use 35 cm thick wall

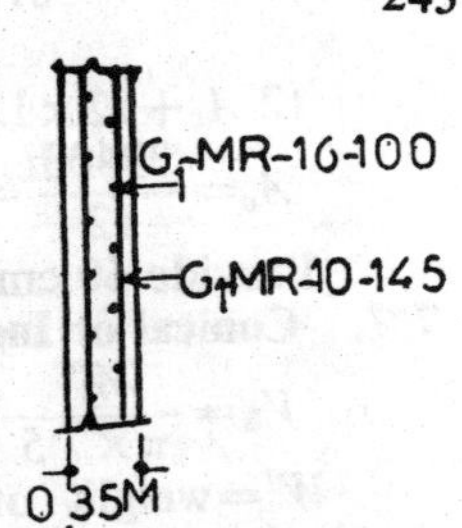

7·5·3. Secondary Reinforcement

$A_t = {\cdot}3\%$ concrete area

$$= \frac{{\cdot}3}{100} \times 35 \times 100 = 10{\cdot}5 \text{ cm}^2$$

Use 10 mm ϕ at 14·5 cm c/c on both faces

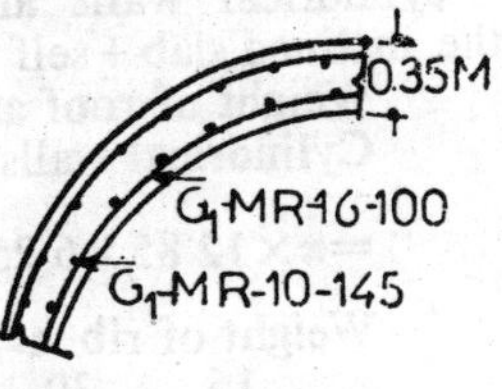

7·5·4. Stresses in the Wall

Tensile stress

$$= \frac{39000}{35 \times 100 + 12 \times 40{\cdot}22}$$

$$= \frac{39000}{3902} = 9{\cdot}8 \text{ kg/cm}^2 < 12 \text{ kg/cm}^2$$

Compressive stress. Vertical load from the dome $= V_1 = 2050$ kg/m

weight of wall

$$= \frac{35}{100} \times 2400 \times 1 \times 6{\cdot}25 = 5250 \text{ kg/m}$$

Weight of rib

$$= \frac{15}{100} \times \frac{30}{100} \times 1 \times 2400 = 108 \text{ kg/m}$$

Total load per metre width of wall

$V_2 = 7408$ kg/m

Compressive stress=

$$= \frac{7408}{100 \times 35} = 2{\cdot}13 \text{ kg/cm}^2$$

Stress is very low.

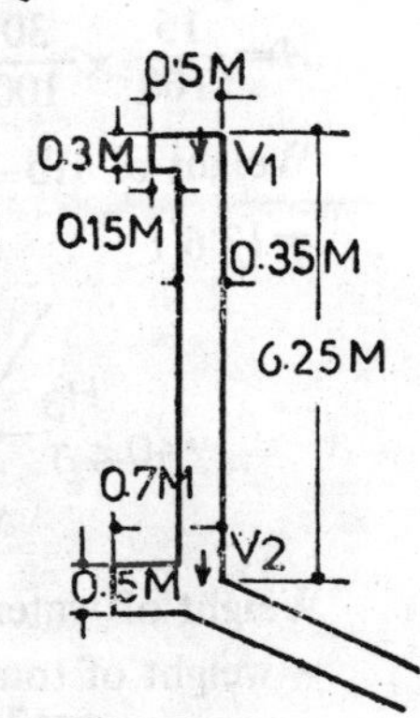

7·6. Ring Beam at the Junction of Cylindrical and Conical Portion

Horizontal thrust H_2 acting outwards at the junction $= H_2 = \frac{V_2}{\tan \alpha} = 7408$ kg/m

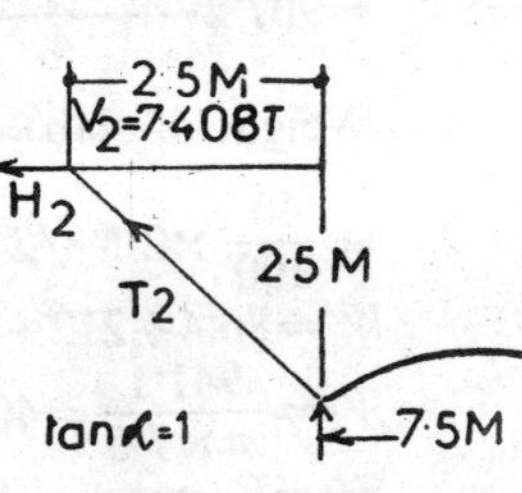

Hoop tension in the rib

$$= H_2 \times \frac{D}{2} = 7408 \times \frac{12{\cdot}5}{2} = 46500 \text{ kg}$$

7·6·1. Main reinforcement in the ring beam

$$A_t = \frac{46500}{1000} = 46{\cdot}5 \text{ cm}^2$$

Use 25 mm ϕ – 10 Nos.

(A_t provided = 49·09 cm²)

7·6·2. Dimensions of the Ring Beam

$$\sigma_{ct} = \frac{T}{A_c + (m-1)A_t}$$

$$12 = \frac{46500}{A_c + 12 \times 49{\cdot}09}$$

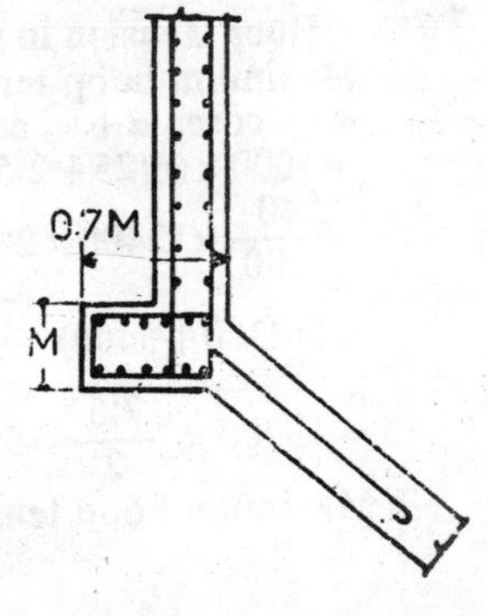

$$12\ A_c + 12 \times 12 \times 49{\cdot}09 = 46500$$

$$A_c = \frac{39400}{12} = 3280\ \text{cm}^2.$$

Provide 50 cm × 70 cm ring beam

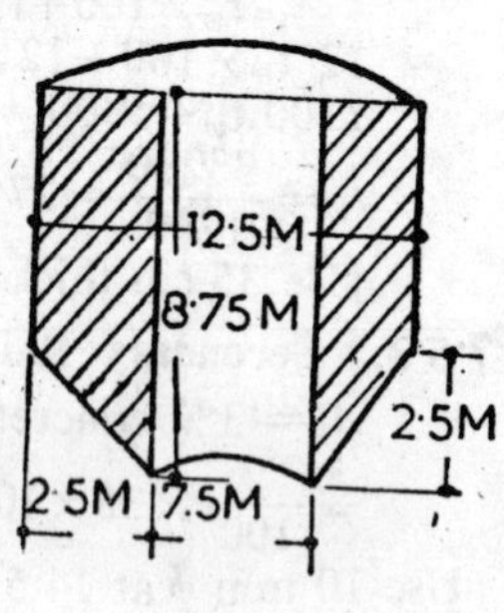

7·7. Conical or Inclined Slab

$$V_3 = \frac{W'}{\pi \times 7{\cdot}5}\ \text{kg/m}$$

W' = weight of roof and its loads + cylindrical walls and ribs + water above the inclined slab + self weight of inclined slab

Weight of roof and its load = 80·4 t

Cylindrical walls

$$= \pi \times 12{\cdot}85 \times 6{\cdot}25 \times \frac{35}{100} \times \frac{2400}{1000} = 212\ \text{t}$$

Weight of rib A

$$A = \frac{15}{100} \times \frac{30}{100} \times \frac{2400}{1000} \times \pi \times 13{\cdot}35 = 4{\cdot}5\ \text{t}$$

$$\text{Weight of rib} = \frac{35}{100} \times \frac{30}{100} \times \frac{2400}{1000} \times \pi \times 13{\cdot}55$$

$= 17{\cdot}6$ t

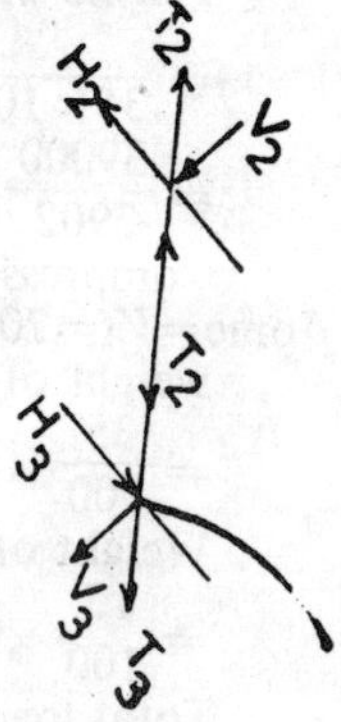

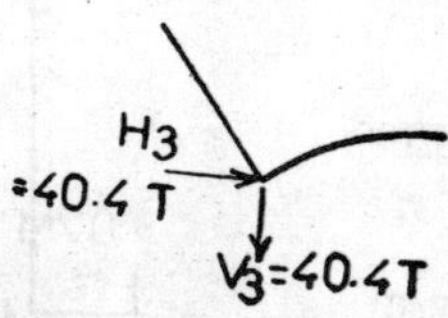

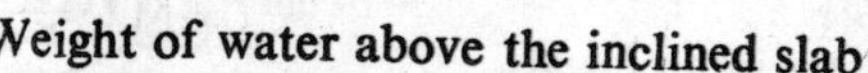

Weight of water above the inclined slab

= weight of total capacity − weight of water above domed floor

$$= 907{\cdot}2 - \frac{\pi \times 7{\cdot}5^2}{4} \times 8{\cdot}75 \times 1 = 907{\cdot}2 - 386 = 521{\cdot}2\ \text{t}$$

$$\text{Weight of slab} = \frac{40}{100} \times 2{\cdot}5\sqrt{2} \times \pi \times \text{mean radius} \times \frac{2400}{1000}$$

$$= \frac{40}{100} \times 2{\cdot}5\ \sqrt{2} \times \pi \times 10 \times \frac{2400}{1000} = 106{\cdot}2\ \text{t}$$

$$W' = 80{\cdot}4 + 212 + 4{\cdot}5 + 17{\cdot}6 + 521{\cdot}2 + 106{\cdot}2 = 941{\cdot}9\ \text{t}$$

$$V_3 = \frac{941{\cdot}1}{\pi \times 7{\cdot}5} = 40{\cdot}4\ \text{t}$$

$$H_3 = V_3 \cot \alpha = 40{\cdot}4 \times 1 = 40{\cdot}4\ \text{t}$$

7·7·1. Hoop Tension in the Inclined Slab

Maximum hoop tension

$$= (p \operatorname{cosec} \alpha + w_s \cot \alpha)\ D/2$$

$$= 1000 \times (6{\cdot}25 + 2{\cdot}5)\sqrt{2}$$

$$+ \frac{40}{100} \times (2400 \times 2{\cdot}5 \times \sqrt{2})\ \frac{7{\cdot}5}{2}$$

$$= (12300 + 3400)\ \frac{7{\cdot}5}{2}$$

$$= 15700 \times \frac{7{\cdot}5}{2} = 59{,}000\ \text{kg}$$

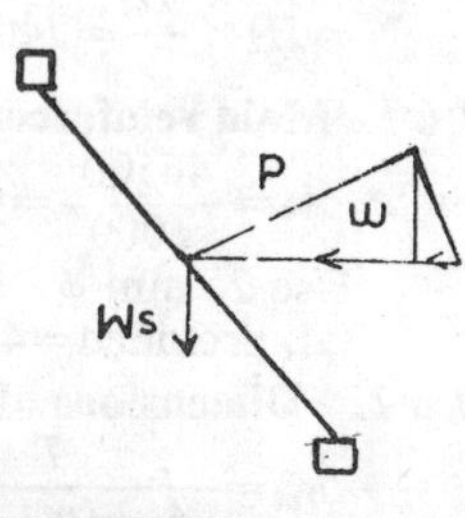

Maximum hoop tension in the inclined slab = 59000 kg

7·7·2. Hoop Steel

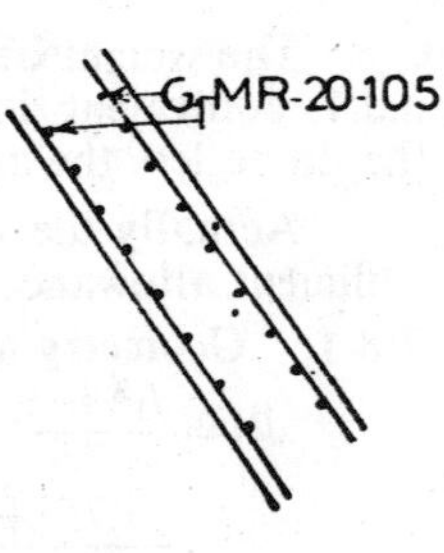

$$A_t=\frac{59000}{1000}=59 \text{ cm}^2$$

Use hoop steel on both the faces.

A_t per face $=29{\cdot}5$ cm²

Use 20 mm ϕ 10·5 cm c/c

(Area provided $=29{\cdot}91\times2=59{\cdot}82$ cm²)

7·7·3. Bending Moment

$$BM=\frac{V_3L}{8}=\frac{40{,}400\times2{\cdot}5}{8}=12600 \text{ m kg}$$

$$=1260{,}000 \text{ cm kg}$$

Axial compression in the inclined slab

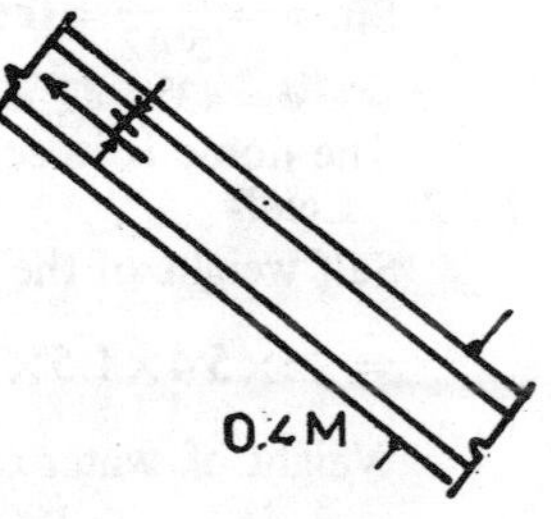

$$T_2=\frac{V_2}{\sin\alpha}=7408\times\sqrt{2}=10{,}500 \text{ kg}$$

Bending moment due to thrust T_2

$=10500\times(35-20)=10500\times15$

$=157{,}500$ cm kg

Total bending moment

$=1260{,}000+157{,}500$

$=1417{,}500$ cm kg

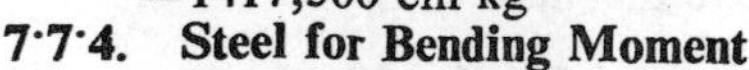

7·7·4. Steel for Bending Moment

$$A_t=\frac{1417500}{1250\times{\cdot}86\times35}=37{\cdot}5 \text{ cm}^2$$

Use 22 mm ϕ at 10 cm c/c

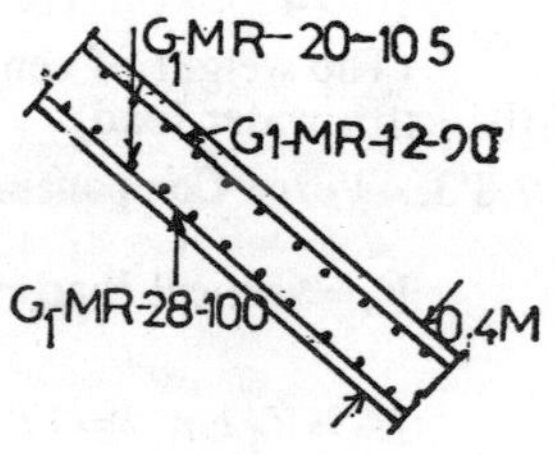

7·7·5. Secondary Steel

$$A_t=\frac{{\cdot}3}{100}\times40\times100=12 \text{ cm}^2$$

Use 12 mm ϕ 9 cm c/c at top

Stresses. $\sigma_{ct}=\dfrac{59000}{40\times100+59{\cdot}82\times12}$

$$=\frac{59000}{4715}-12{\cdot}5 \text{ kg/cm}^2 > 12 \text{ kg/cm}^2$$

However this can be reduced by increasing the thickness of the slab to 45 cm

7·7·6. Depth Required for BM

$$d=\sqrt{\frac{1417500}{12{\cdot}6\times100}}=\sqrt{1120}=33{\cdot}5 \text{ cm}$$

$d=40$ cm, $d_e=35$ cm

7·8. Domed Floor

The bottom dome has to carry :

(*i*) Self weight of dome

(*ii*) Self weight central shaft

(*iii*) Weight of water contained in the cylindrical portion above the dome

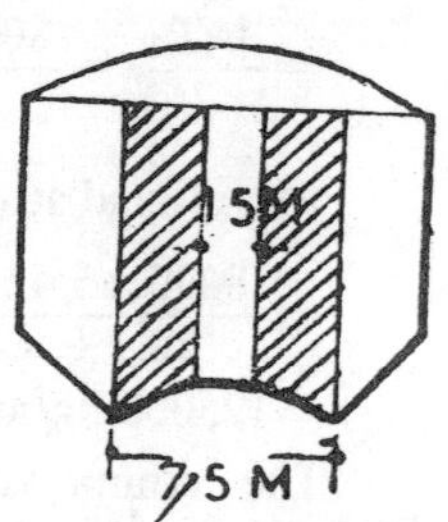

In estimating the weight of water it is safe to take full depth of water as 8·75 m approximately over the whole area of the dome including the opening through which the shaft passes

The weight of the shaft is generally regarded as an additional u.d.l. equivalent in intensity to its weight divided by the area of the dome less the area of the shaft.

Actually the assumption of water load over the wall provides sufficient allowance for the dead load of the shaft wall.

7·8·1. Geometry of the Dome

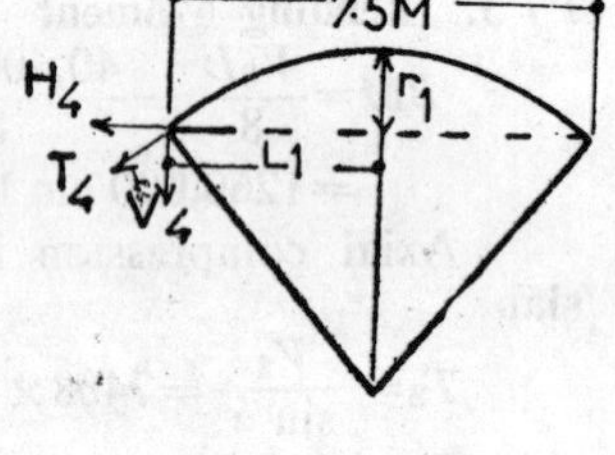

$R_1 = \frac{l_1^2 + r_1^2}{2r_1}$

$= \frac{3{\cdot}75^2 + 1{\cdot}5^2}{2 \times 1{\cdot}5} = \frac{16{\cdot}25}{3}$

$= 5{\cdot}42$ m

$\sin\phi = \frac{3{\cdot}75}{5{\cdot}42} = 0{\cdot}69$, $\cos\phi = {\cdot}7230$

$\phi = 43° - 42' < 51° - 48'$.

The dome is in compression

7·8·2. Loads

Self weight of the dome (assuming 20 cm thick) $= 2\pi R_1 r_1 \times w$

$= 2\pi \times 5{\cdot}4 \times 1{\cdot}5 \times \frac{2400}{1000} \times \frac{20}{{\cdot}100} \times 1 = 24{\cdot}5$ t

Weight of water contained above the floor

$= \frac{\pi d^2}{4} \times h \times \frac{1000}{1000} = \frac{T \times (7{\cdot}5)^2}{4} \times 8{\cdot}75 \times 1 = 386$ t

(The weight of central shaft has been accounted by considering the extra water load)

7·8·3. Force Components at springs

$V_4 = \text{vertical load} = \frac{386 + 24{\cdot}5}{\pi \times 7{\cdot}5} = \frac{410{\cdot}5}{\pi \times 7{\cdot}5} = 17{\cdot}5$ t

$H_4 = V_4 \cot\phi = 17{\cdot}5 \times \frac{(5{\cdot}42 - 1{\cdot}5)}{3{\cdot}75} = \frac{17{\cdot}5 \times 3{\cdot}92}{3{\cdot}75} = 18{\cdot}3$ t

$T_4 = \frac{V_4}{\sin\phi_1} = V_4 \frac{R_1}{l_1} = \frac{17{\cdot}5 \times 5{\cdot}42}{3{\cdot}75} = 25{\cdot}4$ t

7·8·4. Stresses

Intensity of the uniformly distributed load on the domed floor

$w_1 = \frac{410{\cdot}5}{2\pi R_1 r_1} = \frac{410{\cdot}5}{2\pi \times 5{\cdot}42 \times 1{\cdot}5} = 8$ t

Hoop stress

$= \frac{w_1 R_1}{2t} = \frac{8000 \times 5{\cdot}42 \times 100}{2 \times 20} = 108000 \text{ kg/m}^2 = 10{\cdot}8 \text{ kg/cm}^2$

Meridional stress $c = \frac{w_1 R_1}{t(1 + \cos\phi_1)}$

$= \frac{8000 \times 5{\cdot}42 \times 100}{20(1 + {\cdot}7230)}$

$= 125000 \text{ kg/m}^2 = 12{\cdot}5 \text{ kg/cm}^2$

The compressive stresses are within the limits

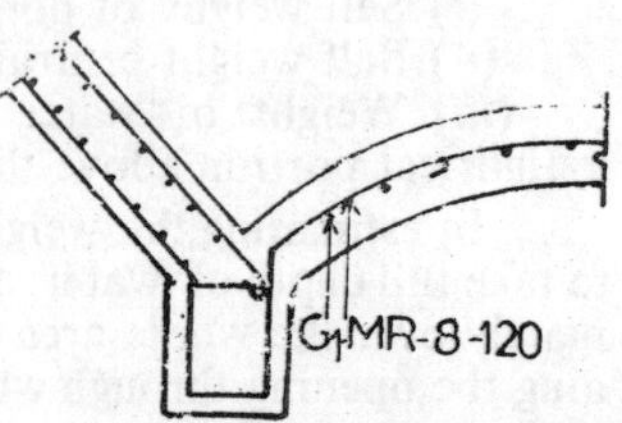

7·8·5. Minimum Reinforcement

$A_t=\frac{\cdot 2}{100}\times 20\times 100=4\text{ cm}^2$

Use 8 mm ϕ 12 cm c/c in both directions

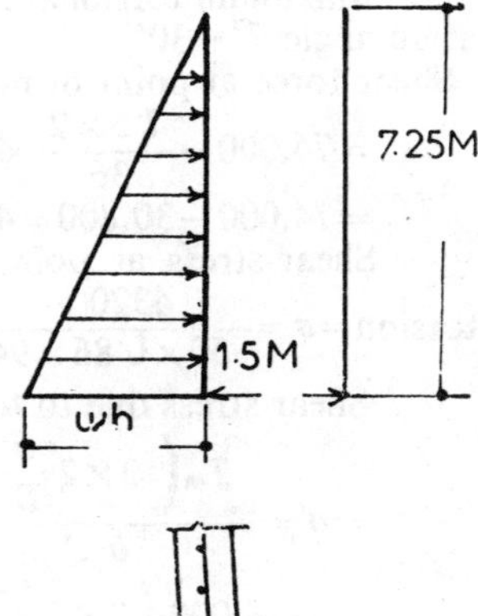

7·8·6. Central Shaft

Maximum pressure due to water
$=1000\times 7{\cdot}25=7250$ kg

Maximum hoop compression

$C=\frac{7250\times 1{\cdot}5}{2}=5425$ kg

assuming 15 cm thick compressive

stress $=\frac{5425}{100\times 15}=3{\cdot}62$ kg/cm²

Stress is very low

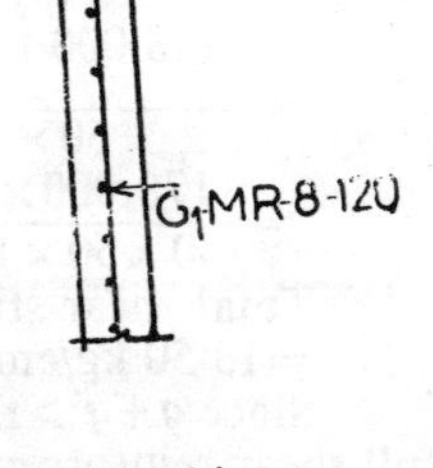

7·8·7. Nominal Reinforcement

$A_t=\frac{\cdot 3}{100}\times 15\times 100=4{\cdot}5\text{ cm}^2$

Use 8 mm ϕ 11 cm c/c both directions

7·9. Ring Beam

Total vertical load per metre on the ring beam $V=V_3+V_4=40{\cdot}4+17{\cdot}5=57{\cdot}9$ t

Net horizontal force $=H_3-H_4$

$H=40{\cdot}4-18{\cdot}3=22{\cdot}1$ t

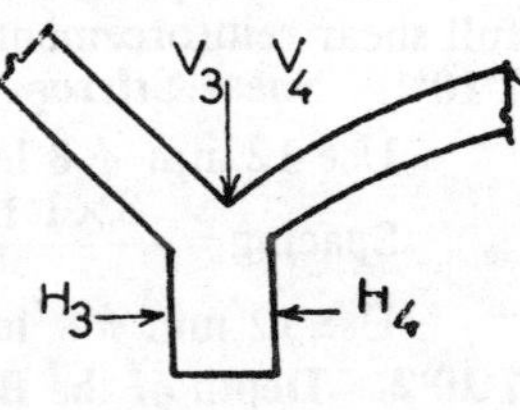

7·9·1. Trial Section

Assume 50 cm × 100 cm beam self weight of the beam/metre

$=1\times\frac{50}{100}\times\frac{100}{100}\times 2400=1200\text{ kg}=1{\cdot}2\text{ t}$

Total vertical load/metre $=57{\cdot}9+1{\cdot}2=59{\cdot}1$ t

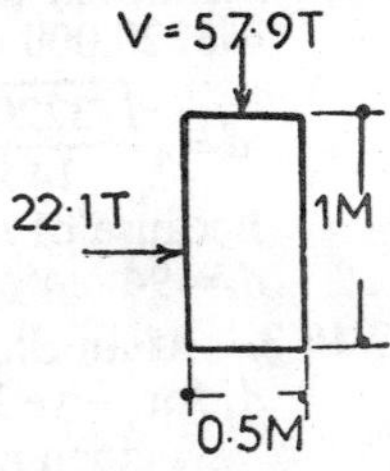

7·10. Number of Columns : Bending Moments

Assume 10 coulmns. $W=$ Total load
$=\pi\times 8\times 59{\cdot}1=1480$ t

Bending moment at support (−ve)
$=\cdot 0054\,Wr=\cdot 0054\times 1480\times 4=32{\cdot}20$ mt
$=3220{,}000$ cm kg

Maximum bending moment at centre
(+ve) $=0{\cdot}0023\times Wr=0{\cdot}0023\times 1480\times 4$
$=13{\cdot}80\text{ mt}=1380{,}000$ cm kg

Shear force $=\frac{\text{Total load}}{2\times\text{number of columns}}$

$=\frac{1480}{2\times 10}=74\text{ t}=74{,}000$ kg

Torsional moment $=\cdot 0003\,Wr$

$T_m=\cdot 003\times 1480\times 4=1{\cdot}78$ mt
$=178{,}000$ cm kg

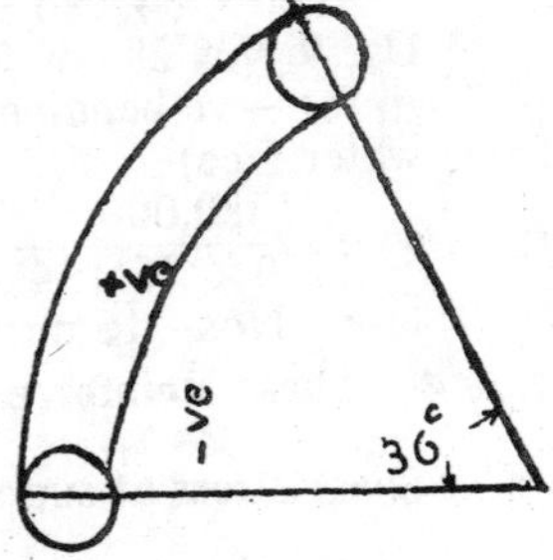

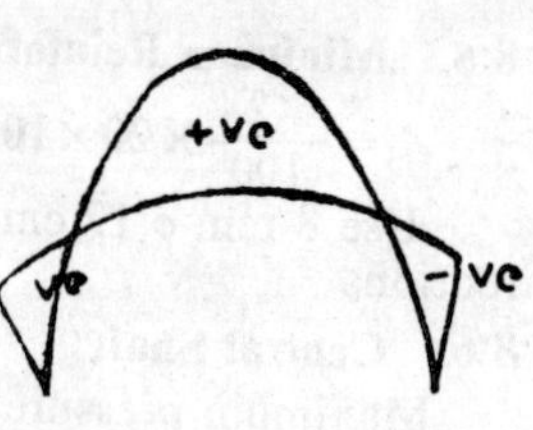

Maximum torsional moment occurs at an angle 7°–30′

Shear force at point of maximun torsion

$$=74{,}000-\frac{7{\cdot}5\times 2}{36}\times 74{,}000$$

$$=74{,}000-30{,}800=43{,}200 \text{ kg}$$

Shear stress at point of maximum

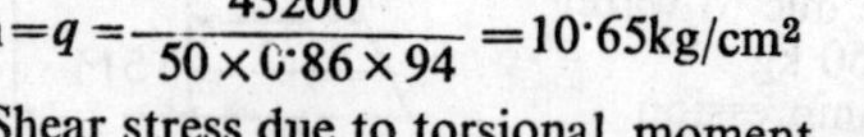

$$\text{torsion}=q=\frac{43200}{50\times 0{\cdot}86\times 94}=10{\cdot}65 \text{kg/cm}^2$$

Shear stress due to torsional moment,

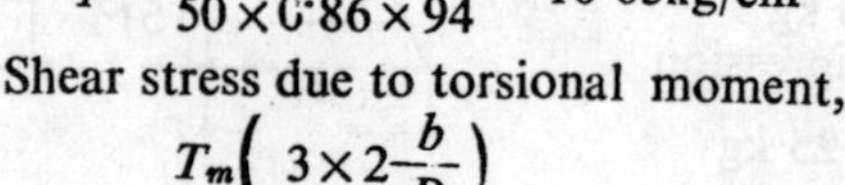

$$=q'=\frac{T_m\left(3\times 2-\frac{b}{D}\right)}{b^2D}$$

$$=\frac{178{,}000\left(3\times 2\times \frac{50}{100}\right)}{50\times 50\times 100}$$

$$=\frac{178{,}000\times 4}{50\times 50\times 100}=2{\cdot}85 \text{ kg/cm}^2$$

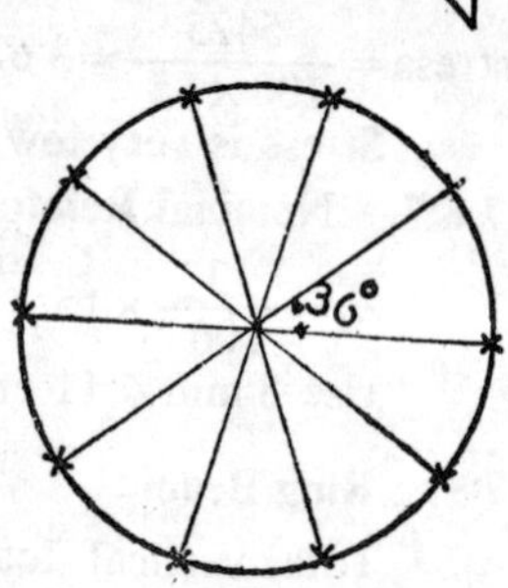

Total shear stress $=q+q'=10{\cdot}65+2{\cdot}85$

$=13{\cdot}50$ kg/cm² >7 kg/cm² <28 kg/cm²

Since $q+q'>$ the allowable shear stress, full shear reinforcement shall be provided.

7·10·1. Shear Stirrups at support

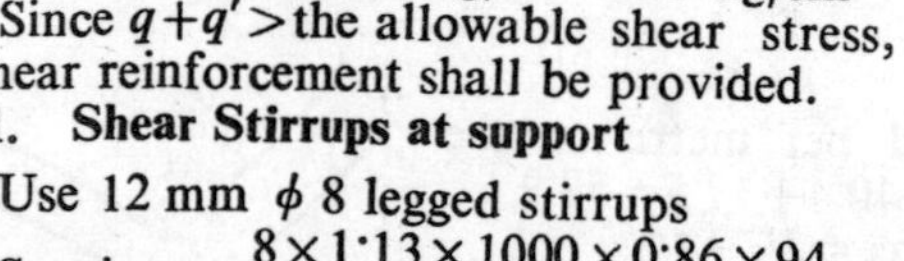

Use 12 mm ϕ 8 legged stirrups

$$\text{Spacing}=\frac{8\times 1{\cdot}13\times 1000\times 0{\cdot}86\times 94}{74{,}000}=9{\cdot}55 \text{ cm}$$

Use 12 mm ϕ 8 legged stirrups at 9·5 cm c/c

7·10·2. Depth of the Beam

Maximum bending moment (at support, water face) $=3220{,}000$ cm kg

$$d=\sqrt{\frac{3220{,}000}{14\times 50}}=\sqrt{4600}=68 \text{ cm}$$

Because of shear and torsional moment, Use $d=100$ cm, $d_e=94$ cm

7·10·3. Main Steel

A_t for –ve bending moment (water face)

$$=\frac{3220{,}000}{1000\times {\cdot}84\times 94}=40{\cdot}6 \text{ cm}^2$$

Use 7 Nos 28 cm ϕ (43·10 cm²)

A_t for +ve bending moment (away from water face)

$$=\frac{1380{,}000}{1250\times {\cdot}86\times 94}=13{\cdot}6 \text{ cm}^2$$

Use 7 Nos – 16 mm ϕ (14·07 cm²)

7-G₁MR-28

5-G₁MR-8

7 G₁MR 16

7·10·4. Shear Reinforcement

$$\text{Shear stress at support}=\frac{74{,}000}{50\times {\cdot}86\times 94}=18{\cdot}3 \text{ kg/cm}^2$$

7·10·5. Shear Stirrups at Point of Maximum Torsion

Using 9·5 cm spacing, Area of steel required for shear [illegible]ce

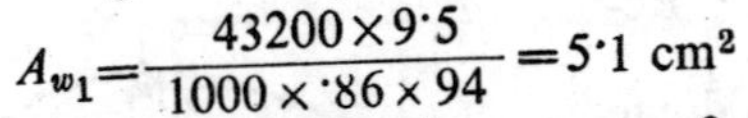

$$A_{w_1}=\frac{43200\times 9\cdot 5}{1000\times \cdot 86\times 94}=5\cdot 1 \text{ cm}^2$$

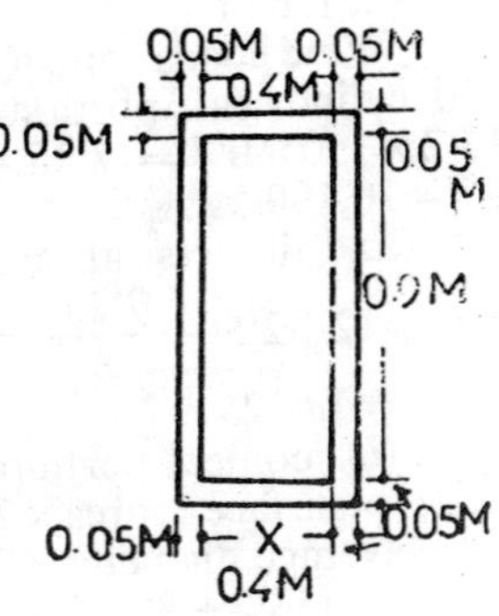

Using 9·5 cm spacing, area of steel required for torsional shear stress

$$A_{w_2}=\frac{T_m\times p}{0\cdot 8\times t_w\times X\times Y}$$

$$=\frac{78{,}000\times 9\cdot 5}{0\cdot 8\times 1000\times 90\times 40}=\cdot 587 \text{ cm}^2$$

Total steel required at point of maximum torsion $=5\cdot 1+\cdot 587=5\cdot 687$ cm²

Using 10 mm ϕ each leg $=0\cdot 79$,

number of legs required $=\dfrac{5\cdot 687}{0\cdot 79}$

$=8$ Nos Use 8 legged -10 mm ϕ stirrups at 9·5 cm c/c

7·10·6. Longitudinal Reinforcement for Torsion

Volume of longitudinal reinforcement per unit length
=Volume of stirrups per unit length

provided for torsion $=\dfrac{\cdot 587\times 90}{9\cdot 5}=5\cdot 55$ cm².

10 cm ϕ -8 Nos

7·10·7. Hoop Compressive Stress

$H=22\cdot 1$ t $=22100$ kg

Hoop compression $=H\times\dfrac{D}{2}$

$=22{,}100\times\dfrac{7\cdot 5}{2}=83{,}000$ kg

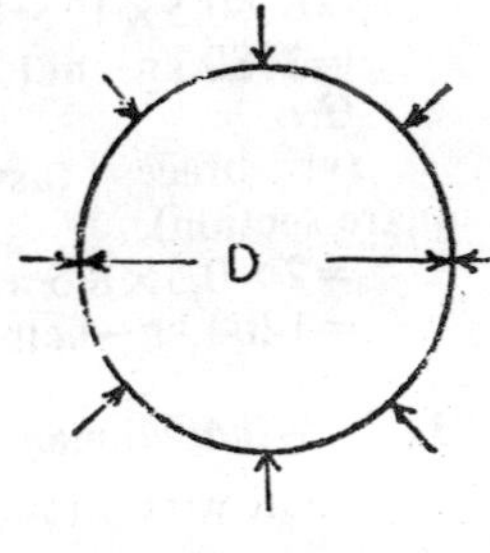

Compression stress $=\dfrac{83000}{100\times 50}$

$=16\cdot 6$ kg/cm²

Its effect is neglected since concrete can take this compressive stress

7·11. Staging or Tower Columns

Loads. (*i*) Tankfull, direct axial load on each column from the tank portion $=74\times 2=148$ t

Self weight of the column (assuming 50 cm dia column)

$=\dfrac{\pi}{4}\times(0\cdot 5)^2\times 12\times\dfrac{2400}{1000}$

$=5\cdot 65$ t

Total vertical load $=153\cdot 65$ t

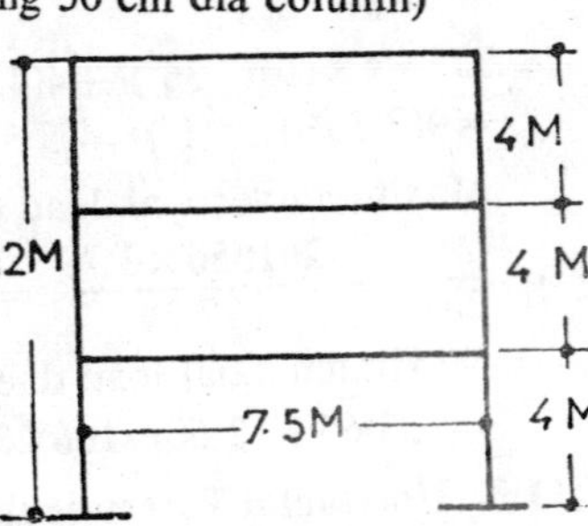

(*ii*) Tank Empty, weight of water in the tank $=907\cdot 2$ t

Axial load per column due to water $=\dfrac{907\cdot 2}{10}=90\cdot 72$ t

Actual load per column when the tank is empty $=153\cdot 65-90\cdot 72$
$=62\cdot 93$ t

7·11·1. Wind Loads

Wind pressure = 150 kg/m^2 Reduction factor = 0·7

Wind loads on, (*i*) top dome and cylindrical portion = (6·25+1) × 13·2 × 150 × 0·7 = 7·25 × 13·2 × 150 × 0·7 = 10,100 kg

and it acts at a height above

$$G.L. = 12+2·5+\frac{7·25}{2} = 14·5+3·625$$

= 18·125 m

(*ii*) conical portion of the tank = surface area × 2·5 × 150 × 0·7 = average diameter × 2·5 × 150 × 0·7

$$=\frac{(13·2+8·5)}{2}\times 2·5\times 150\times 0·7$$

= 10·85 × 2 × 150 × 0·7
= 2860 kg.

This acts at = 12 + 1·25 = 13.25m above *G.L.*

(*iii*) circular beam
= 1 × 8·5 × 150 × 0·7 = 895 kg.
This acts at 12·5m above *G.L.*

(*iv*) Column
= 6 × 0·5 × 12 × 150 × 0·7
= 3780 kg – acts at 6m above *G.L.*

(*v*) braces (assuming 50 cm square section)
= 2 × 0·5 × 8·5 × 150
= 1200 kg – acts at 6m above *G.L.*

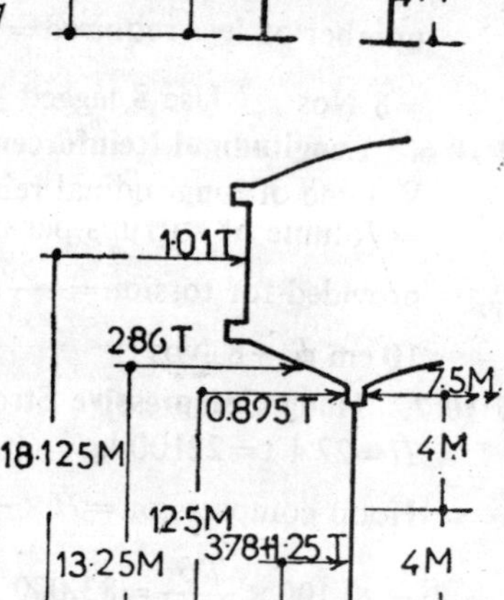

7·11·2. Wind Moment

Total wind moment in the tank = 10,100 × 18·125 × 2860 × 13·25 + 895 × 12·5 + 3780 × 6 + 1250 × 6
= 182,500 + 38,000 + 11,100 + 22750 + 7500 = 261,850m kg.

Maximum axial load on the farthest column due to wind load = $Mr/\Sigma r^2$

$\Sigma r^2 = 4\times(\sin\ 36°)^2 + 4(4\ \sin\ 72°)^2 = 4\times(4\times·5878)^2 + (4\times·9511)^2$
$= 4(2·35^2 + 3·71^2) = 4(5·51 + 13·8) = 19·31\times 4 = 77·24$

Maximum vertical load on the column at 3·71m

$$\frac{Mr}{\Sigma r^2} = \frac{261850\times 3·71}{77·24} = 12600\ \text{kg} = 12.6\ \text{t}$$

Maximum axial load due to wind and dead loads
= 153·65 + 12·60 = 166·25 *t*

7·11·3. Horizontal Forces at the base of the Column

Total horizontal wind loads = 10,100 + 2860 + 895 + 3780 + 1250

$= 18885$ kg

Horizontal shear per column $= \frac{18885}{10}$

$= 1888{\cdot}5$ kg

Maximum bending moment at the base of the column

$= 1888{\cdot}5 \times 2 = 3777{\cdot}0$ m kg.

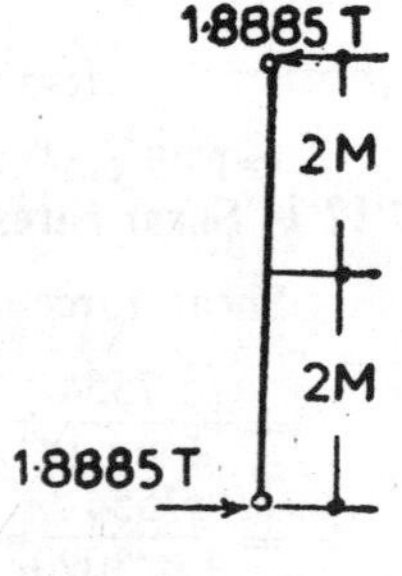

7·11·5. Trial Section

Use 8 bars of 40 mm ϕ ($12{\cdot}57$ cm^2)

Equivalent area

$$= \frac{\pi \times 50^2}{4} + 12 \times 12{\cdot}57 \times 8 = 1960 + 1200$$

$= 3160$ cm^2

Moment of inertia

$$= \frac{\pi \times 50^4}{64} + 12\left(2 \times 12{\cdot}57 \times 19^2 + 4 \times 1257 \times \frac{122}{\sqrt{2}}\right)$$

0.05M
45°
20M

$= 306700 + 12(9100 + 9100)$

$= 306700 + 219000 = 525700$ cm^4

$$\text{Direct stress} = \sigma_c' = \frac{P}{A} = \frac{166250}{2160}$$

$= 52{\cdot}6$ kg/cm^2

$$\text{bending stress} = \sigma_{cb}' = \frac{My}{1} = \frac{277700 \times 25}{525700} = 18 \text{ kg/cm}^2$$

$$\frac{\sigma_c'}{\sigma_c} + \frac{\sigma_{cb}}{\sigma_{cb}} < 1$$

$$\frac{52{\cdot}6}{66{\cdot}7} + \frac{18}{93{\cdot}3} = 0{\cdot}786 + {\cdot}192 = 0{\cdot}978 < 1. \text{ section is safe.}$$

7.11·5. Trial section Increasing the Concrete Section

60 cm dia column and $8 - 25$mmϕ ($4{\cdot}91$ cm^2 each)

$$\text{Equivalent area} \frac{\pi d^2}{4} + (m-1)\,A_{se}$$

$$\frac{\pi \times 60^2}{4} + 12 \times 8 \times 4{\cdot}91 + 2925 + 471 = 3396 \text{ cm}^2$$

Moment of inertia

$$\frac{\pi}{64} \times 60^4 + 12\left(2 \times 4{\cdot}91 \times 19{\cdot}75^2 + 4 \times 4{\cdot}91 \times \left(\frac{19{\cdot}75}{\sqrt{2}}\right)^2\right)$$

$= 635{,}000 + 12(780 + 780)4{\cdot}91 = 635000 + 92{,}000 = 727{,}000$ cm^4

$$\sigma_c' = \frac{166250}{3396} = 496 \text{ kg/cm}^2$$

$$\sigma_{cb}' = \frac{377700 \times 3}{727000} = 15{\cdot}6 \text{ kg/cm}^2$$

$$\frac{\sigma_c'}{\sigma_c} + \frac{\sigma_{cb}'}{\sigma_{cb}} = \frac{47{\cdot}6}{66{\cdot}7} + \frac{15{\cdot}6}{93{\cdot}3} = {\cdot}740 + {\cdot}167 = {\cdot}907 < 1$$

section is safe

7·12. Braces

Maximum moment in the braces $= 2 \times 3777 = 7554$ m kg

Assume a section of 50 cm $\times$ 50 cm

Equal reinforcement in tension and compression flange is provided $A_t \times 42 \times \sigma_{st} = BM$

$$A_t = \frac{755400}{42 \times 1400}$$

$= 13{\cdot}9 \text{ cm}^2$

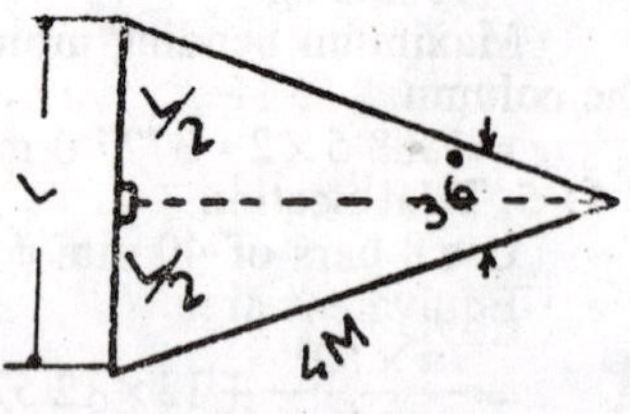

7·12·1. Shear Stress

$$\text{Shear force} = \frac{\text{Bending moment}}{\text{Half span}}$$

$$= \frac{7554}{4 \sin 18^\circ}$$

$$= \frac{7554}{4 \times {\cdot}3090} = 6100 \text{ kg}$$

$$\text{Shear stress} = \frac{6100}{0{\cdot}87 \times 50 \times 46} = 3{\cdot}06 \text{ kg/cm}^2 < 5 \text{ kg/cm}^2$$

Provide nominal stirrups 6 mm ϕ 30 cm c/c

7·13. Foundation

Loads

when tank is full, the total load on the foundation $= 166{\cdot}25 \times 10 = 1662{\cdot}5$ t

Add 10% for foundation self *wt.* $= 166{\cdot}25$t

Total load (say) $= 1830$t

Bearing capacity of the soil $= 40\text{t}/m^2$

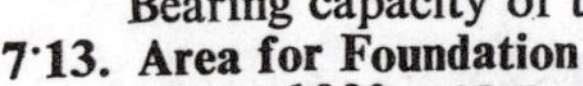

7·13. Area for Foundation

Area $= \frac{1830}{40} = 45{\cdot}7$ sq. m

Assume outside diameter of raft $= 10$ m, inside diameter $= 6$ m

6M

8M

10 M

$$\text{Area provided} = \frac{\pi}{4}(10^2 - 6^2) = \frac{\pi}{4}(100 - 36) = \frac{\pi}{4} \times 62$$

$= 48{\cdot}6$ sq. m

$$\text{Upward pressure on the raft} = \frac{1830}{48{\cdot}6} = 37{\cdot}5 \text{ t/m}^2$$

7·14. Circular Beam : Bending Moment

$$\text{Load on the circular beam per metre length} = \frac{1830}{\pi \times 8} = 73\text{t}$$

Support bending moment ($-ve$) $= {\cdot}0054\, Wr = {\cdot}0054 \times 1662{\cdot}5 \times 4$

$= 36{\cdot}0$ *mt*

Central moment ($+$ve) $= {\cdot}0023 \times 1662{\cdot}5 \times 4 = 15{\cdot}4$ mt

7·14·1. Torsional Moment

$T_m = {\cdot}0003\, Wr = {\cdot}0002 \times 1662{\cdot}5 \times 4 = 2{\cdot}0$ mt

Maximum occurs at $7\frac{1}{2}^\circ$ from support

7·14·2. Shear Force

$$\text{Shear force at support} = \frac{1662{\cdot}5}{10 \times 2} = 83{\cdot}125 \text{ t}$$

Shear force at point of maximum torsion

$$= 83{\cdot}125 - 83{\cdot}125 \times \frac{7{\cdot}5}{18} = 83{\cdot}125 - 34{\cdot}6 = 48{\cdot}525 \text{ t}$$

7·14·3. Trial Section

Use $M-200$, Assume 50 cm $\times$ 100 cm

depth required $= d_e = \sqrt{\dfrac{3600000}{50 \times 12 \cdot 1}} = \sqrt{5920} = 77$ cm.

Use $d_e = 94$ cm, $d = 100$ cm

7·14·4. Area of Steel

At support, $A_t = \dfrac{3600000}{1400 \times \cdot 87 \times 94}$

$= 31 \cdot 4$ cm²

Use 20 mm ϕ—10 Nos

at centre, $A_t = \dfrac{1540000}{1400 \times \cdot 87 \times 94}$

$= 13 \cdot 42$ cm²

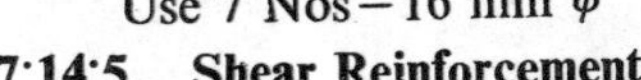

Use 7 Nos—16 mm ϕ

7·14·5. Shear Reinforcement

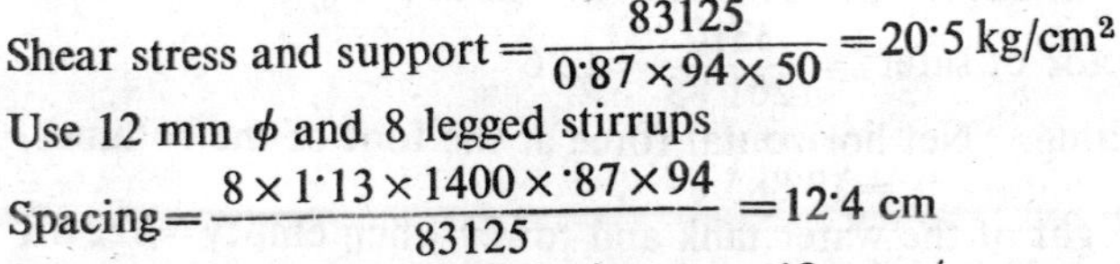

Shear stress and support $= \dfrac{83125}{0 \cdot 87 \times 94 \times 50} = 20 \cdot 5$ kg/cm²

Use 12 mm ϕ and 8 legged stirrups

Spacing $= \dfrac{8 \times 1 \cdot 13 \times 1400 \times \cdot 87 \times 94}{83125} = 12 \cdot 4$ cm

Use 12 mm ϕ and 8 legged stirrups at 12 cm c/c

7·14·6. Shear Reinforcement at Point of Maximum Torsion

Steel for shear force using a spacing of 12 cm

$A_{w_1} = \dfrac{48525 \times 12}{1400 \times \cdot 87 \times 94} = 5 \cdot 1$ cm²

Steel for torsional moment using a spacing of 12 cm

$A_{w_2} = \dfrac{200000 \times 12}{0 \cdot 8 \times 1400 \times 40 \times 90} = \cdot 592$ cm²

$A_w = 5 \cdot 692$ cm²

No. of legs of 12 mm ϕ required

$= \dfrac{5 \cdot 692}{1 \cdot 13} = 6$. Use 6 legged 12 mm ϕ at 12 cm c/c

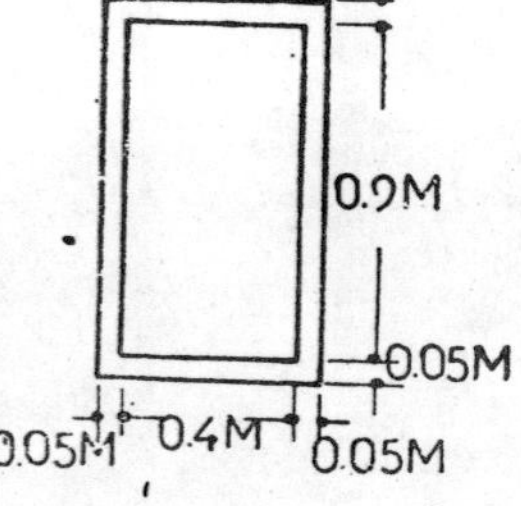

7·14·7. Longitudinal Steel for Torsion

Volume of steel required for torsion unit length $= \dfrac{\cdot 592 \times 90}{12} = 6 \cdot 4$ cm²

Use 6 Nos—12 mm ϕ at 3 Nos per face

7·15. Raft Slab

Projection of the raft slab from the edge of the beam $= 0 \cdot 75$ m

Bending moment $= 37500 \times 0 \cdot 75 \times \dfrac{0 \cdot 75}{2} = 10500$ m kg

$= 1050{,}000$ cm kg

7·15·1. Depth

Effect of depth $d_e = \sqrt{\dfrac{1050{,}000}{12 \cdot 1 \times 100}} = \sqrt{870} = 29 \cdot 5$ cm

Use $d = 35$ cm, $d_e = 30$ cm

7·15·2. Main Steel

$$A_t = \frac{1050,000}{0·87 \times 1400 \times 30} = 27·6 \text{ cm}^2$$

Use 20 mm ϕ at 11 cm c/c

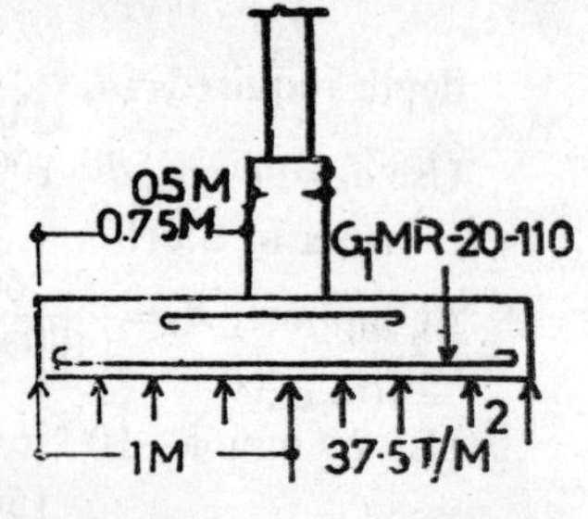

7·15·3. Secondary Steel

$$A_t = \frac{·15}{100} \times 35 \times 100 = 5·25 \text{ cm}^2$$

Use 10 mm ϕ 14·5 cm c/c

7·16. Stability

Overturning moment=261,850 m kg =261·85 mt

Total weight including water and foundations=1830 t

When the tank is empty weight of the structure=1830−907·2 =822·8 t

Stabilizing moment=822·8×4=3310 mt

$$\text{Factor of safety} = \frac{3310}{261·85} = 12·6$$

Sliding. Net horizontal force at the foot of the columns =18·885 t

Weight of the water tank and tower when empty=822·8 t

$$\text{Factor of safety} = \frac{0·5 \times 822·8}{18·885} = 21·6$$

Cylindrical Elevated Tank with Domical Bottom

8

Capacity 275,000 Litres

8·1. Data

Capacity=275 cu m. Elevation of the tank=18 m. Bearing capacity of the soil=5 t/m²

Materials available : Concrete *M* 200 for the tank, *M* 150 for staging, Steel Ribbed—Torsteel.

8·2. Characteristic Strengths

$\sigma_{st}=1300$ kg/cm² (water face) $m=13$

$\sigma_{cb}=70$ kg/cm² $jd=\cdot 86$ $R=12\cdot 43$

$\sigma_{ct}=12$ kg/cm²

$\sigma_{st}=2100$ kg/cm² $m=18$

$\sigma_{cb}=50$ kg/cm² $jd=\cdot 89$ $R=6\cdot 97$

Compression in column bars $\sigma_{st}=1750$ kg/cm², $\sigma_t=40$ kg/cm²

8·3. Trial Dimensions

Required capacity=275 cu m

$D=9\cdot 5$ m, $H=4\cdot 75$ m,

$$h=\frac{D}{6}=1\cdot 5 \text{ m}$$

Volume of the cylindrical portion including bottom dome

$$=\frac{\pi D^2}{4}\times H=\frac{\pi\times 9\cdot 5^2}{4}\times 4\cdot 75$$

$=336$ cu m

Volume of bottom domed portion

$$=\frac{\pi h}{6}(3r^2+h^2)$$

$$=\frac{\pi\times 1\cdot 5}{6}(3\times 4\cdot 75^2+1\cdot 5^2)=55 \text{ cu m}$$

Actual volume provided=336−55=281 cu m>275 cu m

8·4. Domed Roof

Loads : Assume 10 cm thick.

Self weight of the dome

$=\frac{10}{100}\times 1\times 1\times 2400=240$ kg/m²

Wind and accidental loading

=150 kg/m²

Water proofing etc.=110 kg/m²

Total load=500 kg/m²

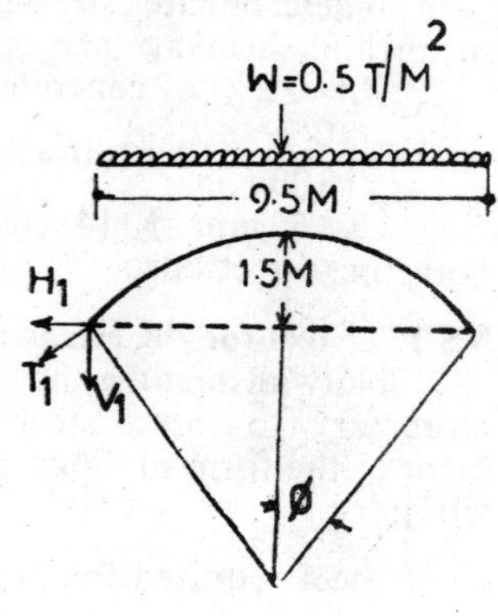

8·4·1. Geometry of the Dome

Rise=$r=1\cdot 5$ m

Span=$2l=9\cdot 5$ m

$$R=\text{Radius}=\frac{l^2+r^2}{2r}=\frac{4\cdot 75^2+1\cdot 5^2}{2\times 1\cdot 5}$$

$$=\frac{22\cdot 5+2\cdot 25}{3}=\frac{24\cdot 75}{3}=8\cdot 25 \text{ m}$$

$\text{Sin } \phi = \frac{4{\cdot}75}{8{\cdot}25} = 0{\cdot}575,\ \phi = 35° < 51° - 48'$

Therefore the entire dome will be subjected to compressive stresses only.

8·4·2. Stresses in the Dome

At crown, Hoop stress ($\phi=0°$) $S = \frac{w_1 R}{2t} = \frac{500 \times 8{\cdot}25}{2 \times {\cdot}10}$

$= 20600 \text{ kg/m}^2 = 2{\cdot}06 \text{ kg/cm}^2$

Meridional stress $c = \frac{w_1 R}{t(1+\cos\phi)} = \frac{500 \times 2{\cdot}25}{{\cdot}10(1+0{\cdot}82)}$

$= 22500 \text{ kg/m}^2 = 2{\cdot}25 \text{ kg/cm}^2$

The stresses are however low.

8·4·3. Force Components

Surface area $= 2\pi Rr = 2\pi \times 8{\cdot}25 \times 1{\cdot}5 = 77{\cdot}5$ sq. m

Total load $W = 77{\cdot}5 \times 500 = 38750 \text{ kg} = 38{\cdot}75$ t

$$V_1 = \frac{W}{2\pi l} = \frac{38{\cdot}75}{2 \times \pi \times 4{\cdot}75} = 1{\cdot}30 \text{ t}$$

$$H_1 = V_1 \text{ col } \phi = 1{\cdot}07\left(\frac{R-r}{l}\right) = \left(\frac{8{\cdot}25 - 1{\cdot}5}{4{\cdot}75}\right)$$

$$= \frac{6{\cdot}75}{4{\cdot}75} \times 1{\cdot}30 = 1{\cdot}85 \text{ t}$$

$$T_1 = \frac{V_1}{\text{Sin } \phi} = V_1 \frac{R}{l} = \frac{1{\cdot}85 \times 8{\cdot}25}{4{\cdot}75} = 3{\cdot}21 \text{ t}$$

8·4·4. Shear Stress at the Edge

Shear stress along the perimeter of the dome

$$= \frac{V_1}{\text{area}} = \frac{1300}{100 \times 10} = 1{\cdot}30 \text{ kg/cm}^2 < 5 \text{ kg/cm}^2$$

A 10 cm thick $M-150$ concrete slab will do as the stresses are low.

8·5. Main Reinforcement

Since there are no tensile stresses anywhere in the dome, only a nominal reinforcement is required, to allow for such indeterminate stresses due to wind, distortion, shrinkage and temperature etc.

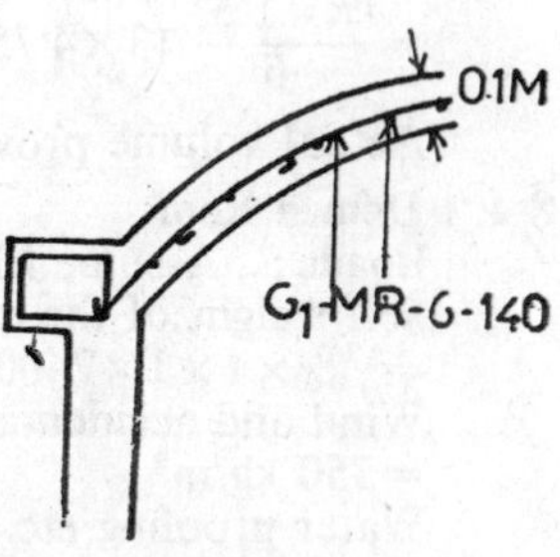

$A_t = 0{\cdot}2\%$ of concrete area

$$= \frac{0{\cdot}2}{100} \times 10 \times 100 = 2 \text{ cm}^2$$

Use 6 mm ϕ 14 cm c/c ($<3d$) in both the directions.

8·5·1. Steel for the Shear at the Junction of the Rib and Dome

Notwithstanding the low shear stress along the perimeter, it is customary to insert steel to resist all the shear, the reinforcement being in the form of links continuous around the main bars in the rib.

$$\text{Steel required for shear} = \frac{V_1}{1750} = \frac{1300}{1750} = {\cdot}74 \text{ cm}^2$$

Use 6 mm ϕ at 38 cm c/c

8·6. Rib Reinforcement

Maximum hoop tension$=T=H_1\times\dfrac{D}{2}=\dfrac{1850\times 9{\cdot}5}{2}=8800$ kg

$A_t=\dfrac{8800}{2300}=3{\cdot}83$ cm^2. Use 4—12 mm ϕ, A_t provided (4·52 cm^2)

8·6·1. Rib Dimensions

$\sigma_{ct}=11$ kg/cm^2

$$\sigma_{ct}=\frac{T}{A_e+(m-1)A_t}$$

$$11=\frac{8800}{A_e+17\times 4{\cdot}52}$$

$11A_e+11\times 17\times 4{\cdot}52=8800$

$11A_e=8800-845=7955$

$$A_e=\frac{7955}{11}=724 \text{ cm}^2$$

Use 25 cm × 30 cm

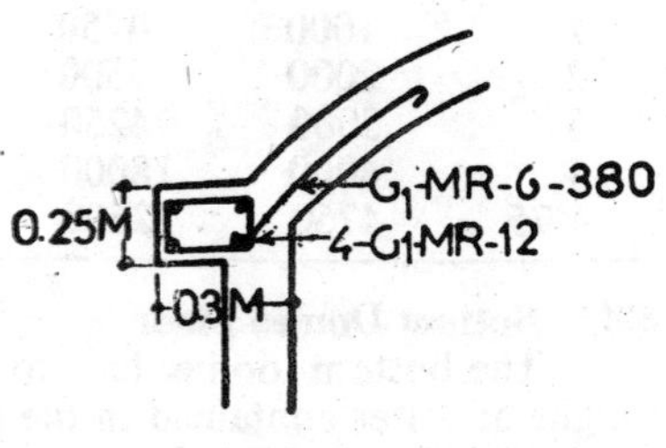

8·7. Cylindrical Wall Hoop Tension

The cylindrical wall is designed to resist hoop tension

Maximum water pressure at the bottom of the cylindrical wall.

$p=wh=1000\times 4{\cdot}75=4750$ kg/m^2

Maximum hoop tension in the wall

$$=T=p\times\frac{D}{2}=4750\times\frac{9{\cdot}5}{2}$$

$=27500$ kg

8·7·1. Main Steel

Area of steel required at the bottom

1 m height$=\dfrac{22500}{1300}=17{\cdot}3$ cm^2

Use 16 mm ϕ—11 cm c/c

(A_t provided$=18{\cdot}28$ cm^2)

8·7·2. Thickness of the Wall

$$\sigma_{ct}=\frac{T}{A_c+(m-1)A_t}$$

$$12=\frac{22500}{A_c+12\times 18{\cdot}28}$$

$12\times A_c+12\times 12\times 18{\cdot}28=22500$

$12A_c+2710=22500$

$12A_c\ 22500-2710=19790$

$$A_c=\frac{19790}{12}=1650 \text{ cm}^2$$

$A_c=t_w\times 100=1650$ cm^2

$t_w=16{\cdot}5$ cm

Use 20 cm thick wall

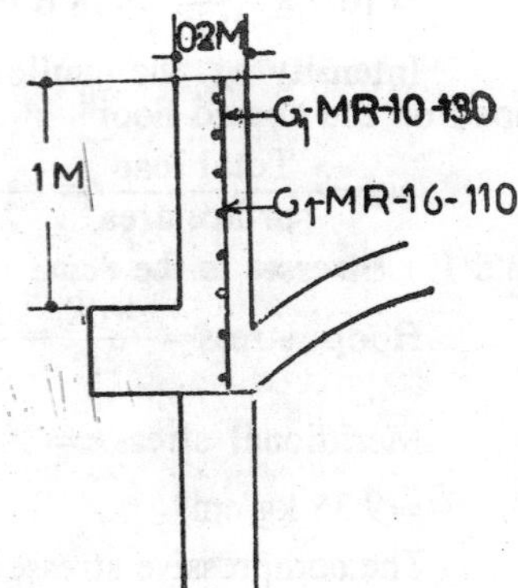

8·7·3. Secondary Steel

$A_t=0{\cdot}3\%$ of concrete area, $=\dfrac{{\cdot}3}{100}\times 20\times 100=6$ cm^2

Use 10 mm ϕ 13 cm c/c

8·7·4. Main Reinforcement at Different Levels of the Wall

depth *m*	*p* = *wh*	*hoop tension* $p\frac{D}{2}$ *kg*	A_t *cm²*	*spacing* *cm*	
0	—	—	—	—	
1	1000	4750	3·65	10 mm ϕ 21	0 – 1 m
2	2000	9500	7·3	12 ,, 15	1 to 2 m
3	3000	14250	10·9	16 ,, 18	2 to 3 m
4	4000	18000	13·9	16 ,, 14	3 to 4 m
4·75	4750	22500	17·3	16 ,, 11	4 to 4·75 m

8·8. Bottom Domed Floor

The bottom dome has to carry: (1) Self weight of dome (2) weight of water contained in the cylindrical portion above the dome.

Geometry of the dome as per roof dome.

Capacity provided = 281 cu m, Weight of water = 281 t

Assuming 20 cm thick slab self weight

$= 2\pi Rr \times w$

$= 2\pi \times 8{\cdot}25 \times 1{\cdot}5 \times \frac{20}{100} \times 1 \times 1 \times \frac{24000}{10000}$

$= 37{\cdot}2$ t

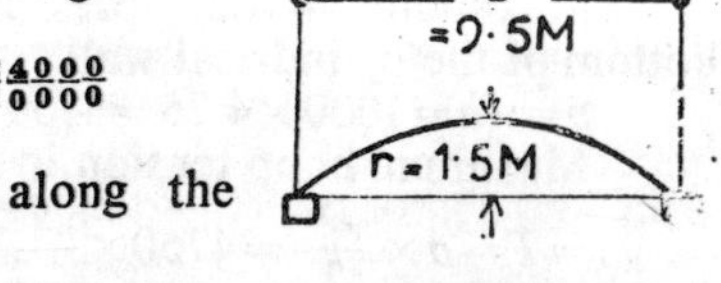

Total weight per metre length along the circumference of the dome

$$V_2 = \frac{281 + 37{\cdot}2}{\pi \times 9{\cdot}5} = \frac{318{\cdot}2}{\pi \times 9{\cdot}5} = 10{\cdot}7 \text{ t}$$

$$H_2 = V_2 \cot \phi$$

$$= 10{\cdot}7 \times \left(\frac{8{\cdot}25 - 1{\cdot}5}{47{\cdot}5}\right) = 15{\cdot}3 \text{ t}$$

$$T_2 = \frac{V_2}{\sin \phi} = V_2 \frac{R}{l}$$

$$= 10{\cdot}7 \times \frac{8{\cdot}25}{4{\cdot}75} = 18{\cdot}6 \text{ t}$$

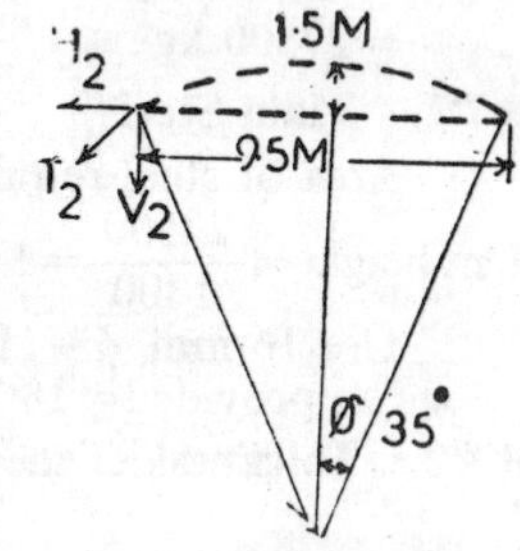

Intensity of the uniformly distributed load on the domed floor

$$w_1 = \frac{\text{Total load}}{\text{surface area}} = \frac{318{\cdot}2}{2\pi \times 8{\cdot}25 \times 1{\cdot}5} = 4{\cdot}1 \text{ t/m}^2 = 4100 \text{ kg/m}^2$$

8·8·1. Stresses in the dome

$$\text{Hoop stress} = \frac{w_1 R}{2t} = \frac{4100 \times 8{\cdot}25 \times 100}{2 \times 20} = 8{\cdot}5 \text{ kg/cm}^2$$

$$\text{Meridional stress } c = \frac{w_1 R}{t(1 + \cos \phi)} = \frac{4100 \times 8{\cdot}25 \times 100}{20(1 + 0{\cdot}82)}$$

$= 9{\cdot}35$ kg/cm²

The compressive stresses are however within the limit.

8·8·2. Reinforcement

$$A_t = {\cdot}2\% \text{ of concrete area} = \frac{{\cdot}2}{100} \times 20 \times 100 = 4 \text{ cm}^2$$

Use 10 mm ϕ 19 cm c/c in both the directions.

8·8·3. Steel for the Shear at Junction of the Floor and Side Wall

Steel required for shear

$$=\frac{V_2}{1750}=\frac{10700}{1750}=6{\cdot}15 \text{ cm}^2$$

Use 10 mm ϕ 12 cm c/c

8·9. Rib Dimension

Maximum hoop tension $T=H_2\times D/2$

$$=15300\times\frac{9{\cdot}5}{2}=72{,}500 \text{ kg}$$

$$A_t=\frac{72500}{1300}=56 \text{ cm}^2$$

18 bars of 20 mm ϕ

(A_t provided$=56{\cdot}54$ cm²)

$$\sigma_{ct}=\frac{T}{A_c+(m-1)A_t}$$

Using M 200, $12=\dfrac{72500}{A_c+12\times 56{\cdot}54}$

$$12A_c+12\times12\times56{\cdot}54=72500$$

$$12A_c=72500-8120=64380$$

$$A_c=\frac{64380}{12}=5365 \text{ cm}^2$$

Use $90\times60=5400$ cm²

8.10. Ring Beam

Total vertical load on the ring beam per metre$=V_1$ (weight of domed roof)+rib+V_2 (weight of water+weight of bottom domed floor)+rib+weight of the side wall+self weight of the beam.

$$V_1=1{\cdot}30 \text{ t}$$

$$\text{Top rib}=\frac{25}{100}\times\frac{30}{100}\times\frac{2400}{1000}=0{\cdot}18 \text{ t}$$

$$V_2=10{\cdot}7 \text{ t}$$

$$\text{Bottom rib}=\frac{90}{100}\times\frac{60}{100}\times\frac{2400}{1000}=1{\cdot}3 \text{ t}$$

Weight of side wall$=\frac{20}{100}\times1\times4{\cdot}75\times2{\cdot}4=2{\cdot}28$ t

8·10·1. Trial Section

Assuming 50 cm × 100 cm beam self weight of thc beam

$$=\frac{50}{100}\times\frac{100}{100}\times\frac{2400}{1000}=1{\cdot}20 \text{ t}$$

Total vertical load per metre on the ring beam

$$=1{\cdot}30+0{\cdot}18+10{\cdot}7+1{\cdot}3+2{\cdot}28+1{\cdot}20=16{\cdot}96 \text{ t}$$

Total load$=W=\pi\times9{\cdot}5\times16{\cdot}96=502{\cdot}5$ t

No. of columns$=6$

Max. bending moment at support

$$=\cdot 0148\,\frac{WD}{2}=\cdot 0148\times 50\cdot 25\times\frac{9\cdot 5}{2}$$

$=35\cdot 5$ mt$=35{,}50{,}000$ cm kg

Maximum bending moment at centre

$$=\cdot 0075\times 502\cdot 5\times\frac{9\cdot 5}{2}=17\cdot 9\text{ mt}$$

$=17{,}90{,}000$ cm kg

$$\text{Torsional moment}=\cdot 0015\,\frac{WD}{2}$$

$$=0\cdot 0015\times 502\cdot 5\times\frac{9\cdot 5}{2}=355{,}000\text{ cm kg occurs at }12^\circ-44'$$

$$\text{Shear force}=\frac{\text{Total load}}{2\times\text{No. of column}}=\frac{502\cdot 5}{2\times 6}=41\cdot 9\text{ t}=41900\text{ kg}$$

8·10·2. Depth of Beam Required

Maximum bending moment (at support)$=35{,}50{,}000$ cm kg

$$d=\sqrt{\frac{35{,}50{,}000}{9\cdot 07\times 50}}=\sqrt{7800}=88\text{ cm.}$$ Hence 50 cm$\times$100 cm beam is sufficient.

8·10·3. Shear Force at Point of Maximum Torsion

Shear force at point of max. torsion

$$=41900-\frac{12\cdot 735\times 2}{60}\times 41900=41900-17800=24{,}100\text{ kg}$$

$$\text{Shear stress at point of maximum torsion}=q=\frac{24100}{50\times\cdot 86\times 88}$$

$=6\cdot 35$ kg/cm²

$$\text{Shear stress due to torsional moment}=q'=\frac{T_m\left(3+2\frac{b}{D}\right)}{b^2D}$$

$$=\frac{355000(3+2\frac{50}{100})}{50\times 50\times 100}=\frac{355000\times 4}{50\times 40\times 100}=5\cdot 68\text{ kg/cm}^2$$

Total shear stress$=6\cdot 35+5\cdot 68=12\cdot 03>5$ kg/cm²<20 kg/cm²

$$\text{Shear stress at support}=\frac{41900}{50\times\cdot 86\times 88}=11\cdot 1\text{ kg/cm}^2$$

8·10·4. Shear Stirrups Supports

$$\text{Use 12 mm }\phi-8\text{ legged stirrups}=s=\frac{A_w\,t_w\,jd}{Q}$$

$$=\frac{8\times 1\cdot 13\times 1750\times 0\cdot 86\times 94}{41900}=30\cdot 5\text{ cm c/c Say 30 cm c/c}$$

8·10·5. Main Steel

$d_e=88$ cm $\quad d=100$ cm

$$A_t\text{ for }-\text{ve }BM\text{ water face}=\frac{35{,}50{,}000}{1300\times\cdot 86\times 88}=36\cdot 1\text{ cm}^2$$

Use 8 Nos of 25 mm ϕ ($A_t=39\cdot 27$ cm²)

A_t for +ve bending moment (away from water face)

$$=\frac{17{,}90{,}000}{2100\times\cdot 86\times 86}=11\cdot 2\text{ cm}^2$$

Use 6 Nos of 16 mm ϕ ($A_t=12\cdot 06$ cm²)

8·10·6. Shear Stirrups at Point of Maximum Torsion

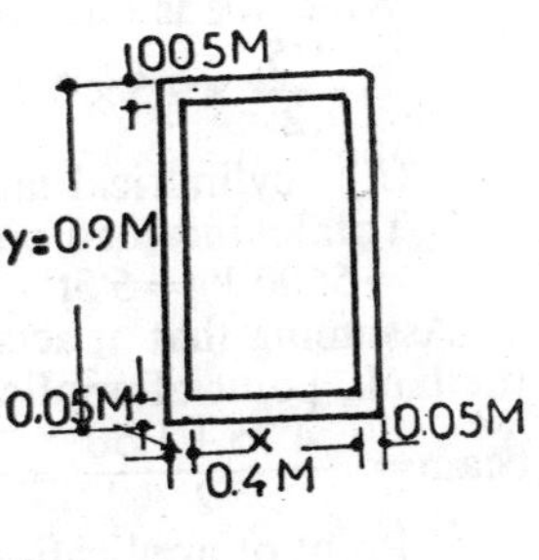

Using 30 cm spacing, area of steel required for shear force of 24100 kg

$$Aw_1=\frac{Q\times s}{t_w-jd}=\frac{24100\times 30}{1750\times \cdot 86\times 88}$$

$=5{\cdot}45\ cm^2$

Using 30 cm spacing area of steel required for torsional shear stress

$$Aw_2=\frac{Tm\times s}{0{\cdot}8\times t_w\times XY}$$

$$=\frac{355000\times 30}{0{\cdot}8\times 1750\times 40\times 90}=2{\cdot}1\ cm^2$$

Total steel required at point of maximum torsion

$=5{\cdot}45+2{\cdot}1=7{\cdot}55\ cm^2$

Using 12 mm ϕ area per leg$=1{\cdot}13\ cm^2$

No. of legs required$=\dfrac{7{\cdot}55}{1{\cdot}13}=6{\cdot}7$

Use 8 legged 12 mm ϕ stirrups at 30 cm c/c

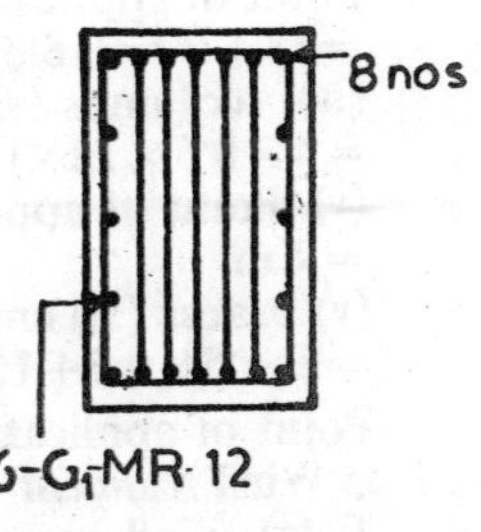

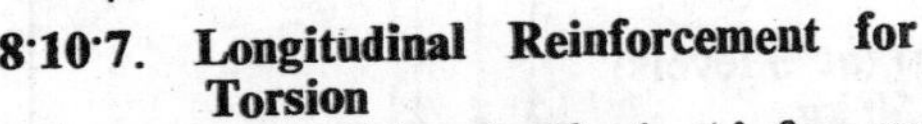

8·10·7. Longitudinal Reinforcement for Torsion

Volume of longitudinal reinforcement per unit length=Volume of stirrups per unit length for torsion$=\dfrac{2{\cdot}1}{3}\times 90=6{\cdot}3\ cm^2$

Use 6 Nos-12 mm ϕ

8·11. Staging Columns

Loads : (*i*) Tank full. Direct axial load on each column from the tank portion

$=\dfrac{502{\cdot}5}{6}=83{\cdot}8$ t

Self weight of the column (50 cm dia)

$=\dfrac{\pi}{4}(0{\cdot}5)^2\times 18\times 2{\cdot}4=8{\cdot}47$ t

Assuming 50 cm$\times$30 cm brace wt. of braces crried by each column$=\dfrac{3{\cdot}08}{5}=6{\cdot}16$ t

Total vertical load$=83{\cdot}8+8{\cdot}47+6{\cdot}16$ $=98{\cdot}43$ t

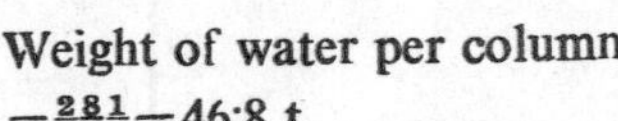

(*ii*) Tank empty. Total weight of water 281 t

Weight of water per column

$=\frac{281}{6}=46{\cdot}8$ t

Weight on each column including self weight of column

$=98{\cdot}43-46{\cdot}8=51{\cdot}63$ t.

8·11·1. Wind Loads

Wind pressure$=150\ kg/m^2$, Reduction factor$=0{\cdot}7$

Wind loads on: (*i*) top dome

$=\frac{1\cdot5}{2}\times9\cdot5\times0\cdot7\times150=750$ kg,

(*ii*) cylindrical tank portion $=9\cdot5\times4\cdot75\times0\cdot7\times150=4750$ kg

Total wind load on the tank

$=5500$ kg $=5\cdot5$t

Assuming that it acts at mid height of the tank, point of application above ring beam $=\frac{4\cdot75+1\cdot50}{2}=\frac{6\cdot25}{2}=3\cdot125$ m

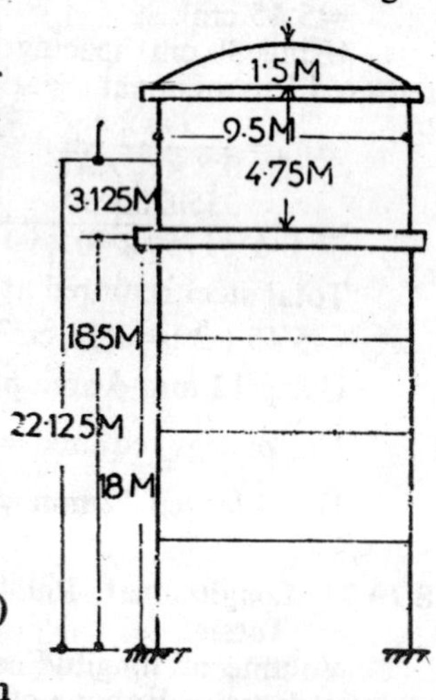

Point of application above *GL*

$=3\cdot125+1+18=22\cdot125$ m

(*iii*) wind on circular ring beam

$1\times9\cdot5\times0\cdot7\times150=1000$ kg

Point of application above *G.L.*

$=18+0\cdot5=18\cdot5$

(*iv*) columns (assume 50 cm dia)

$=4\times0\cdot5\times18\times0\cdot7\times150=3800$ kg

(*v*) Point of application above *G.L.*

$=9$ m

(*v*) braces (50 cm depth) (at 3 levels)

$=3\times\cdot5\times9\cdot5+150=2150$ kg

Point of application above $G.L.=9$ m

8·11·2. Wind Moment

Total wind moment $=5\cdot5\times22\cdot125+1\times18\cdot5$
$+3\cdot8\times9+2\cdot15\times9$

$=122+18\cdot5+34\cdot2+18\cdot3=19300$ mt

Maximum axial load in the farthest column due to wind load $=\frac{Mr}{\Sigma r^2}$

$\Sigma r^2=4\times(4\cdot75\sin60^\circ)^2=4\times(4\cdot75\times\cdot866)^2$

$=4\times16\cdot9=67\cdot6$

Maximum vertical load due to wind loads

$=\frac{194\cdot05\times4\cdot1}{67\cdot6}=11.8$t

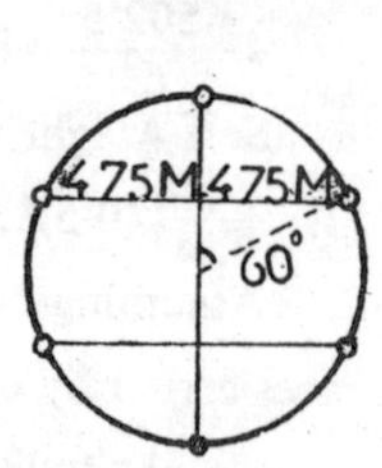

Maximum axial load on the column due to wind and other loads $=98\cdot43+11\cdot80=110\cdot23$ *t*

8·11·3. Total Horizontal Force at the base of the column

Total horizontal wind loads on the tank

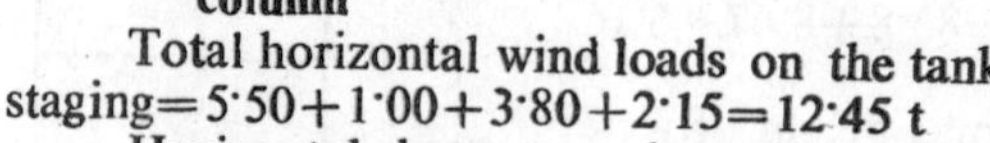

staging $=5\cdot50+1\cdot00+3\cdot80+2\cdot15=12\cdot45$ t

Horizontal shear per column

$=\frac{12\cdot45}{6}=2\cdot075$t

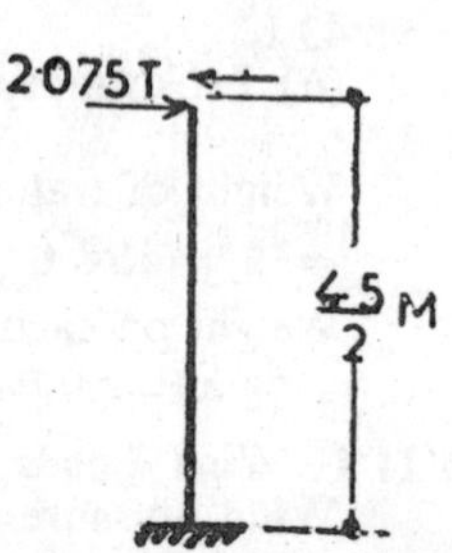

Maximum bending moment at the base of the column

$=2\cdot075\times\frac{4\cdot5}{2}=4\cdot7$ mt

8·11·4. Trial Section for the Column

Axial load on the column $=98\cdot43+11\cdot8$

$=110\cdot23$ t

Bending moment$=4{\cdot}7$ t

Assume 50 cm dia with 8 bars of 25 mm $\phi=(39{\cdot}27\ \text{cm}^2)$

Using M 150, $P=\sigma_c \times A_c+\sigma_{sc}\ A_{sc}$

$$=50\times\left(\frac{\pi\times 50^2}{4}-A_{sc}\right)+1750\ A_{sc}$$

$=40\times(1960-39{\cdot}27)+1750\times 39{\cdot}27$

$=40\times 1920{\cdot}73+1750\times 39{\cdot}27=76{,}800+69{,}000$

$=145{\cdot}8$ t$>110{\cdot}23$ t (overdesigned because of bending stresses)

8·11·5. Stresses in the Column section

M 150 concrete, Area$=\dfrac{\pi\times 50^2}{4}+17\times 39{\cdot}27=1960+630$

$=2590\ \text{cm}^2$

Equivalent moment of inertia

$$=\frac{\pi}{64}50^4+17\left[4{\cdot}91\times 2\times 20^2+4{\cdot}91\times 4\left(\frac{20}{\sqrt{2}}\right)^2\right]$$

$=306{,}700+17(3940+3940)=440{,}000\ \text{cm}^4$

Direct stress $=P/A=\sigma'i=110230/2590=42{\cdot}8\ \text{kg/cm}^2$

Bending stress$=My/I=\sigma_b=470.000\times 25/440{,}700=26{\cdot}5$ kg/cm²

When wind loads are considered the allowable stress can be increased by 332%

Combined stress

$$=\frac{\sigma_c'}{\sigma_c}+\frac{\sigma_{cb}'}{\sigma_{cb}}=\frac{42{\cdot}8}{50\times 1{\cdot}33}+\frac{26{\cdot}5}{50\times 1{\cdot}33}$$

$=0{\cdot}85+{\cdot}4=1{\cdot}205>1$ Unsafe

GrMR-8-300

8 GMR-25

Try $8-32$ mm ϕ $(64{\cdot}34\ \text{cm}^2)$

equivalent area=

$$\frac{\pi\times 50^2}{4}+17\times 64{\cdot}34$$

$=1960+1100=3060\ \text{cm}^2$

Moment of inertia

$$\frac{\pi\times 50^4}{60}+17(2\times 8{\cdot}04\times 19{\cdot}5^3\times 4\times 8{\cdot}04\left(\frac{19{\cdot}5}{\sqrt{2}}\right)^2$$

$=306700+17(6100+6100)=306{,}700+207{,}000=513{,}700\ \text{cm}^4$

$$\sigma_c'=\frac{110230}{3060}=36\ \text{kg/cm}^2$$

$$\sigma_{cb}'=\frac{470{,}000\times 25}{513{,}700}=22{\cdot}8\ \text{kg/cm}^2$$

$$\text{Combined stress}=\frac{36}{40\times 1{\cdot}33}+\frac{22{\cdot}8}{50\times 1{\cdot}33}$$

$=675+{\cdot}342=1{\cdot}017$ very nearly 1

Maximum moment in the brace

$=2\times$col moment$=2\times 4{\cdot}7=9{\cdot}4$ mt

Try 50 cm$\times$30 cm self wt

$$BM=\frac{6{\cdot}16}{2}\times\frac{4{\cdot}75}{8}=1{\cdot}84\ \text{mt}$$

Total $BM=9{\cdot}4+1{\cdot}84=11{\cdot}24$ mt

$A_t\times 42\times\sigma_{st}=BM$

$A_t\times 42\times 2100=1124000$ cm kg

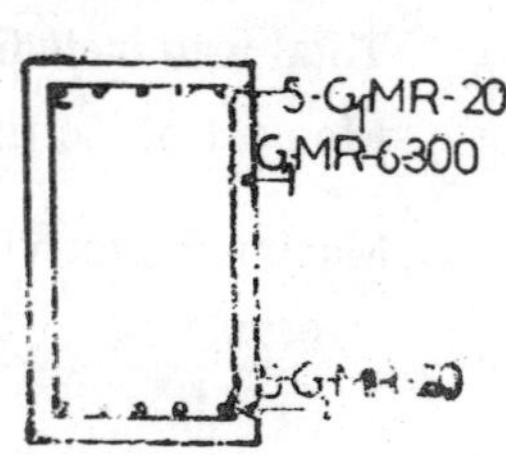

$$At = \frac{1124000}{2100 \times 42} = 12{\cdot}6 \text{ cm}^2$$

Use 5Nos $-$ 20 mm ϕ on each face

8·12. Shear Stress

$$\text{Shear face} = \frac{BM}{l/2} = \frac{940000}{4{\cdot}75 \times \sin 30}$$

$$= \frac{940000 \times 2}{4{\cdot}75 \times 100} = 3950 \text{ kg}$$

$$\text{Shear Stress} = \frac{3950}{0{\cdot}86 \times 46 \times 30}$$

$= 3{\cdot}32$ kg/cm² < 5 kg/cm² Use nominal stirrups 6 mm @ 30 cm c/c

8·13. Foundation size of the raft

Total load on the foundation when the tank is full including wind $= 104{\cdot}07 \times 6$

$= 624{\cdot}42$ t

wt. of braces $= 30{\cdot}80$ t

Self wt. of foundation at 10% $= 65{\cdot}52$ *t*

Total load (including vertical load due to wind) $= 720{\cdot}74$ t

Bending capacity of soil $= 5$ t/m²

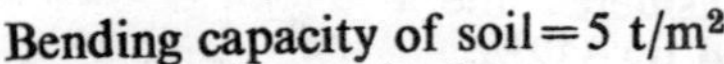

Area of raft required

$$= \frac{720{\cdot}74}{5} = 144{\cdot}15 \text{ m}^2 \text{ say } 145 \text{ sq m}$$

$$\frac{\pi D^2}{4} = 145$$

$$D = \sqrt{\frac{145 \times 4}{\pi}} = \sqrt{185}$$

$= 13{\cdot}6$ m say 14 m

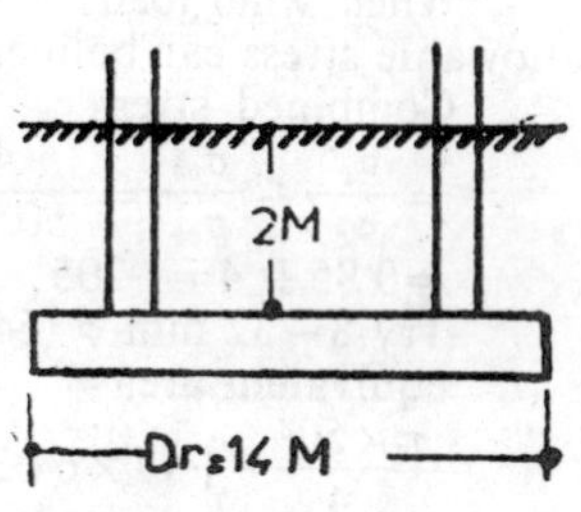

8·13·1. Pressure Distribution Below the Raft

Wind load moment about the basis of the raft $= M_w$

Wind on tank+dome $= 5{\cdot}5 \times 24{\cdot}125 = 134$ mt

Ring beam $= 1 \times 20{\cdot}5 = 20{\cdot}5$ mt

Columns $= 3{\cdot}8 \times 11 = 41{\cdot}8$ mt

Braces $= 2{\cdot}15 \times 11 = 23{\cdot}65$ mt

Total moment $M_w = 219{\cdot}95$ mt

Axial load on the raft.

Tank+water $= 502{\cdot}5$ t

Colums $= 8{\cdot}47 \times 6 = 50{\cdot}82$ t

braces $= 30{\cdot}80$ t

Total load $= 584{\cdot}12$ t

10% for the self wt. of foundation 58·40

Total load including self wt. $= 642{\cdot}62$ t

$$\text{Moment of inertia of raft } I = \frac{\pi D^4}{64} = \frac{\pi \times 14^4}{64} = 1880 \text{ m}^4$$

$$\text{Maximum pressure} = \frac{P}{A} + \frac{My}{I}$$

$$= \frac{642{\cdot}62 \times 4}{\pi \times 14^2} + \frac{219{\cdot}95 \times 7}{1880} = 4{\cdot}15 + 0{\cdot}81 = 4{\cdot}96 \text{ t/m}^2 < 5 \text{ t/m}^2$$

Minimum pressure $= 4{\cdot}15 - 0{\cdot}81 = 3{\cdot}34$ t/m²
Upward pressure due to self wt. of foundation raft

$$= \frac{58400 \times 4}{\pi \times 14^2} = 376 \text{ kg/m}^2$$

Net upward max. pressure
$= 4960 - 376 = 4584$ kg/m²
Net min. upward pressure
$3340 - 376 = 2964$ kg/m²

Average net upward pressure

$$= \frac{7548}{2} = 3774 \text{ kg/m}^2 = 3{\cdot}8 \text{ t/m}^2 \text{ (say)}$$

8·14. Ring Girder

Radius of the ring girder $= \dfrac{9{\cdot}5}{2} = 4{\cdot}75$ m

$$W = \frac{3{\cdot}8 \times \pi \times 14}{4} = 585 \text{ t}$$

BM at support (+ve) $= {\cdot}0148\, Wr = {\cdot}0148 \times 585 \times 4{\cdot}75 = 41$ mt
BM at centre (−ve) $= {\cdot}0075 \times Wr = {\cdot}0075 \times 585 \times 4{\cdot}75 = 21$ mt
Torsional moment $= {\cdot}00015 \times Wr = {\cdot}0015 \times 585 \times 4{\cdot}75 = 4{\cdot}16$ mt
Using M_{150} concrete and torsteel
$\sigma_{cb} = 50$ kg/cm², $\sigma_{st} = 1300$ kg/cm², $R = 9{\cdot}07$

8·14·1. Depth of Beam

Assuming 50 cm width for the beam
$9{\cdot}07 \times 50 \times d^2 = 41000{,}000$

$$d = \sqrt{\frac{4100{,}000}{9{\cdot}07 \times 50}} = \sqrt{9050} = 95 \text{ cm}$$

Use 50 cm × 105 cm with 10 cm cover, $d_e = 95$ cm

8·14·2. Shear force at point of Max. Torsion

Shear force at support

$$= \frac{\text{Total load}}{2 \times \text{No. of supports}} = \frac{585}{2 \times 6} = 46{\cdot}6 \text{ t} = 48600 \text{ kg}$$

Shear force at point of max. torsion

$$= 48600 - \frac{12{\cdot}735 \times 2}{60} \times 48600 = 48600 - 20600 = 28000 \text{ kg}$$

Shear stress at support

$$q_s = \frac{48600}{{\cdot}86 \times 95 \times 50} = 11{\cdot}7 \text{ kg/cm}^2 > 5 \text{ kg/cm}^2$$

Shear stress at point of maximum shear

$$q = \frac{28000}{50 \times {\cdot}86 \times 95} = 6{\cdot}85 \text{ kg/cm}^2$$

Shear stress due to torsional moment

$$q' = \frac{T_m\left(3 + 2\frac{b}{d}\right)}{b^2 D}$$

$$= \frac{416000\,(3 + 2 \times \frac{50}{115})}{50 \times 50 \times 105} = \frac{416000 \times 3{\cdot}91}{50 \times 50 \times 105} = 5{\cdot}91 \text{ kg/cm}^2$$

Total shear stress $= q + q'$
$= 6{\cdot}85 + 5{\cdot}91 = 12{\cdot}76$ kg/cm² > 5 kg/cm² < 20 kg cm²

8·14·3. Reinforcement Main Steel

A_t for +ve BM at bottom d_e=95 cm

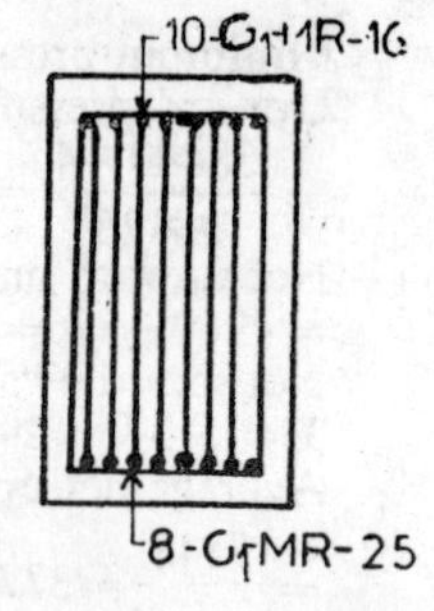

$$A_t=\frac{4100000}{1300\times\cdot86\times95}=38\cdot6\ \text{cm}^2$$

Use 8 bars of 25 mm ϕ

A_t=39·26 cm²

A_t for −ve BM at top

$$=\frac{2100000}{1300\times\cdot86\times95}=19\cdot6\ \text{cm}^2$$

Use 10−16 mm ϕ (A_t=21·1 cm²)

8·14·4. Shear Stirrups at Supports

Use 12 mm ϕ−8 legged stirrups spacing

$$s=\frac{A_w\ t_w\ j_d}{Q}=\frac{8\times1\cdot13\times1750\times0\cdot86\times95}{48600}=26\cdot5\ \text{cm}$$

Use 12 mm ϕ−8 legged at 25 cm c/c

8·14·5. Shear Stirrups at point of Maximum Torsion

Using 25 cm spacing area of steel required for shear force of

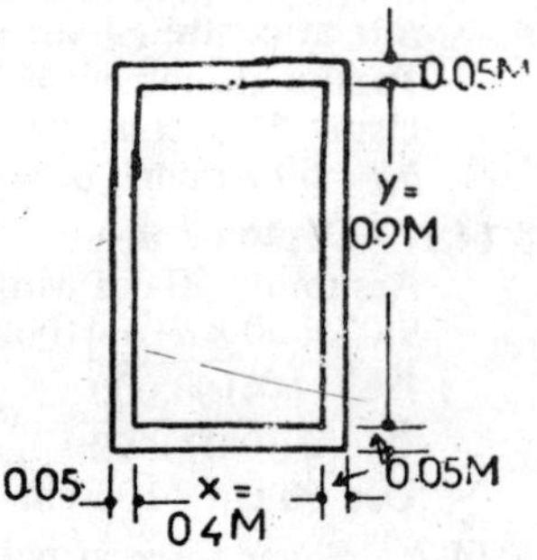

$$Aw_1=\frac{s\times Q}{t_w\times j_d}=\frac{27000\times25}{1750\times\cdot86\times95}$$

$=4\cdot89$ cm²

Using 25 cm spacing, area of steel required for torsional shear stress,

$$As_2=\frac{T_m\times s}{0\cdot8\times t_w\times X\times Y}$$

$$=\frac{416000\times25}{0\cdot8\times1750\times40\times100}$$

$=1\cdot86$ cm²

Total steel required at point of maximum torison
$=4\cdot89+1\cdot86=6\cdot75$ cm²

Using 12 mm ϕ area per leg=1·13 cm³

Number of legs required $=\frac{6\cdot75}{1\cdot13}=5\cdot8$ Nos−6 Nos (say)

Use 6 legged −12 mm ϕ stirrups at 25 cm c/c

8·14·6. Longitudinal Reinforcement for Torsion

Volume of longitudinal reinforcement per unit length for torsion

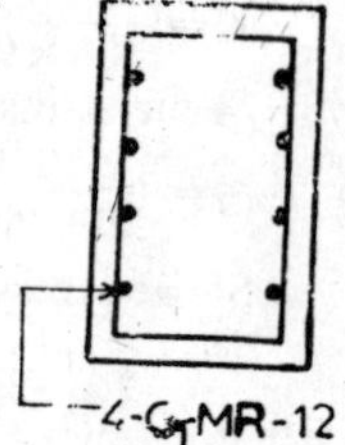

$$=1\cdot86\times\frac{100}{25}=7\cdot4\ \text{cm}^2$$

Area=7·4 cm² per unit length.

Use 8−12 mm ϕ - 4 Nos per face
(A_t−9·04 cm²)

8·14·7. Raft Slab

Projection of the slab from the centre line of the support=2·25m

$$\text{Bending moment}=\frac{3\cdot8\times2\cdot25^2}{2}=9\cdot65\ \text{mt}$$

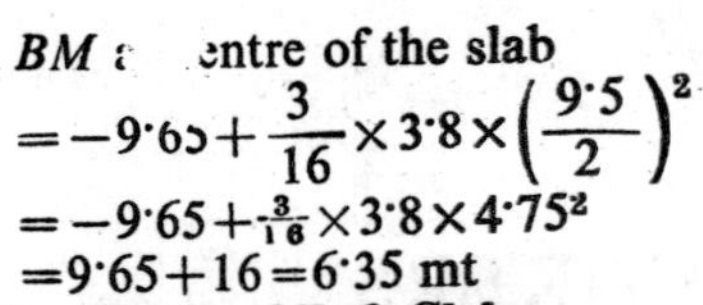

BM at centre of the slab

$$= -9{\cdot}65 + \frac{3}{16} \times 3{\cdot}8 \times \left(\frac{9{\cdot}5}{2}\right)^2$$

$$= -9{\cdot}65 + \tfrac{3}{16} \times 3{\cdot}8 \times 4{\cdot}75^2$$

$$= 9{\cdot}65 + 16 = 6{\cdot}35 \text{ mt}$$

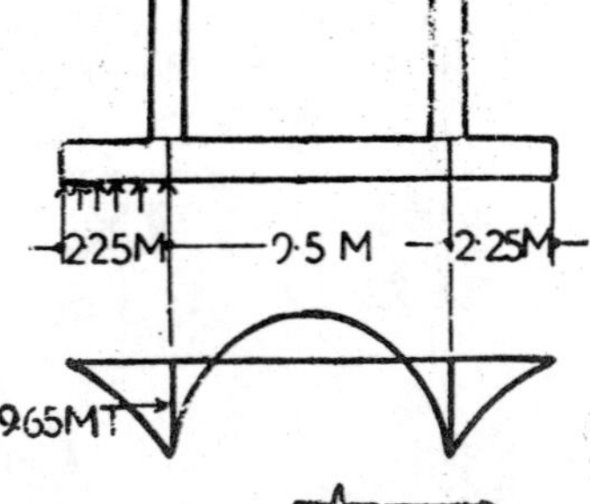

8·14·8. Depth of Raft Slab

$$9{\cdot}07 \times 100 \times d^2 = 965{,}000$$

$$d = \frac{965{,}000}{9{\cdot}07 \times 100} = 1050 = 32{\cdot}5 \text{ cm}$$

Use $d = 50$ cm $d_e = 40$ cm

$$A_t = \frac{965{,}000}{1300 \times {\cdot}86 \times 40} = 21{\cdot}6 \text{ cm}^2$$

Use 25 mm ϕ 22 cm c/c both ways in cantilever portion and for central portion

$$\frac{9{\cdot}65}{6{\cdot}35} \times 22 = 33{\cdot}5 \text{ cm}$$

8·15. Stability

Weight of tank and staging when tank is empty

Tank $= 502{\cdot}5 - 281 = 221{\cdot}5$ t

Columns $= 50{\cdot}82$ t

Braces $= 39{\cdot}80$ t

$$\text{Bottom ring beam} = \frac{110}{100} \times \frac{50}{100} \times \pi \times 9{\cdot}5 \times 2{\cdot}4 = 39{\cdot}5 \text{ t}$$

$$\text{Raft } \frac{\pi \times 14^2}{4} \times \frac{50}{100} \times 2{\cdot}4 = 185{\cdot}0 \text{ t}$$

Total wt. $= 527{\cdot}62$ t

Total horizontal wind force on the tank and staging $= 12{\cdot}45$ t

Coefficient of friction for black cotton soil $= 0{\cdot}2$

Factor of safety against sliding

$$= \frac{0{\cdot}2 \times 527{\cdot}62}{12{\cdot}45} = 8{\cdot}5$$

Factor of safety against overturning

$$= \frac{\text{stabilizing moment}}{\text{wind moment}} = \frac{527{\cdot}62 \times \frac{9{\cdot}5}{2}}{219{\cdot}95} = 11{\cdot}4$$

IX. BUNKERS

General Information I

1·1. General

A shallow bin is called a bunker. Bunkers are used to store ,rains, cement coal etc. A container in which the plane of rupture comes out on the surface of material is said to be shallow. Bunkers are used for storing coal. The inside surface is lined with anti attrition lining of concrete or blue brick tiles.

1·2. Limiting Condition

$\frac{d}{b} = \tan\frac{90° + \phi}{2}$ if ϕ = the angle of repose of the material = 25° (say)

$\frac{d}{b} = 1·5697$, that is for a bunker the dimensions are such that depth to breadth ratio shall be less than 1·5

1·3. Types

(*i*) Hoppers with sloping bottoms on all sides. (*ii*) Trough bunkers with two sides sloping (*iii*) Bunkers in which only one side sloping (*iv*) Circular bunkers with conical bottom

These bunkers may be constructed in single compartment or more than one compartment in which case it is called batteries of bunkers.

1·4. Side Walls

For a small bunker the maximum height of side walls is normally 2·5 to 3 m.

1·5. Bottom Hoppers

Bottom hopper sides should be steep enough to discharge the materials effectively. A slope of 1 : 1½ is used. Hopper is a term sometimes applied to a structure used for storing bulk materials loaded at the top and discharged at the bottom. Sloping sides of the bottom hopper facilitates the gravity discharged of the material.

1·6. Shape

Bunkers are mostly rectangular or square in plan, but circular ones are not uncommon.

1·7. Pressures in a Bunker

It is assumed that the horizontal pressure on the vertical side of a bunker is caused by the wedge of the filling. It is common to use one of the methods used to compute pressures on retaining walls with a filling of unlimited extent. The objections, however in applying these theories, to compute pressures are :

(*i*) that they do not take into account the effect of jamming or corner effect which occurs in bunkers and

(*ii*) that these theories do not consider the impact effect of materials while loading.

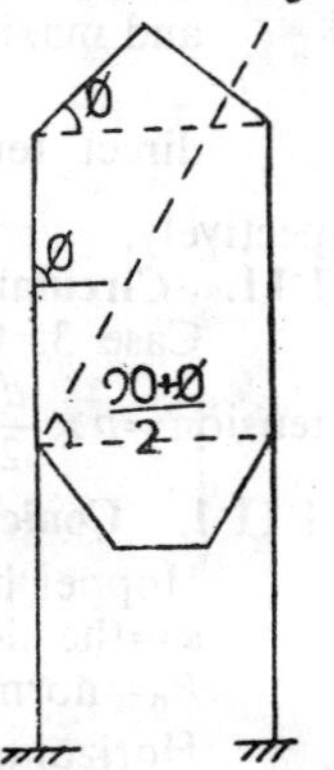

Rankines theory, however gives satisfactory results and therefore it is used to compute the pressures when containers are filled with material, the surcharge takes the slopes equal to the angle of repose of the material.

(*i*) With surcharge pressure on the vertical wall

$$p = wh\cos\delta \times \frac{\cos\delta - \sqrt{\cos^2\delta - \cos^2\phi}}{\cos\delta + \sqrt{\cos^2\delta - \cos^2\phi}}$$

If the surcharge angle $\delta = \phi$, angle of repose, $p = wh\cos\phi$ and it acts parallel to surcharge. The horizontal pressure $= wh\cos^2\phi$

(*ii*) Level fill $p = wh\left(\dfrac{1-\sin\phi}{1+\sin\phi}\right)$

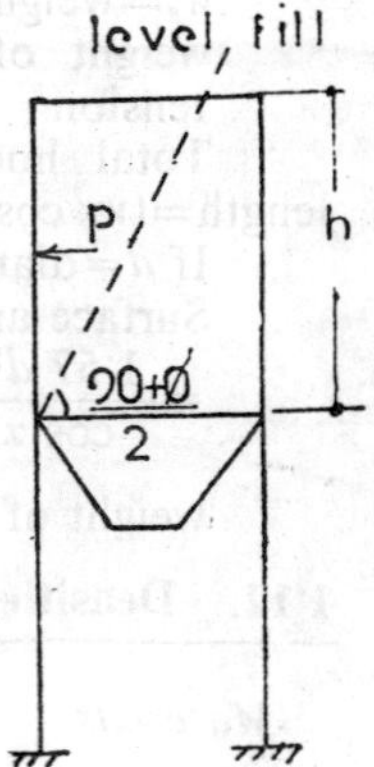

1·8. Probing Holes

Special tapered iron castings with covers shall be provided to attend to jamming of the material. These are known as probing holes and are provided down in the sloping bottom.

1·9. Square-Bending Moments in side Walls

Case 1 (*i*) Square bunker $h < L$

Bending moment al corner$= \dfrac{pL^2}{12}$

Bending moment at the centre of the wall

$$= \frac{pL^2}{24}$$

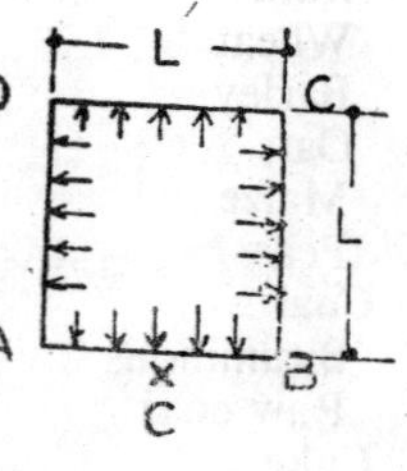

1·10. Rectangulaı

(*ii*) Rectangular –ve bending moment at

corners$= \dfrac{p}{12}(B^2 - BL + L^2)$

+ve bending moment at centre of shorter side B

$$= \frac{pB^2}{8} - \frac{p}{12}(B^2 - BL + L^2) = \frac{p}{24}(B^2 + 2BL - 2L^2)$$

+ve bending moment at the centre of the longer side L

$$= \frac{p}{24}(L^2 + 2BL - 2B^2)$$

Case 2. $h>L$ Maximum +ve bending moment$=\frac{Ph^2}{15}$

and maximum −ve bending moment$=\frac{Ph^2}{33{\cdot}5}$

direct tension$=\frac{pL}{2}$ and $\frac{pB}{2}$ on shorter and longer wall respectively.

1·11. Circular

Case 3. Circular walls are designed for hoop tension$=p\times\frac{d}{2}$

1·11·1. Conical Hopper Bottoms

Hopper bottom is designed as a conical dome.

α=the slope of the wall of the hopper.

P_n=normal pressure due to contents/sq. m

Horizontal pressure causing hoop tension$=p_n \sin\alpha$/sq. m

w_s=weight of the sloping slab/sq. m

weight of the sloping slab causing hoop tension $=w_s \cos\alpha$

Total hoop tension in the wall per metre length$=(w_s \cos\alpha+p_n \sin\alpha)\ d/2$

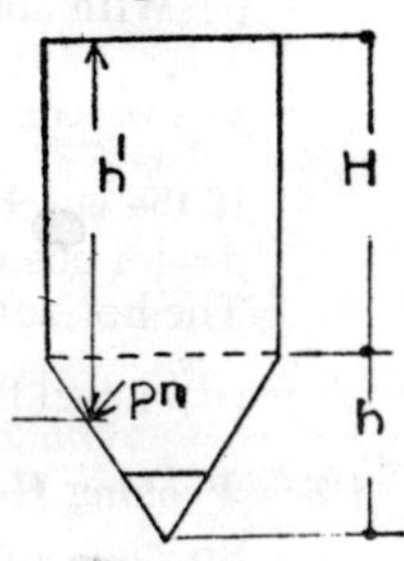

If d=diameter of the cone

Surface area of conical walls is approximately

$$=\frac{1{\cdot}57\ d^2}{\cos\alpha}$$

weight of conical walls$=\frac{1{\cdot}57\ w_s\ d^2}{\cos\alpha}$

1·12. Densities of Materials

Materials	*kg/cu m*	ϕ	$\frac{1-\sin\phi}{1+\sin\phi}$	$\frac{1+\sin\phi}{1-\sin\phi}$
Grains				
Wheat	800-850	25°	0·406	2·46
Barley	650	27°	0·375	2·66
Oats	450	28°		
Maize	725	28°		
Peas	850	25°		
Coal				
Bituminons	900	40°	0·217	4·61
Raw coal	800			
Coke	300-500	35°	0·271	3·69
Iron ore / Haematite	1950	45°	0·172	5·83
Lime Stone	1300-1800	35°	0·271	3·69
Dolomite	1200	—	—	—
Cement	1450	—	—	—
Clinker	1350-1450	30°	0·5	2
Slurry	1650	—	—	—
Sand	1450-1600	34°	0·284	3·52

Circular Coal Bunker 2
Capacity 30 tonnes

2·1. Data

Capacity=30 t, Density of coal=900 kg/m³, Angle of repose =40°, Elevation of the bottom of the hoppcr to the G.L.=3 m

Bunker carries a surcharge at angle equal to the angle of repose opening=50 cm×50 cm, Wind load=150 kg/m², Reduction coefficient=0·7.

Materials available, concrete=*M* 150, Steel—grade—*I*

Bearing capacity=10 t/m²

2·2. Characteristic Strengths

$\sigma_{cb}=50$ kg/cm² $\quad m=18$

$\sigma_{st}=1400$ kg/cm² $\quad jd=0{\cdot}87\,d \quad R=8{\cdot}7$

2·3. Trial Dimensions

Required capacity$=\dfrac{30000}{900}=33{\cdot}5$ cu m

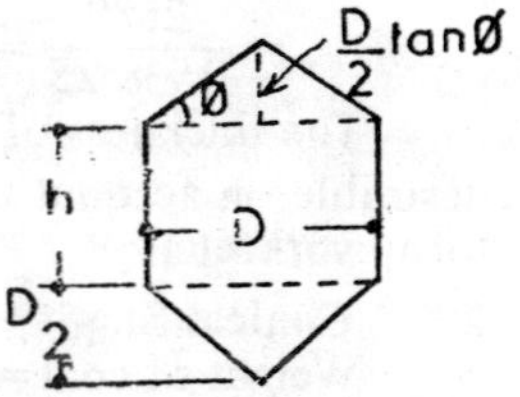

Approximately 25% of the capacity is provided by the surcharge and conical portion. Cylindrical portion accommodates the the remaining.

$\dfrac{\pi D^2 h}{4}=\cdot 75\times 33{\cdot}5=25{\cdot}2$ cu m. Say 26 cu m

$D^2=\dfrac{26\times 4}{\pi\times h}$

Assuming 3 m depth, $D^4=\dfrac{26\times 4}{\pi\times 3}=11$

$D=3{\cdot}3$ m, Use $D=3{\cdot}25$ m

2·4. Actual Capacity provided

Surcharge$=\frac{1}{3}\pi r^2 h$

$=\frac{1}{3}\times\pi\left(\dfrac{3{\cdot}25}{2}\right)^2\times\dfrac{3{\cdot}25}{2}\tan 40°$

$=\frac{1}{3}\times\pi(1{\cdot}625)^2\times 1{\cdot}625\times{\cdot}8391=3{\cdot}4$ cu m

Cylindrical portion$=\pi r^2 h$

$=\pi\times 1{\cdot}625^2\times 3=24{\cdot}9$ cu m

Conical portion$=\dfrac{\pi h_2}{12}(D^2+D_1^2+DD_1)$

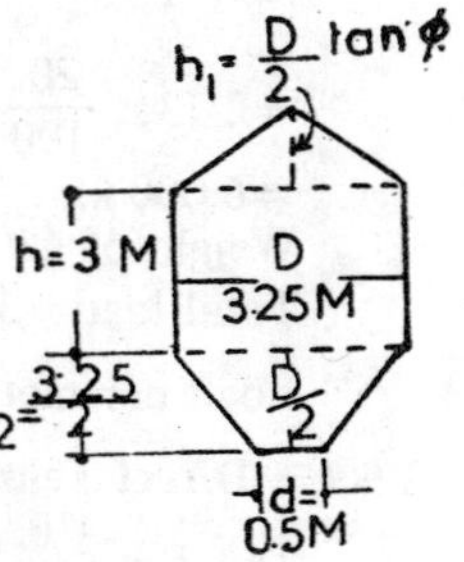

$=\dfrac{\pi\times 1{\cdot}625}{12}(3{\cdot}25^2+{\cdot}5^2+3{\cdot}25\times{\cdot}5)$

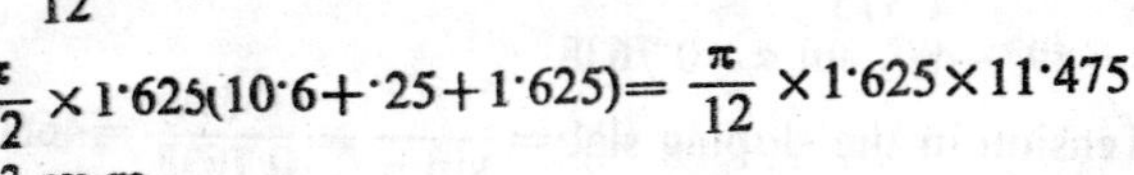

$=\dfrac{\pi}{12}\times 1{\cdot}625(10{\cdot}6+{\cdot}25+1{\cdot}625)=\dfrac{\pi}{12}\times 1{\cdot}625\times 11{\cdot}475$

$=5{\cdot}3$ cu m.

Actual capacity$=3{\cdot}4+24{\cdot}9+5{\cdot}3=33{\cdot}6$ cu m

2·4·1. Pressure Calculations

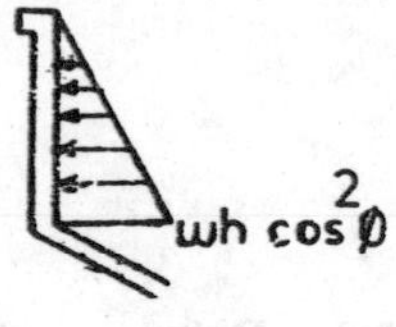

Maximum horizontal pressure at the bottom of the side wall, $p = wh \cos^2 \phi = 900 \times 3 \times (\cdot 766)^2$ $= 1585\ \text{kg/m}^2$

2·4·2. Hoop Tension

$$T = p \times \frac{D}{2} = 1585 \times \frac{3 \cdot 25}{2} = 2680\ \text{kg}$$

2·4·3. Hoop Steel

$$A_t = \frac{2680}{1400} = 1 \cdot 92\ \text{cm}^2$$

2·4·4. Thickness of the Wall. Tensile Stress in the Wall

Assume 15 cm thick wall

$$\text{Minimum reinforcement} = \frac{0 \cdot 15}{100} \times 15 \times 100$$

$= 2 \cdot 25\ \text{cm}^2$. Use 8 mm ϕ at 22 cm c/c both directions

G₁-MR-8-220

$$\sigma_{st} = \frac{2680}{100 \times 15 + 17 \times 2 \cdot 25} = \frac{2680}{1500 + 38 \cdot 25}$$

$$= \frac{2680}{1538 \cdot 25} = 1 \cdot 75\ \text{kg/cm}^2$$

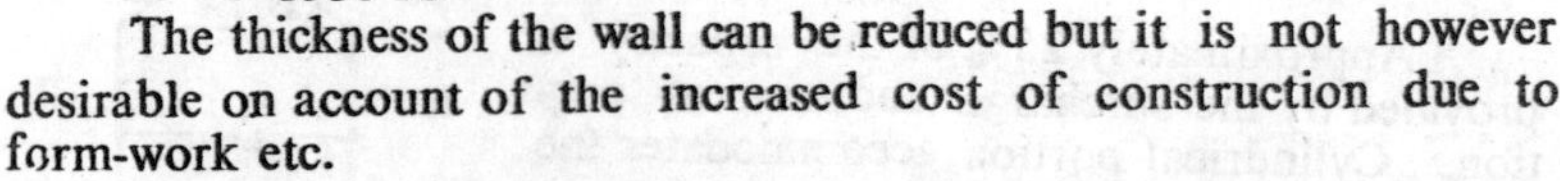

The thickness of the wall can be reduced but it is not however desirable on account of the increased cost of construction due to form-work etc.

2·5. Conical Sloping Bottom

Weight of coal $= 33 \cdot 6 \times 900 = 30240$ kg

Weight of sloping bottom including lining (assuming 20 cm thick)

0·572M
3M
D
1·375
·25
1·625

$$= \sqrt{1 \cdot 375^2 + (1 \cdot 625)^2} \times \frac{20}{100} \times 2\pi \left(\frac{1 \cdot 625 + \cdot 25}{2} \right) \times 2400$$

$$= \sqrt{1 \cdot 88 + 2 \cdot 65} \times \frac{20}{100} \times 2\pi \times \cdot 9375 \times 2400$$

$$= 2 \cdot 13 \times \frac{20}{100} \times 2\pi \times \cdot 9375 \times 2400$$

$= 6{,}000$ kg

Weight of gate $= 100$ kg

Total load $= 30240 + 6000 + 100 = 36340$ kg

$$\text{Load per metre of circumference} = T = \frac{36340}{\pi \times 3 \cdot 25} = 3570\ \text{kg}$$

2·5·1. Direct Tension

$$\tan \alpha = \frac{1 \cdot 625}{1 \cdot 375} = 1 \cdot 1850$$

$\alpha = 49° - 48'$, $\sin \alpha = 0 \cdot 7638$

$$\text{Tension in the sloping slab} = \frac{T}{\sin \alpha} = \frac{3570}{0 \cdot 7638} = 4680\ \text{kg}$$

$$\text{Steel required} = \frac{4680}{1400} = 3 \cdot 42\ \text{cm}^2.$$ Use 8 mm ϕ 14·5 cm c/c

2·5·2. Hoop Steel for Conical Portion

The normal pressure at depth h_1

$$p_n = wh_1\left(\cos^2\alpha + \frac{1-\sin\phi}{1+\sin\phi}\sin^2\alpha\right)$$

$$=900\times 4{\cdot}384\left(\cos^2 49°-48' + \frac{1-\sin 40°}{1+\sin 40°}\sin^2 49°\ 18'\right)$$

$$=900\times 4{\cdot}384({\cdot}6587^2+{\cdot}217\times{\cdot}7638^2)$$

$$=900\times 4{\cdot}384({\cdot}432+{\cdot}127)$$

$$=900\times 4{\cdot}384({\cdot}559)=2200\ \text{kg/m}^2$$

$p_r = p_n \operatorname{cosec}\alpha$

Radial reaction$=p_r=2200\times\operatorname{cosec}\alpha=\dfrac{2200}{{\cdot}7638}=2870\ \text{kg/m}^2$

Radial component of the weight of sloping slab$=W_s \operatorname{Cot}\alpha$

$$=\frac{20}{100}\times 2200\times\operatorname{Cot}\alpha=480(\cot 49°\text{-}48')=\frac{480}{\tan 49°\ 48'}$$

$$=\frac{480}{1{\cdot}185}=405\ \text{kg}$$

Hoop tension$=T=p\times\dfrac{D}{2}$

$$=(2870+405)\left(\frac{3{\cdot}25+0{\cdot}5}{2}\right)\times\frac{1}{2}$$

$$=3275\times\frac{3{\cdot}75}{2}\times\frac{1}{2}=3060\ \text{kg}$$

$$A_t=\frac{3060}{1400}=2{\cdot}2\ \text{cm}^2$$

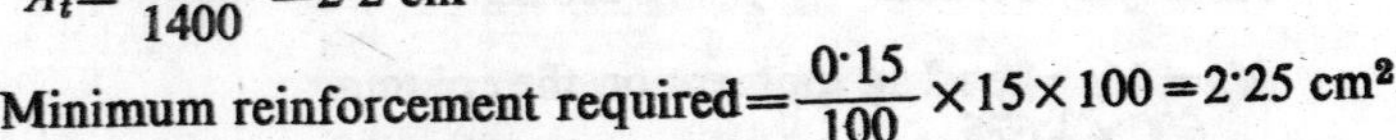

Minimum reinforcement required$=\dfrac{0{\cdot}15}{100}\times 15\times 100=2{\cdot}25\ \text{cm}^2$

Use 8 mm ϕ at 22 cm c/c

2·5·3. Stresses in the Conical Hopper

$$\sigma_{ct}=\frac{T}{b\times t+(m-1)A_s}=\frac{3060}{100\times 15+17\times 2{\cdot}25}=\frac{3060}{1500+38{\cdot}3}$$

$=2\ \text{kg/cm}^2<11\ \text{kg/cm}^2$

The thickness, however cannot be reduced below 15 cm due to possible impact of material while loading.

2·6. Columns

Loads: Weight of coal$=30240$ kg

Sloping conical bottom$=6000$ kg

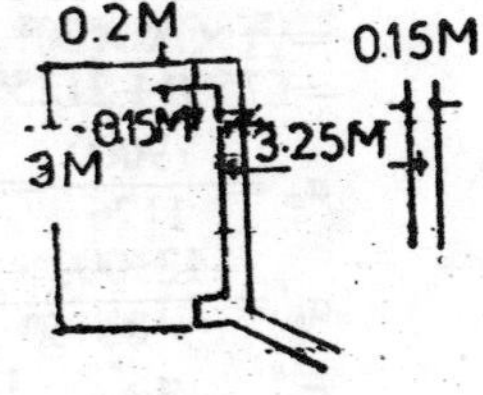

Side wall$=\pi\times 3{\cdot}4\times 3\times\dfrac{15}{100}\times 2400$

$=11{,}500$ kg

Top rib (15 cm$\times$20 cm)

$=\pi\times 3{\cdot}70\times\dfrac{15}{100}\times\dfrac{20}{100}\times 2400$

$=840$ kg

Bottom rib (15 cm$\times$40 cm)$=\pi\times 3{\cdot}70\times\dfrac{15}{100}\times\dfrac{40}{100}\times 2400$

$=1680$ kg

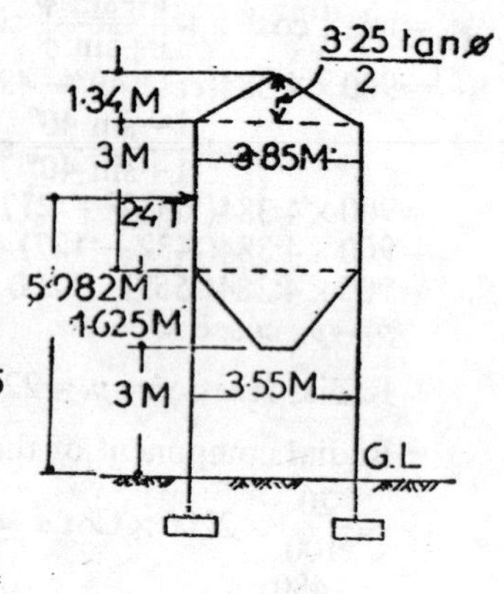

Weight of gate = 100 kg
Pilaster's etc. add = 800 kg
Total load = 52,000 kg.
Axial Load on each column
$=\frac{52}{4}=13$ t
Self weight of column
$=\frac{30}{100}\times\frac{30}{100}\times 5\times 2400$
$=1080$ kg $\doteq 1{\cdot}08$ t
Total load $=14{\cdot}08$ t say 15 t

2·6·1. Wind Loads

Wind load on the bunker
$=\cdot 7\times 150\times(1{\cdot}34+3+1{\cdot}625)\times 3{\cdot}85$
$=\cdot 7\times 150\times 5{\cdot}965\times 3{\cdot}85$
$=2400$ kg
Acts at a height
$=3+\frac{5{\cdot}965}{2}=3+2{\cdot}982=5{\cdot}982$ m

Wind moment $=2400\times 5{\cdot}982=14210$ m kg
$=1421000$ cm kg
Axial load in the column due to wind
moment $=\frac{Mr}{\Sigma r^2}=\frac{14{\cdot}21\times 1{\cdot}775}{2\times 1{\cdot}775^2}$
$=\frac{14{\cdot}21}{2\times 1{\cdot}775}=4$ t

r = 1.775M
3.55M

Shear per column $=\frac{2400}{4}=600$ kg

$M_A=M_B=$ bending moment on the column

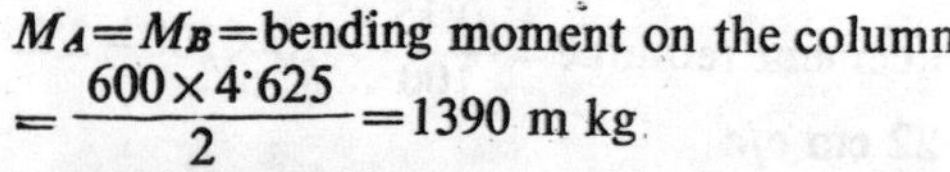

$=\frac{600\times 4{\cdot}625}{2}=1390$ m kg

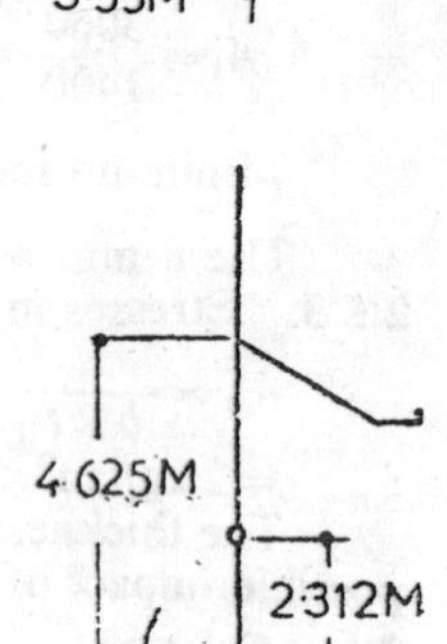

2·6·2. Column Trial Section

Total axial load $=15+4=19$ t
Bending moment $=1390$ m kg
Try 30 cm × 30 cm with 4—20 mm ϕ
Equivalent area $=A_c+(m-1)\,A_t$
$=30\times 30+17\times 3{\cdot}14\times 4=900+226$
$=1126\ \text{cm}^2$
$I=$ moment of inertia
$=\frac{1}{12}\times 30\times 30^3+4\times 3{\cdot}14\times 17\times(10)^2$
$=67500+21250=88750\ \text{cm}^4$

$\sigma_c'=\frac{19000}{1126}=16{\cdot}9\ \text{kg/cm}^2$

$\sigma_b'=\frac{139000\times 15}{88750}=23{\cdot}5\ \text{kg/cm}^2$

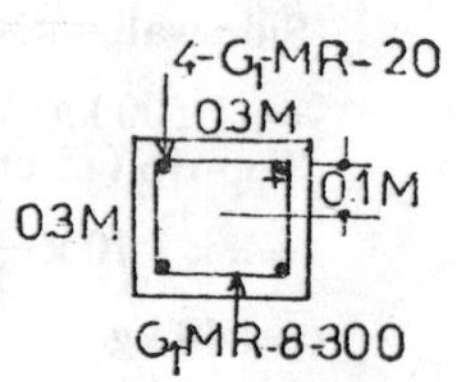

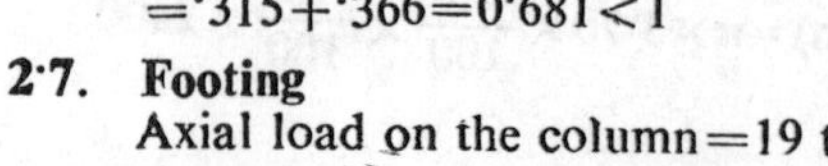

$\frac{\sigma_c'}{\sigma_c}+\frac{\sigma_{cb}'}{\sigma_{cb}}=\frac{16{\cdot}9}{53{\cdot}3}+\frac{23{\cdot}5}{66{\cdot}7}$
$=\cdot 315+\cdot 366=0{\cdot}681<1$

2·7. Footing

Axial load on the column $=19$ t

Weight of footing (10%)=1·9 t
Total axial load on the soil=20·9 t
Maximum bending moment on the column=139000 cm kg

Eccentricity of the load $e=\frac{M}{P}=\frac{139000}{20900}=6{\cdot}7$ cm $={\cdot}067$ m

Area required$=\frac{20{\cdot}9}{10}=2{\cdot}09$ sq. m

To allow for eccentricity, Use 1·75 m×1·75 m footing.

2·7·1. Pressure Under Footing

Maximum pressure$=\frac{P}{A}\left(1+\frac{6e}{B}\right)$

$=\frac{20{\cdot}9}{1{\cdot}75\times1{\cdot}75}\left(1+\frac{6\times{\cdot}067}{1{\cdot}75}\right)=8{\cdot}4$ t/m² < 10 t/m²

Maximum pressure

$\frac{P}{A}\left(1-\frac{6e}{B}\right)=5{\cdot}25$ t/m²

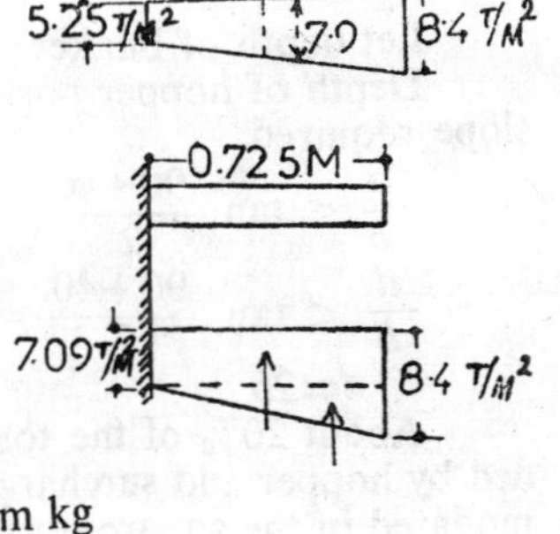

Maximum pressure at the edge of the column

$=5{\cdot}25+(8{\cdot}4-5{\cdot}25)\times\frac{102{\cdot}5}{175}$

$=5{\cdot}25+3{\cdot}15\times\frac{102{\cdot}5}{175}=7{\cdot}09$

Maximun bending moment

$=7{\cdot}09\times{\cdot}725\times\frac{725}{2}$

$+\frac{1}{2}(8{\cdot}4-7{\cdot}09)\times{\cdot}725\times\frac{2}{3}\times{\cdot}725$

$=7{\cdot}09\times{\cdot}725\times\frac{{\cdot}725}{2}$

$+\frac{1}{2}\times1{\cdot}31\times{\cdot}725\times\frac{2}{3}\times{\cdot}725$

$=1{\cdot}841+0{\cdot}23=2{\cdot}071$ mt$=207100$ cm kg

2·7·2. Depth of Footing

$d_e=\sqrt{\frac{207100}{8{\cdot}7\times100}}=\sqrt{238}=15{\cdot}4$

Use $d=25$ cm and $d_e=20$ cm

$A_t=\frac{207100}{1400\times{\cdot}07\times20}=8{\cdot}55$ cm²

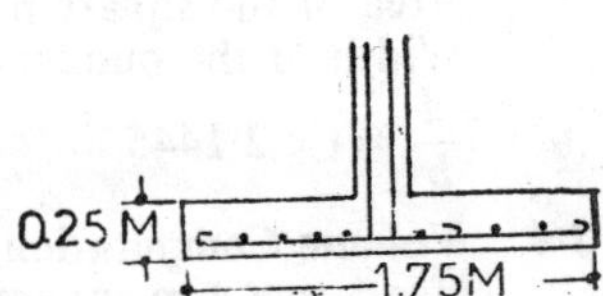

Use 12 mm ϕ at 13 cm c/c both ways.

Square Coal Bunker 3
Capacity 30 Tonnes

3·1. Data

Capacity=30 t,
Density of coal=900 kg/m^3
Angle of repose=40°
Elevation of the bottom of the hopper to the G.L.=3 m,
Bunker is expected to carry surcharge at an angle of repose opening=50 cm×50 cm

3·2. Characteristic Strengths

M 150 concrete and grade–1 steel
σ_{cb}=50 kg/cm^2, σ_c=40 kg/cm^2, j_d=·87
σ_{st}=1400 kg/cm^2, m=18, R=8·7

3·3. Trial Dimensions

$$\text{Required capacity}=\frac{30000}{900}=33{\cdot}5 \text{ cu m}$$

Let depth of bunker=d=3 m
Depth of hopper portion depends on the slope required

$$\frac{d}{b}<\tan\frac{90+\phi}{2}$$

$$\frac{d}{b}<\tan\frac{90+40}{2}=<\tan 65°=<2{\cdot}1445$$

$$d \simeq 2b$$

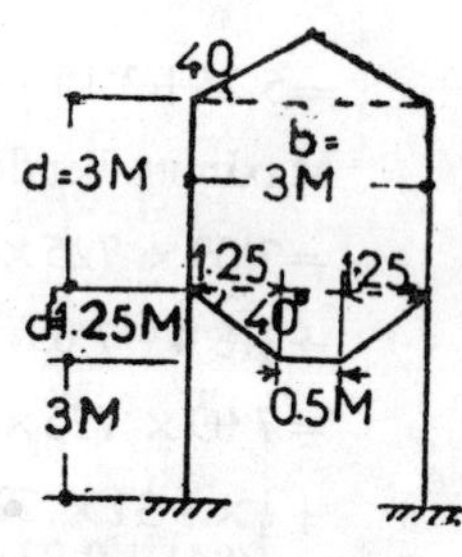

About 20% of the total capacity is provided by hopper and surcharge. The rest is accommodated in the square portion.

Area of the square portion=$\frac{27}{3}$=9 sq. m,
Width of the bunker=3 m,

$$\frac{d}{b}=1<2{\cdot}1445$$

3·4. Volume Computation

Use 3 m×3 m square bunker with 3 m depth and a hopper of 1 m

Surcharge=3×3×$\frac{1}{3}$×1·5 tan 40°
=4·5 tan 40°=4·5×·8391=3·77 cu m
Square portion=3×3×3=27 cu m
Hopper portion

$$=\frac{d'}{3}\left(a_1+a_2+\sqrt{a_1\times a_2}\right)$$

Where a_1=area of cross section of the square portion and a_2=area at the opening

$$=\frac{1{\cdot}25}{3}\left(3\times3+\tfrac{1}{2}\times\tfrac{1}{2}+\sqrt{9\times\tfrac{1}{4}}\right)$$

$$=\frac{1\cdot25}{3}\,(9+\tfrac{1}{4}+\sqrt{2\cdot25})$$

$$=\frac{1\cdot25}{3}\,(9\cdot25+1\cdot5)$$

$$=\frac{1\cdot25}{3}\,(10\cdot75)=4\cdot87 \text{ cu m}$$

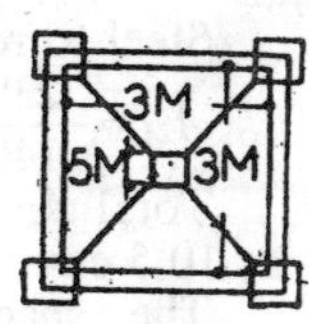

Total volume = 3·77 + 27·00 + 4·87 = 35·64 cu m

Actual capacity provided = 34·64 cu m

3·5. Pressure Calculations

Maximum horizontal pressure at the bottom of the side wall

$p = \text{w.d. } \cos^2\phi = 900\times3\times\cos^2 40^\circ$

$= 900\times3\times(\cdot766)^2 = 1585 \text{ kg/m}^2$

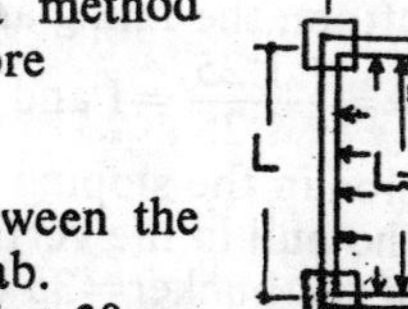

Some times the pressure is also computed on the basis of the average depth of fill material

$$p = wh_a \times \frac{1-\sin 40^\circ}{1+\sin 40^\circ}$$

$= 900\times3\cdot625\times0\cdot217 = 710 \text{ kg/m}^2$

This on the low side and in the absence of a theory, the safest rule is to use a method which errs on the high side and therefore $p = wd \cos^2\phi$ is preferred.

3·6. Side Walls Bending Moments

The side walls span horizontally between the columns and is treated as a continuous slab.

Horizontal span of the wall (assuming 30 cm bearing) = 3·0 + ·3 = 3·3 m

Maximum bending moment $= \frac{pl^2}{12} = \frac{1585\times3\cdot3^2}{12} = 1440$ m kg

3·6·1. Direct Tension in the Wall

$$T = \frac{pL}{2} = 1585\times\times\frac{3}{2} = 2380 \text{ kg}$$

3·6·2. Thickness of the Wall

Trial thickness of the wall

$$d_e = \sqrt{\frac{144000}{100\times8\cdot7}} = \sqrt{166} = 12\cdot9 \text{ cm}$$

Taking 12·5 cm effective depth and 15 cm total depth, net bending moment $= M - T_x$

$= 144000 - 2380(12\cdot5 - 7\cdot5) = 144000 - 2380\times5 = 144000 - 11900$

$= 132100$ cm kg

$8\cdot7\times100\times d_e^2 = 132100$

$$d_e = \sqrt{\frac{132100}{8\cdot7\times100}} = \sqrt{152} = 12\cdot4 \text{ cm}$$

Use $d_e = 12\cdot5$ cm and $d = 15$ cm

3·6·3. Main Steel

Steel to resist bending

$$A_{tb} = \frac{132100}{1400\times\cdot87\times12\cdot5} = 8\cdot7 \text{ cm}^2$$

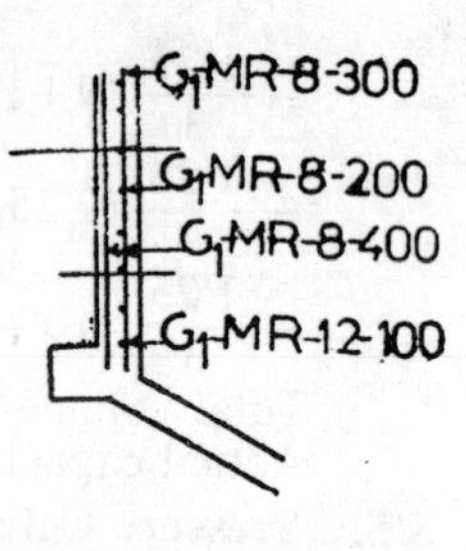

Steel to resist direct tension

$$A_{tt}=\frac{2380}{1400}=1.7 \text{ cm}^2$$

Total $A_t=8.7+1.7=10.4$ cm² Use 12 mm ϕ at 10.5 cm c/c at bottom one metre height.

The spacing of these bars can be reduced towards top as the pressure p reduces.

3.6.4. Secondary Reinforcement

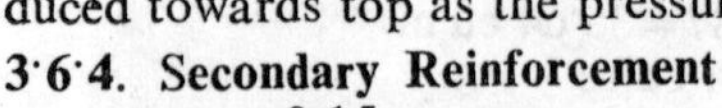

$$A_t=\frac{0.15}{100}\times 15\times 100=2.25 \text{ cm}^2$$

Use 8 mm ϕ at 20 cm c/c.

3.6.5. Top Rib

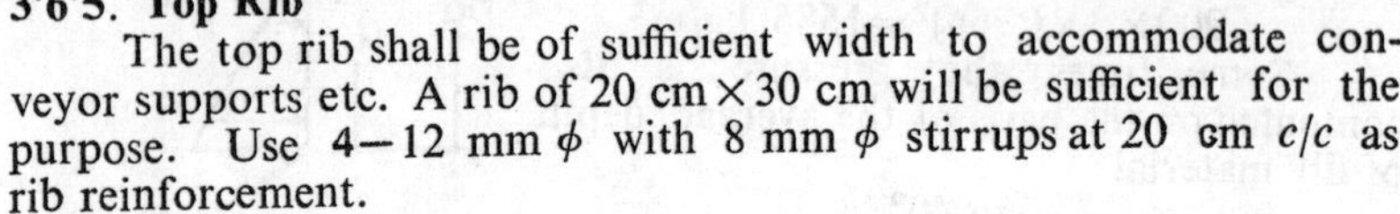

The top rib shall be of sufficient width to accommodate conveyor supports etc. A rib of 20 cm×30 cm will be sufficient for the purpose. Use 4—12 mm ϕ with 8 mm ϕ stirrups at 20 cm c/c as rib reinforcement.

3.6.6. Hopper Bottom Direct Tension

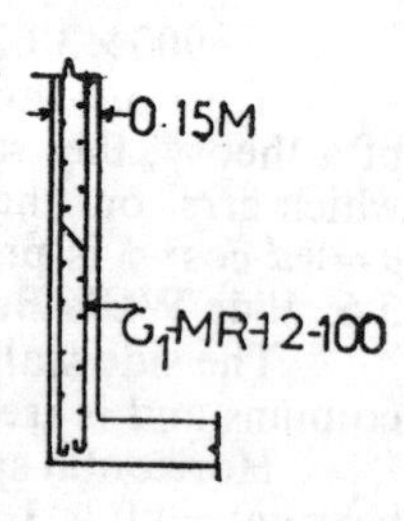

There is a direct pull T at the bottom of the side wall due to material contained in it, weight of sloping bottom, the lining and the gate.

$$\tan\alpha=\frac{1.25}{1.25}=1 \text{ and } \alpha=45°$$

Tension in the sloping bottom=T cosec α

T=the pull in the vertical wall. Weight of the contents of the bunker=35.64×900=32150 kg

self weight of sloping bottom assuming 20 cm thick including lining

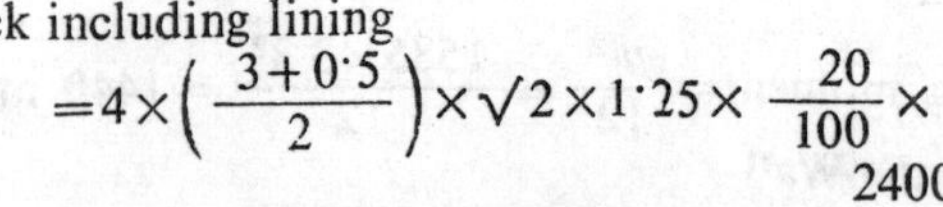

$$=4\times\left(\frac{3+0.5}{2}\right)\times\sqrt{2}\times 1.25\times\frac{20}{100}\times 2400$$

$$=4\times 1.75\times 1.77\times 480=5950 \text{ kg}$$

Gate and other fittings say $=100$ kg

Total load$=32150+5950+100$

$=38,200$ kg

Load per side of hopper

$$=\frac{38,200}{4}=9550 \text{ kg}$$

Load per metre run at the junction

$$T=\frac{9550}{3}=3183.3 \text{ kg}$$

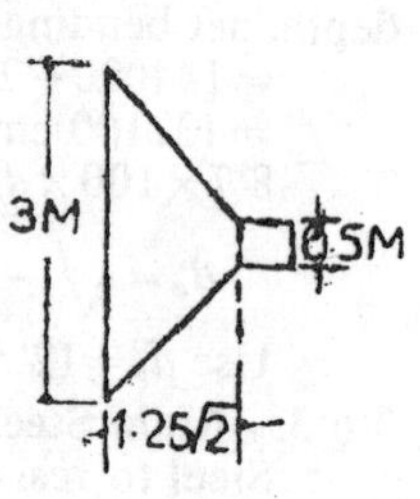

Direct pull in the sloping bottom

$T_s=3183.3\times\text{cosec}\,\alpha=3183.3\times\sqrt{2}$

$=4500$ kg

3.6.7. Reinforcement

At the top of the sloping bottom steel required to resist the direct tension

$$=\frac{4500}{1400}=3.21 \text{ cm}^2 \quad \text{Use 10 mm } \phi \text{ at 24 cm } c/c$$

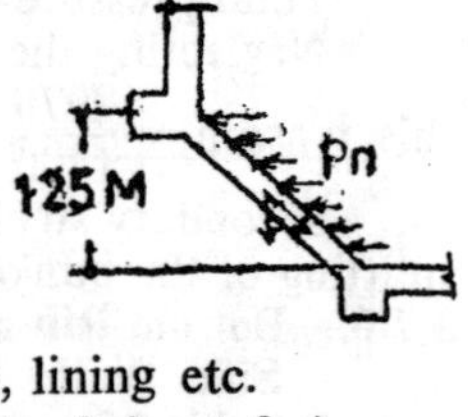

The direct tension decreases gradually towards bottom. Two 12 mm ϕ hanging bars may be provided per side to provide monolithic action between the side wall and hopper bottom.

3·7. Sloping Bottom as Slab

Loads: (*i*) Normal component of the pressure due to the contents of the bunker (*ii*) Normal component of the self weight of slab, lining etc.

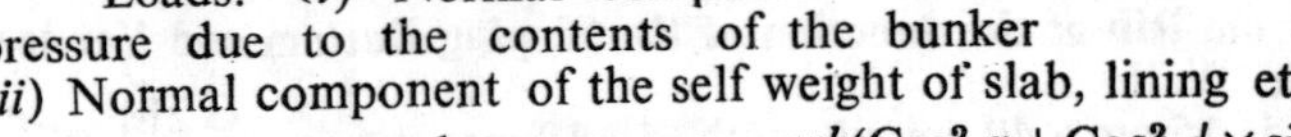

Average normal pressure $=p_n=wh(\text{Cos}^2\,\alpha+\text{Cos}^2\,\phi\times\sin^2\alpha)$

$$=900\left(3+\frac{1{\cdot}25}{2}+\frac{1{\cdot}25}{2}\times\tan 40^\circ\right)(\cos^2 45^\circ+\cos^2 40^\circ\times\sin^2 45^\circ)$$

$$=900(3+0{\cdot}625+{\cdot}625\times0{\cdot}8391)(\tfrac{1}{2}+{\cdot}766^2\times\tfrac{1}{2})$$

$$=900\times4{\cdot}153\times0{\cdot}794=2980\text{ kg/m}^2$$

Normal component due to self weight of slab including lining etc. $=\dfrac{20}{100}\times2400\cos 45^\circ=480\times\dfrac{1}{\sqrt{2}}=340\text{ kg/m}^2$

Total normal load $=2980+340$
$=3320\text{ kg/m}^2$

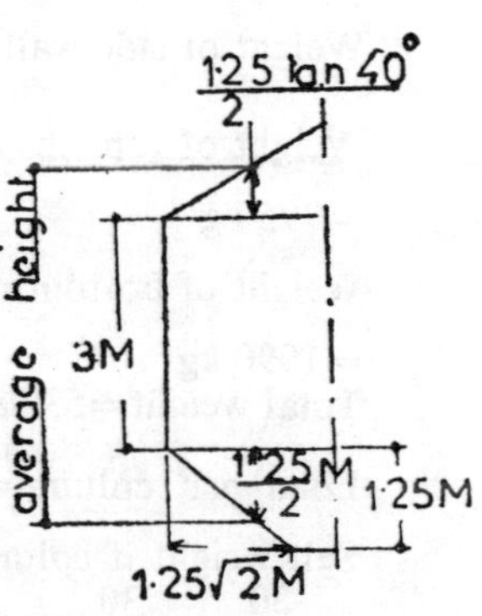

3·7·1. Bending Moment

$$M=\frac{wl^2}{12}=\frac{3320}{12}\times(\sqrt{2}\times1{\cdot}25)^2$$

$$=\frac{3320}{12}\times(1{\cdot}77)^2=870\text{ m kg}$$

Maximum bending moment
$=87000$ cm kg

3·7·2. Depth of slab

$$8{\cdot}7\times100\times d_e^2=87000$$

$$d_e=\sqrt{\frac{87000}{8{\cdot}7\times100}}=\sqrt{100}$$

$=10$ cm. Use $d=15$ cm $d_e=10$ cm and lining 5 cm

3·7·3. Main Steel

Steel to resist bending

$$=A_t=\frac{87000}{1400\times{\cdot}87\times12}=5{\cdot}92\text{ cm}^2$$

Use 10 mm ϕ at 13 cm c/c

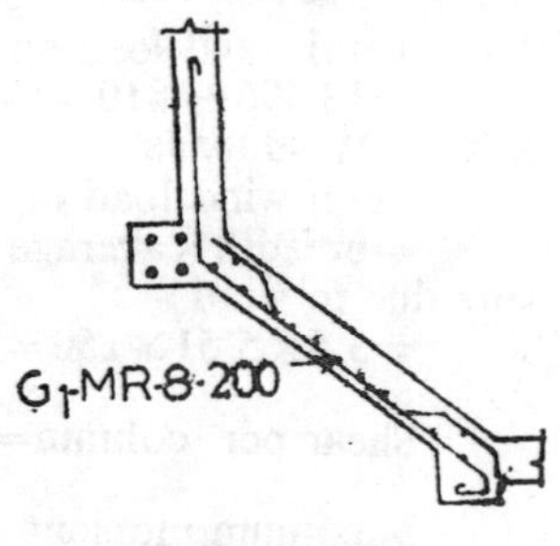

3·7·4. Secondary Reinforcement

$A_t=0{\cdot}15\%$ of concrete

$$=\frac{0{\cdot}15}{100}\times15\times100=2{\cdot}25\text{ cm}^2$$

3·7·5. Tendency to Split the Bunker

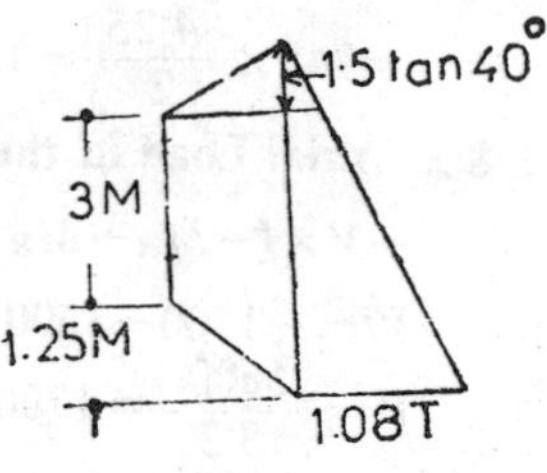

Maximum pressure acting on a vertical plane through the middle of the bunker

$$p=900\times(3+1{\cdot}5\tan 40^\circ+1{\cdot}25)\frac{1-\sin 40^\circ}{1+\sin 40^\circ}$$

$$=900\times(3+1{\cdot}5\times{\cdot}8391+1{\cdot}25)\times{\cdot}0{\cdot}217$$

$$=900(3+1{\cdot}26+1{\cdot}25)\times{\cdot}217$$

$$=900\times5{\cdot}51\times{\cdot}217=1080\text{ kg}$$

Total pressure $=\frac{1}{2}\times 1080\times 5{\cdot}51\times 3=2990\times 3=7970$ kg

Neglecting the concrete tensile strength steel required to resist this tension $=\dfrac{7970}{1400}=5{\cdot}7\ \text{cm}^2$

Secondary steel however would take this tension and prevent splitting of the bunker.

3·7·6. Bottom Rib at the Junction of the Sloping Bottom and Vertical Side Wall

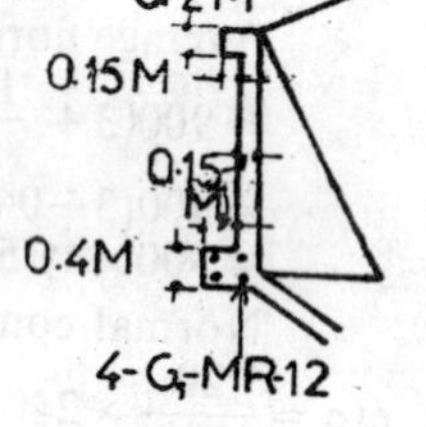

Provide 15 cm × 40 cm rib with 4 – 12 mm ϕ with 8 mm ϕ at 20 cm c/c stirrups.

3·8. Columns

Loads : Weight of the contents of the bunker $=32150$ kg

Self weight of sloping bottom including lining $=5950$ kg

Weight of gate $=100$ kg

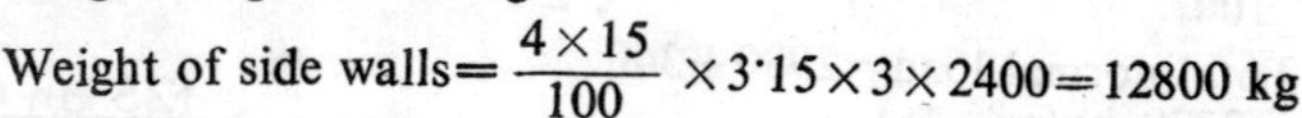

Weight of side walls $=\dfrac{4\times 15}{100}\times 3{\cdot}15\times 3\times 2400=12800$ kg

Weight of top ribs (4 Nos) $=3{\cdot}45\times 4\times \dfrac{20}{100}\times\dfrac{15}{100}\times 2400$ $=995$ kg

Weight of bottom ribs (4 Nos) $=3{\cdot}45\times 4\times\dfrac{40}{100}\times\dfrac{15}{100}\times 2400$ $=1990$ kg

Total weight $=53{\cdot}985$ t.

Load per column $=\dfrac{53{\cdot}985}{4}=13{\cdot}496\ t$

Self weight if column (assuming 30 m × 30 cm)

$=\dfrac{30}{100}\times\dfrac{30}{100}\times 3{\cdot}75\times 2400=810$ kg

Total axial load on the column $=13{\cdot}496+{\cdot}810=14{\cdot}306$ t

3·8·1. Wind loads

Total wind load on the bunker $=$ breadth × average height × intensity of pressure due to wind

$=3{\cdot}6\times 5{\cdot}51\times 150=2980$ kg

Shear per column $=\dfrac{2980}{4}=745$ kg

Maximum moment at the column base

$=745\times\dfrac{4{\cdot}25}{2}=1500$ m kg

3·8·2. Axial Load in the Leeward Column due to wind

$$-V\times l-M_A-M_B+\frac{2980}{2}\times 5{\cdot}75=0$$

$$Vl=-1500-1500+8600=5600$$

$$V=\frac{5600}{3{\cdot}3}=1700\ \text{kg}=1{\cdot}7\ t$$

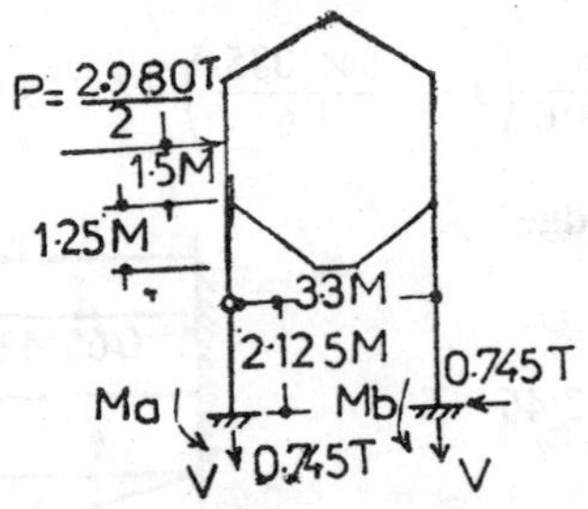

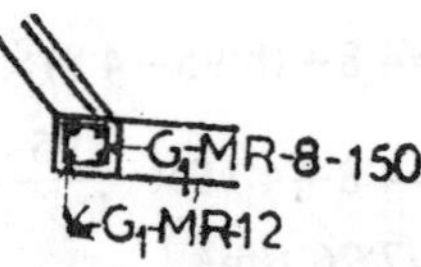

Total axial load on the column $=16{\cdot}06$ t $=16\ T$ (say)
Maximum bending moment in the column $=1500$ m kg

3·8·3. Trial Section

Use 30 cm × 30 cm with 4—20 mm ϕ

Equivalent area $=A$

$=30\times30+17\times3{\cdot}14\times4=900+226$

$=1126$ cm²

Moment of inertia

$I=\frac{1}{12}\times30\times30^3+4\times3{\cdot}14\times17(10)^2$

$=67{,}500+21250=88{,}750$ cm⁴

$$\sigma_c'=\frac{P}{A}=\frac{16000}{1126}=14{\cdot}2\ \text{kg/cm}^2$$

$$\sigma_b'=\frac{M_y}{I}=\frac{150000\times15}{88750}=25{\cdot}4\ \text{kg/cm}^2$$

$$\frac{\sigma_c'}{\sigma_c}+\frac{\sigma_{cb}'}{\sigma_{cb}}=\frac{14{\cdot}2}{53{\cdot}3}+\frac{25{\cdot}4}{66{\cdot}7}={\cdot}255+{\cdot}382={\cdot}637<1$$

3·9. Foundation Footing

Total direct load on the column $=16000$ kg

Add 10% for weight of footing $=1600$ kg

Total load $=17600$ kg

Moment at the base of the column

$=1500$ m kg $=150000$ cm kg

$$\text{Eccentricity}=\frac{150000}{17600}=8{\cdot}5\ \text{cm}={\cdot}085\ \text{m}$$

$$\text{Area required}=\frac{17{,}600}{10}=1{\cdot}76\ \text{sq. m}$$

Use 1·35 m × 1·35 m (1·825 sq. m)

3·9·1. Pressure below the Footing

Maximum pressure

$$=\frac{P}{A}\left(1+\frac{6e}{B}\right)$$

$$=\frac{17{\cdot}6}{1{\cdot}825}\left(1+\frac{6\times{\cdot}085}{1{\cdot}35}\right)$$

$=13{\cdot}2$ t/m² >10 t/m²

Increase the footing size.

Try 1·6 × 1·6 m square

Maximum pressure

$$=\frac{17{\cdot}6}{1{\cdot}6\times1{\cdot}6}\left(1+\frac{6\times{\cdot}085}{1{\cdot}6}\right)$$

$=6{\cdot}75\,(1+{\cdot}30)=8{\cdot}95$ t/m²

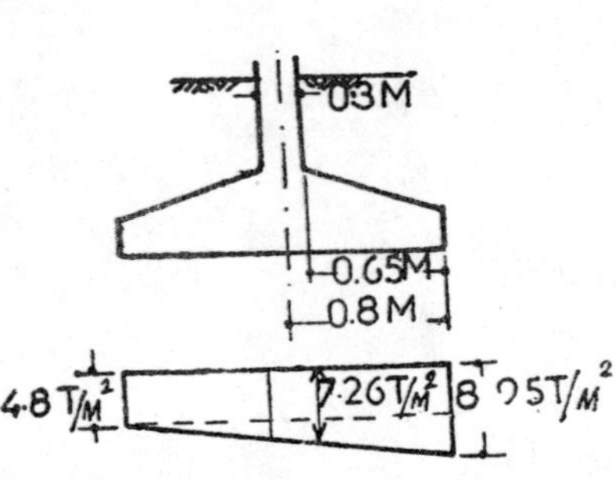

Minimum pressure$=\frac{17{\cdot}6}{1{\cdot}6\times1{\cdot}6}\left(1-\frac{6\times{\cdot}085}{1{\cdot}6}\right)=6{\cdot}75\times0{\cdot}7$

$=4{\cdot}8$ t/m²

Maximum pressure at the edge

$=4{\cdot}8+(8{\cdot}95-4{\cdot}8)\times\frac{0{\cdot}95}{1{\cdot}6}$

$=4{\cdot}8+4{\cdot}15\times\frac{0{\cdot}95}{1{\cdot}6}=4{\cdot}8+2{\cdot}46$

$=7{\cdot}26$ t/m²

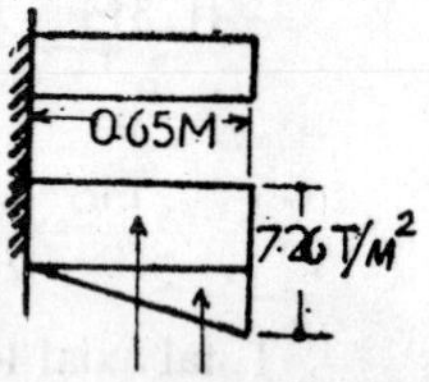

Bending moment

$=7{\cdot}26\times{\cdot}65\times\frac{{\cdot}65}{2}+\frac{1}{2}\times{\cdot}65\times(8{\cdot}95-7{\cdot}26)$

$\times\frac{2}{3}\times{\cdot}65$

$=1{\cdot}52+\frac{1}{2}\times{\cdot}65\times1{\cdot}69\times\frac{2}{3}\times{\cdot}65=1{\cdot}52+{\cdot}237=1{\cdot}757$ mt

$=175700$ cm kg

3·9·2. Depth of Footing

$$d_e=\sqrt{\frac{175700}{8{\cdot}7\times100}}=\sqrt{202}=14{\cdot}6 \text{ cm}$$

Use $d=20$ cm $\quad d_e=16$ cm

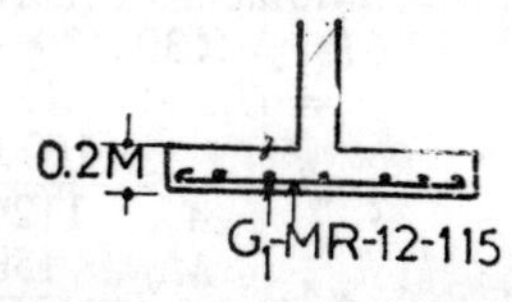

$$A_t=\frac{176700}{1400\times{\cdot}87\times15}=9{\cdot}6 \text{ cm}^2$$

Use 12 mm ϕ at 11·5 cm c/c at the bottom.

X. SILOS

General Information 1

1·1. General

A deep bin in which the plane of rupture cuts the side wall, is called a silo. A large proportion of the weight of the filling is supported on the side walls by friction between the concrete and the material stored. Usually they are circular in shape.

1·2. Limiting Condition

$$\frac{d}{b} > \tan \frac{90+\phi}{2}$$

For a value of $\phi=25°$, $d/b>1·5697$

The depth of a silo is $>1·5$ times its width or diameter.

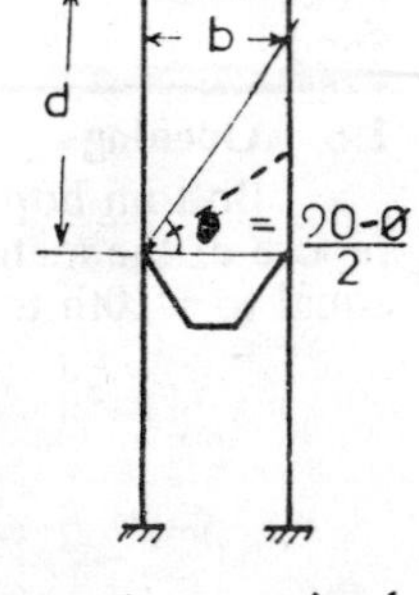

1·3. Theoretical Minimum Depth

The theoretical minimum depth of a silo is

$$d=b\times\left[\tan\phi+\sqrt{\tan\phi}\left(\frac{1+\tan^2\phi}{\tan\phi+\tan\mu}\right)\right]$$

Where ϕ=angle of surcharge or angle of repose.

μ=angle of internal friction between the wall and the contained material.

1·4. Theories for Pressure Computations

(*i*) Janssen's theory, the horizontal pressure

$$=P_h=\frac{wR}{\mu'}\left(1-e^{\frac{\mu' k}{R}h}\right)$$

where P_h=intensity of the horizontal pressure, w=density of material, $R=A/U$, the hydraulic mean depth

A=area, U=perimeter, $\mu'=\tan\mu$ and $k=\left(\frac{1-\sin\phi}{1+\sin\phi}\right)$

h=depth below top at which pressures are being calculated

1·5. Airy's Theory

Plane of rupture is given by

$$\tan\theta=\sqrt{\frac{2h}{b}\,\frac{1+\mu^2}{\mu+\mu'}+\frac{1+\mu^2}{\mu+\mu'}\times\frac{(1-\mu\mu')}{\mu+\mu'}}-\frac{1-\mu\mu'}{\mu+\mu'}$$

and the horizontal pressure

$$P=\frac{wb}{2}\,(2h-b\tan\theta)\frac{\tan\theta-\mu}{1-\mu\mu'+(\mu+\mu')\tan\theta}$$

where, θ=angle between the plane of separation of the wedge-shaped mass causing maximum pressurc and the horizontal, h=depth of silo, b=breadth or diameter of silo, μ=coefficient of friction of material on material, μ'=coefficient of friction on the wall and the material, w=density of the stored material.

Coefficient of friction of various materials

Material	*Density*	*Material on material* μ	*Material on concrete* μ'
Wheat	850	0·466	0·444
Maize	700	0·521	0·423
Barley	625	0·507	0·452
Beans	750	0·616	0·442
Peas	800	0·472	0·296
Soya bean	750	—	0·27
Cement	1450	0·316	0·700
Coal	800-900	0·700	0·700
Coke	450-500	0·839	0·839
Sand	1600	0·674	0·577

1·6. Opening

Bottom hopper should have a slope steeper than the angle of repose of the material and the opening at the bottom is generally kept equal to 1/10th to 1/15th times the diameter of the silo.

A Cement Silo Capacity 3,000 Tonnes 2

2·1. **Data**

Capacity=3000 t, Weight=1450 kg/m^3 Angle of repose=35°, μ'=0·70

Elevation of the bottom of the opening above G.L.=3 m

Wind load=150 kg/m^2, Reduction coefficient=0·7

Bearing capacity of soil=45 t/m^2

Use Janseen's method

Materials available, *M* 150 concrete Grade—1 steel

2·2. **Characteristic Strengths**

σ_{cb}=50 kg/cm^2 $\quad$ m=18

σ_{st}=1400 kg/cm^2 $\quad$ jd=0·87 d

σ_{ct}=11 kg/cm^2 $\quad$ R=8·7

2·3. **Trial Dimensions**

Capacity required=3000 t=$\dfrac{3000000}{1450}$=2060 cu m

Let the diameter of the silo D=10 m

h=depth of conical portion=5 m

Diameter of the opening=$\dfrac{1}{10}$ to $\dfrac{1}{15}D$

d=1 m (say)

Volume of the conical hopper

$$=\frac{\pi h}{12}(D^2+d^2+Dd)$$

$$=\frac{\pi\times 5}{12}(10^2+1^2+10\times 1)$$

$$=\frac{\pi\times 5}{12}\times 111=146 \text{ cu m}$$

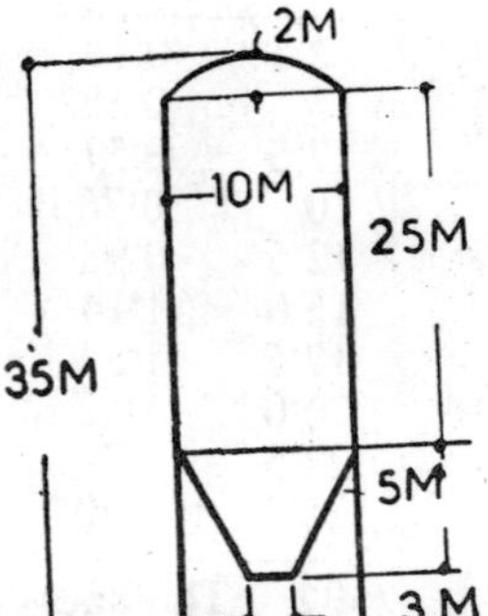

Remaining capacity=2060−146=1924

H=Depth of cylindrical portion of the silo

$$\frac{\pi D^2}{4}\times H=V$$

$$H=\frac{4V}{\pi D^2}=\frac{4\times 1924}{\pi\times 10^2}=24{\cdot}5 \text{ m Use } H=25 \text{ m}$$

Actual volume provided, conical portion=146 cu m

Cylindrical portion=$\dfrac{\pi\times 10^2}{4}\times 25$=1960 cu m

Actual volume provided=1960+146=2106 cu m>2060 cu m

Capacity in tonnes=2106×1450=3070 t

2·4. **Side Wall Computation of Pressure**

$$R=\frac{\pi D^2}{4\times\pi D}=\frac{D}{4}=\frac{10}{4}=2{\cdot}5 \text{ m}$$

$$K=\frac{1-\sin\phi}{1+\sin\phi}=\frac{1-\sin 35^\circ}{1+\sin 35^\circ}=0{\cdot}271$$

$\mu'=0{\cdot}7$

Intensity of the horizontal pressure $p_h=\frac{wR}{\mu'}\left(1-e^{\frac{\mu' kh}{R}}\right)$

$$\frac{\mu' k}{R}=\frac{0{\cdot}7\times0{\cdot}271}{25}={\cdot}076$$

$$\frac{wR}{\mu'}=\frac{1450\times2{\cdot}5}{0{\cdot}7}=5190$$

2·4·1. Sample Calculations

$p_h=2{\cdot}5$ m

$$p_h=5190\left(1-e^{-{\cdot}076\times2{\cdot}5}\right)=5190\left(1-e^{-{\cdot}19}\right)$$

$$=5190\times0{\cdot}173=900\text{ kg/m}^2$$

$$T=\frac{p_h D}{2}=\frac{900\times10}{2}=4500\text{ kg}$$

$$A_t=\frac{T}{\sigma_{st}}=\frac{4500}{1400}=3{\cdot}22\text{ cm}^2$$

h Depth m	$x=\frac{\mu' k}{R}h$	$(1-e^{-x})$	p_h	$T=\frac{p_h D}{2}$ kg	A_{st} cm²	Spacing cm
2·5	0·19	0·173	900	4500	3·22	24
5	0·38	0·316	1640	8200	5·88	19—12 mm ϕ
7·5	0·57	0·435	2250	12250	8·86	12·5
10	0·76	0·532	2760	13800	9 9	20
12·5	0·95	0·613	3180	15900	11·4	17—16 mm ϕ
15·0	1·14	0·680	3520	17600	12·6	15·5
17·5	1·33	0·735	3810	19050	13·61	14·5
20·0	1·52	0·781	4060	20300	14·5	13·6
22·5	1·71	0·819	4240	21200	15·2	13
25·0	1·90	0·860	4600	23000	16·5	12

2·4·2. Thickness of Wall On Cracking Stress Consideration

Maximum hoop tension $T=23000$ kg

$$\sigma_{ct}=\frac{23000}{100\times t_w+A_t(m-1)}$$

$$11=\frac{23000}{100\times t_w+16{\cdot}5\times17}$$

$$1100\,t_w+11\times16{\cdot}5\times17=23000$$

$$1100\,t_w+3100=23000$$

$$1100\,t_w=23000-3100=19900$$

$$t_w=\frac{19900}{1100}=18{\cdot}1\text{ cm}$$

Use 20 cm thick wall

2·4·3. Secondary Vertical Steel

Usually ·2% of concrete are$=\frac{{\cdot}2}{100}\times20\times100=4\text{ cm}^2$

Use 10 mm ϕ at 19 cm c/c

2·5. Hopper Bottom

Steel must be provided in the hopper to resist the hoop tension and steel will also be required to hang the hopper to side walls.

Horizontal pressure/sq. m $= p_h$

projected area of 1 sq. m $= 1 \times \sin \alpha$ sq. m

Horizontal pressure on projected area

$= p_h \sin \alpha = 4600 \times \cdot 7431 = 3420$ kg/m²

due to self weight (assuming 20 cm thick)

$= w_s \cos \alpha = \frac{20}{100} \times 2400 \times \cos \alpha$

$= 480 \times 0{\cdot}6691 = 322$ kg

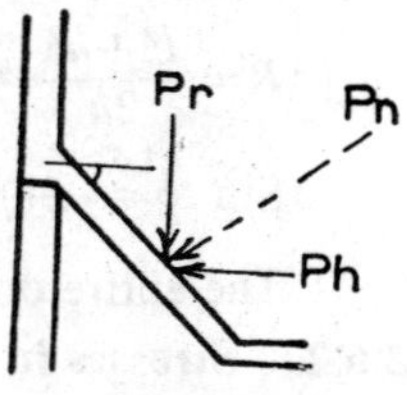

Hoop tension/metre $= (3420 + 322)\frac{10}{2} = 3742 \times 5 = 18{,}710$ kg

2·5·1. Hoop Steel

$$A_t = \frac{18710}{1400} = 13{\cdot}4 \text{ cm}^2$$

Use 16 mm ϕ at 14·5 cm c/c and thickners of slab 20 cm as in the side wall.

2·5·2. Hanging Steel

The total vertical weight of the contents of the silo and the weight of the hopper bottom must be hung from cylindrical wall.

Weight of the contents $= 3070000$ kg

Weight of conical hopper approximately

$$= \frac{1{\cdot}52\, W_s\, D^2}{\cos \alpha}$$

$$= \frac{1{\cdot}52 \times 20 \times 2400 \times 10^2}{\cos \alpha \;\; 100}$$

$$= \frac{1{\cdot}52 \times 480 \times 100}{\cdot 6691} = 108{,}500 \text{ kg}$$

Total vertical load $= 307{,}0000 + 108{,}500$

$= 3{,}178{,}500$ kg

Load per metre of the circumference

$$= \frac{3178500}{\pi \times 10} = 101{,}000 \text{ kg}$$

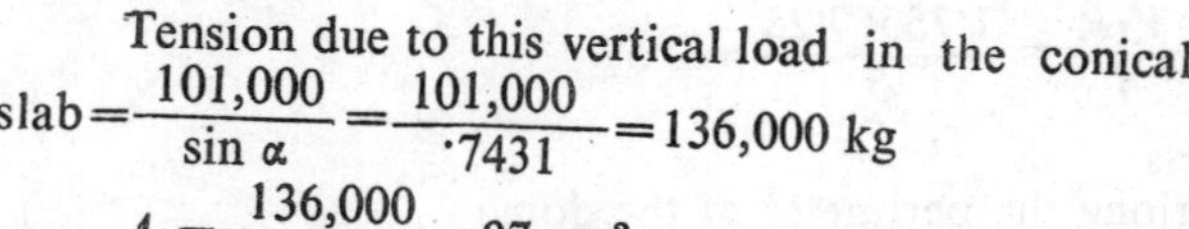

Tension due to this vertical load in the conical

$$\text{slab} = \frac{101{,}000}{\sin \alpha} = \frac{101{,}000}{\cdot 7431} = 136{,}000 \text{ kg}$$

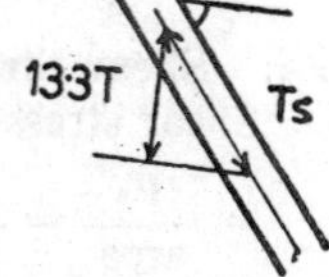

$$A_t = \frac{136{,}000}{1400} = 97 \text{ cm}^2$$

Use 32 mm ϕ at 10 cm c/c

2·6. Domed Roof

Diameter $= 10$ m

Loads : Self weight of dome (15 cm)

$= \frac{15}{100} \times 2400 \times 1 = 360$ kg/m²

Wind and accidental loads $= 150$ kg/m²

Water proofing etc. $= 90$ kg/m²

$w =$ Total load $= 600$ kg/m²

2·6·1. Geometry of the Dome

Rise of the dome $h=\frac{1}{5}\times 10=2$ m

t_s thickness of dome $=15$ cm

$(R-h)^2+l^2=R^2$

$$R=\frac{l^2+h^2}{2h}=\frac{5^2+2^2}{2\times 2}=\frac{25+4}{4}=\frac{29}{4}=7{\cdot}25 \text{ m}$$

$$\text{Sin } \phi=\frac{5}{7{\cdot}25}=0{\cdot}67,\ \phi=44^\circ<51^\circ-48'$$

The entire dome is in compression

2·6·2. Stresses in the Dome

Hoop stresses at $\phi=0$

$$s=\frac{-wR}{2t}\left(\frac{1-\cos\phi-\cos^2\phi}{1+\cos\phi}\right)$$

$$\frac{-wR}{2t}=\frac{600\times 7{\cdot}25}{2\times{\cdot}15}=14500 \text{ kg/m}^2=1{\cdot}45 \text{ kg/cm}^2$$

$$\text{Meridional stress } c=\frac{wR}{t(1+\cos\phi)}$$

$$=\frac{600\times 7{\cdot}25}{{\cdot}15(1+0{\cdot}7193)}=17000 \text{ kg/m}^2=1{\cdot}7 \text{ kg/cm}^2$$

The compressive stresses are very low, the slab can be reduced to 10 cm thick. However there is not much advantage in reducing it below 10 cm

2·6·3. Force Components

Surface area of dome $=2\pi Rh=2\pi\times 7{\cdot}25\times 2=91{\cdot}25$ sq. m

Total load on the dome $=W=600\times 91{\cdot}25$

$=55000$ kg $=55$ t

H1 T1 V1

$$V_1=\frac{W}{2\pi l}=\frac{55}{2\pi\times 5}=1{\cdot}75 \text{ t}$$

$$H_1=V_1\cot\phi=V_1\left(\frac{R-h}{l}\right)=1{\cdot}75\left(\frac{7{\cdot}25-2}{5}\right)$$

$$=1{\cdot}75\times\frac{5{\cdot}25}{5}=1{\cdot}84 \text{ t}$$

$$T_1=\frac{V}{\sin\phi}=\frac{V_1 R}{l}=\frac{1{\cdot}75\times 7{\cdot}25}{5}=2{\cdot}57 \text{ t}$$

2·6·4. Shear Stress

Shear stress along the perimeter of the dome

$$=\frac{V_1}{\text{area}}=\frac{1750}{15\times 100}=1{\cdot}17 \text{ kg/cm}^2<5 \text{ kg/cm}^2$$

2·6·5. Minimum Reinforcement

$A_t=\frac{0{\cdot}2}{100}\times 15\times 100=3$ cm^2. Use 8 mm ϕ 16 cm c/c in both the directions

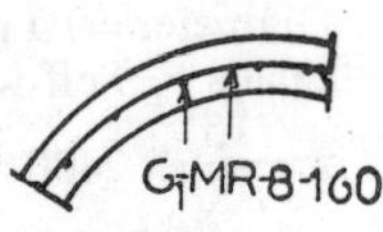

2·7. Rib Dimensions

$$\text{Maximum hoop tension } T=H_1\times\frac{D}{2}=1840\times\frac{10}{2}=9200 \text{ kg}$$

$$A_t = \frac{9200}{1400} = 6{\cdot}55 \text{ cm}^2$$

Use 6 Nos – 12 mm ϕ ($6{\cdot}78$ cm²)

$$\sigma_{ct} = \frac{T}{A_c + (m-1)A_t}$$

$$11 = \frac{9200}{A_c \ \ 17 \times 6{\cdot}78} = \frac{8200}{A_c + 115}$$

$$11A_c + 1260 = 9200$$

$$11A_c = 9200 - 1260 = 7940$$

$A_c = 625$ cm². Use 25 cm × 30 cm rib

2·8. Columns

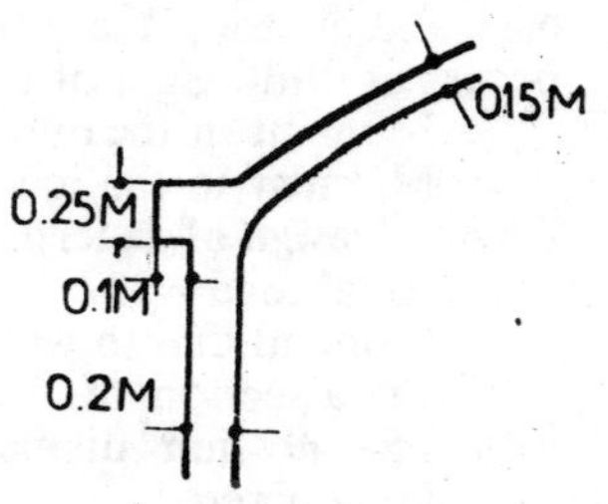

Loads : Roof load = 55 t
Cement = 3070 t
Conical hopper = $108{\cdot}5$ t
Side walls

$$= \pi \times 10{\cdot}2 \times 25 \times \frac{20}{100} \times \frac{2400}{1000}$$

$= 384$ t

Total load = $3617{\cdot}5$ t

$$\text{Load per column} = \frac{3617{\cdot}5}{12} = 301 \text{ t}$$

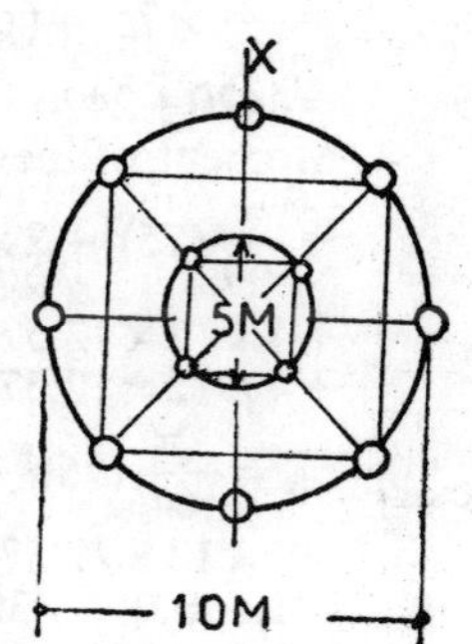

Assuming 75 cm circular columns
Self weight of the column

$$= \frac{\pi \times (0{\cdot}75)^2}{4} \times 8 \times \frac{2400}{1000} = 8{\cdot}45 \text{ t}$$

Total dead load on each column
$= 301 + 8{\cdot}45 = 309{\cdot}45$ t

2·8·1. Wind Loads

Wind load on silo

$$= 0{\cdot}7 \times 32 \times 10 \times \frac{150}{1000} = 33{\cdot}6 \text{ t}$$

Point of application = 16 + 3 = 19 m
Wind load moment = $33{\cdot}6 \times 19 = 640$ mt

2·8·2. Axial Load in the Column Due to Wind Moment

Axial load due to wind moment in the extreme column $= \dfrac{Mr}{\Sigma r^2}$

$$\Sigma r^2 = 2 \times 5^2 + 4 \times \left(\frac{5}{\sqrt{2}}\right)^2 + 4 \times \left(\frac{2{\cdot}5}{\sqrt{2}}\right)^2$$

$$= 50 + 50 + 12{\cdot}56 = 112{\cdot}56 \text{ m}^2$$

$$\text{Axial load} = \frac{640 \times 5}{112{\cdot}56} = 28{\cdot}5 \text{ t}$$

Total load per column due to dead and wind loads
$= 309{\cdot}45 + 28{\cdot}5 = 337{\cdot}75$ t

2·8·3. Shear in the Columns.

Inner columns can take twice as much shear as the outer columns.

If S is the shear force in the outer column

Then, $8 \times S + 4 \times 2S = 33{\cdot}6$ t

$16S = 33{\cdot}6$ t, $S = \frac{33{\cdot}6}{16} = 2{\cdot}11$ t

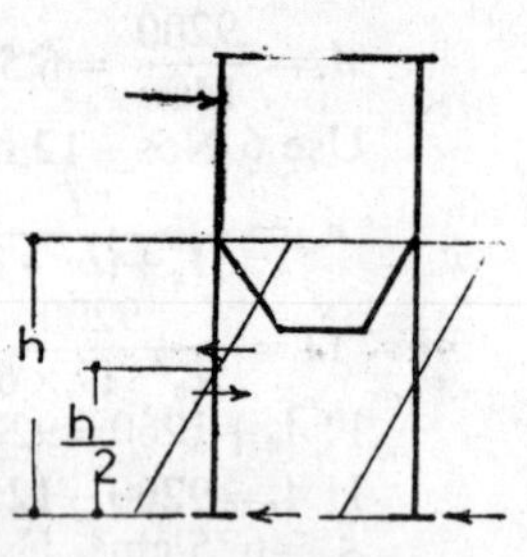

Therefore shear force in the inner column $= 2{\cdot}11 \times 2 = 4{\cdot}22$ t

2·8·4. Moment in the Column

Assuming the columns to be fixed at base and at top, the point of contraflexure occurs at mid height of the column.

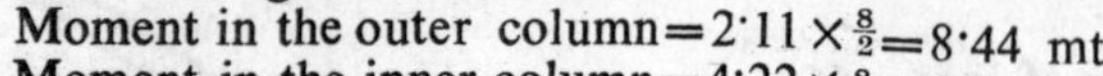

Moment in the outer column $= 2{\cdot}11 \times \frac{8}{2} = 8{\cdot}44$ mt

Moment in the inner column $= 4{\cdot}22 \times \frac{8}{2} = 16{\cdot}88$ mt

2·8·5. Design of Inner Column

Axial load $= 337{\cdot}75$ t

Moment due to wind $= 16{\cdot}88$ t

Try a section of 75 cm diameter with 15—40 mm diameter bars with M—200 concrete

Equivalent area $= A_c + (m-1)A_t$

$= \frac{\pi}{4} \times 75^2 + (13-1) \times 12{\cdot}57 \times 16$

$= 4420 + 2420 = 6840$ cm^2

Moment of inertia $= I_{xx}$

$= \frac{\pi}{64}(75)^4 + 2 \times 12 \times 12{\cdot}57 \times (32)^2$

$+ 4 \times 12 \times 12{\cdot}57 \times (32 \sin 22{\cdot}5°)^2 + 4 \times 12 \times 12{\cdot}57(32 \sin 45°)^2$

$+ 4 \times 12 \times 12{\cdot}57(32 \sin 67{\cdot}5°)^3$

$I_{xx} = \frac{\pi}{64}(75)^4 + 2 \times 12 \times 12{\cdot}57(32)^2$

$+ 4 \times 12 \times 12{\cdot}57[(32 \times 0{\cdot}3827)^2 + (32 \times {\cdot}707)^2 + (32 \times {\cdot}9239)^2]$

$= 1{,}530{,}000 + 309{,}000 + 600[149 + 506 + 871]$

$= 1{,}530{,}000 + 309{,}000 + 600 \times 1526$

$= 1{,}530{,}000 + 309{,}000 + 915{,}000 = 2{\cdot}754 \times 10^6$ cm^4

2·8·6. Stresses in the Column

$\sigma_c' = \frac{337750}{6840} = 49{\cdot}2$ kg/cm^2, $\sigma_b' = \frac{1688000 \times 75}{2 \times 2{,}754{,}000} = 23$ kg/cm^2

$\frac{\sigma_c'}{\sigma_c} + \frac{\sigma_{cb}'}{\sigma_{cb}} < 1.$

$\frac{49{\cdot}2}{1{\cdot}33 \times 50} + \frac{23}{70 \times 1{\cdot}33} = 0{\cdot}74 + 0{\cdot}245 = 0{\cdot}985 < 1$

2·9. Foundations Raft

Axial load on each column $= 337{\cdot}75$ t

Total load on the outer ring beam $= 337{\cdot}75 \times 8 = 2700$ t

Total load on the inner ring beam $= 337{\cdot}75 \times 4 = 1350$ t

Total load on the foundation $= 2702 + 1350 = 4050$ t

Self weight of the foundation at 10% $= 450$ t

Total load $= 4455$ t

Area of the raft required $= \frac{\text{load}}{\text{B.C. of soil}} = \frac{4455}{45} = 99$ m^2

Try outer diameter of the raft =12 m, inner diameter of the raft=3 m, and the outer columns on a diameter of 10 m, and the inner columns on a diameter of 5 m

Area of the raft provided$=\frac{\pi}{4}(12^2-3^2)$

$=\frac{\pi}{4}(144-9)=\frac{3{\cdot}14}{4}\times 135=106 \text{ m}^2>99 \text{ m}^2$

Upward pressure$=\frac{4455}{106}=42 \text{ t/m}^2$

2·9·1. Approximate Bending Moment in the Raft

Maximum cantilever moment $=\frac{wl^2}{2}=\frac{42\times 1\times 1}{2}=21$ mt

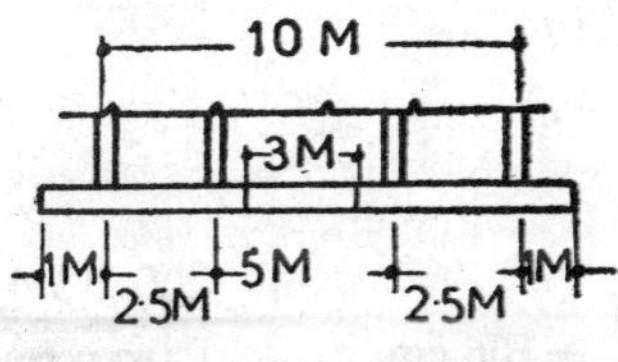

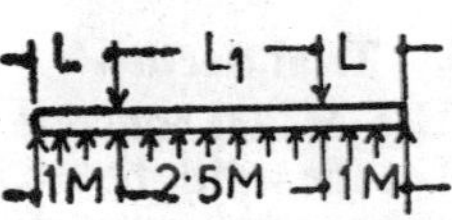

Maximum +ve BM at centre

$=\frac{wl_1^2}{8}-\frac{wl^2}{2}=\frac{42\times 2{\cdot}5^2}{8}-21=32{\cdot}7-21=11{\cdot}7$ mt

depth of slab$=\sqrt{\frac{2100{\cdot}000}{100\times 12{\cdot}13}}=41{\cdot}6$cm

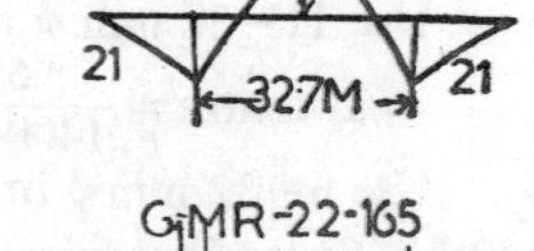

Use $d=45$ cm, $d_e=42$ cm

Area of steel required for −ve BM

$=\frac{2100{,}000}{1400\times{\cdot}87\times 42}=41 \text{ cm}^2$

Use 22 mm ϕ at 9 cm c/c

Area of steel for +ve BM

$=\frac{1170{,}000}{1400\times{\cdot}87\times 42}=22{\cdot}9 \text{ cm}^2$

Use 22 mm at 16·5 cm c/c

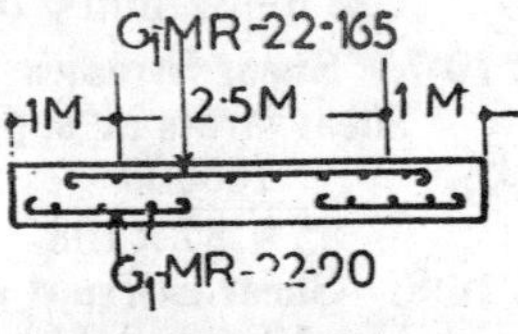

2·10. Ring Beam

Load on the outer ring beam=2700 t

Load per metre run of circumference of the girder

$=w=\frac{2700}{\pi\times 10}=86$ t

Load on the inner girder per meter run of circumference

$=w=\frac{1350}{\pi\times 5}=86$ t

2·10·1. Design of Inner Ring Beam Bending Moments

Number of supports=4, $\theta=90°$ $w=86$ t

Maximum −ve BM at supports$=0{\cdot}137\times w\times r^2\ \theta$

$=0{\cdot}137\times 86\times 2{\cdot}5^2\ \frac{\pi}{190}\times 90=116$ mt

Maximum +ve BM at centre$={\cdot}07\times w\times r^2\ \theta$

$={\cdot}07\times 86\times(2{\cdot}5)^2\times\frac{\pi}{100}\times 90=59$ mt

2·10·2. Shear Force

Maximum shear force at support$=\frac{wr\,\theta}{2}$

$$=\frac{86\times2{\cdot}5}{2}\times\frac{\pi}{180}\times90=168 \text{ t}$$

2·10·3. Shear Force at Maximum Point of Torsion

Torsional moment is maximum at 19·25°

SF at point of maximum torsion

$$=168-\frac{168\times19{\cdot}25}{45}=168-72=96 \text{ t}$$

2·10·4. Torsional moment

Maximum torsional moment at 19·25°$=\cdot021\times wr^2\,\theta$

$$=\cdot021\times86\times(2{\cdot}5)^2\,\frac{\pi}{100}\times90=17{\cdot}8 \text{ mt}$$

2·10·5. Trial Section

Use M 200 concrete

Let the width be$=$85 cm

$$\text{depth of beam}=\sqrt{\frac{11600{,}000}{12{\cdot}13\times85}}=106 \text{ cm}$$

Use $d=115$ cm $d_e=108$

2·10·6. Area of Steel

$$A_t \text{ at support}=\frac{11600{,}000}{1400\times{\cdot}87\times108}=88 \text{ cm}^2$$

Use 11$-$32 mm ϕ in one row at top. $A_t=88{\cdot}47$ cm²

$$A_t \text{ at centre}=\frac{5900{,}000}{1400\times{\cdot}87\times108}=44{\cdot}7 \text{ cm}^2$$

Use 6$-$32 mm ϕ in one row at bottom

2·10·7. Shear Stresses

Shear stress at support

$$=\frac{168000}{85\times{\cdot}87\times108}=21{\cdot}1 \text{ kg/cm}^2>5 \text{ kg/cm}^2$$

2·10·8. Shear Stirrups at Support

Use 12 mm ϕ 11$-$legged stirrups

$$\text{spacing}=\frac{11\times1{\cdot}13\times1400\times0{\cdot}87\times108}{168{,}000}=9 \text{ cm c/c}$$

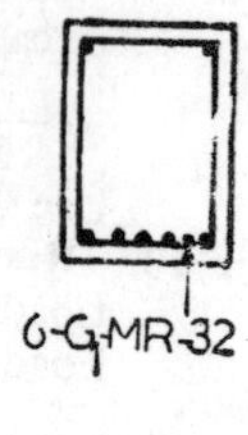

2·10·9. At Point of Maximum Torsion

Use 9 cm spacing for stirrups

Area of steel required

$$=\frac{96000\times9}{1400\times{\cdot}87\times108}=6{\cdot}7 \text{ cm}^2$$

Area of steel required to resist torsional moment at that section (19·25°)

$$=\frac{17{\cdot}8\times100\times1000}{1400\times{\cdot}8\times80\times105}=1{\cdot}9 \text{ cm}^2$$

Total steel required at point of maximum torsion

$=6{\cdot}7+1{\cdot}9=8{\cdot}6$ cm²

Use 11$-$legged 10 mm ϕ

2·10·10. Longitudinal Steel for Torsion

Volume of steel provided in the legs of stirrups for torsion/cm length

$$= \frac{1{\cdot}9 \times 105}{9} = 22{\cdot}2 \text{ cm}^2$$

Use 8—20 mm ϕ bars 4 on each face

4-G,MR-20

2·11. Outer Ring Beam Bending Moments

Number of supports = 8

Total load = 2700 t, $w = 86$ t, $\theta = 45°$

Maximum —ve BM at support $= 0{\cdot}066 \times wr^2\ \theta$

$$= 0{\cdot}066 \times 86 \times 5^2 \frac{\pi}{180} \times 45 = 111 \text{ mt}$$

Maximum +ve BM at centre $= 0{\cdot}03 \times wr^2\ \theta$

$$= {\cdot}03 \times 86 \times 5^2 \times \frac{\pi}{180} \times 45 = 50{\cdot}5 \text{ mt}$$

2·11·1. Shear Force

$$\text{Maximum SF at support} = \frac{wr\ \theta}{2} = \frac{86 \times 5}{2} \times \frac{\pi}{180} \times 45 = 161 \text{ t}$$

SF at point of maximum torsion (at 9·5°)

$$= 168 - \frac{168 \times 9{\cdot}5}{22{\cdot}5} = 168 - 71 = 97 \text{ t}$$

2·11·2. Torsional Moment

Maximum torsional moment $= {\cdot}005 \times wr^2\ \theta$

$$= {\cdot}005 \times 86 \times 5^2\ \frac{\pi}{180} \times 45 = 8{\cdot}45 \text{ mt}$$

2·11·3. Design of Outer Ring Beam

Same section 85 cm × 115 cm of inner ring beam can be used, as the bending moments are nearly the same.

Square Wheat Silo 3

3·1. Data

Capacity=600 t, Weight of wheat=850 kg/m³, Angle of repose=25°

For wheat, coefficient of friction of grain on grain $\mu = 0{\cdot}466$

Coefficient of friction between the grain and the concrete wall $\mu' = 0{\cdot}444$

Elevation of the bottom of the opening above G.L.=3 m

Wind load=150 kg/m²

Use AIRY's theory for pressure computation

Materials available, Concrete – M 150, Steel=grade=1

Bearing capacity of soil=40 t/m²

3·2. Characteristic Strengths

$\sigma_{cb} = 50$ kg/cm² $\quad m = 18$

$\sigma_{st} = 1400$ kg/cm² $\quad j_d = 0{\cdot}87\, d,\ R = 8{\cdot}7$

3·3. Trial Dimensions

$$\text{Capacity of the silo} = 600 \text{ t} = \frac{600{,}000}{850} = 705 \text{ cu m}$$

Let the side of the bunker =5 m

Volume of the hopper

$$\tfrac{1}{3} = h\,(b^2 + b_1{}^2 \div \sqrt{b^2 b_1{}^2})$$

$$= \frac{2{\cdot}5}{3}(5^2 + {\cdot}5^2 + \sqrt{5^2 \times {\cdot}5^2})$$

$$= \frac{2{\cdot}5}{3}(25 + {\cdot}25 + \sqrt{25 \times {\cdot}25})$$

$$= \frac{2{\cdot}5}{3} \times 27{\cdot}73 = 22{\cdot}9 \text{ cum}$$

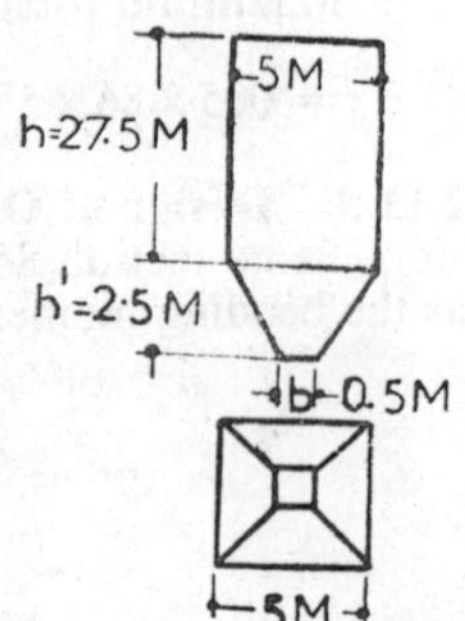

h=depth required for the square portion to accommodate the remaining capacity

$$5 \times 5 \times h = 705 - 22{\cdot}9 = 682{\cdot}1$$

$$h = \frac{682{\cdot}1}{25} = 27{\cdot}3 \text{ m, Use } 27{\cdot}5 \text{ m}$$

Actual capacity provided, hopper=22·9 cum

Square=5×5×27·5=687·5 cum

Total=710·4 cum>705 cum

3·4. Pressure Computations

Plane of rupture

$$\tan\theta = \sqrt{\frac{2h}{b} \times \frac{1+\mu^2}{\mu+\mu'} + \frac{1+\mu^2}{\mu+\mu'} \times \frac{1-\mu\mu'}{\mu+\mu'}} - \frac{1-\mu\mu'}{\mu+\mu'}$$

$$= \sqrt{\frac{2\times 27{\cdot}5}{5} \times \frac{1+{\cdot}466^2}{{\cdot}466+{\cdot}444} + \frac{1+{\cdot}466^2}{{\cdot}466+{\cdot}444} \times \frac{1-{\cdot}466\times{\cdot}444}{{\cdot}466+{\cdot}444}} - \frac{1-{\cdot}466\times{\cdot}444}{{\cdot}466+{\cdot}444}$$

$$=\sqrt{11\times1\cdot337+1\cdot165}-0\cdot871=\sqrt{14\cdot7+1\cdot165}-0\cdot871$$
$$=3\cdot981-0\cdot871=3\cdot11$$
$$\theta=72^\circ-12'$$
$$p=\frac{wb}{2}(2h-b\tan\theta)\times\frac{\text{taa }\theta-\mu}{1-\mu\mu'+(\mu+\mu')\tan\theta}$$
$$=\frac{850\times5}{2}(2\times27\cdot5-5\times3\cdot11)$$
$$\times\frac{3\cdot11-0\cdot466}{1-0\cdot466\times\cdot444+(0\cdot466+\cdot444)3\cdot11}$$
$$=2125(55-15\cdot55)\times\frac{2\cdot644}{0\cdot793+0\cdot910\times3\cdot11}$$
$$=2125\times39\cdot45\times\frac{2\cdot644}{3\cdot643}=60500 \text{ kg/sq. m}$$

Total pressure on the sides $=60{,}500\times5\times4=1{,}220{,}000$ kg $=1220$t
Weight of grain held by friction $=1{,}20{,}000\times\mu'$
$=1{,}220{,}000\times0\cdot444=542000$ kg $=542$ t.
Weight of grain in the side upto 27·5 m depth

$$=5\times5\times27\cdot5\times\frac{850}{1000}=584 \text{ t}$$

Pressure on the bottom of the silo $=584-542=42$t

3·4·1. Pressure Computations General Expression Smaller Depths

$$\tan\theta=\sqrt{\tfrac{2}{5}h\times1\cdot337+1\cdot65}-0\cdot871$$
$$p=\frac{850\times5}{2}(2h-5\tan\theta)\times\frac{\tan\theta-0\cdot466}{0\cdot793+0\cdot91\tan\theta}$$
$$=2125(2h-5\tan\theta)\times\frac{\tan\theta-0\cdot466}{0\cdot793+\cdot91\tan\theta}$$

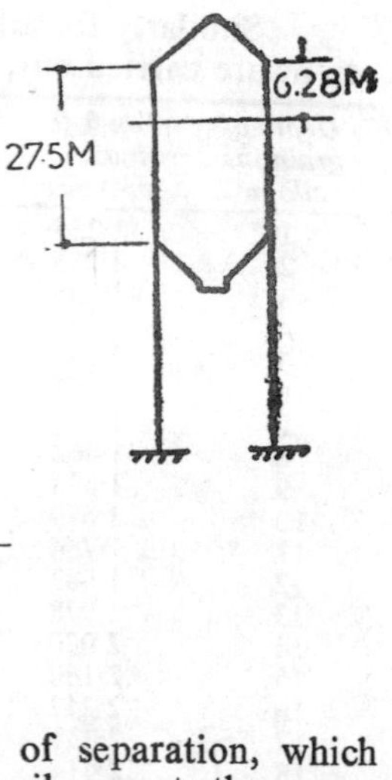

For smaller depths,

$$\tan\theta=\mu+\sqrt{\mu\frac{1+\mu^2}{\mu+\mu'}}$$
$$=0\cdot466+\sqrt{0\cdot466\times\frac{1+0\cdot466^2}{0\cdot91}}=1\cdot255$$
$$p=\frac{wh^2}{2\tan\theta}\times\frac{\tan\theta-\mu}{(1-\mu\mu')+(\mu+\mu')\tan\theta}$$
$$=\frac{850h^2}{2\times1\cdot255}\times\frac{1\cdot255-0\cdot466}{(1-\cdot466\times\cdot444)+0\cdot91\times1\cdot255}$$
$$=\frac{425h^2}{1\cdot255}\times\frac{0\cdot789}{0\cdot793+1\cdot142}=140h^2$$

where $h=0$ to $b\tan\theta=5\times1\cdot255=6\cdot28$ m

This is the depth for which the plane of separation, which causes maximum pressure on one side of the silo. meets the opposite side at the surface of the grain.

3·4·2. Larger depths >6·28 m

For larger depths, where the plane of separation meets the opposite side within the grain.

$$\tan\theta=\sqrt{\frac{2h}{b}\times\frac{1+\mu^2}{\mu+\mu'}+\frac{1+\mu^2}{\mu+\mu'}\cdot\frac{1-\mu\mu'}{\mu+\mu'}}-\frac{1-\mu\mu'}{\mu+\mu'}$$

For wheat, $\dfrac{1+\mu^2}{\mu+\mu'}=1{\cdot}337,\quad \dfrac{1-\mu\mu'}{\mu+\mu'}=0{\cdot}871$

$$\frac{1-\mu^2}{\mu+\mu'}\times\frac{1-\mu\mu'}{\mu+\mu'}=1{\cdot}337\times0{\cdot}871=1{\cdot}165$$

$$\tan\theta=\sqrt{\frac{2\times h}{b}\times1{\cdot}337+1{\cdot}165}-{\cdot}871$$

$$p=\frac{wb}{2}(2h-b\tan\theta)\times\frac{\tan\theta-\mu}{1-\mu\mu'+(\mu+\mu')\tan\theta}$$

$$=\frac{850}{2}\times5(2\times h-5\times\tan\theta)\times\frac{\tan\theta-0{\cdot}466}{0{\cdot}793+{\cdot}910\tan\theta}$$

$$=2125\,(2h-5\tan\theta)\times\frac{\tan\theta-0{\cdot}466}{0{\cdot}793+{\cdot}910\tan\theta}$$

Calculations for 10 m depth, $h=10$ m

$$\tan\theta=\sqrt{\frac{2\times10}{5}\times1{\cdot}337+1{\cdot}165}-0{\cdot}871$$

$$=\sqrt{4\times1{\cdot}337+1{\cdot}165}-0{\cdot}871=\sqrt{5{\cdot}35+1{\cdot}165}-0{\cdot}871$$

$$=\sqrt{6{\cdot}515}-0{\cdot}871=2.55-{\cdot}871=1{\cdot}679$$

$$p=2125(2\times10-5\times1{\cdot}679)\times\frac{1{\cdot}679-0{\cdot}466}{0{\cdot}793+{\cdot}91\times1{\cdot}679}$$

$$=2125(20-8{\cdot}4)\times\frac{1{\cdot}213}{2{\cdot}313}=2125\times11{\cdot}6\times\frac{1{\cdot}213}{2{\cdot}313}=12900\text{ kg}$$

Similarly for all other depths at an interval of 1 m computations are carried out,

Depth of grain in silo m	*Tan θ for max. pressure*	*Weight of grain in the silo*	*Pressure P/m run of wall*	*The intensity of side pressure kg/m²*
1	1·255	21.250	140	—
2	1·255	42,500	560	420
3	1·255	63,750	1260	700
4	1·255	85,000	2240	980
5	1·255	106,250	3600	1360
6	1·255	127,500	5050	1450
7	1·345	148,750	6750	1700
8	1·462	170,000	8650	1900
9	1·574	191 250	10,700	2050
10	1·679	212,500	12,900	2200
11	1·784	233,750	15,100	2200
12	1·883	255,000	17.500	2400
13	1·978	276,250	19,900	2400
14	2·070	297,500	22,400	2500
15	2·160	318,750	25,000	2600
16	2·247	340 000	27,800	2800
17	2·331	367,250	30,500	2700
18	2·414	382,500	33,100	2600
19	2·494	403.75)	36,100	3000
20	2·573	425,000	39,100	3000
21	2·650	446,250	41,600	2500
22	2·725	467,500	44,600	3000
23	2·798	488,750	47,500	2900
24	2·871	510,000	50,400	2900
25	2·941	531,250	53,500	3100
26	3·011	552,500	56,800	3300
27·5	3.113	584,000	61,000	4200

3·4·3. Bending moments in the Soil Walls

Maximum corner moment in the wall at any depth $h = M_c = \frac{-pl^2}{12}$

Maximum moment at the centre of he wall

$$= M_m = \frac{pl^2}{8} - \frac{pl^2}{12} = \frac{pl^2}{24}$$

Sample calculation for 27·5 m depth

Horizontal pressure p at depth 5 m=4200 kg/m

P
5M×5M

$$M_c = \frac{-4200 \times 5 \times 5}{12} = -8750 \text{ m kg}$$

$$M_m = \frac{+4200 \times 5 \times 5}{24} = 4375 \text{ m kg}$$

3·4·4. Thickness of the Slab

$$t = \sqrt{\frac{8750 \times 100}{87 \times 100}} = \sqrt{762} = 27{\cdot}6 \text{ cm}$$

Us 30 cm thick side walls and effective thickness 27·5 cm

3·4·5. Main Steel

$$A_t = \frac{8750 \times 100}{1400 \times {\cdot}87 \times 27{\cdot}5} = 26{\cdot}1 \text{ cm}^2$$

Use 20 mm at 15 cm c/c at corners.

A_t centre use 20 mm ϕ at 22 cm c/c since +veBM is half of −ve BM.

3·4·6. Secondary Steel

A_t=0·3% of concrete section

$= \frac{0{\cdot}3}{100} \times 30 \times 100 = 9 \text{ cm}^2$. Use 10 mm ϕ at 17 cm c/c on both faces.

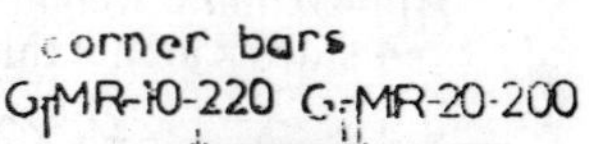

Similarly for various depth the reinforcement can be calculated, keeping the same thickness of wall throughout.

Depth from top m	*Maximum pressure P kg/m*	*Max. B.M m kg*	A_t *cm²*	*Diameter and spacing mm*
0−5	1450	3030	6·9	12−160
5−10	2200	4570	10·4	12−105
10−15	2800	5810	13·2	16−150
15−20	3000	6250	14·2	16−140
20−25	3300	6850	15·5	16−125
25−27·5	4200	8750	26·1	20−150

3·5. Hopper Bottom

Horizontal pressure at level C, that is, at a depth =27·5+1·1=28·6 m

$$\text{Tan } \theta = \sqrt{\left[\frac{2 \times 28{\cdot}6}{5} \times 1{\cdot}337 + 1{\cdot}165\right]} - 0{\cdot}871$$

$$= \sqrt{15{\cdot}3 + 1{\cdot}165} - 0{\cdot}871 = 4{\cdot}05 - {\cdot}871 = 3{\cdot}179$$

$$P=\frac{wb}{2}\,(2h-b\tan\theta)\,\frac{\tan\theta-\mu}{1-\mu\mu'+(\mu+\mu')\tan\theta}$$

$$P=2125(2\times28{\cdot}6-5\times3{\cdot}179)\times\frac{3{\cdot}179-0{\cdot}466}{0{\cdot}793+{\cdot}0{\cdot}91\times3{\cdot}179}$$

$$=2125(57{\cdot}2-15{\cdot}895)\,\frac{2{\cdot}713}{3{\cdot}683}$$

$$=2125\times41{\cdot}305\times\frac{2{\cdot}713}{3{\cdot}683}=65500 \text{ kg}$$

The intensity of horizontal pressure

$P_h=65500-61{,}000=4500 \text{ kg/m}^2$

$P_v=9000 \text{ kg/m}^2$

3·5·1. Bending Moment

Maximum bending moment $=\cdot03125\ P_n\ D^2$

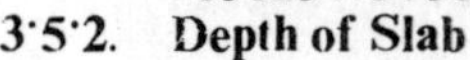

$=\cdot03125\times9750\times(2{\cdot}62)^2=2090$ m kg

3·5·2. Depth of Slab

$$\text{depth of slab } d=\sqrt{\frac{209000}{8{\cdot}7\times100}}$$

$=\sqrt{215}=14{\cdot}6$ cm. Use $d=20$ cm and $d_e=15$ cm

3·5·3. Main Steel

$$A_t=\frac{209000}{1400\times{\cdot}87\times15}=11{\cdot}4 \text{ cm}^3$$

The approximate total weight of the concrete hopper

$$=4\times\frac{(5+0{\cdot}5)}{2}\times3.4\times\frac{20}{100}\times2400=18000 \text{ kg}=18 \text{ t}$$

Approximate weight of hopper below the level C

$=4$ sides $\times$ area $\times$ thickness $\times$ 2400

$$=4\times\frac{(3{\cdot}2+0{\cdot}5)}{2}\times2\times\frac{20}{100}\times2400$$

$$=4\times\frac{3{\cdot}7}{2}\times2\times480=7100 \text{ kg}=7{\cdot}1 \text{ t}$$

Total vertical pressure at the top of slope

$=P_v\times\text{area}=9000\times5\times5=226000$ kg$=226$ t

Total weight over the section C

$=9000\times3{\cdot}2\times3{\cdot}2=92{,}000$ kg$=92$ t

Total approximate weight of grain in the hopper

$$=5\times5\times\frac{2{\cdot}5}{3}\times250=13600 \text{ kg}=13{\cdot}6 \text{ t}$$

Approximate weight of grain in the hopper below the level C

$$=3{\cdot}2\times3{\cdot}2\times\frac{(2{\cdot}5-1{\cdot}1)}{3}\times850$$

$$=3{\cdot}2\times3{\cdot}2\times\frac{1{\cdot}4}{3}\times850=4100 \text{ kg}=4{\cdot}1 \text{ t}$$

Total weight acting at the top of the slope

$=18+226+13{\cdot}6=257{\cdot}6$ t

Total weight acting at the level C

$=7{\cdot}1+92+4{\cdot}1=103{\cdot}2$ t

The perimeter at the top of the slope$=5\times4=20$ m

Vertical load per metre of perimeter$=\frac{257{\cdot}6}{20}=12{\cdot}88$ t

The perimeter at level $C=3{\cdot}2\times4=12{\cdot}8$ m

Vertical load per metre of the perimeter$=\frac{103{\cdot}2}{12{\cdot}8}=10{\cdot}3$ t

Co-efficient K_t from graph for $\theta=48°$, $K_t=0{\cdot}74$

Component of the force acting above the top of slope which acts down$=\frac{12{\cdot}88}{0{\cdot}74}=17{\cdot}5$ t

Component of the force acting at level C which acts down $=\frac{10{\cdot}3}{0{\cdot}74}=13{\cdot}9$ t

3·5·4. Hanging Steel

Hanging steel required at the top of the slope

$=\frac{17500}{1400}=12{\cdot}5\text{ cm}^2$

Hanging steel required at level $C=\frac{13900}{1400}=9{\cdot}9\text{ cm}^2$

3·5·5. Hoop Tension

Hoop tension in the sides at the top of the slope
$=0{\cdot}5\,ph\,K_t\times L=0{\cdot}5\times4500\times0{\cdot}74\times5=8320$ kg

Area of steel required$=\frac{8320}{1400}=5{\cdot}95\text{ cm}^2$

Hoop tension in the sides at the level C
$=0{\cdot}5\times4500\times0{\cdot}74\times3{\cdot}2=5320$ kg

Area of steel required$=\frac{5320}{1400}=3{\cdot}8\text{ cm}^2$

3·6. Arrangement of Steel

Longitudinal steel in the outside face of slab for bending moment $=11{\cdot}4\text{ cm}^2$

Steel for direct tension at level $C=9{\cdot}9\text{ cm}^2$

Total steel$=11{\cdot}4+9{\cdot}9=21{\cdot}3\text{ cm}^2$

For the length of 3·2 m steel required$=21{\cdot}3\times3{\cdot}2=68\text{ cm}^2$

Use 22 bars of 20 mm ϕ (68·42 cm²) at a spacing of 15·2 cm at level C

At the top of slope bending moment is zero and the steel required for the direct tension is$=12{\cdot}5\text{ cm}^2$

For the entire width total steel required$=12{\cdot}5\times5=62{\cdot}5\text{ cm}^2$

Which is less than the steel required at level C. Therefore, the same steel is used for the entire slab as per at level C.

The steel required at the top of the slope to cover the negative moment$=11{\cdot}4\text{ cm}^2$

For the entire width of 5 m steel required$=11{\cdot}4\times5=57{\cdot}0\text{ cm}^2$

Use 20 bars—20 mm ϕ (62·84 cm²)

Horizontal steel at top of slope

Area required for bending moment$=11{\cdot}4\text{ cm}^2$

A_t required for direct tension$=12{\cdot}5\text{ cm}^2$

Total $11{\cdot}4+5{\cdot}95=17{\cdot}35\text{ cm}^2$

Similarly horizontal steel at level $C=11{\cdot}4+3{\cdot}8=15{\cdot}2\text{ cm}^2$

Use 16 mm ϕ at 11·5 cm c/c (17·49 cm²)

XI. REINFORCED CONCRETE SLAB BRIDGES

General Information 1

1·1. General

Highway bridges, are usually of reinforced concrete, because of their durability, economy and ease in construction.

1·2. Width of Roadway

Minimum width for (*i*) Single lane bridge=3·8 m (*ii*) two-lane =6·8 m (*iii*) four-lane=12·8 m with 1·2 m central verge (*iv*) Multi-lane=3·8 m for the first lane and 3 m for every additional lane with 1·2 m central verge

1·3. Width of Footway

Minimum width of footway=1·5 m

1·4. Culvert

Structure whose length between the inner-faces of abutments is equal to or less than 6 m

1·5. Free Board

Minimum free board=60 cm

1·6. Kerb

Safety kerbs shall not be less than 60 cm. Road kerb shall not be less than 22·5 cm

1·7. Reinforcement-cover

Minimum cover at the ends—2·5 cm or twice the diameter of the bar

Minimum cover for longitudinal main reinforcement=40 mm or the diameter of the bar

Minimum cover for concrete in direct contact with earth=6·5 cm

1·7·1. Size

(*i*) Maximum size of the bar=45 mm (*ii*) Minimum size of the bar=6 mm

1·7·2. Spacing

Minimum distance between the two bars=diameter of the bar or 6 mm more than the nominal maximum size of the coarse aggregate.

Vertical spacing between the layers of steel shall not be less than 12 mm or diameter of the bar or maximum size of the aggregate whichever is greater.

Spacing of the main bars shall not exceed 30 cm or twice the effective depth, whichever is smaller in slabs.

Spacing of the secondary reinforcement shall not exceed 45 cm or thrice the effective depth whichever is smaller.

1·8. Effective Span

(*i*) Centre io centre of bearings (*ii*) clear span+effective depth or centre to centre of supports whichever is smaller

1·8·1. Secondary Reinforcement

Minimum reinforcement to resist 0·3 times *LL* bending moment and 0·2 times dead load moment shall be provided as distribution steel.

1·8·2. Skin Reinforcement

If the depth of web of a beam is greater than 1·5 m, skin longitudinal reinforcement of an area of at least 0·5% of the web area shall be provided on each face. The spacing of the bars shall not exceed 20 cm

1·9. Effective Flange For-T-Beams

The effective flange width *B* for a *T* Beam shall be the least of

(*i*) $\frac{1}{4}$ of the effective span of the beam (*ii*) centre to centre of ribs. (*iii*) $br+12ds$

1·9·1. Effective Flange Width For L-Beams

(*i*) $\frac{1}{10}$ of the effective span of the beam (*ii*) Breadth of rib plus $\frac{1}{2}$ the clear distance between the rib (*iii*) $br+4ds$

1·10. Load Classification

(*i*) Heavy loading – *I - R.C.* class '*AA*' (*ii*) standard loading —*I.R.C.* class '*A*' (*iii*) light loading – *I.R.C.* class '*B*'

1·11. Types of R.C.C. Bridges

Single span : (*i*) Slab bridges and culverts upto 6 m (*ii*) Continuous slab bridges upto 8 m (*iii*) Slab and *T*-beam bridges for spans between 5 m to 20 m (*iv*) Balanced cantilever bridges upto 20 m (*v*) Arched bridges upto 70 m (*vi*) Bow string girder over 20 m

1·12. Live Loads on Deck Slabs

For slabs spaning in one direction

(*i*) For single concentrated load the effective

$$\text{width} = e = kx\left(1-\frac{x}{l}\right)+W$$

where k = a constant depending on $\frac{B}{l}$

where l = effective span, B = width of the slab

W = the breadth of concentration area of the load that is the dimension of the tyre or track contact area over the road surface of the slab in a direction at right angles to the span plus the thickness of the wearing coat.

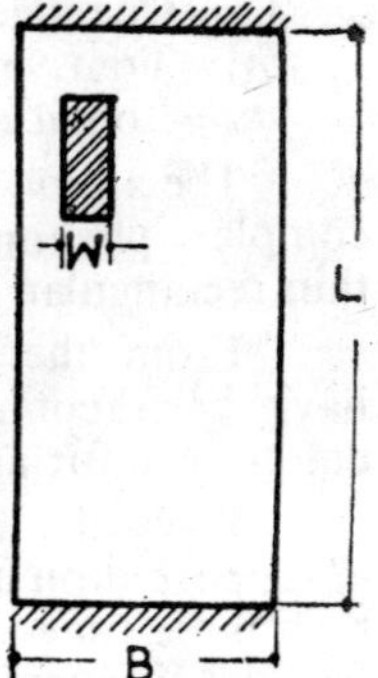

$\frac{B}{l}$	*K for* simply supported slab	*K for* continuous slab	$\frac{B}{l}$	*K for* simply supported slab	*K for* continuous slab
0·1	0·40	0·40	1·1	2·60	2·28
0·2	0·80	0·80	1·2	2·64	2·36
0·3	1·16	1·16	1·3	2·72	2·40
0·4	1·48	1·44	1·4	2·80	2·48
0·5	1·72	1·68	1·5	2·84	2·48
0·6	1·96	1·84	1·6	2·88	2·52
0·7	2·12	1·96	1·7	2·92	2·56
0·8	2·24	2·08	1·8	2·96	2·60
0·9	2·36	2·16	1·9	3·00	2·60
1·0	2·48	2·24	2 and over	3·00	2·60

1·12·1. Pigeaud or Marcus Methods B/L>3

The effect of concentrated loads on slabs spanning in one or two directions may be calculated from either Pigeaud or Marcus methods. The ratio of width of the slab to effective span exceeds 3, this method is mostly used.

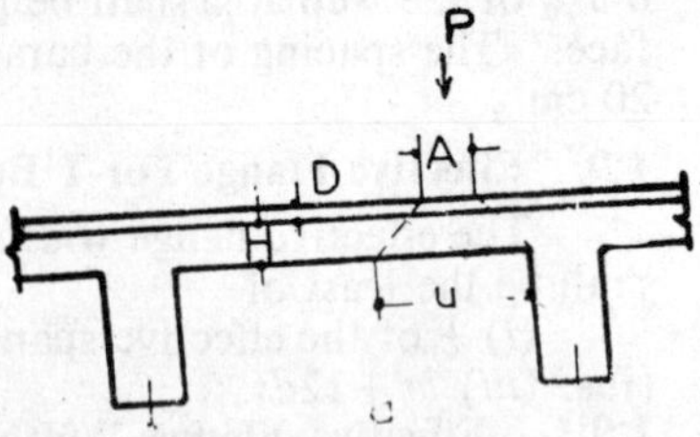

(*a*) Concentrated Loads :

Notation

A=contact width of load in the direction of transverse span.

B=Contact length in the direction of longitudinal span.

u=Load spread in transverse direction

v=Load spread in longtudinal direction

H=Thickness of slab

D=Wearing coat thickness

P=Wheel Load

a=Transverse or short span

b=Longitudinal or long span

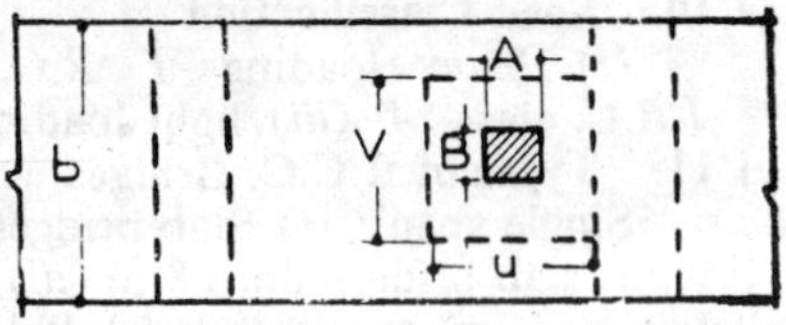

M_a=Coefficient for transverse moment

M_b=Coefficient for longitudinal moment

m_a=Transverse moment/metre width of slab

m_b=Longitudinal moment/metre width of slab

The action of concetrated load on a thin plate or slab is a complex phenomenon. Pigeaud analysed the effect of these loads on thin rectangular plates supported along their edges.

From the formulae, developed by Pigeaud a nnmber of curves have been compiled, from which the maximum moments can be determined for any rectangular contact of the load.

Pigeaud's method is particularly more rational, where the slab is supported on all the four sides and where the width of slab is more than three times the span as in the case of slabs supported on two parallel sides only.

In the case of slabs having parallel supports in one direction and where the width of slab is more than three times the span, the Pigeaud's method enables the longitudinal moments to be computed for various conditions of loading met with and as particularly applicable to bridge design.

By making suitable allowances for the effects of fixity or continuity at the supports, reasonable accuracy can be obtained in the design of slabs.

1·12·2. Bending moment

The values from the graphs give $M_a\times 10^2$ and $M_b\times 10^2$ and the coefficients are therefore to be divided by 100 to get the actual coefficients.

(*i*) The concentrated load is assumed to spread through the filling such that

$$u=\sqrt{(A+2D)^2+H^2} \text{ or } A+2D \text{ and } v=\sqrt{(B+2D)^2+H^2} \text{ or } B+2D$$

(*ii*) The ratios for u/a and v/b are calculated.

(*iii*) Obtain the coefficients corresponding to these ratios $M_a \times 10^2$ and $M_b \times 10^2$ from the appropriate graph of a/b ratio.

(*iv*) Maximum bending moments in the transverse direction $= m_a = P(M_a + 0{\cdot}15\ M_b)$ and in the longitudinal direction $= m_b = P({\cdot}15\ M_a + M_b)$ and 0·15 is the Poisson's ratio

1·12·3. Shear Force

Shear intensity at middle of longer side $= \dfrac{P}{2b+a}$

at middle of shorter side $= P/3b$

1·12·4. More Concentrated Loads case (i) $\dfrac{b}{a} = \infty$

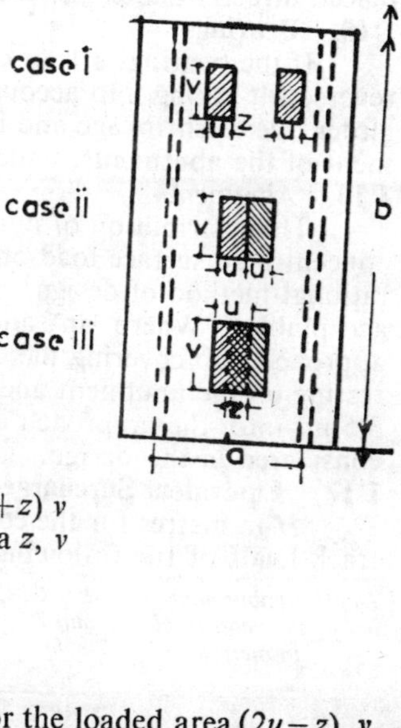

Different conditions :

(*i*) Load spread may touch one another

(*ii*) Load spread may be separate

(*iii*) Load spread may overlap

Find u/a ratio

Substitute $2u/a$ for u/a and $2\,P$ for P

Find M_a and M_b corresponding to v/a and $2u/a$ ratios from $b/a = \infty$ curve

case i

case ii

case iii

1·12·5. Case (ii)

$$p = \frac{P}{uv}$$

Find M_a and M_b for the whole area $(2\,u+z)\,v$

$$m_a = M_a\,p\,(2\,u+z)\,v,\quad m_b = M_b\,p\,(2\,u+z)\,v$$

Similarly find M_a and M_b for the area z, v

$$m_{a_1} = M_{a_1}\,pzv,\quad m_{b_1} = M_{b_1}\,pzv$$

The resultant bending moment

$$m'_a = m_a - m_{a_1},\quad m'_b = m_b - m_{b_1},$$

1·12·6. Case (iii)

Similarly the two bending moments for the loaded area $(2u-z).\,v$ and the other two moments for the area zv are found. The addition of these moments gives the resultant bending moments.

1·12·7. Final Moments

Poisson's ratio=0·15 shall be taken into account in the final moments.

$$m_a = (M_a + 0{\cdot}15\ M_b)\,P,\qquad m_b = (0{\cdot}15\ M_a + M_b)\,P$$

1·12·8. Reduction Factors

80% for maximum moment near the centre of spans and negative moment at interior supports and 50% for negative moments at end supports.

1·13. Impact Factors Class A and B

For class '*A*' and Class '*B*' bridges $= \dfrac{4{\cdot}5}{6+L}$ where L=span

1·13·1. Class AA

For class *AA* bridges: (*i*) For spans less than 9 m

(*a*) For tracked vehicles 25% for spans upto 5 m linearly

reducing to 10 percent for spans of 9 m (*b*) For wheeled vehicles 25%

(*ii*) For spans of 9 m or more

(*a*) Tracked vehicles : 10% upto 40 m span and as in the curve for spans exceeding 40 m (*b*) Wheeled vehicles : 25% for span upto 12 m and as in the curve for spans exceeding 12 m

1·14. Pier Cap

When bearings are used minimum thickness of pier cap =22·5 cm for spans upto 25 m and 30 cm for spans exceeding 25 m

1·14·1. Reinforcement

The cap shall be reinforced with 2% of steel distributed equally at top and bottom in two directions. Additional reinforcement of a mesh consisting of 6 mm at 7·5 cm c/c in both directions shall be placed directly under the bearings.

1·15. Bearings

If the bearings are of cut-rollers, they shall be given an initial reverse tilt taking into account, the possible change in length of the girder due to shrinkage and if resting on abutments the tilt and movement of the abutments, under earth pressure.

1·16. Abutments

The distribution of normal pressure on a retaining wall due to a concentrated surface load on the backfill shall be obtained by any rational method of design. Equations such as Spangler's equation is acceptable. Where an adequately designed reinforced concrete approach slab covering the centre width of the roadway, with one end resting on the abutment and extending for a length of not less than 3·5 m into the approach is provided, no live load surcharge need be considered in the design.

1·17 Equivalent Surcharge Heights

H in metres for the concentrated surface loads due to wheel or track loads of the following I.R.C. standard loading

Depth of abutment below the road level in metres	*I.R.C. Class 'AA' and 70 loadings*		*I RC Class 'A' loading*		*I RC Class 'B' loading*	
	Single lane	*Multilane bridges*	*Single lane*	*Multi lane*	*Single lane*	*Multi lane*
0·2	26·0	15·4	14·3	17·2	8·3	10·0
1·0	15·0	9·1	8·5	10 0	5·1	5·8
2·0	8·0	5·5	5·1	6·1	3·0	3·7
3·0	6·8	4·1	3·8	4·6	2·3	2·7
4·0	5·5	3·3	3·0	3·5	1·8	2·1
6·0	3·8	2·3	2·2	2·6	1·3	1·5
8·0	3·0	1·8	1·7	2·0	1·0	1·2
10·0 and above	2·6	1·5	1·4	1·7	0·9	1·0

Note : Length of abutment (*i*) single lane $L=4·5$ m

(*ii*) multi lane $L=7·5$ m

Angle of internal friction $\phi-30°$,

Weight of fill$=W=1600$ kg/m^3

Pressure acts in the horizontal direction

For different values of L_1, ϕ_1 and W_1, multiply the figures in the table by

$$\frac{(4·5 \text{ or } 7·6 \text{ m})}{L_1} \times \frac{(1+\sin\phi_1)}{3(1-\sin\phi_1)} \times \frac{1600}{W_1}$$

to get corresponding figures.

Deck Slab Bridge Single Lane Class A Loading 2

2·1. Data

Clear span=3·5 m Loading=*IRC*—Class *A*
Number of lanes=Single Road width=3·8 m
Safety kerbs=60 cm wide
Average thickness of wearing coat=8 cm
Materials available, Concrete *M* 200 grade, Steel grade—1

2·2. Relevant Codes

IRC standard specifications and code of practice for road bridges —sec (*i*) to (*iv*)

2·3. Over all Width of Road Way

Width of roadway=3·8+1·2=5 m
Thickness of the slab ($\frac{1}{12}$ of span)=30 cm
Effective thickness of the slab=30−4=26 cm

2·4. Effective Span

Effective span=clear span+effective depth
=3·5+·26=3·76 m

2·5. Impact Factor Allowed

$$\text{Impact factor allowed}=\frac{4{\cdot}5}{6{\cdot}0+L}$$

2·6. Characteristic Strengths

M 200 concrete

$\sigma_{ct}=70$ kg/cm^2	$m=13$
$\sigma_{st}=1400$ kg/cm^2	$j_d=0{\cdot}87$ d
$q=7$ kg/cm^2	$R=12{\cdot}1$
$\sigma_{bt}=13$ kg/cm^2	

2·7. Loads

Dead loads : Self weight of slab (30 cm assumed)

$$=\frac{30}{100}\times1\times1\times2400=720 \text{ kg/m}^2$$

Wearing coat (8 cm average thickness)

$$=\frac{8}{100}\times1\times1\times2200=176 \text{ kg/m}^2$$

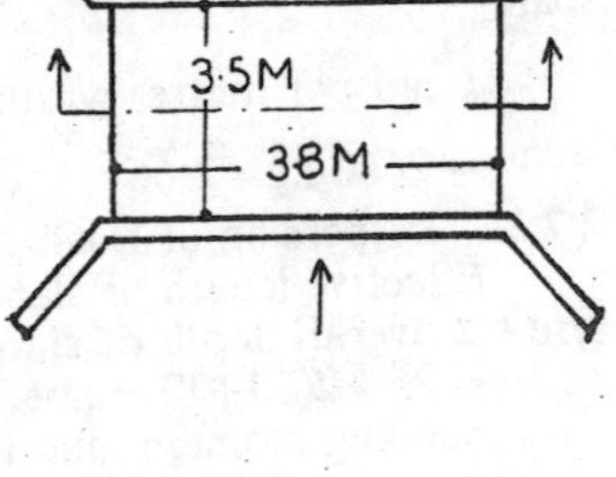

Total dead load=896 kg/m²
Live loads Class A Loading

2·7·1. Dead Load Bending Moment

Effective span=L=3·76 m
Dead load bending moment

$$=\frac{wL^2}{8}=\frac{896\times3{\cdot}76^2}{8}$$

=1500 m kg=150,000 cm kg

Live loads : Ratio of width of slab to effective span

$$=\frac{B}{L}=\frac{5}{3{\cdot}76}=1{\cdot}33<3$$

In this case the slab is spanning in one direction only since the width of slab is less than thrice the effective span and therefore Pigeaud's method of distribution will not apply.

The effective width of dispersion of wheel loads in the direction at right angles to the span is given by the method suggested by I.R.C. specifications.

$$e = Kx\left(1-\frac{X}{L}\right)+w$$

$$\frac{B}{L}=\frac{5}{3{\cdot}76}=1{\cdot}33$$

For simply supported slab : $K=2{\cdot}72$ for $B/L=1{\cdot}3$ and $K=2{\cdot}80$ for $B/L=1{\cdot}4$. K for $1{\cdot}33=2{\cdot}72+\frac{(2{\cdot}80-2{\cdot}72)(1{\cdot}33-1{\cdot}3)}{(1{\cdot}4-1{\cdot}3)}$

$$=2{\cdot}72+0{\cdot}08\times\frac{{\cdot}03}{{\cdot}1}=2{\cdot}72+{\cdot}024=2{\cdot}744 \text{ m}$$

Maximum bending moment is produced when the heaviest axle loads are on the span. For such position of the loads,

$$x=\frac{3{\cdot}76}{2}-\frac{1{\cdot}2}{2}$$

$$=1{\cdot}88-0{\cdot}6=1{\cdot}28 \text{ m}$$

W= width of which contact + 2 × wearing coat thickness

$$=50+2\times 8=66 \text{ cm}=0{\cdot}66 \text{ m}$$

$$e=2{\cdot}744\times 1{\cdot}28\left(1-\frac{1{\cdot}28}{3{\cdot}76}\right)+0{\cdot}66$$

$$=2{\cdot}744\times 1{\cdot}28\times(1-{\cdot}34)+{\cdot}66=2{\cdot}744\times 1{\cdot}28\times 0{\cdot}66+0{\cdot}66$$

$$=2{\cdot}3+0{\cdot}66=2{\cdot}96 \text{ m}$$

Effective widths overlap and $e=1+1{\cdot}8+\frac{2{\cdot}96}{2}=4{\cdot}28$ m

Effective width of dispersion for the axial load = 4·28 m

To obtain maximum bending moment the heaviest axle loads shall be placed symmetrically with respect to span. For this position all other loads are off the span.

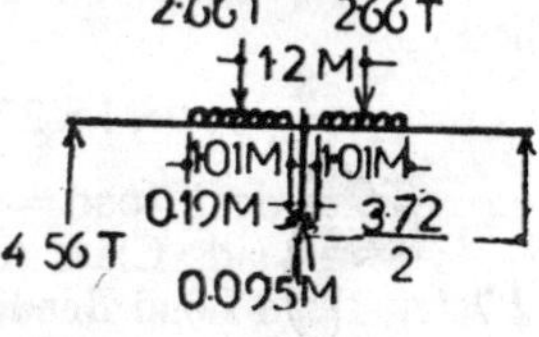

$$\text{Load per metre width}=\frac{11400}{4{\cdot}28}$$

$$=2660 \text{ kg}=2{\cdot}66 \text{ t}$$

2·7·2. Dispersion of Load Along the Span

Effective length of slab along the span = width of contact of the tyre + 2 overall depth of slab including the wearing coat

$$=25+2(30+8)=25+76=101 \text{ cm}=1{\cdot}01 \text{ m}$$

bending moment due to live load

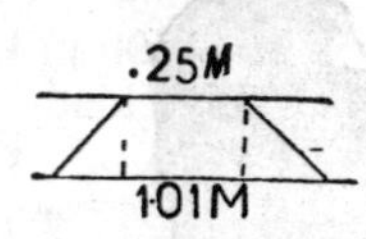

$$=2{\cdot}66\times\frac{3{\cdot}76}{2}-2{\cdot}82\times\frac{1{\cdot}2}{2}$$

$$=2{\cdot}66(1{\cdot}88-0{\cdot}6)=2{\cdot}66\times 1{\cdot}25=3{\cdot}36 \text{ mt}$$

$$=336{,}000 \text{ cm kg}$$

2·8. Impact Factor

$$\text{Impact factor}=\frac{4{\cdot}5}{6{\cdot}0+L}=\frac{4{\cdot}5}{6+3{\cdot}76}=\frac{4{\cdot}5}{9{\cdot}76}=0{\cdot}46$$

Bending moment due to impact factor

$=0{\cdot}46\times366{,}000=155{,}000$ cm kg

Total bending moment $=150{,}000+336{,}000+155{,}000$

$=641{,}000$ cm kg

2·8·1. Depth of Slab

Using M 200 concrete, $12{\cdot}1\ bd_e^2=641{,}000$

$$d_e=\sqrt{\frac{641{,}000}{12{\cdot}1\times100}}=\sqrt{531}=23{\cdot}1 \text{ cm}$$

Use $d_e=26$ cm and $d=30$ cm

2·8·2. Main Reinforcement

$$A_t=\frac{641{,}000}{1400\times{\cdot}87\times26}=20{\cdot}5 \text{ cm}^2$$ Use 16 mm ϕ at 9·5 cm c/c

A_t provided 21·16 cm²

2·8·3. Secondary Steel

Steel required to resist 0·3 times the L.L. bending moment and, 0·2 times the Dead load bending moment, shall be provided as distribution steel

$BM=0{\cdot}3\times(336{,}000+155{,}000)+0{\cdot}2\times150{,}000$

$=0{\cdot}3\times491{,}000+30{,}000=147{,}300+30{,}000=177300$ cm kg

Assuming 12 mm ϕ effective depth available

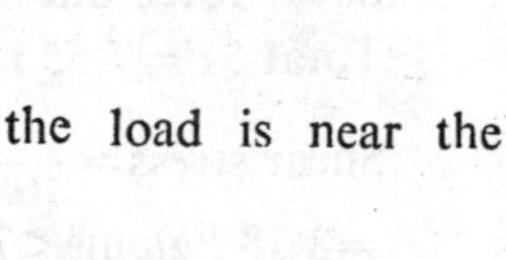

$=30-\left(3{\cdot}2+1{\cdot}6+\dfrac{1{\cdot}2}{2}\right)$

$=30-5{\cdot}4=24{\cdot}6$ cm

$$A_t=\frac{177{,}300}{1400\times24{\cdot}6\times0{\cdot}87}=5{\cdot}9 \text{ cm}^2$$

Use 12 mm ϕ 19 cm c/c

G₁-MR-12-190
0.15L
G₁-MR-16-95

2·8·4. Shear Force

Maximum shear force occurs when the load is near the support.

For the load position, $x_1=0{\cdot}505$ m

Effective width of slab for the 1st wheel load

$e_1=Kx_1\left(1-\dfrac{x_1}{2}\right)=W$

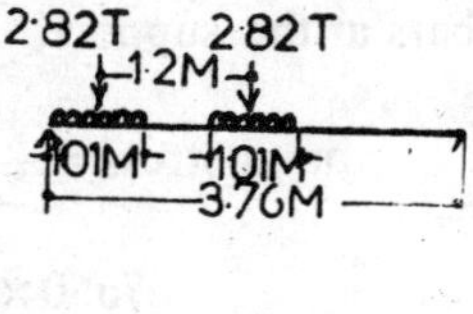

$=2{\cdot}744\times0{\cdot}505\left(1-\dfrac{0{\cdot}505}{3{\cdot}76}\right)+\dfrac{(50+2\times8)}{100}$

$=2{\cdot}744\times0{\cdot}505\times(1-{\cdot}134)+{\cdot}66$

$=2{\cdot}744\times0{\cdot}505\times0{\cdot}866+0{\cdot}66=1{\cdot}2+10{\cdot}660=1{\cdot}86$ m

2·82T 2·82T
1·2M
·101M ·101M
3·76M

Effective widths overlap

$$e=1+1\cdot8+\frac{1\cdot86}{2}=3\cdot73 \text{ m}$$

Load per metre width $=\frac{11\cdot4}{3\cdot73}=3\cdot06$ t

Effective width for second wheel loads

$$x_2=1\cdot2+\cdot505=1\cdot705 \text{ m}$$

$$e_2=Kx_2\left(1-\frac{x_2}{2}\right)+W$$

$$=2\cdot744\times1\cdot705\left(1-\frac{1\cdot705}{3\cdot76}\right)+0\cdot66$$

$$=2\cdot744\times1\cdot705\times(1-\cdot454)+0\cdot66$$

$$=2\cdot744\times1\cdot705\times\cdot546+0\cdot66=2\cdot53+0\cdot66=3\cdot19 \text{ m}$$

Effective widths overlap:

$$\text{Effective width}=0\cdot6+0\cdot4+1\cdot8+\frac{3\cdot19}{2}=2\cdot8+1\cdot595=4\cdot395 \text{ m}$$

Load per metre width

$$=\frac{11\cdot4}{4\cdot395}=2\cdot60 \text{ t}$$

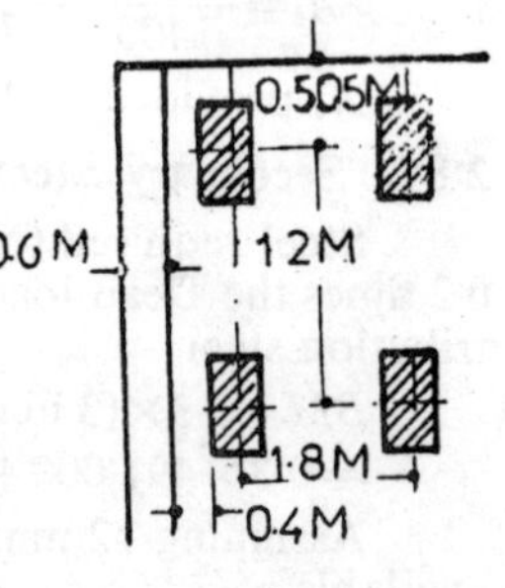

Shear force for the worst load position,

$$R=3\cdot06\times\left(\frac{3\cdot76-\cdot505}{3\cdot76}\right)+2\cdot60\left(\frac{3\cdot76-1\cdot705}{3.76}\right)$$

$$=3\cdot06\times\frac{3\cdot255}{3\cdot76}+\frac{2\cdot60}{3\cdot76}\times2\cdot055$$

$$=2\cdot66+1\cdot42=4\cdot08 \text{ t}$$

Shear force due to impact on live load $=4\cdot08\times0\cdot46=1\cdot88$ t

Shear force due to Dead Load $=0\cdot896\times\frac{3\cdot76}{\cdot2}=1\cdot69$ t

Total shear force: Shear force due to L.L $=4\cdot08$ t

Shear force due to impact factor $=1\cdot88$ t

Shear force due to dead load $=1\cdot69$ t

Total SF $=7\cdot65$ t

$$\text{Shear stress}=\frac{7650}{100\times\cdot87\times26}$$

$$=3\cdot38 \text{ kg/cm}^2<7 \text{ kg/cm}^2$$

2·8·5. Bond Stress

Bond stress: Assuming 50% bent up bars at the support,

$$\text{bond stress}=\frac{Q}{\Sigma Ojd}=\frac{7650}{5\cdot03\times\frac{100}{9\cdot5\times2}\times.87\times26}$$

$$=\frac{7650\times19}{5\cdot03\times100\times\cdot87\times26}=12\cdot8 \text{ kg/cm}^2<13 \text{ kg/cm}^2$$

2·9. Kerbs

Safety kerbs are designed for a live load of 400 kg/m² and a horizontal load of 750 kg/m.

2·10. Vertical Loads

Live load on the Kerb

$=0{\cdot}6\times1\times1\times400=240$ kg/m

Self weight of the Kerb

$=\frac{60}{100}\times\frac{60}{100}\times2400=865$ kg/m

Weight of hand rail (assumed)

$=70$ kg/m

Total vertical load$=240+865+70=1175$ kg/m²

Bending moment due to vertical loads

$=1175\times\frac{3{\cdot}76^2}{8}=2090$ m kg$=209{,}000$ cm kg

Bending moment due to class A loading per metre width including impact$=336{,}000+155{,}000=491{,}000$ cm kg

bending moment on 60 cm wide kerb

$=\frac{491{,}000\times60}{100}=294{,}600$ cm kg

Total bending moment in the Kerb

$=294600+209{,}000=503{,}600$ cm kg

2·11. Depth of Kerb

$d_e=\sqrt{\frac{503{,}600}{12{\cdot}1\times60}}=\sqrt{700}=26{\cdot}40$ cm<60 cm provided

effective depth available$=60-4=56$ cm

Steel required in the Kerb$=\frac{503{,}600}{1400\times{\cdot}87\times56}=7{\cdot}40$ cm²

Use 16 mm ϕ 4 Nos (A_t provided 8·04 cm²)

2·11·1. Shear Stress

Maximum shear force due to kerb loading

$=\frac{1175\times3{\cdot}76}{2}=2200$ kg

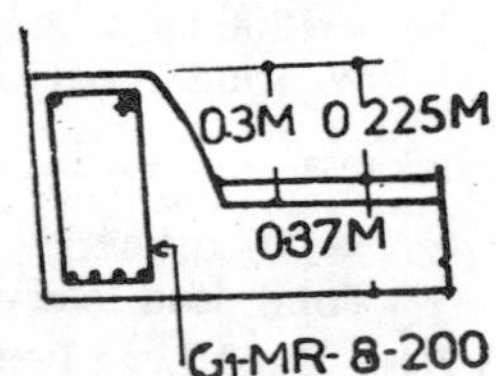

due to class A live load and impact

$=4080+1880=5960$ kg

Total SF$=2200+5960=8160$ kg

Shear stress$=\frac{8160}{60\times{\cdot}87\times56}$

$=2{\cdot}8$ kg cm²<7 kg/cm²

Use nominal stirrups 8 mm ϕ at 20 cm c/c

2·11·2. Horizontal Loads

Horizontal load on the kerb$=750$ kg/m

Bending moment$=750\times30=22{,}500$ cm kg

2·11·3. Steel Required

$A_t=\frac{22500}{1400\times{\cdot}87\times56}=0{\cdot}33$ cm²

This is a small quantity vertical leg of stirrup reinforcement at 20 cm c/c will resist the bending moment and therefore no extra reinforcement is needed.

R.C.C. Deck Slab Bridge Two Lane Class A Loading 3

3·1. Data

Clear span = 6 m. Loading = IRC--Class A. Number of lanes = two. Road width = 6·8m. Safety kerbs = 60 cm wide. Average thickness of wearing coat = 8 cm.

Materials available, M 150 grade concrete, grade – I steel

3·2. Relevant Codes

IRC standard specifications and code of practice for railway bridges sections one to four

3·3. Overall Width of Road Way

Width of roadway = 3·8 + 3 + 1·2 = 8 m

3·4. Thickness of the Slab

Thickness of the slab ($\frac{1}{12}$ span) = 50 cm. Effective depth = (50 – 4) = 46 cm

3·4·1. Effective Span

Effective span = clear span + effective depth = 6 + ·46 = 6·46 m

3·5. Impact Factor Allowed

$$\text{Impact factor} = \frac{4 \cdot 5}{6 \cdot 0 + 6 \cdot 46} = 0 \cdot 36$$

3·6. Characteristic Strengths

$\sigma_{cb} = 50$ kg/cm² $\quad m = 18$

$\sigma_{st} = 1400$ kg/cm² $\quad jd = 0 \cdot 87\, d$

$q = 5$ kg/cm² $\quad R = 8 \cdot 7$

$\sigma_{bl} = 10$ kg/cm²

3·7. Loads

Dead loads : Self weight of slab (50 cm assumed)

$$= \frac{50}{100} \times 1 \times 1 \times 2400$$

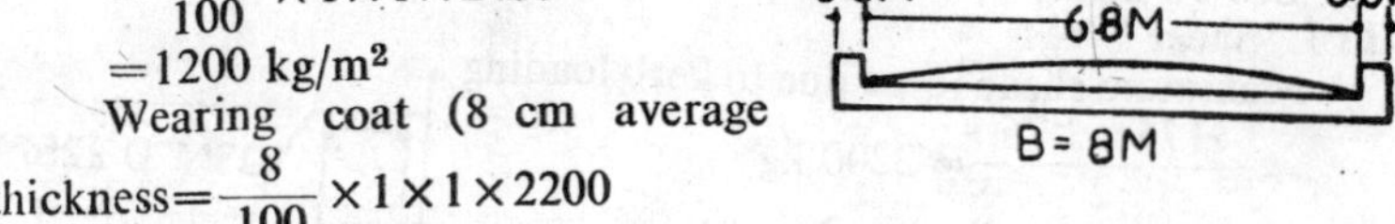

= 1200 kg/m²

Wearing coat (8 cm average thickness = $\frac{8}{100} \times 1 \times 1 \times 2200$

= 176 kg/m²

Total dead load = 1376 kg/m²

3·7·1. Dead Load Bending Moment

Effective span = 6·46 m

$$\text{Bending moment} = \frac{1376 \times 6 \cdot 46^2}{8}$$

= 5720 m kg = 572,000 cm kg

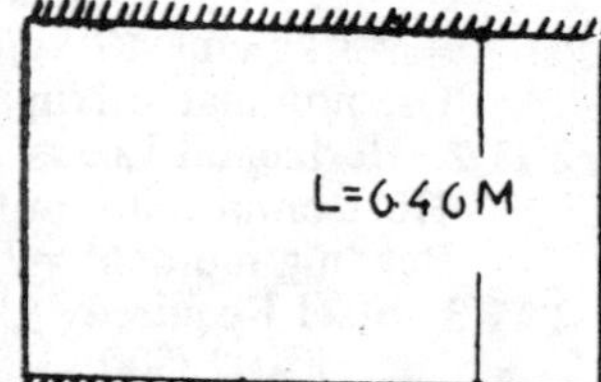

Live loads : Ratio of width of slab to effective span

$$= \frac{B}{L} = \frac{8}{6 \cdot 46} = 1 \cdot 24 < 3$$

3·7·2. Effective Width of Dispersion Wheel Loads

According to IRC specification, $e = Kx\left(1 - \frac{x}{L}\right) + W$

x is determined for the position of axle loads that produces maximum bending moment in the span. Obviously the maximum bending is produced when the heavier loads are on the span. Heaviest axle load in class $-A$ Vehicle is 11·4 t. x shall be determined for this axle load. When these loads are placed symmetrically from the centre of the span, maximum bending moment will occur. For this arrangement all other loads will be off the span since the load in front of 1st 11·4 t axle load is 3·2 m from it and the rear one from 2nd 11·4 t axle load is 4·3 m behind it.

Therefore $x=\frac{6{\cdot}46}{2}-\frac{1{\cdot}2}{2}=3{\cdot}23-0{\cdot}60=2{\cdot}63$ m

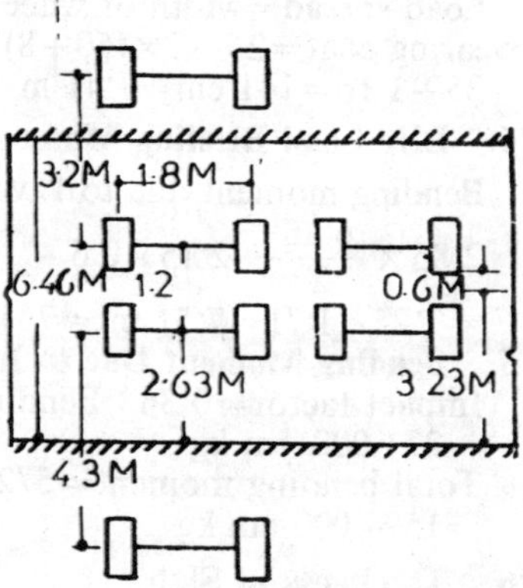

$\frac{B}{L}=\frac{8{\cdot}00}{6{\cdot}46}=1{\cdot}24$

$K=2{\cdot}64$ for $\frac{B}{L}=1{\cdot}2$,

and $K=2{\cdot}72$ for $\frac{B}{L}=1{\cdot}3$

For $\frac{B}{L}=1{\cdot}24$,

$K=2{\cdot}64+\frac{{\cdot}08\times0{\cdot}4}{0{\cdot}1}$

$=2{\cdot}672$

W=width of contact of wheel $+2\times$wearing coat

$=50+(2\times8)$

$=66$ cm$=0{\cdot}66$ m

3·2M, 1·8M, 6·46M, 1·2, 0·6M, 2·63M, 3 23M, 4·3M

$e=Kx\left(1-\frac{x}{L}\right)+W=2{\cdot}672\times2{\cdot}63\left(1-\frac{2{\cdot}63}{6{\cdot}46}\right)+0{\cdot}66$

$=2{\cdot}672\times2{\cdot}63\times{\cdot}594+0{\cdot}66=4{\cdot}7\times0{\cdot}66=5{\cdot}36$ m

Effective widths of wheel loads over lap case (i) Two vehicles on the span

The effective width of dispersion. When two vehicles pass or cross (since two lane bridge)

g=clear distance between the two vehicles=0·4 m to 1·2 m uniformly increasing for clear road widths from 5·5 m to 7·5 m

Therefore for a clear road width of 6·8 m

$g=\frac{0{\cdot}4+(1{\cdot}2-0{\cdot}4)}{(7{\cdot}5-5{\cdot}5)}\times(6{\cdot}8-5{\cdot}5)$

$=0{\cdot}4+\frac{0{\cdot}8\times1{\cdot}3}{2}=0{\cdot}92$ m

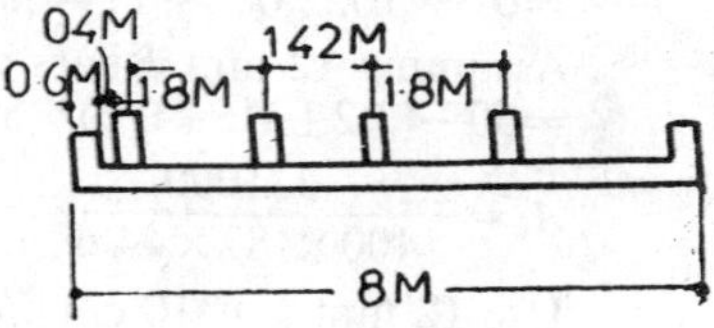

Distance between the centre to centre of the two adjacent vehicles$=0{\cdot}92+\frac{0{\cdot}50}{2}+\frac{0{\cdot}50}{2}$

$=1{\cdot}42$ m

04M, 142M, 0·6M, 1·8M, 1·8M, 8M

The effectiva width$=0{\cdot}6+0{\cdot}4+1{\cdot}8+1{\cdot}42+1{\cdot}8+\frac{5{\cdot}36}{2}$

$=8{\cdot}7>8$ m. Therefore effective width$=8$ m

Load per metre width$=\frac{2\times 11{\cdot}4}{8}=2{\cdot}85$ t

Case (*ii*) When only one vehicle is on the span

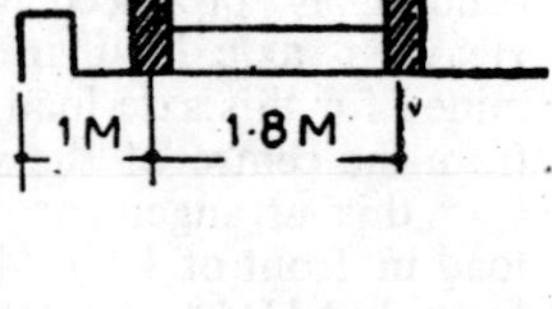

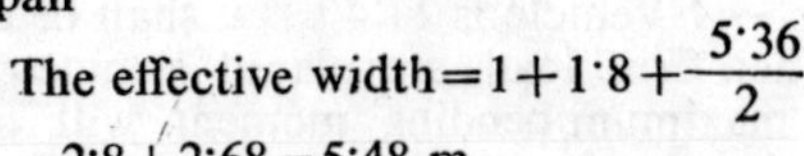

The effective width$=1+1{\cdot}8+\frac{5{\cdot}36}{2}$

$=2{\cdot}8+2{\cdot}68=5{\cdot}48$ m

Load per metre width$=\frac{11{\cdot}4}{5{\cdot}48}=2{\cdot}08$ t

Therefore case (*i*) is critical

3·7·3. Dispersion of Load Along the Span

Load spread=width of wheel contact+2×depth of slab including wearing coat$=25+2\times(50+8)$

$25+1{\cdot}16=141$ cm$=1{\cdot}41$ m

3·7·4. Live Load Bending Moment

2·85T 2·85T
1·2M
0·705M
6·46M
2·85T

Bending moment due to Live Load

$2{\cdot}85\times\frac{6{\cdot}46}{2}-2{\cdot}85\times 0{\cdot}6$

$=9{\cdot}22-1{\cdot}71=7{\cdot}51$ mt$=751{,}000$ cm kg

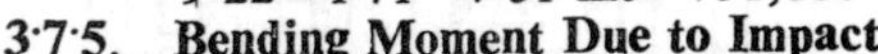

3·7·5. Bending Moment Due to Impact

Impact factor$=0{\cdot}36$: Bending moment$=0{\cdot}36\times 751{,}000$
$=271{,}000$ cm kg

Total bending moment$=572{,}000+751{,}000+271{,}000$
$=1594{,}000$ cm kg

3·7·6. Thickness of Slab

Using *M* 150, $8{\cdot}7\,bd_e^2=1594{,}000$

$$d_e=\sqrt{\frac{1954{,}000}{8{\cdot}7\times 100}}=\sqrt{1830}=42{\cdot}7 \text{ cm}$$

Use $d=50$ cm $d_e=46$ cm

3·7·7. Main Reinforcement

$$A_t=\frac{1594{,}000}{1400\times{\cdot}87\times 46}=28{\cdot}4 \text{ cm}^2$$

Use 20 mm ϕ 11 cm c/c

G1 MR 12 100
G1 MR 20 110
1M

3·7·8. Secondary Reinforcement

Steel required to resist 0·3 times the L.L., B.M. and 0·2 times the dead load bending moment shall be provided as distribution steel

Bending moment$=0{\cdot}3(751{,}000+271{,}000)+0{\cdot}2\times 572{,}000$
$=0{\cdot}3\times 1022{,}000+114400=306600+114400=421000$ cm kg

Assuming 12 mm ϕ bars effective depth available
$=50-(3{\cdot}2+1{\cdot}6+0{\cdot}6)=50-5{\cdot}4=44{\cdot}6$ cm

$$A_t=\frac{421000}{1400\times{\cdot}87\times 44{\cdot}6}=10{\cdot}9 \text{ cm}^2$$

Use 12 mm ϕ at 10 cm c/c

3·7·9. Shear Force Distance of Loads from the Support

Maximum shear force occurs when the load is near the support.

If the heaviest axle load of 11·4 t is at a distance

$$x_1=\frac{1\cdot41}{2}=0\cdot705 \text{ m}$$

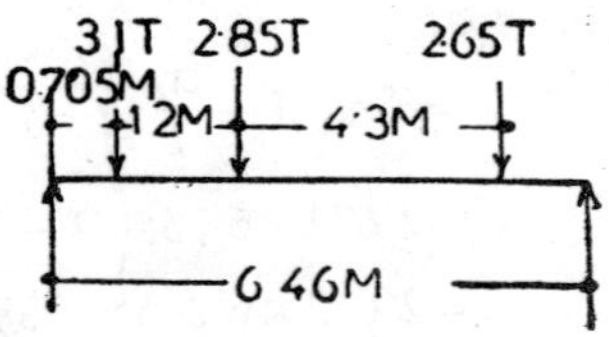

The load behind it will be at a distance

$x_2=0\cdot705+1\cdot2=1\cdot905$ m

For this arrangement the axle load behind this axle, will also be on the span.

Distance $x_3=0\cdot705+1\cdot2+4\cdot3=6\cdot205$ m

3·7·10. Effective Widths and Load/metre 1st Axle Load

e_1=effective width for 1st axle load on the span

$$=Kx_1\left(1-\frac{x_1}{L}\right)+W$$

$$=2\cdot672\times0\cdot705\left(1-\frac{0\cdot705}{6\cdot46}\right)+0\cdot66$$

$$=2\cdot672\times0\cdot705\times\cdot891+0\cdot66$$

$$=1\cdot66+0\cdot66=2\cdot32 \text{ m}$$

The effective width overlap.

The effective width

$$=0\cdot6+0\cdot4=1\cdot8+1\cdot42+1\cdot8+\frac{2\cdot32}{2}$$

$$=7\cdot18 \text{ m}$$

Load per meter width for the Ist axle load on the span

$$=\frac{2\times11\cdot4}{7\cdot18}=3\cdot19 \text{ t}$$

e_1 when only one vehicle is on the span

$$=0\cdot6+0\cdot4+1\cdot8+\frac{2\cdot32}{2}=3\cdot92 \text{ m}$$

Load per metre$=\frac{11\cdot4}{3\cdot92}=2\cdot90$ t which is less than the previous case

Second axle Load e_2=effective width for the second axle loads on the span

$$=Kx_2\left(1-\frac{x_2}{L}\right)+W=2\cdot672\times1\cdot905\left(1-\frac{1\cdot905}{6\cdot46}\right)+0\cdot66$$

$$=2\cdot672\times1\cdot905\times\cdot704+0\cdot66=3\cdot59+0\cdot66=4\cdot25 \text{ m}$$

Effective widths overlap

effective width$=0\cdot6+0\cdot4+1\cdot8+1\cdot42+1\cdot8+\frac{4\cdot25}{2}$

$=8\cdot045$ m>8 m

Load per metre width$=\frac{2\times11\cdot4}{8}=2\cdot85$ t

When only one vehical is on the span

$$e_2=0\cdot6+0\cdot4+1\cdot8+\frac{4\cdot25}{2}=4\cdot925 \text{ m}$$

Load per metre$=\frac{11\cdot4}{4\cdot925}=2\cdot32$ t which is less

Third axle Load—e_3 = effective width for the third axle loads on the span

$$=Kx_3\left(1-\frac{x_3}{L}\right)+W$$

$$=2{\cdot}672\times6{\cdot}205\left(1-\frac{6{\cdot}205}{6{\cdot}46}\right)+0{\cdot}66$$

$$=2{\cdot}672\times6{\cdot}205\times{\cdot}038+0{\cdot}66$$

$$=0{\cdot}632+0{\cdot}660=1{\cdot}292 \text{ m}$$

0·705M, 1·2M, 4·3M, 6·46M

Effective widths do not overlap

$$\text{Load per metre}=\frac{6{\cdot}8}{1{\cdot}292\times2}=2{\cdot}65 \text{ t}$$

Shear force at the support

$$R=3{\cdot}19\times\left(\frac{6{\cdot}46-0{\cdot}705}{6{\cdot}46}\right)+2{\cdot}85\left(\frac{6{\cdot}46-1{\cdot}905}{6{\cdot}46}\right)$$

$$+2{\cdot}85\left(\frac{6{\cdot}46-6{\cdot}205}{6{\cdot}46}\right)$$

$$=3{\cdot}19\times\frac{5{\cdot}755}{6.46}+\frac{2{\cdot}85\times4{\cdot}555}{6{\cdot}46}+\frac{2{\cdot}65\times{\cdot}255}{6{\cdot}46}$$

$$=2{\cdot}94+1{\cdot}94+0{\cdot}106=4{\cdot}986 \text{ t}$$

Shear force due to impact own live load $=4{\cdot}986\times0{\cdot}36=1{\cdot}79$ t

$$\text{Shear force due to dead load}=\frac{1{\cdot}376\times6{\cdot}46}{2}=4{\cdot}42 \text{ t}$$

Total shear force $=4{\cdot}986+1{\cdot}79+4{\cdot}420=11{\cdot}196$ t $=11196$ kg

3·7·11. Shear Stress

$$\text{Shear stress}=\frac{11196}{100\times{\cdot}87\times46}=2{\cdot}82 \text{ kg/cm}^2<5 \text{ kg/cm}^2$$

3·7·12. Bond Stress

Assuming that alternate bars are bent up at support, bond stress

$$=\frac{11196}{\dfrac{6{\cdot}28\times100\times0{\cdot}87\times46}{11\times2}}=\frac{11196\times22}{6{\cdot}28\times{\cdot}87\times46\times100}=9{\cdot}9 \text{ kg/cm}^2$$

3·8. Kerbs

Safety kerbs are designed for a live load of 400 kg/m² and a horizontal load of 750 kg/m

3·8·1. Vertical Loads

Live load on the kerb $=0{\cdot}6\times1\times1\times400=240$ kg/m

$$\text{Self weight of concrete}=\frac{80}{100}\times\frac{60}{100}\times1\times2400=1150 \text{ kg/m}$$

Weight of hand rails = 60 kg/m

Total vertical load = 1450 kg/m

$$\text{Bending moment}=\frac{1450\times6{\cdot}46^2}{8}=5630 \text{ m kg}=563{,}000 \text{ cm kg}$$

Bending moment due to class—A loading per metre including impact = 751,000 + 271,000 = 1022,000 cm kg

$$\text{Bending moment on 60 cm wide kerb}=\frac{60}{100}\times1022{,}000$$

$=615{,}000$ cm kg

Total bending moment on the kerb $=563{,}000+615{,}000$ $=1178{,}000$ cm kg

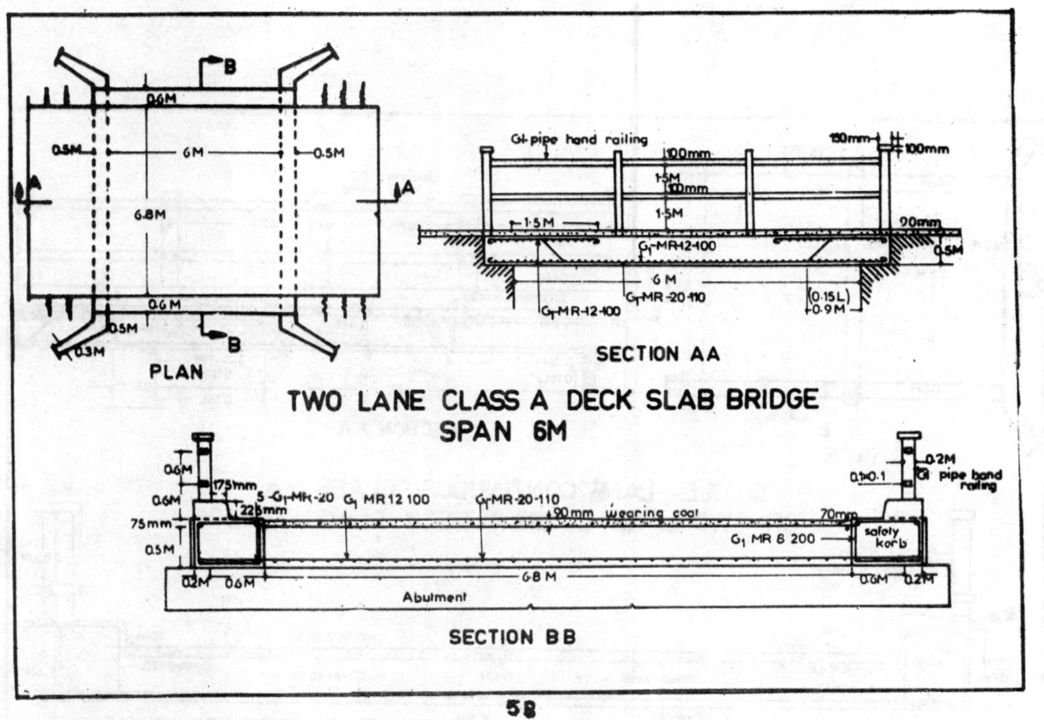

TWO LANE CLASS A DECK SLAB BRIDGE
SPAN 6M

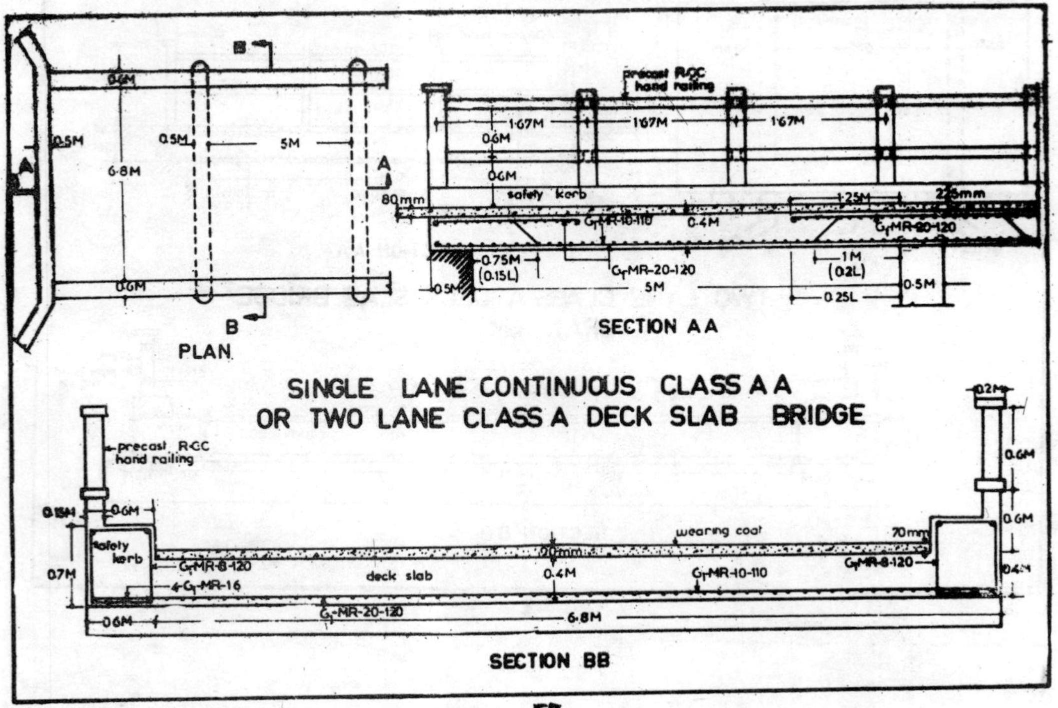

PLAN

SECTION AA

SINGLE LANE CONTINUOUS CLASS AA
OR TWO LANE CLASS A DECK SLAB BRIDGE

precast RCC hand railing
1·67M
1·67M
1·67M
0·6M
0·6M
safety kerb
80mm
225mm
1·25M
G_1-MR-10-110
0·4M
G_1-MR-20-120
0·75M (0.15L)
G_1-MR-20-120
1M (0.2L)
0·5M
5M
0·25L
0·5M
0·6M
0·5M
6·8M
0·5M
5M
0·6M

precast RCC hand railing
0·15M
0·6M
safety kerb
0·7M
G_1-MR-8-120
4-G_1-MR-16
deck slab
70mm
0·4M
wearing coat
70mm
G_1-MR-10-110
G_1-MR-8-120
G_1-MR-20-120
0·6M
6·8M
0·2M
0·6M
0·6M
0·4M

SECTION BB

3·8·2. Depth of Kerb

$$d_e=\sqrt{\frac{1778,000}{8{\cdot}7\times60}}=\sqrt{1360}=37 \text{ cm}$$

effective depth available=80—5=75 cm

$$A_t=\frac{1178,000}{1400\times{\cdot}87\times75}\quad \text{Use 5 Nos} -20 \text{ mm } \phi$$

3·8·3. Shear Stress

Maximum shear force due to kerb loading

$$=\frac{1450\times6{\cdot}46}{2}=4700 \text{ kg}$$

due to class A loading live load including impact=4986+1820 =6806 kg

Total shear stress=4700+6806=11506 kg

$$\text{Shear stress}=\frac{11506}{60\times{\cdot}87\times75}=1{\cdot}75 \text{ kg/cm}^2<5 \text{ kg/cm}^2$$

Use nominal stirrups 8 mm ϕ at 20 cm c/c

3·8·4. Horizontal Load

Horizontal load on the kerb=750 kg/m

Bending moment=750×30=22500 cm kg

3·8·5. Steel Required

$$A_t=\frac{22500}{1400\times{\cdot}87\times75}={\cdot}247 \text{ cm}^2$$

This is very small and the bending moment will be resisted by the vertical legs of the nominal stirrups and therefore no extra reinforcement is necessary.

R.C.C. Deck Slab Bridge Single lane Class-AA or two lane Class-A Loading 4

4.1. Data

Clear span = 6 m, Loading = class-AA Tracked vehicle,
Number of lanes = two, Road width = 6·8 m,
Safety kerbs = 60 cm wide, Average thickness of wearing coat = 8 cm
Materials available, Concrete *M* 150 grade, steel grade−1

4·2. Relevant Codes

IRC standard specifications and code of practice for highway bridges—section one to four.

4·3. Overall Width of the Bridge

Width of road way = 3·8 + 3 + 1·2 = 8 m

4·4. Slab Thickness

Slab thickness (at 1/12 span) = 50 cm
Effective depth = (50 − 4) = 46 cm

4·5. Effective Span

Effective span = clear span + effective depth = 6 + 0·46 = 6·46 m

4·6. Impact Factor Allowed

For spans < 9 m for tracked vehicles, 25% upto 5 m and is reduced linearly to 10% for spans upto 9 m

$$\text{Impact factor for } 6{\cdot}46 \text{ m} = 25\% - \frac{(25\% - 10\%) \times (6{\cdot}46 - 5)}{(9-5)}$$

$$= 25\% - 15\% \times \frac{1{\cdot}46}{4} = 25\% - 5{\cdot}5\% = 19{\cdot}5\% = 0{\cdot}195$$

4·7. Characteristic Strengths

$\sigma_{cb} = 50$ kg/m² $\quad m = 18$
$\sigma_{st} = 1400$ kg/cm² $\quad j_d = 0{\cdot}87\, d$
$q = 5$ kg/cm² $\quad R = 8{\cdot}7$
$\sigma_{bl} = 10$ kg/cm²

4·8. Loads

Dead loads : Self weight of slab (50 cm assumed)

$$= \frac{50}{100} \times 1 \times 1 \times 2400 = 1200 \text{ kg/m}^2$$

Wearing coat (8 cm average thickness)

$$= \frac{8}{100} \times 1 \times 1 \times 2200 = 176 \text{ kg/m}^2$$

Total dead load = 1376 kg/m²

4·8·1. Dead Load Bending Moment

Effective span = 6·46 m

$$\text{Bending moment} = \frac{1376 \times 6{\cdot}46^2}{8} = 5720 \text{ m kg} = 572{,}000 \text{ cm kg}$$

Live loads : Ratio of width of slab to the effective span.

$$=\frac{B}{L}=\frac{8}{6{\cdot}46}=1{\cdot}24<3$$

Therefore the slab spans in one direction only. Pigeaud's method of distribution will not apply

4·8·2. Effective width of Dispersion of Loads

According to IRC specification

$$e=Kx\left(1-\frac{x}{L}\right)+W$$

$$\frac{B}{L}=\frac{L'}{L}=\frac{8}{6{\cdot}46}=1{\cdot}24$$

For $\frac{B}{L}=1{\cdot}2,\ K=2{\cdot}64,$

For $\frac{B}{L}=1{\cdot}3,\ K=2{\cdot}72$

Therefore $\frac{B}{L}=1{\cdot}24$

$$K=2{\cdot}64+\frac{0{\cdot}08\times{\cdot}04}{0{\cdot}1}=2{\cdot}672$$

The distance x from C.G. of the load to the support is determined for the condition, such that the load position should produce maximum bending moment on the span. Maximum bending moment is produced, when the vehicle is at the centre of the span.

$$x=\frac{6{\cdot}46}{2}=3{\cdot}23\text{ m}$$

W=breadth of concentration area of the load+twice the thickness of wearing coat.

$$=85+2\times8=101\text{ cm}=1{\cdot}01\text{ m}$$

$$e=Kx\left(1-\frac{x}{L}\right)+W=2{\cdot}672\times3{\cdot}23\left(1-\frac{3{\cdot}23}{6{\cdot}46}\right)+1{\cdot}01$$

$$=2{\cdot}672\times3{\cdot}23\times0{\cdot}5+1{\cdot}01=4{\cdot}34+1{\cdot}01=5{\cdot}35\text{ m}$$

Effective widths overlap :

Therefore the effective width

$$=2{\cdot}225+2{\cdot}05+\frac{5{\cdot}35}{2}$$

$$=2{\cdot}225+2{\cdot}05+2{\cdot}675=6{\cdot}95\text{ m}$$

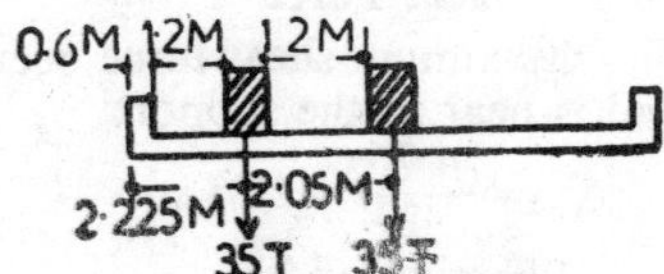

4·8·3. Live Load Per Metre Width

Live load per metre width due to class AA vehicle

$$=\frac{2\times35}{6{\cdot}95}=10{\cdot}2\text{ t}$$

4·8·4. Dispersion of Load Along the Span

Load spread=width of track contact+2×depth of slab including wearing coat=360+2(50+8)=360+116=476 cm=4·76 m

4·8·5. Liveload Bending Moment

Maximum bending moment due to

$$\text{Live load}=\frac{10{\cdot}2}{2}\left(\frac{6{\cdot}46}{2}-\frac{4{\cdot}76}{4}\right)$$

$=5{\cdot}1(3{\cdot}23 - 1{\cdot}19)$

$=5{\cdot}1\times2{\cdot}04=9{\cdot}6$ mt

$=960{,}000$ cm kg

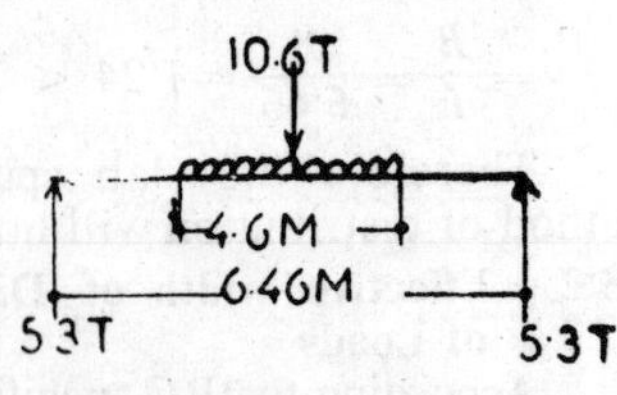

4·8·6. Bending Moment Due to Impact

Bending moment due to impact on live load=Impact factor × Live load moment=0·195 × 960,000

$=187{,}000$ cm kg

Total bending moment due to dead, live and impact loads

$=572{,}000+960{,}000+187{,}000=1719{,}000$ cm kg

4·9. Slab Thickness

$8{\cdot}7\ bd_e^2=1719{,}000$

$$d_e=\sqrt{\frac{1719{,}000}{8{\cdot}7\times100}}=\sqrt{1980}=44{\cdot}5 \text{ cm}$$

Use $d=50$ cm, $d_e=46{\cdot}2$ cm

Using 22 m ϕ bars this leaves a clear cover of 2·7 cm

4·9·1. Main Reinforcement

$$A_t=\frac{1719{,}000}{1400\times{\cdot}87\times46{\cdot}2}=30{\cdot}5 \text{ cm}$$

Use 22 mm ϕ at 12 cm c/c

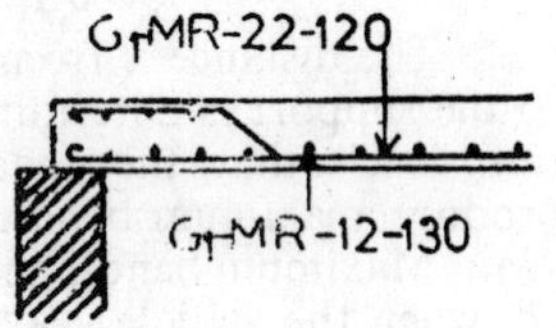

4·9·2. Secondary Reinforcement

Steel required to resists 0·3 times the L.L. bending moment and 0·2 times the D.L. bending moment shall be provided as distribution steel.

Bending moments$=0{\cdot}3\ (960{,}000+187{,}000)+0{\cdot}2\times572{,}000$

$=344{,}100+114{,}400=458{,}500$ cm kg

Using 12 mm ϕ, Effective depth available

$$=50-\left(2{\cdot}7+2{\cdot}2+\frac{1{\cdot}2}{2}\right)=50-5{\cdot}5=44{\cdot}5 \text{ cm}$$

$$A_t=\frac{458{,}500}{1400\times{\cdot}87\times44{\cdot}5}=8{\cdot}5 \text{ cm}^2$$

Use 12 mm ϕ at 13 cm c/c

4·9·3. Shear Force

Maximum shear force occurs when the load is near to the support

$$x_1=\frac{4{\cdot}76}{2}=2{\cdot}38 \text{ m}$$

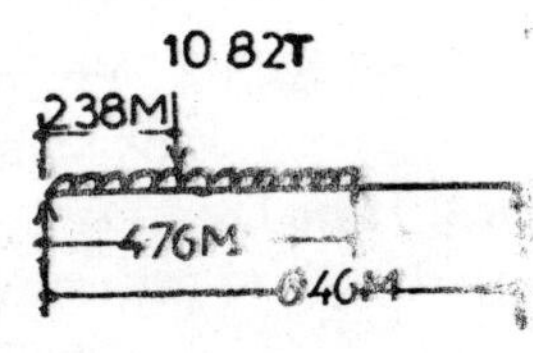

Effective width

$$e_1=K\,x_1\left(1-\frac{x_1}{L}\right)+W$$

$$=2{\cdot}672\times2{\cdot}38\left(1-\frac{2{\cdot}38}{6{\cdot}46}\right)+1{\cdot}01$$

$=2{\cdot}672\times2{\cdot}38\times0{\cdot}63+1{\cdot}01=4{\cdot}02+1{\cdot}01=5{\cdot}03$

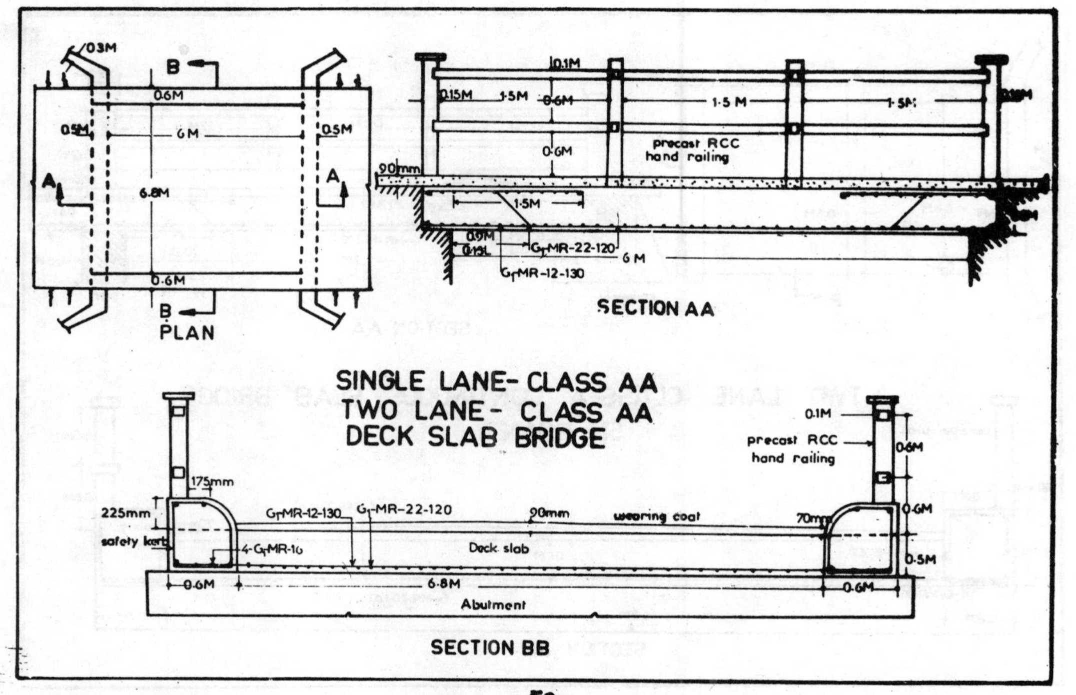
0.3M
B
0.6M
0.5M
6M
0.5M
A
6.8M
A
0.6M
B
PLAN
0.1M
0.15M
1.5M
0.6M
1.5 M
1.5M
0.6M
precast RCC
hand railing
90mm
1.5M
0.9M
0.15L
G1-MR-22-120
6 M
G1-MR-12-130
SECTION AA
SINGLE LANE- CLASS AA
TWO LANE- CLASS AA
DECK SLAB BRIDGE
0.1M
precast RCC
hand railing
0.6M
175mm
225mm
G1-MR-12-130
G1-MR-22-120
90mm
wearing coat
70mm
0.6M
safety kerb
4-G1-MR-16
Deck slab
0.5M
0.6M
6.8M
0.6M
Abutment
SECTION BB

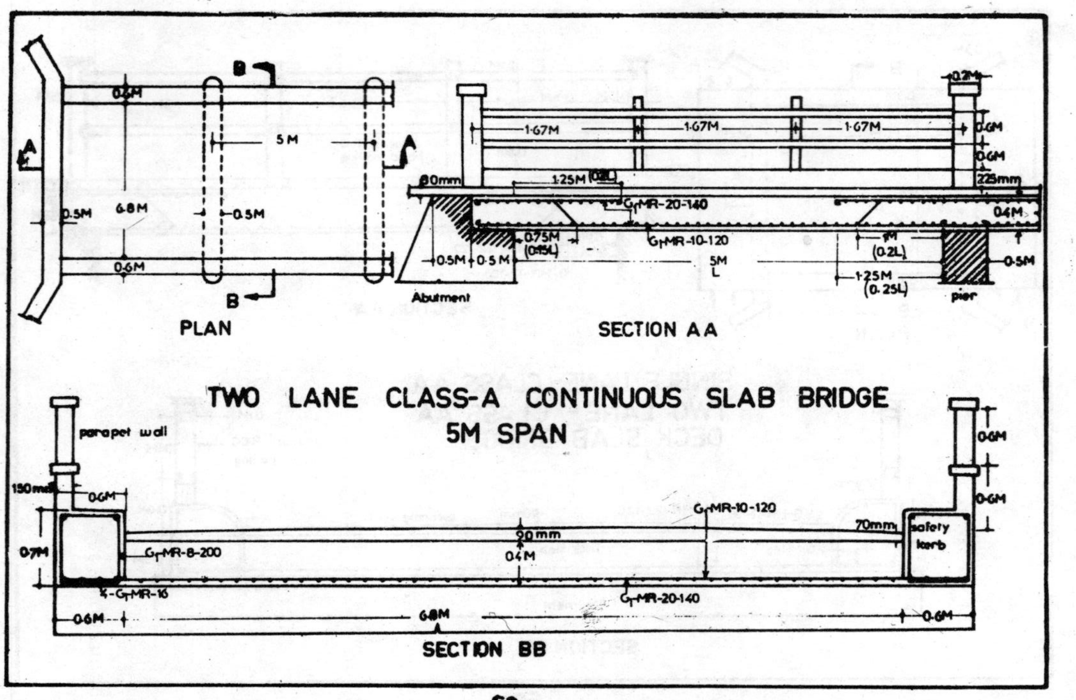

TWO LANE CLASS-A CONTINUOUS SLAB BRIDGE
5M SPAN

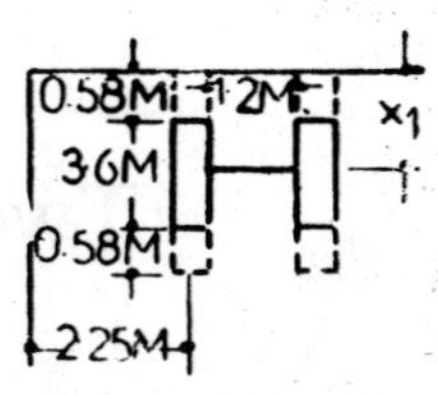

Effective widths overlap, Effective width

$$=2{\cdot}05+\frac{5{\cdot}03}{2}+2{\cdot}225$$

$$=2{\cdot}225+2{\cdot}05+2{\cdot}515=6{\cdot}79 \text{ m}$$

Load per meter width

$$=\frac{70}{6{\cdot}79}=10{\cdot}35\text{t}$$

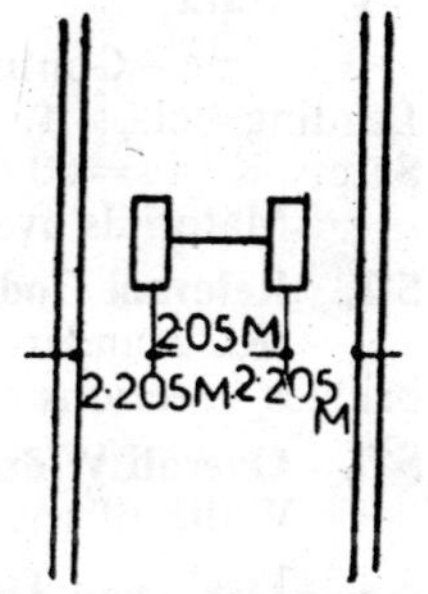

Maximum shear force

$$=\frac{10{\cdot}35\times(6{\cdot}46-2{\cdot}38)}{6{\cdot}46}$$

$$=\frac{9{\cdot}9}{6{\cdot}46}\times 4{\cdot}08=6{\cdot}55 \text{ t}=6550 \text{ kg}$$

Shear force due to impact on live load
$=0{\cdot}195\times 6550=1280$ kg

Shear force due to dead load

$$=1376\times\frac{6{\cdot}46}{2}=4420 \text{ kg}$$

Total shear force $=6550+1280+4420$
$=12250$ kg

4·9·4. Shear Stress

$$\text{Stress}=\frac{12250}{100\times{\cdot}87\times 46{\cdot}2}$$

$=3{\cdot}06$ kg/cm² < 5 kg/cm²

4·9·5. Bond Stress

Assuming that alternate bars are bent up at support and bond stress

$$=\frac{12250\times 11\times 2}{6{\cdot}91\times 100\times{\cdot}87\times 46{\cdot}2}=9{\cdot}65 \text{ kg/cm}^2 < 10 \text{ kg/cm}^2$$

Continuous Slab Bridge Two Lane Class-A Loading 5

5·1. Data

Type—Continuous slab of 5 equal spans, clear span=5 m, Leading=class-*A*, Number of lanes=Two, Road width=6·8 m, Safety Kerbs=60 cm wide, Average thickness of wearing coat=8 cm.

Materials available=Concrete *M* 150 grade, Steel grade—*I*.

5·2. Relevant Codes

IRC standard specifications and code of practice for highway bridges—sections one to four.

5·3. Overall Width of Roadway

Width of roadway=3·8+3+1·2=8 m

5·4. Trial Slab Thickness

Slab thickness ($\frac{1}{12}$ span)=40 cm

Effective depth=36 cm

5·5. Effective Span

Effective span=clear span+effective depth=5+·36=5·36 m

5·6. Impact Factor Allowed

$$\text{Impact factor} = \frac{4{\cdot}5}{6{\cdot}0+5{\cdot}36} = 0{\cdot}396$$

5·7. Characteristic Strengths

$\sigma_{cb}=50$ kg/cm² $\quad m=18$

$\sigma_{st}=1400$ kg/cm² $\quad jd=0{\cdot}87\,d$

$q=5$ kg/cm² $\quad R=8{\cdot}7$

$\sigma_{bt}=10$ kg/cm²

5·7·1. Loads

Dead loads : Self weight of slab (40 cm assumed)

$$=\frac{40}{100}\times1\times1\times2400=960 \text{ kg/m}^2$$

$$\text{Wearing coat (8 cm thick)}=\frac{8}{100}\times1\times1\times2200=176 \text{ kg/m}^2$$

Total dead load=1036 kg/m²

5·7·2. Dead Load Bending Moment

Effective span=5·36 m

$$\text{Dead load bending moment}=\frac{1036\times5{\cdot}36^2}{10}=2970 \text{ m kg}$$

=297,000 cm kg

Live loads: Ratio of width of slab to effective span

$$=\frac{B}{L}=\frac{8}{5{\cdot}36}=1{\cdot}49<3$$

Therefore Pigeaud's method of distribution will not apply.

According to IRC specification effective width of dispersion is

$$e=Kx\left(1-\frac{x}{L}\right)+W$$

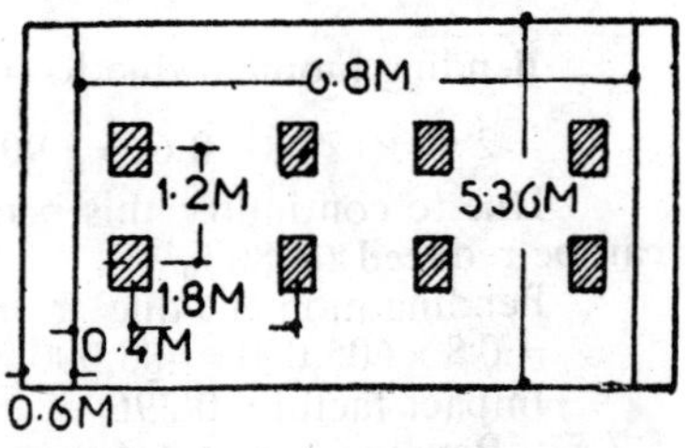

x should be determined for that load position which gives the maximum bending moment. For class A loading heaviest axle loads being 11.4 t, these axle loads produce the maximum bending moment.

$$x=\frac{5{\cdot}36}{2}-\frac{1{\cdot}2}{2}=2{\cdot}68-0{\cdot}6=2{\cdot}08 \text{ m} \qquad \frac{B}{L}=\frac{8}{5{\cdot}36}=1{\cdot}49$$

For a continuous slab, For $\frac{B}{L}=1{\cdot}4$, $K=2{\cdot}48$, and for $\frac{B}{L}=1{\cdot}5$, $K=2{\cdot}48$, $\frac{B}{L}=1{\cdot}49, K=2{\cdot}48$.

W=width of contact of wheel $+2\times$wearing coat thickness

$=50+2\times 8=\cdot 66$ m

0.4M
1 M 1.8M 1.42M 1.8M 1M
0.6M

$$e=2{\cdot}48\times 2{\cdot}08\left(1-\frac{2{\cdot}08}{5{\cdot}36}\right)+0{\cdot}66$$

$=2{\cdot}48\times 2{\cdot}08\times 0{\cdot}61+0{\cdot}66=3{\cdot}15+0{\cdot}66=3{\cdot}81$ m

The effeetive widths of wheel loads overlap.

Case (*i*). Two vehicles on the span

The effective width of dispersion when two vehicles pass or cross (since two lane bridge)

g=clear distance between the two vehicles $=0{\cdot}4$ m to $1{\cdot}2$ m uniformly increasing for clear road widths from $5{\cdot}5$ m to $7{\cdot}5$ m

Therefore for a clear road width of $6{\cdot}8$ m

$$g=0{\cdot}4+\left(\frac{1{\cdot}2}{7{\cdot}5}\ \frac{0{\cdot}4}{5{\cdot}5}\right)\times(6{\cdot}8-5{\cdot}5)=0{\cdot}4+\frac{0{\cdot}8\times 1{\cdot}3}{2}=0{\cdot}92 \text{ m}$$

Distance between the centre to centre of the two adjacent vehicles $=0{\cdot}92+\frac{0{\cdot}5}{2}+\frac{0{\cdot}5}{2}=1{\cdot}42$ m

$$\text{The effective width}=0{\cdot}6+0{\cdot}4+1{\cdot}8+1{\cdot}42+1{\cdot}8+\frac{3{\cdot}81}{2}=7{\cdot}97 \text{ m}$$

$$\text{Load per metre width}=\frac{2\times 11{\cdot}4}{7{\cdot}97}=2{\cdot}90 \text{ t}$$

Case (*ii*). When only one vehicle is on the span

$$\text{The effective width}=0{\cdot}6+0{\cdot}4+1{\cdot}8+\frac{3{\cdot}81}{2}=4{\cdot}75 \text{ m}$$

$$\text{Lood per metre width}=\frac{11{\cdot}4}{4{\cdot}75}=2{\cdot}4 \text{ t}$$

Therefore case (*i*) is critical.

5·7·3. Dispersion of the Load along the Span

Load spread = width of wheel contact in the direction of span $+2\times$depth of slab including wearing coat

$=25+2(40+8)=25+96=121$ cm $=1{\cdot}21$ m

5·7·4. Live Load Bending Moment

Bending moment due to live load $=2·90\times\left(\frac{5·36}{2}-\frac{1·2}{2}\right)$

$=2·90\times(2·68-0·6)=2·90\times2·08=6·05$ mt $=605,000$ cm kg

Due to continuity this bending moment can be reduced to 80%

Bending moment due to live load
$=0·8\times605,000=485,000$ cm kg

Impact factor $=0·396$

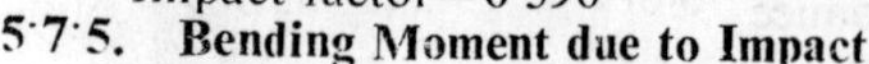

5·7·5. Bending Moment due to Impact

Bending moment due to impact factor on live load
$=0·396\times485,000 \quad 192,000$ cm kg

Total bending moment $=297,000+485,000+192,000$
$=954,000$ cm kg

5·7·6. Slab Thickness

$8·7\ bd_e^2=954,000$ cm kg

$$d_e=\sqrt{\frac{954,000}{100\times8·7}}=\sqrt{1130}$$

$=33·1$ cm. Use $d_e=36$ cm,
$d=40$ cm.

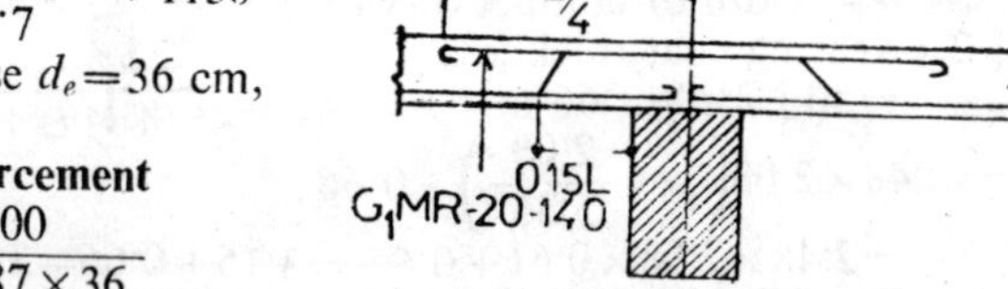

5·7·7. Main Reinforcement

$$A_t=\frac{954,000}{1400\times·87\times36}$$

$=21·8\ \text{cm}^2$. Use 20 mm ϕ at 14 cm c/c

5·7·8. Secondary Reinforcement

Steel required to resist 0·3 times the *L.L* bending moment and 0·2 times the *D.L* bending moment shall be provided as distribution steel.

Bending moment $=0·3\ (485,000+192,000)+0·2\times297,000$
$=203,100+59,400=262,500$ cm kg

Assuming 10 mm ϕ, Effective depth available
$=40-(3+2+0·5)=34·5$ **cm**

$$A_t=\frac{262,500}{1400\times8·7\times34·5}=6·25\ \text{cm}^2.$$ Use 10 mm ϕ at 12 cm c/c

5·7·9. Shear Force

Maximum shear force occurs when the load is near the supports

The first axle load is at a distance $x_1=\frac{1·21}{2}=0·605$ m

The second axle load is at a distance $x_2=·605+1·2=1·805$ m

For this load position, all other loads will be off the span

5·7·10. Effective Widths

e_1 = effective width for 1st axle load on the span
$=Kx_1(1-x_1/L)+W$

$=2·48\times0·6051-\left(\frac{·605}{5·36}\right)+0·66=2·48\times0·605\times0·887+0·66$

$=1·31+0·66=1·97$ m

Effective width

$$e_1=\frac{1·81}{2}+1·8+1·42+1·8+\frac{1·97}{2}=6·91\ \text{m}$$

Load per metre width $= \frac{2\times 11{\cdot}4}{6{\cdot}91} = 3{\cdot}30$ t

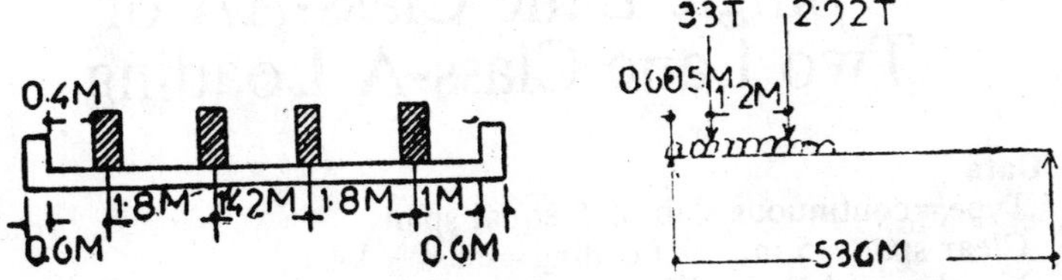

When only one vehicle is on the span, load per metre width will be less

Effective width for 2nd axle load

$$e_2 = 2{\cdot}48\times 1{\cdot}805\left(1-\frac{1{\cdot}805}{5{\cdot}36}\right)+0{\cdot}66 = 2{\cdot}48\times 1{\cdot}805\times 0{\cdot}664+0.66$$

$= 2{\cdot}98+0{\cdot}66 = 3{\cdot}64$ m

Effective width $= 0{\cdot}6+0{\cdot}4+1{\cdot}8+1{\cdot}42+1{\cdot}8+\frac{3{\cdot}64}{2} = 7{\cdot}84$ m

Load per metre width $= \frac{2\times 11{\cdot}4}{7{\cdot}84} = 2{\cdot}92$ t

Shear force at the support

$$= \frac{3{\cdot}30}{5{\cdot}36}(5{\cdot}36-0{\cdot}605)+\frac{2{\cdot}92}{5{\cdot}36}(5{\cdot}36-1{\cdot}805)$$

$$= \frac{3{\cdot}35}{5{\cdot}36}\times 4{\cdot}755+\frac{2{\cdot}99}{5{\cdot}36}\times 3{\cdot}55 = 2{\cdot}90+1{\cdot}94 = 4{\cdot}84 \text{ t} = 4840 \text{ kg}$$

Shear force due to impact on live load $= 4840\times 0{\cdot}396 = 1905$ kg

Shear force due to dead load $= 1036\times \frac{5{\cdot}36}{2} = 2775$ kg

Total shear force $= 4840+1905+2775 = 9520$ kg

5·7·11. Shear Stress

Shear stress $= \frac{9520}{100\times {\cdot}87\times 36} = 3{\cdot}05 \text{ kg/cm}^2 < 5 \text{ kg/cm}^2$

5·7·12. Bond Stress

Assuming that alternate bars are bent up at support

$=$ bond stress $= \frac{9520\times 27}{6{\cdot}28\times 100\times {\cdot}87\times 36} = 13{\cdot}2 \text{ kg/cm}^2 > 10 \text{ kg/cm}^2$

Provide additional bars at end support. Use 20 mm ϕ at 13·5 cm c/c so that bond stress is reduced to half. These additional bars also will resist negative bending moment at support.

Continuous Slab Bridge Single Lane Class-AA or Two Lane Class-A Loading 6

6·1. Data

Type = continuous slab of 5 equal spans
Clear span = 5 m, Loading = class – AA
Number of lanes = single Road width = 6·8 m
Safety kerbs = 60 cm wide
Average thickness of wearing coat = 8 cm
Materials available Concrete M 150 grade, steel—grade - 1

6·2. Relevant Codes

IRC standard specifications and code of practice for railway bridges - sections one and four

6·3. Overall width of Roadway

Width of roadway = 3·8 + 3 + 1·2 = 8 m

6·4. Trial Slab Thickness

Slab thickness ($\frac{1}{12}$ span) = 40 cm
Effective depth = 36 cm

6·5. Effective Span

Effective span = Clear span + effective depth = 5 + 0·36 = 5·36 m

6·6. Impact Factor Allowed

For spans < 9 m for tracked vehicles impact factor allowed is 25% upto 5 m and is reduced linearly to 10% for spans upto 9 m

Impact factor for 5·36 m

$$= 25\% - \frac{(25\% - 10\%) \times (5 \cdot 36 - 5)}{(9-5)}$$

$$= 25\% - 15\% \times \frac{0 \cdot 36}{4} = 25\% - 1 \cdot 35\% = 23 \cdot 65\% = \cdot 2365$$

6·7. Characteristic Strengths

M 150 concrete

$\sigma_{cb} = 50$ kg/cm^2 $m = 18$
$\sigma_{st} = 1400$ kg/cm^2 $jd = 0 \cdot 87\ d$
$q = 5$ kg/cm^2 $R = 8 \cdot 7$
$\sigma_{bt} = 10$ kg/cm^2

6·8. Loads

Dead loads: Self weight of slab (40 cm assumed)

$$= \frac{40}{100} \times 1 \times 1 \times 2400 = 960 \text{ kg/m}^2$$

Wearing coat (8 cm thick) $= \frac{8}{100} \times 1 \times 1 \times 2200 = 176$ kg/m^2

Total dead load = 1036 kg/m^2

6·8·1. Dead Load Bending Moment

Effective span = 5·36 m
Dead load bending moment since continuous

$$= \frac{1036 \times 5 \cdot 36^2}{10} = 2970 \text{ m kg} = 297000 \text{ cm kg}$$

Live loads: Ratio of width of slab to effective span

$$\frac{B}{L}=\frac{8}{5\cdot36}=1\cdot49<3$$

Therefore Pigeaud's method of distribution will not apply

According to IRC specification effective width of dispersion is

$$e=\text{K}x\left(1-\frac{x}{\text{L}}\right)+\text{W}$$

x should be determined for that load position which gives the maximum bending moment. For class AA loading with tracked vehicle, the maximum bending moment occurs when the vehicle is symmetrically placed about the centre of span.

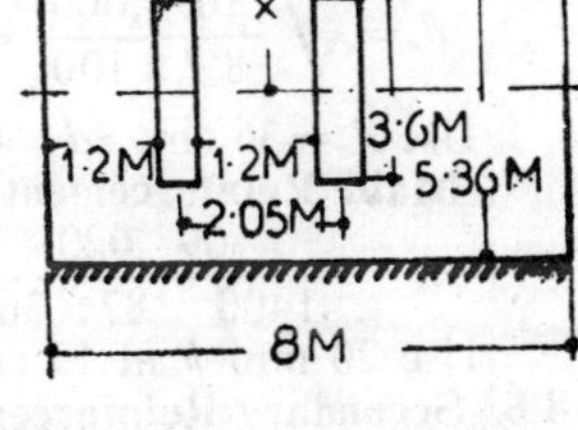

x = the distance of the centre of gravity of the concentrated load from the near support

$$x=\frac{5\cdot36}{2}=2\cdot68 \text{ m}$$

W = breadth of concentration area of the load + twice the thickness of wearing coat

$= 85+2\times8=101$ cm $=1\cdot01$ m

$$\frac{B}{L}=\frac{8}{5\cdot36}=1\cdot49$$

For a continuous slab:

For $\frac{B}{L}=1\cdot4$, $K=2\cdot48$ For $\frac{B}{L}=1\cdot5$, $K=2\cdot48$

For $\frac{B}{L}=1\cdot49$, $K=2\cdot48$

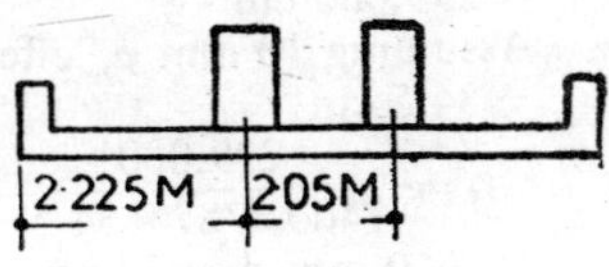

Effective width = e

$$=Kx\left(1-\frac{x}{L}\right)+W$$

$$=2\cdot48\times2\cdot68\left(1-\frac{2\cdot68}{5\cdot36}\right)+1\cdot01$$

$=2\cdot48\times2\cdot68\times0\cdot5+1\cdot01=3\cdot32+1\cdot01=4\cdot33$ m

Effective widths overlap: Therefore the effective width

$$=0\cdot6+1\cdot2+0\cdot85+1\cdot2+\frac{0\cdot85}{2}+\frac{4\cdot33}{2}$$

$=2\cdot225+2\cdot05+2\cdot165=6\cdot48$ m

6·8·2. Live Load Per Metre Width

Live load per metre width due to class-AA

$$\text{Vehicle}=\frac{2\times35}{6\cdot48}=10\cdot80 \text{ t}$$

6·8·3. Dispersion of load along the span

Load spread = width of track contract
+2×depth of slab including wearing coat

$=360+2\,(50+8)=360+116=476$ cm $=4\cdot76$ m

6·8·4· Live Load Bending

Maximum bending moment due to live load

$$=\frac{10\cdot80}{2}\left(\frac{5\cdot36}{2}-\frac{4\cdot76}{4}\right)=\frac{10\cdot80}{2}\,(2\cdot68-1\cdot19)$$

$$=\frac{10{\cdot}80}{2}\times 1{\cdot}49=8{\cdot}05 \text{ mt}=805{,}000 \text{ cm kg}$$

Due to continuity, since the slab is continuous this bending moment can be reduced to 80%

Bending moment $=805{,}000\times{\cdot}8=644{,}000$ cm kg

Impact factor $=0{\cdot}2365$

6·8·5. Bending Moment due to Impact

Bending moment due lo impact factor on Live load

$=0{\cdot}2365\times 644{,}000=153{,}000$ cm kg

Total bending moment $=297{,}000+644{,}00+153{,}000$

$=1094{,}000$ cm kg

6·8·6. Slab Thickness

$8{\cdot}7\ bd_e^2=1094{,}000$

$$d_e=\sqrt{\frac{1094{,}000}{8{\cdot}7\times 100}}=\sqrt{1250}=35{\cdot}5 \text{ cm}$$

Use $d_e=36$ cm, $d=40$ cm

6·8·7. Main Reinforcement

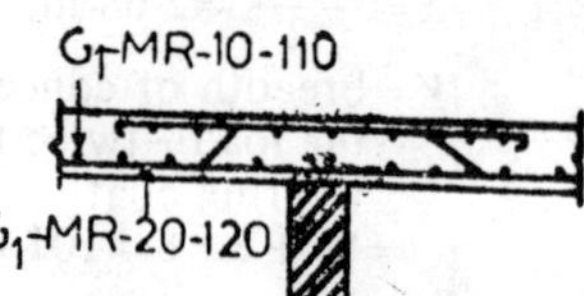

$$A_t=\frac{109\text{-},020}{1400\times{\cdot}87\times 36}=25{\cdot}2 \text{ cm}^2$$

Use 20 mm ϕ at 12 cm c/c

6·8·8. Secondary Reinforcement

Steel required to resist 0·3 times the L.L bending moment and 0·2 times the D.L bending moment shall be provided as distribution steel.

Bending moment$=0{\cdot}3(644{,}000+153{,}000)+0{\cdot}2\times 297{,}000$

$=0{\cdot}3\times 279{,}000+0{\cdot}2\times 297{,}000=239{,}600+59{,}400$

$=299{,}000$ cm kg

Assuming 10 mm ϕ, effective depth available$=40-(3+2+0{\cdot}5)$

$=34.5$ cm

$$A_t=\frac{299{,}000}{1400\times{\cdot}87\times 34{\cdot}5}=7{\cdot}1 \text{ cm}^2$$

Use 10 mm ϕ at 11 cm c/c

6·8·9. Shear Force

Maximum shearforce occurs when the load is near to the support

$$x_1=\frac{4{\cdot}76}{2}=2{\cdot}38 \text{ m}$$

Effective widths

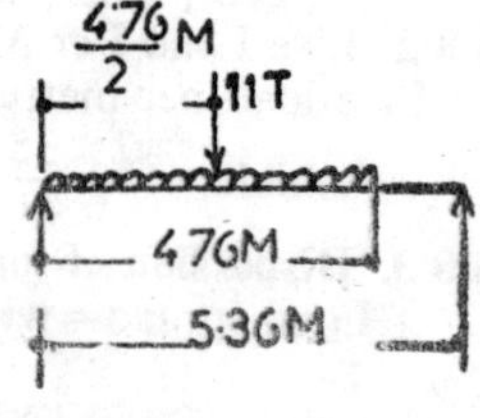

$$=e_1=Kx_1\left(1-\frac{x_1}{2}\right)+W$$

$$=2{\cdot}48\times 2{\cdot}38\left(1-\frac{2{\cdot}38}{5{\cdot}36}\right)+1{\cdot}01$$

$=2{\cdot}48\times 2{\cdot}38\times{\cdot}555+1{\cdot}01$

$=3{\cdot}27+1{\cdot}01=4{\cdot}28$ m

Effective widths overlap:

$$\text{Effective width}=\frac{4{\cdot}28}{2}+2{\cdot}05+\frac{4{\cdot}28}{2}$$

$=2{\cdot}05+4{\cdot}28=6{\cdot}33$ m

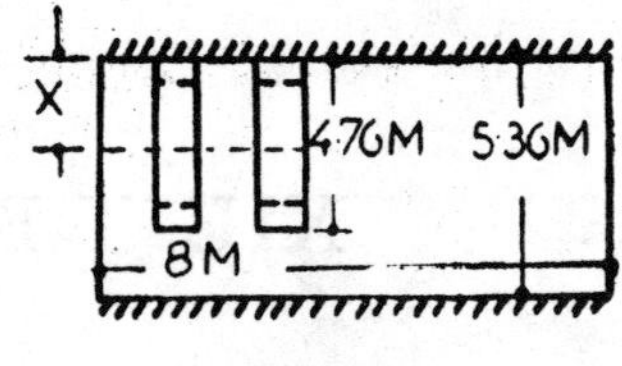

Lead per metre width

$$=\frac{2\times 35}{6\cdot 33}=11\cdot 0 \text{ t}$$

Maximum shear force

$$11\cdot 0\times\left(\frac{5\cdot 36-2\cdot 28)}{5\cdot 36}\right)$$

$$=11\cdot 0\times\frac{2\cdot 98}{5\cdot 36}=6\cdot 15 \text{ t}$$

$=6150$ kg

Shear force due to impact on live load $=0\cdot 2365\times 6150$
$=1450$ kg

Shear force due to dead load

$$=\frac{1036\times 5\cdot 36}{2}=2775 \text{ kg}$$

Total shear force
$=6150+1450+2775=10{,}375$ kg

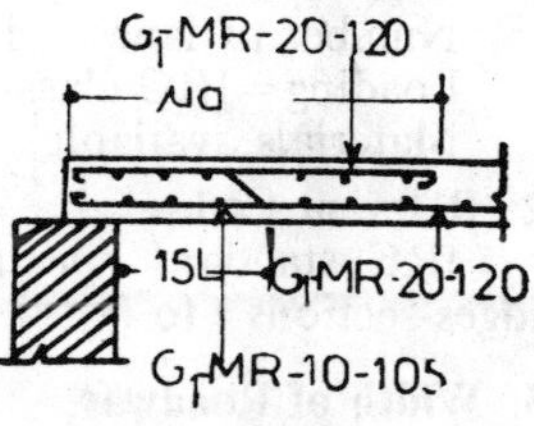

6·8·10. Shear Stress

$$\text{Shear stress}=\frac{20{,}375}{200\times\cdot 87\times 396}$$

$=3\cdot 3$ kg/cm$^2<5$ kg/cm^2

6·8·11. Bond Stress

Assuming that alternate bars are bent up at support, bond stress

$$=\frac{10\cdot 375\times 24}{6\cdot 25\times 100\times\cdot 87\times 36}=12\cdot 8 \text{ kg/cm}^2>10 \text{ kg/cm}^2$$

6·8·12. Additional Bars at end Support

Provide additional bars at end support to reduce bond stress and to take up negative moment, use 20 mm ϕ at 12 cm c/c

XII. T-BEAM BRIDGES

Single-Lane IRC-Class-A T-Beam Bridge 1

1.1. Data

Clear span=10 m, Bearing on either side=60 cm,
Number of lanes= single Width of safety kerbs=60 cm,
Loading=IRC class-A: Wearing coat=8 cm (average),
Materials available, Concrete M 150 grade, steel grade—1

1·2. Relevant Codes

IRC standard specifications and code of practice for road bridges-sections *i* to *iv*

1·3. Width of Roadway

Overall width of roadway$=3{\cdot}8+1{\cdot}2=5$ m

1·4. Impact Factor Allowed

$$\text{Impact factor allowed}=\frac{4{\cdot}5}{6{\cdot}0+L}$$

For spans<3 m Impact factor$=0{\cdot}5$

1·5. Characteristic strengths

$\sigma_{cb}=50$ kg/cm^2 $m=18$
$\sigma_{st}=1400$ kg/cm^2 $jd=0{\cdot}87\,d$
$q=5$ kg/cm^2 $R=8{\cdot}7$
$\sigma_{bl}=10$ kg/cm^2

1·6. Deck slab Trial Arrangement of Ribs

Use three ribs with 1·75 m c/c and a overhang of 0·75 m on eitherside. Let, rib width=30 cm

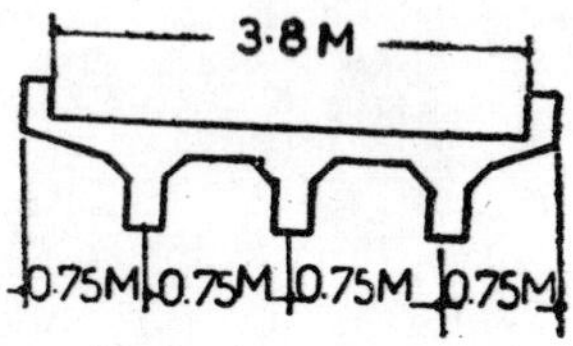

Clear span$=1{\cdot}75-{\cdot}30=1{\cdot}45$ mt between ribs

Using fillets, the clear span can be taken as the effective span,

Effective span=1·45 m

3·8 M
0.75M 0.75M 0.75M 0.75M

1·7. Slab Thickness

Thickness of slab ($\frac{1}{8}$ span)$=\frac{145}{8}=18$ cm

1·8. Dead Loads

$$\text{Weight of slab (18 cm)}=\frac{18}{100}\times1\times1\times2400=432\text{ kg/m}^2$$

$$\text{Weight of wearing coat}=\frac{8}{100}\times1\times1\times2200=176\text{ kg/m}^2$$

Total dead load=608 kg/m^2

1·9. Dead load Bending Moment

Dead load bending moment $=\frac{wl^2}{10}=\frac{608\times1{\cdot}45^2}{10}=128$ m kg $=12{,}800$ cm kg

Live loads: The slab spans in one direction only, the width of slab is more than three times the span

Pigeaud's method of distribution of live load can be applied to determine the dispersion of the wheel loads. Maximum BM will occur when the heaviest axle load are on the span.

A=wheel contact in the direction of transverse span, a=50 cm

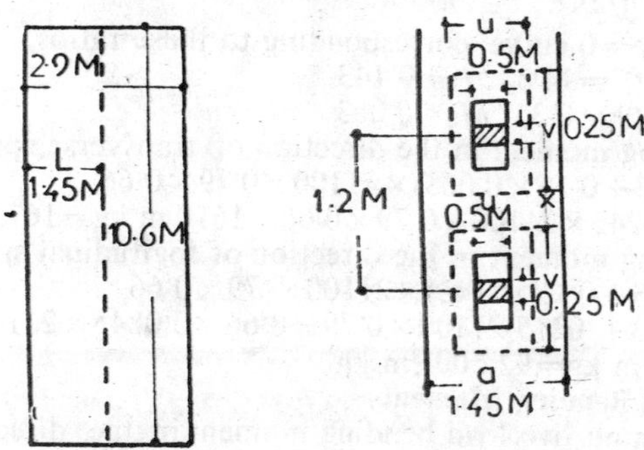

B=wheel contact in the direction of longitudinal span, b=25 cm

D=thickness of wearing coat=8 cm

$$\rho=\frac{a}{b}=\frac{1{\cdot}45}{\infty}=0$$

u=width of load spread in transverse direction$=A+2D$ $=50+2\times8=66$ cm$=0{\cdot}66$ m

v=width of load spread in longitudinal direction$=B+2D$ $=25+2\times8=41$ cm$=0{\cdot}41$ m

Magnitude of the wheel load $=\frac{11{\cdot}4}{2}=5{\cdot}7$ t$=5700$ kg

Intensity of load $=\frac{5700}{u\times v}=\frac{5700}{0{\cdot}66\ \ 0{\cdot}41}=21{,}100$ kg/m^2

1·10. Moments on the whole Area of u and 2v+x

$$\frac{u}{a}=\frac{0{\cdot}66}{1{\cdot}45}=0{\cdot}455$$

$$\frac{2v+x}{a}=\frac{2\times0{\cdot}41+0{\cdot}79}{1{\cdot}45}=\frac{1{\cdot}61}{1{\cdot}45}=1{\cdot}11$$

From $\rho=0$ curve, corresponding to these ratios

$M_1\times10^2=11{\cdot}9,\quad M_1=0{\cdot}119$

$M_2\times10^2=2{\cdot}5,\quad M_2=0{\cdot}025$

Bending moment on the direction of transverse span

$\doteq(M_1+0{\cdot}15\,M_2)\times$ intensity of load $\times$ area

$=(0{\cdot}119+0{\cdot}15\times{\cdot}025)\times21100\times(2v+x)\times u$

$=(0{\cdot}119+0{\cdot}15\times{\cdot}025)\ 21100\times1{\cdot}61\times0{\cdot}66$

$=(0{\cdot}119+{\cdot}00375)\times21100\times1{\cdot}61\times0{\cdot}66$

$=0{\cdot}12275\times21100\times1{\cdot}61\times0{\cdot}66=2750$ m kg$=275{,}000$ cm kg

Bending moment in the direction of longitudinal span
$=(M_2+0{\cdot}15\ M_2)\times$ intensity of load $\times$ area
$=(\cdot 025+0{\cdot}15\times\cdot 119)\ 2100\times 1{\cdot}61\times 0{\cdot}66$
$=(\cdot 025+\cdot 0178)21100\times 1{\cdot}61\times 0{\cdot}66=\cdot 0428\times 21100\times 1{\cdot}61\times 0{\cdot}66$
$=955$ m kg$=95{,}500$ cm kg

1·10·1. Moments on the Area x and u

$$\frac{u}{a}=\frac{0{\cdot}66}{1{\cdot}45}=0{\cdot}455$$

$$\frac{x}{a}=\frac{0{\cdot}79}{1{\cdot}45}=0{\cdot}545$$

From $\rho=0$ curve, corresponding to these ratios,
$M_1\times 10^2=14{\cdot}3,\ M_1=0{\cdot}143$
$M_2\times 10^2=6{\cdot}3,\quad M_2=0{\cdot}063$
Bending moment in the direction of transverse span
$=(\cdot 143+0{\cdot}15\times 0{\cdot}063)\times 21100\times 0{\cdot}79\times 0{\cdot}66$
$=0{\cdot}15245\times 21100\times 0{\cdot}79\times 0{\cdot}66=1670$ m kg$=167{,}000$ cm kg
Bendihg moment in the direction of logitudinal span
$=(\cdot 063+0{\cdot}15\times\cdot 143)\times 21100\times\cdot 79\times 0{\cdot}66$
$=(\cdot 063+\cdot 0215)21100\times 0{\cdot}79\times 0{\cdot}66\quad 0{\cdot}0845\times 21100\times 0{\cdot}79\times 0{\cdot}66$
$=925$ m kg$=92{,}500$ cm kg

1·10·2. Net Bending Moments

Maximum live load bending moment in the direction of transverse span$=275000-167000=108{,}000$ cm kg

Maximum live load bending moment in the direction of longitudinal span$=95{,}500-92{,}500$ cm kg$=3000$ cm kg

1·10·3. Bending Moment Due to Impact

Impact factor$=0{\cdot}5$ for spans less than 3 m

Bending moment due to impact on live load in the direction of transverse span$=0{\cdot}5\times 108{,}000=54{,}000$ cm kg

Since the slab is continuous over *T*-beam ribs, the bending moment can be reduced to 80%. Bending moment due to live load including impact$=(108{,}000+54000)\times 0{\cdot}8=162{,}000\times 0{\cdot}8$
$=129{,}600$ cm kg

Total bending moment due to DL, LL and impact
$=12{,}800+129{,}600=142{,}400$ cm kg

1·10·4. Thickness of Deck Slab

$8{\cdot}7\ bd_e^2=142{,}400$

$$d_e\ \sqrt{\frac{142{,}400}{100\times 8{\cdot}7}}=\sqrt{164}=12{\cdot}8\text{ cm.}$$

For highway bridges, 17·5 cm is considered to be a minimum for deck slab subjected to impact loads.

Use $d_e=14$ cm, $d=18$ cm

1·10·5. Main Steel

$$A_t=\frac{142{,}400}{1400\times .87\times 14}$$
$=8{\cdot}4$ cm^2

Use 12 mm ϕ at 13 cm c/c

Negative bending moment at end rib$=\dfrac{wl^2}{16}$

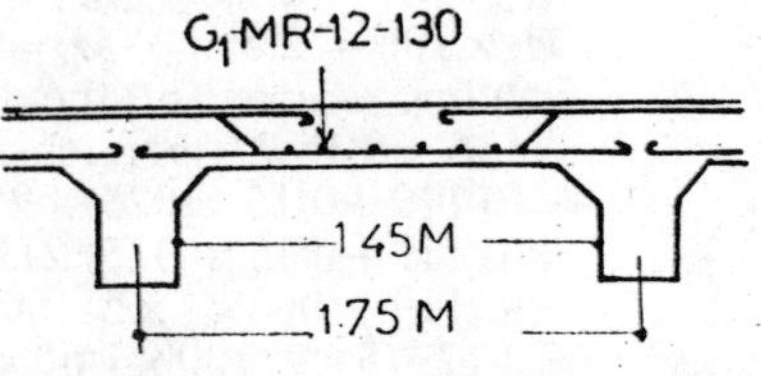

If half the bars are bent up, bending moment resisted by the bent up bars $= \frac{wl^2}{20}$

Remaining bending moment $= \frac{wl^2}{16} - \frac{wl^2}{20} = \frac{wl^2}{80}$

Additional steel required to resist this bending moment

$\frac{A_t}{8} = \frac{8{\cdot}4}{8} = 1{\cdot}05\ cm^2$

Use 6 mm ϕ at 13 cm c/c

1·10·6. Secondary Steel

Secondary reinforcement is provided to resist a moment equal to
$= 0{\cdot}3$ live load moment $+ 0{\cdot}2 \times$ dead load moment
$= 0{\cdot}3 \times 129{,}600 + 0{\cdot}2 \times 12{,}800 = 38{,}880 + 2560 = 41{,}440$ cm kg

Using 8 mm ϕ bars

Effective depth available $= 18 - (3{\cdot}4 + 1{\cdot}2 + 0{\cdot}4) = 13$ cm

$A_t = \frac{41440}{1400 \times {\cdot}87 \times 13} = 2{\cdot}62\ cm^2$

Use 8 cm ϕ at 19 cm c/c

Bending moment in the direction of longitudinal span is very small and the secondary steel provided is more than ample.

1·11. T-Beam

Clear span $= 10$ m

Effective span (centre to centre of bearing) $= 10{\cdot}6$ m

1·11·1. Trial Section

Depth of beam ($\frac{1}{10}$ of span)

$= \frac{1060}{10} = 106$ cm

Depth of rib $= 106 - 18 = 88$ cm

Use 15 cm × 15 cm fillets

Width of rib ($\frac{1}{3}$ depth of rib)

$= 30$ cm

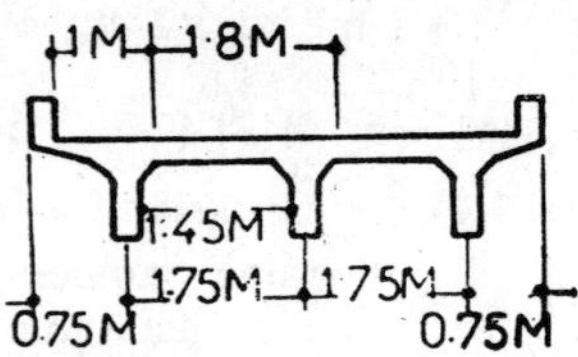

1·11·2. Dead Loads

Deck slab (18 cm) $= \frac{18}{100} \times 1{\cdot}75 \times 2400 = 755$ kg/m

Wearing coat (8 cm) $= \frac{8}{100} \times 1{\cdot}75 \times 2200 = 306$ kg/m

T-beam rib $= \frac{30}{100} \times \frac{88}{100} \times 1 \times 2400 = 632$ kg/m

Fillets $= \frac{1}{2} \times \frac{15}{100} \times \frac{15}{100} \times 2 \times 2400 = 54$ kg/m

Total dead loads $= 755 + 306 + 632 + 54 = 1747$ kg/m

1·11·3. Dead Load Bending Moment

Maximum bending moment due to dead loads

$= \frac{1747 \times 10{\cdot}6^2}{8} = 24{,}600$ m kg

1·11·4. Live Loads

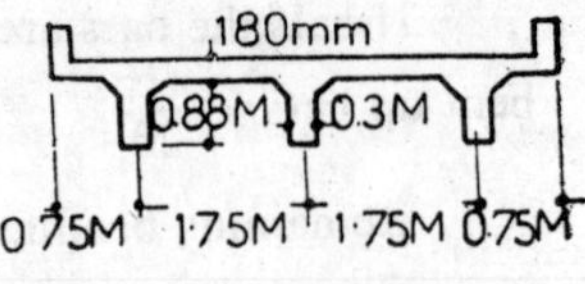

With minimum clearance f=15 cm and with the proposed arrangement of ribs the wheel loads will not come directly on the ribs. The intermediate beam is stressed slightly more when compared to the outer beams, as the reaction due to wheel loads on the intermediate beam is slightly greater.

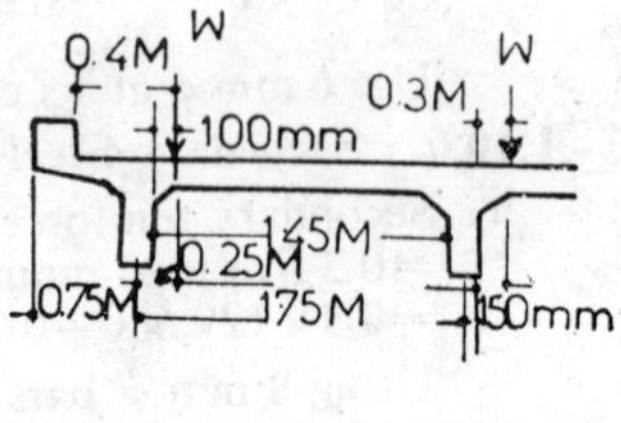

Let W be the magnitude of the wheel load, for the arrangement, the reaction due to these loads on intermediate

$$\text{beam}=\frac{W\times(1{\cdot}75-{\cdot}30)}{1{\cdot}75}\times W\times\frac{{\cdot}25}{1{\cdot}75}$$

$$=W\left(\frac{1{\cdot}45}{1{\cdot}75}+\frac{{\cdot}25}{1{\cdot}75}\right)$$

$$=\frac{W}{1{\cdot}75}\times 1{\cdot}70=0{\cdot}97\,W$$

Each axle load is $=2\,W$

Reaction due to axle load $=0{\cdot}485$ timse the axle load

Class A Load System

(A)	(B)	(C)	(D)	(E)	(F)	(G)	(H)
2·7 t	2·7 t	11·4 t	11·4 t	6·1 t	6·8 t	6·8 t	6·8 t
1·1 m	3·2 m	1·2 m	4·3 m	3 m	3 m	3 m	

Load System to be Considered

(A)	(B)	(C)	(D)	(E)	(F)	(G)	(H)
1·1 m	3·2 m	1·2 m	4·3 m	3 m	3 m	3 m	
1·31 t	1·31 t	5·54 t	5·54 t	3·3 t	3·3	3·3 t	3·3 t

1·11·5. Criterion for Maximum Bending Moment

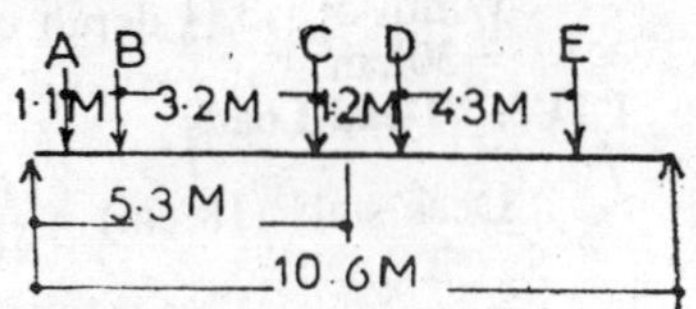

Bending moment under any load X is maximum when that load and the C.G. of the whole system of loads on the span are equidistant from mid-span.

If it is assumed that maximum bending moment occurs under load D only loads B, C, D and E will be on the span.

B C D E
3.2M 1.2M 4.3M
5.3M
10.6M

C.G. of the load system taking moments about B

$$=\frac{5{\cdot}54\times3{\cdot}2+5{\cdot}54\times4{\cdot}4+3{\cdot}3\times8{\cdot}7}{1{\cdot}31+5{\cdot}54+5{\cdot}54+3{\cdot}3}$$

$$=\frac{70{\cdot}7}{15{\cdot}69}=4{\cdot}5\text{ m}$$

$$\text{Reaction } R_2=\frac{15{\cdot}69\times5{\cdot}35}{10{\cdot}5}=7{\cdot}94\text{ t}$$

Bending moment under load $D=R_2\times5{\cdot}35-3{\cdot}3\times4{\cdot}3$

$=7{\cdot}94\times5{\cdot}35-3{\cdot}3\times4{\cdot}3=42{\cdot}5-14{\cdot}2=28{\cdot}3$ mt

1·11·6. Bending Moment Due to Impact on live Load

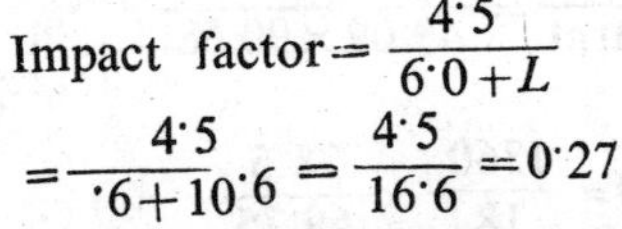

$$\text{Impact factor} = \frac{4{\cdot}5}{6{\cdot}0+L}$$

$$= \frac{4{\cdot}5}{{\cdot}6+10{\cdot}6} = \frac{4{\cdot}5}{16{\cdot}6} = 0{\cdot}27$$

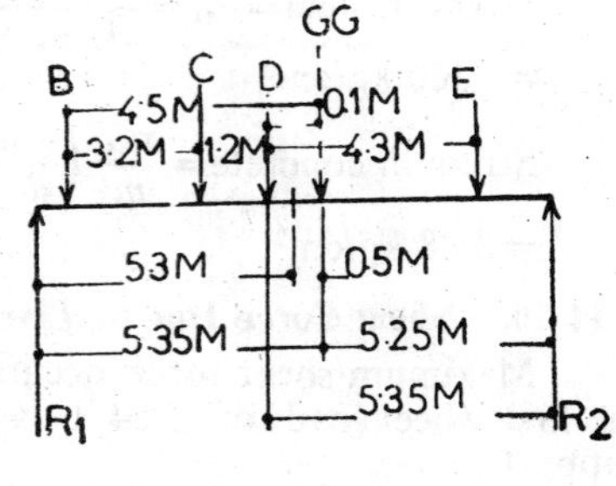

Bending moment due to impact

$= 0{\cdot}27 \times 28{\cdot}3 = 7{\cdot}62$ mt

Total bending moment due to *DL LL* and impact

$= 24{\cdot}6 + 28{\cdot}3 + 7{\cdot}62$

$= 60{\cdot}52$ mt

1·11·7. Design of T-Beam

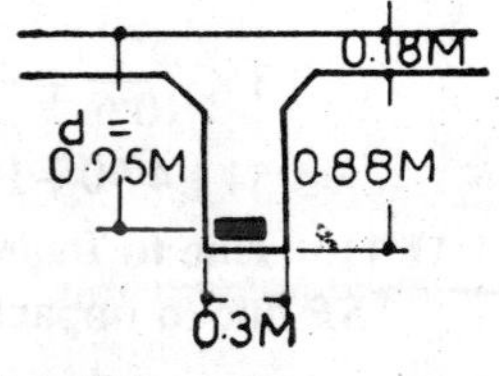

Total deapth of assumed section = 106 cm

Approximate effective depth = 95 cm

Approximate lever arm

$$= d - \frac{ds}{2} = 95 - \frac{18}{2} = 95 - 9 = 86 \text{ cm}$$

$$\text{Approximate } A_t = \frac{6052{,}000}{1400 \times 86} = 50 \text{ cm}^2$$

Use 10 bars of 25 mm ϕ (49·07 cm²) in two layers

Effective depth available

$$= 106 - \left(4 + 2{\cdot}5 + \frac{2{\cdot}5}{2}\right) = 106 - 7{\cdot}75 = 98{\cdot}25 \text{ cm}$$

1·11·8. Effective Width of Flange

The effective width is the least of

(*i*) $\frac{1}{3}$ span $= \frac{10{\cdot}6}{3} = 3{\cdot}53$ m $= 353$ cm,

(*ii*) centre to centre of ribs $= 1{\cdot}75$ m $= 175$ cm

(*iii*) $12\, d_s + br = 12 \times 18 + 30 = 216 + 30 = 246$ cm

Therefore the effective width of flange = 175 cm

Assuming *NA* to be below the flange

$$B \times d_s\left(n - \frac{d_s}{2}\right) = m \times A_t\,(d-n)$$

$$175 \times 18(n-9) = 18 \times 49{\cdot}09 \times (98{\cdot}25 - n)$$

$$175 \times 18n - 175 \times 18 \times 9 = 18 \times 49{\cdot}09 + 98{\cdot}25 - 18 \times 49{\cdot}09\, n$$

$$3150\, n - 28.400 = 86700 - 885\, n$$

$$4035\, n = 114{,}100$$

$$n = 28{\cdot}5 \text{ cm}$$

$$\sigma_{cb}' = \frac{n-ds}{n} \times \sigma_{cb} = \frac{(28{\cdot}5-18)}{28{\cdot}5} \times \sigma_{cb} = \frac{10{\cdot}5}{28{\cdot}5}\sigma_{cb} = {\cdot}368\ \sigma_{cb}$$

$$\bar{y} = \frac{\sigma_{cb} + 2\,\sigma_{cb}'}{\sigma_{cb} + \sigma_{cb}'}\ \frac{ds}{3} = \frac{\sigma_{cb} + 2 \times 0{\cdot}368\ \sigma_{cb}}{\sigma_{cb} + 0{\cdot}368\ \sigma_{cb}} \times 6$$

$$= \frac{1{\cdot}736}{1{\cdot}368} \times 6 = 7{\cdot}67 \text{ cm}$$

Actual lever arm $= d - \bar{y} = 98{\cdot}25 - 7{\cdot}67 = 90{\cdot}58$ cm

1·11·9. Stresses

Stress in steel $=\sigma_{st}=\dfrac{BM}{A_t \times \text{lever arm}}=\dfrac{6052000}{49{\cdot}09 \times 90{\cdot}58}$

$=1360$ kg/cm²

Stress in concrete $=\dfrac{\sigma_{st}}{m}\left(\dfrac{n}{d}\ \dfrac{}{n}\right)=\dfrac{1360}{18}\times\dfrac{28{\cdot}5}{69{\cdot}75}$

$=30{\cdot}8$ kg/cm²

1·11·10. Shear Force Due to Live Loads

Maximum shear force occurs when the first wheel load of 5·54 t is on the support

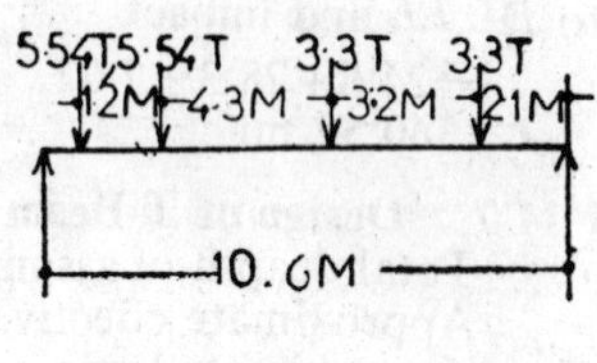

$SF=5{\cdot}54+5{\cdot}54\times\dfrac{(10{\cdot}6-1{\cdot}2)}{10{\cdot}6}$

$+3{\cdot}3\times\dfrac{5{\cdot}1}{10{\cdot}6}+\dfrac{3{\cdot}3\times 2{\cdot}1}{10{\cdot}6}$

$=5{\cdot}54+4{\cdot}90+1{\cdot}585+0{\cdot}655=12{\cdot}7$ t

1·11·11. Due to Impact

SF due to impact on Live Load $=12{\cdot}7\times 0{\cdot}27=3{\cdot}43$ t

1·11·12. Due to Dead Load

SF due to dead load $=\dfrac{1{\cdot}747\times 10{\cdot}6}{2}=9{\cdot}3$ t

Total $SF=12{\cdot}7+3{\cdot}43+9{\cdot}30=25{\cdot}300$ t

1·11·13. Shear Stress

Shear stress $=\dfrac{Q}{b\times ja}=\dfrac{25300}{30\times 90{\cdot}58}=9{\cdot}25$ kg/cm² $<$ 5 kg/cm³

1·11·14. Shear Force that Concrete Can Resist

Shear force that concrete can take

$=5\times 90{\cdot}58\times 30=13600$ kg $=13{\cdot}60$ t

1·11·15. Shear Force at 0·15 S (1·59 m)

Live load : SF at $(0{\cdot}15\ L=0{\cdot}15\times 10{\cdot}6)=1{\cdot}59$ m

$=\dfrac{5{\cdot}14}{10{\cdot}6}(10{\cdot}6-1{\cdot}59)+\dfrac{5{\cdot}14}{10{\cdot}6}\times(10{\cdot}6-1{\cdot}59-1{\cdot}2)$

$+\dfrac{3{\cdot}3}{10{\cdot}6}\times(10{\cdot}6-1{\cdot}59-1{\cdot}2-4{\cdot}3)+\dfrac{3{\cdot}3\times 0{\cdot}51}{10{\cdot}6}$

$=\dfrac{5{\cdot}14\times 9{\cdot}01}{10{\cdot}6}+\dfrac{5{\cdot}14}{10{\cdot}6}\times 7{\cdot}81+\dfrac{3{\cdot}3}{10{\cdot}6}\times 3{\cdot}51+\dfrac{3{\cdot}3\times 0{\cdot}51}{10{\cdot}6}$

$=4{\cdot}38+3{\cdot}80+1{\cdot}1+{\cdot}16=9{\cdot}44$ t

SF due to impact on live load

$=0{\cdot}27\times 9{\cdot}44=2{\cdot}56$ t

SF due to dead load

$=9{\cdot}3-1{\cdot}747\times 1{\cdot}59=6{\cdot}51$ t

Total SF at 1·59 m from support

$=9{\cdot}44+2{\cdot}56+6{\cdot}51=18{\cdot}51$ t

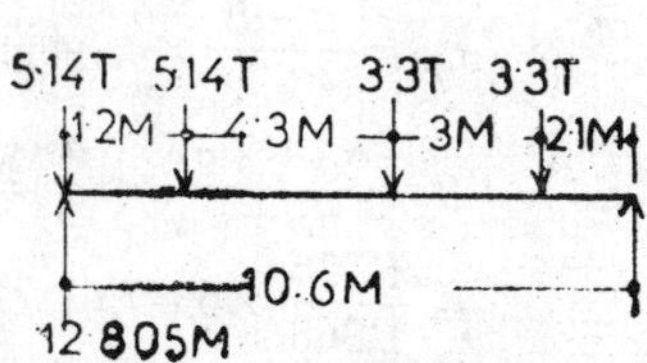

Shear Force at $(0{\cdot}15\ L \times \frac{3}{4}\ d = 2{\cdot}25\ m)$

Distance $= 0{\cdot}15 \times 10{\cdot}6 + \frac{3}{4}$ effective depth

$= 1{\cdot}59 + \frac{3}{4} \times 98{\cdot}25$

$= 1{\cdot}59 + {\cdot}737 = 2{\cdot}327$ m

Say at 2·25 m

SF due to live load

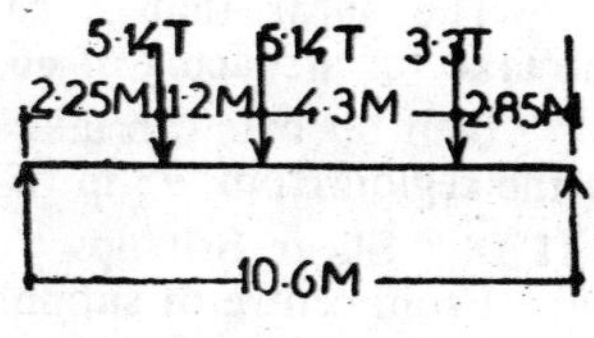

$$= \frac{5{\cdot}14}{10{\cdot}6} \times 8{\cdot}35 + \frac{5{\cdot}14}{10{\cdot}6} \times 7{\cdot}15 + \frac{3{\cdot}3}{10{\cdot}6} \times 2{\cdot}85$$

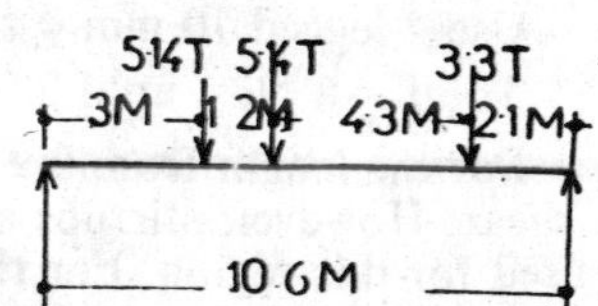

$= 4{\cdot}05 + 3{\cdot}47 + {\cdot}885 = 8{\cdot}405$ t

SF due to impact on live load

$= 0{\cdot}27 \times 8{\cdot}405 = 2{\cdot}27$ t

SF due to dead load $= 9{\cdot}44 - 1{\cdot}747 \times 2{\cdot}25 = 9{\cdot}43 - 3{\cdot}94 = 5{\cdot}50$ t

Total shear Force $= 8{\cdot}405 + 2{\cdot}27 + 5{\cdot}50 = 16{\cdot}175$ t

Shear Foree at $(0{\cdot}15\ L + 1 \times \frac{3}{4}\ d = 3\ m)$

Distance $= 0{\cdot}15\ L + 2 \times \frac{3}{4}\ d$

$= 1{\cdot}59 + 0{\cdot}737 \times 2 = 3{\cdot}06$ m

Say at 3 m

SF due to live load

$$= \frac{5{\cdot}14}{10{\cdot}6} \times 7{\cdot}60 + \frac{5{\cdot}14}{10{\cdot}6} \times 5{\cdot}4 + \frac{3{\cdot}3}{10{\cdot}6} \times 2{\cdot}10$$

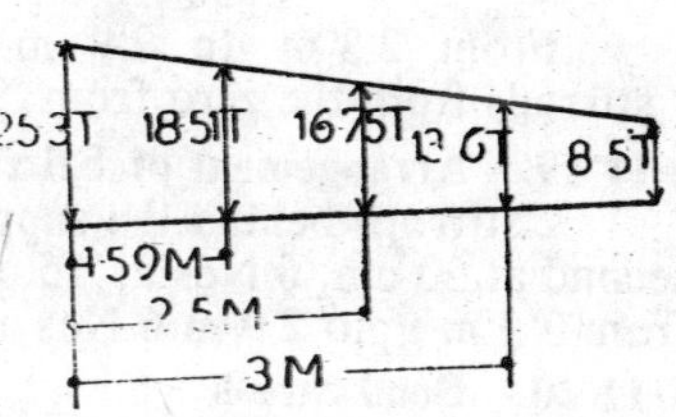

$= 3{\cdot}68 + 2{\cdot}62 + 0{\cdot}65 = 6{\cdot}95$ t

SF due to impact on live load

$= 0{\cdot}27 \times 6{\cdot}95 = 1{\cdot}875$ t

SF due to dead load

$= 9{\cdot}3 - 1{\cdot}747 \times 3 = 9{\cdot}3 - 5{\cdot}24$

$= 4{\cdot}06$ t

Total shear force $= 12{\cdot}885$ t

1·11·16. **Distanee from the end beyond which stirrups are not required**

Assuming uniform variation of shear between 2·25 m and 3 m from the support the shear of 13·67 t occurs at

$$= 2{\cdot}25 + \frac{16{\cdot}750 - 13{\cdot}61}{16{\cdot}750 - 12{\cdot}885} \times 0{\cdot}75 = 2{\cdot}25 + \frac{3{\cdot}14}{3{\cdot}865} \times 0{\cdot}75$$

$= 2{\cdot}25 + {\cdot}596 = 2{\cdot}846$ m

Theoratically stirrups are required over a length of 2·846 m from support, beyond this concrete can resist the shear. However, in practice stirrups are provided throughout the length of span.

1·11·17. Shear Resisted By Bentup Bars

The top layer of main reinforcement bars are bent up.

Three bars of 25 mm ϕ are bent up at 45° at a distance 1·59 m.

Two bars of 25 mm ϕ are bent up at 45° at a distance 2·25 m.

The spacing of the bent up bars along the length of the beam is 66 cm.

These bent up bars will take shear in the region 0·93 to 2·25 m from the support.

The shear that 3 bars of 25 mm ϕ cantake for a spacing of 66 cm

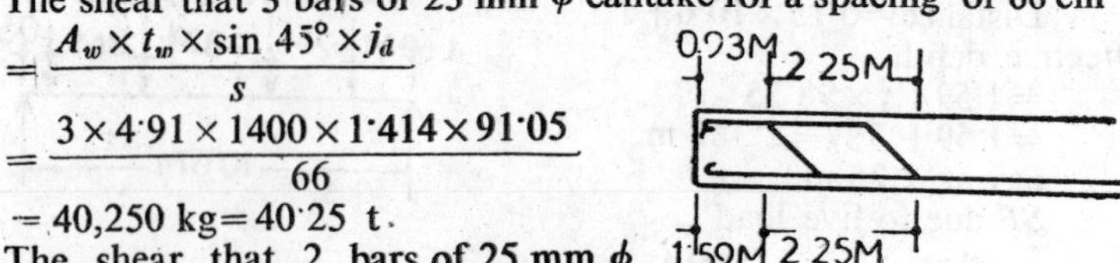

$$= \frac{A_w \times t_w \times \sin 45° \times j_d}{s}$$

$$= \frac{3 \times 4{\cdot}91 \times 1400 \times 1{\cdot}414 \times 91{\cdot}05}{66}$$

$= 40{,}250$ kg $= 40{\cdot}25$ t.

The shear that 2 bars of 25 mm ϕ can take for a spacing of 66 cm $= \frac{2}{3} \times 40{\cdot}25 = 26{\cdot}8$ t

Bent up bars can take all the shear in the region from ·93 m to 2·25 m

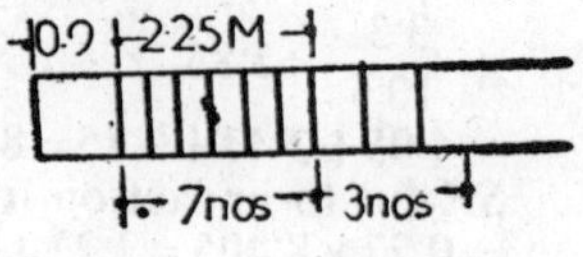

1·11·18. Shear Stirrups

From centre of support to 0·93 m

Using 4 legged – 10 mm ϕ stirrups

$$\text{Spacing} = \frac{A_w \, t_w \, j_d}{Q} = \frac{4 \times 0{\cdot}79 \times 1400 \times 9105}{25300} = 15{\cdot}9 \text{ cm}$$

Use 4-legged 10 mm ϕ stirrups at 15 cm c/c

Total —8 Nos, upto 0·9 m

For the length from 0·9 m to 2·25 m bent up bars will take all the shear. However; stirrups at spacing required beyond 2·25 m shall be used for this region. For the length from 2·25 m to 2·85 m

Spacing of 4 legged 10 mm ϕ stirrups

$$= \frac{4 \times 0{\cdot}79 \times 1400 \times 91{\cdot}05}{16750} = 24{\cdot}1 \text{ cm}$$

From 2·3 m to 2·9 m use three stirrups at 20 cm c/c and 7 stirrups from the zero from ·9 m to 2·3 m.

1·11·19. Arrangement of Stirrups

2 stirrups behind the support, first one at 5 cm from end and the second at 20 cm, 6 Nos at 15 cm c/c upto 0·9 m, 10 Nos at 20 cm c/c from 0·9 m upto 2·9 m 6 Nos at 40 cm c/c from 2·9 m upto 5·3 m

1·11·20. Bond Stress

$$\text{Bond Stress} = \frac{25300}{5 \times 7{\cdot}85 \times 91{\cdot}05} = 7{\cdot}07 \text{ kg/cm}^2 < 10 \text{ kg/cm}^2$$

1·11·21. Negative BM

$$\text{Negative bending moment} = \frac{wl^2}{16}$$

$$\text{Amount of steel} = \frac{49{\cdot}09}{2} = 24{\cdot}54 \text{ cm}^2$$

Five bent up bars give $A_t = 24{\cdot}54$ cm²

Four hanger bars 12 mm $\phi = 1{\cdot}13 \times 4 = 4{\cdot}52$ cm²

Total steel provided $= 25{\cdot}54 + 4{\cdot}52 = 30{\cdot}06$ cm²

Therefore no additional steel is required at top.

T-Beam Bridge Two Lane IRC Class-A Loading 2

2·1. Data

Clear span = 10 m, Bearing for the beam = 0·60 cm
Number of lanes = two, Width of safety kerbs = 60 cm
Loading = IRC class-A, Wearing coat = 8 cm (average)
Materials available, Concrete M 150 grade, steel grade—1

2·2. Relevant Codes

IRC standard specifications and code of practice for road bridges sections (*i*) to (*iv*)

2·3. Width of Roadway

Roadway = 3·8 + 3 + 1·2 = 8 m

2·4. Impact Factor

$$\text{Impact factor allowed} = \frac{4\cdot5}{6\cdot0 + L}$$

For spans < 3 m, the impact factor is = 0·5

2·5. Characteristic strengths

$\sigma_{cb} = 50$ kg/cm², $\sigma_{st} = 1400$ kg/cm²
$m = 18$, $j_d = 0\cdot87$ d
$q = 5$ kg/cm², $R = 8\cdot7$
$\sigma_{bt} = 10$ kg/cm²

2·6. Deck slab trial arrangement of Ribs

Use 4 ribs with 2 m c/c and an overhang of 1 m on either side

Assume width of rib = 35 cm,
Clear span = 2·00 0·35 = 1·65 m

Using 15 cm × 15 cm fillets clear span can be taken as effective span

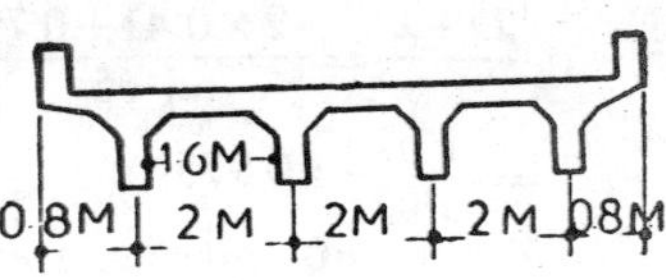

2·6·1. Slab Thickness

Assumed thickness of slab ($\frac{1}{8}$ span) = 20 cm

2·6·2. Dead Loads

Weight of slab (20 cm)

$$= \frac{20}{100} \times 1 \times 1 \times 2400 = 480 \text{ kg/m}^2$$

Weight of wearing coat (8 cm)

$$\frac{8}{100} \times 1 \times 1 \times 2200 = 176 \text{ kg/m}^2$$

Total dead load = 656 kg/m²

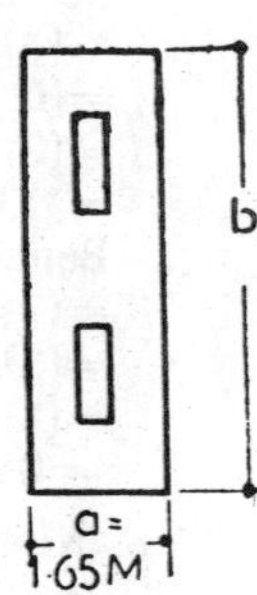

2·6·3. Dead load Bending Moment

Bending moment due to load

$$= \frac{wl^2}{10} = \frac{656 \times 1\cdot65^2}{10} = 178 \text{ m kg} = 17{,}800 \text{ cm kg}$$

2·6·4. Live Loads

IRC Loading class – A: The slab spans in one direction only since the width of slab is more than thrice the span

$=\dfrac{10·6}{1·65}>3$

Pigeaud's method of distribution of live load can be applied to determine the dispersion of the wheel loads. Maximum BM will occur when the heaviest axle loads are on the span

A = wheel contact in the direction of transverse span, $a=50$ cm
B = wheel contact in the direction of longitudinal span $b=25$ cm
D = thickness of wearing coat = 8 cm,

$\rho=\dfrac{a}{b}=\dfrac{1·65}{\infty}=0,$

u = width of load spread in transverse direction $=A+2D$
$=50+2\times8=66$ cm $=0·66$ m
v = width of load spread in longitudinal direction $=B+2D$
$=25+2\times8=41$ cm $=0·41$ m

Magnitude of the wheel load

$=\dfrac{11·4}{2}=5·7$ t = 5700 kg

Intensity of load

$=\dfrac{5700}{u\times v}=\dfrac{5700}{0·66\times0·41}$

$=21100$ kg/m²

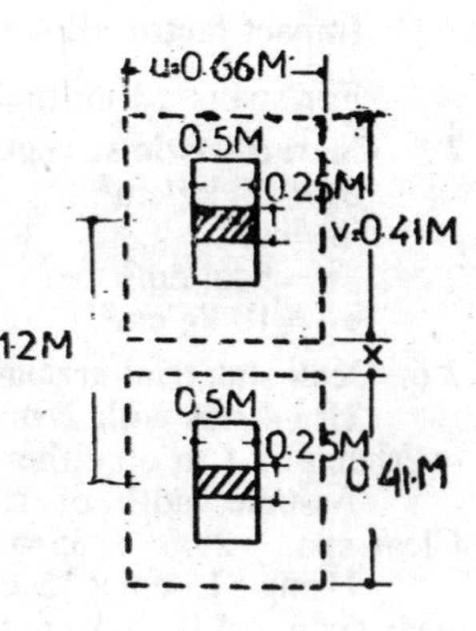

2·6·5. Moments on the whole area of u and 2v+x

$\dfrac{u}{a}=\dfrac{0·66}{1·65}=0·40$

$\dfrac{2v+x}{a}=\dfrac{2\times0·41+0·79}{1·65}$

$=\dfrac{1·61}{1·65}=0·975$

From $\rho=0$ curve corresponding to these ratios

$M_1\times10^2=12·8,\qquad M_1=0·128$
$M_2\times10^2=3·3.\qquad M_2=0·033$

Bending moment in the direction of transverse span
$=(M_1+0·15\,M_2)\times$ intensity of load $\times$ area
$=(0·128+·15\times0·033)\times21100\times(2v+x)\times u$
$=(0·128+·00495)\times21100\times(2\times·41+0·79)\times0·66$
$=·13295\times21100\times1·61\times0·66=2980$ m kg $=298000$ cm kg

Bending moment in the direction of longitudinal span
$=(M_2+0·15\,M_2)\times$ intensity of load $\times$ area
$=(·033+0·15\times0·128)\times21100\times1.61\times0·66$
$=(·033+·0192)\times21100\times1·61\times0·66$
$=·0522\times21100\times1·61\times0·66=1180$ m kg $=118,000$ cm kg

2·6·6. Moments on the area x and u

$\dfrac{u}{a}=\dfrac{0·66}{1·65}=0·40,\quad \dfrac{x}{a}=\dfrac{0·79}{1·65}=0·478$

From $\rho = 0$ curve, corresponding to these ratios
$M_1 \times 10^2 = 15{\cdot}4$, $M_1 = 0{\cdot}154$
$M_2 \times 10^2 = 7{\cdot}3$, $M_2 = 0{\cdot}073$
Bending moment in the direction of transverse span
$= (0{\cdot}154 + 0{\cdot}15 \times {\cdot}073) 21100 \times 0{\cdot}79 \times 0{\cdot}66$
$= 0{\cdot}165 \times 21100 \times 0{\cdot}79 \times 0{\cdot}66 = 1820$ m kg $= 182{,}000$ cm kg
Bending moment in the direction of longitudinal span
$= (0{\cdot}073 + 0{\cdot}15 \times 0{\cdot}154) \times 21100 \times 0{\cdot}79 \times 0{\cdot}66$
$= 1060$ m kg $= 106{,}000$ cm kg

2·6·7. Net Bending Moments

Maximum live load bending moment in the direction of transverse span $= 298{,}000 - 102{,}000 = 116{,}000$ cm kg

Maximum live load bending moment in the direction of longitudinal span $= 118{,}000 - 106{,}000 = 12{,}000$ cm kg

2·6·8. Bending Moments Due to Live Load on Impact

Impact factor upto 3 m span $= 0{\cdot}5$

Bending moment due to impact on live load in the direction of transverse span $= 0{\cdot}5 \times 116000 = 58{,}000$ cm kg

Since the slab is continuous over T-beam ribs, the bending moment can be reduced to 80%

Bending moment due to live load, including impact
$= (116{,}000 + 58{,}000) \times 0{\cdot}8 = 174{.}000 \times 0{\cdot}8 = 139200$ cm kg
Total bending moment due to *DL*, *LL* and impact
$= 17{,}800 + 139{,}200 = 157{,}000$ cm kg

2·6·9. Thickness of Deck Slab

$8\ 7\ bd_e^2 = 157{,}000$

$$d_e = \sqrt{\frac{157{,}000}{100 \times 8{\cdot}7}} = \sqrt{180} = 13{\cdot}4 \text{ cm}$$

For highway bridges 17·5 cm is considered to be a minimum for deck slab subjected to impact loads.

Use $d_e = 14$ cm and $d = 18$ cm
Which is less than the assumed thickness of 20 cm

2·6·10. Main Steel

$$A_t = \frac{157{,}000}{1400 \times 0{\cdot}87 \times 14} = 9{\cdot}2 \text{ cm}^2$$

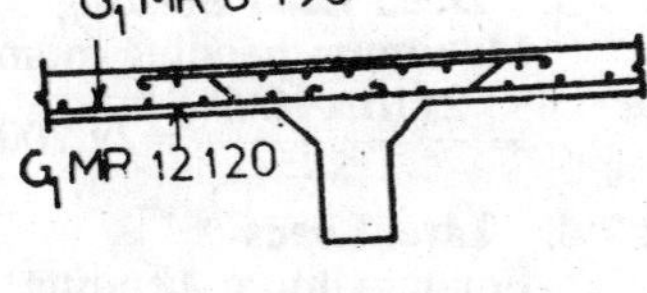

Use 12 mm at 12 cm c/c

Negative bending moment at end ribs $= \dfrac{wl^2}{16}$

If half the bars, are bent up, the bending moment resisted by these bars $= \dfrac{wl^2}{20}$

Additional steel required $= \dfrac{A_t}{8} = \dfrac{9{\cdot}2}{8} = 1{\cdot}15 \text{ cm}^2$

Use 6 mm ϕ bar in between the two bent up bars

2·6·11. Secondary Steel

Secondary reinforcement is provided to resist a moment equal to
$= 0{\cdot}3 \times$ live load moment $+ 0{\cdot}2 \times$ dead load moment
$= 0{\cdot}3 \times 139{,}200 + 0{\cdot}2 \times 17{,}800 = 38760 + 3560 = 42{,}320$ cm kg

Using 8 mm ϕ bars, effective depth available
$=18-(3{\cdot}4+1{\cdot}2+0{\cdot}4)=13$ cm

$$A_t=\frac{42{,}320}{1400\times{\cdot}87\times13}=2{\cdot}68\text{ cm}^2$$

Use 8 mm ϕ at 19 cm c/c

Bending moment in the direction of longitudinal span is very small and the secondary steel provided would take care of this bending moment.

2·7. T-beam

Clear pan=10 m

Effective span (centre to centre of bearing)=10·60 m

2·7·1. Trial Section

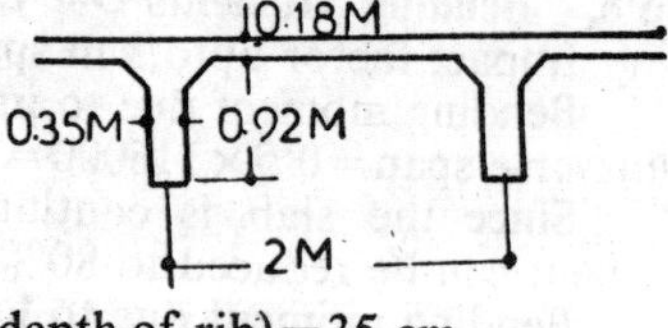

Depth of beam (approximately $\frac{1}{9}$ of span)$=\frac{1060}{9}=118$ cm,

Use depth of beam=118 cm,

Depth of rib=118−18=100 cm

Use fillets 15 cm×15 cm

Width of rib (approximately $\frac{1}{3}$ of depth of rib)=35 cm

Loads on intermediate *T*-Beam : Centre to centre of ribs=2 m

2·7·2. Dead Loads

Deck slab (18 cm)$=\frac{18}{100}\times2{\cdot}0\times2400=864$ kg/m

Wearing coat (8 cm)$=\frac{8}{100}\times2{\cdot}0\times2200=352$ kg/m

T-beam rib$=\frac{35}{100}\times\frac{100}{100}\times1\times2400=840$ kg/m

Fillets$=\frac{1}{2}\times\frac{15}{100}\times\frac{15}{100}\times2\times2400=54$ kg/m

Total dead loads=2110 kg/m

2·7·3. Dead Load Bending Moment

Maximum bending moment due to dead loads

$$=\frac{2110\times10{\cdot}6^2}{8}=29{,}700\text{ m kg}=29{\cdot}7\text{ mt}$$

2·7·4. Live Loads

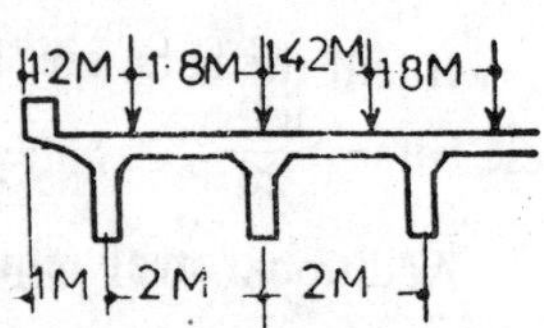

For maximum bending moment to occur in the intermediate *T*-beams, the wheel loads of one vehicle is considered to be on the centre of the *T*-beam and the wheel loads of the other passing vehicle to be closest to the beam.

Minimum clearance between the two crossing or passing vehicles for clear road width of 6·8 m

$$=0{\cdot}4+\left(\frac{1{\cdot}2-0{\cdot}4}{7{\cdot}5-5{\cdot}5}\right)(6{\cdot}8-5{\cdot}5)=0{\cdot}4+0{\cdot}52=0{\cdot}92\text{ m}$$

Distance between centre to centre of wheels=0·92+0·5=1·42 m

Let W=the wheel load of the vehicle. For the arrangement shown the reaction due to all wheel loads on the intermediate girder

$$=W+W\times\frac{(2-1{\cdot}8)}{2}+W\frac{(2-1{\cdot}42)}{2}=W\left(1+\frac{0{\cdot}2}{2}+\frac{{\cdot}58}{2}\right)$$

$=1{\cdot}39\ W$

Magnitude of each axle load$=2\ W$

The reaction due to axle loads$=\dfrac{1{\cdot}39}{2}=0{\cdot}695$

To obtain the load system to be considered on the beam, the axle loads are multiplied by a factor 0·695.

Class A Loading

(*A*)		(*B*)		(*C*)		(*D*)		(*E*)		(*F*)		(*G*)		(*H*)	
2·7 t	1·1 m	2·7 t	3·2 m	11·4 t	1·2 m	11·4 t	4·3 m	6·8 t	3 m	6·8 t	3 m	6·8 t	3 m	6·8 t	20 m

Loading system to be considered

(*A*)	(*B*)	(*C*)	(*D*)	(*E*)	(*F*)	(*G*)	(*H*)
1·88 t	1·88 t	7·93 t	7·73 t	4·73 t	4·73 t	4·73 t	4·73 t

2·7·5. Criterion For Maximum Bending Moment

Bending moment under any load is maximum when that load and the C.G. of the whole system of loads on the span are equidistant from mid span. Assuming that

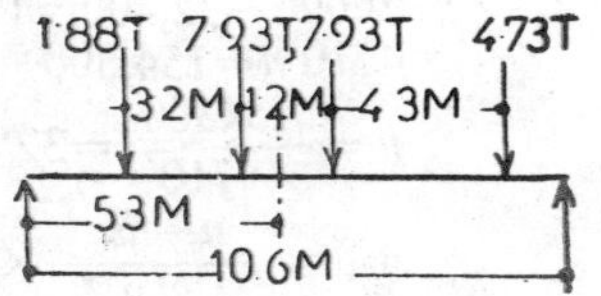

The maximum bending moment occurs under load D, loads B, C, D and E will be on the span.

C-G of the load system taking moments about B

$$=\frac{7{\cdot}93\times3{\cdot}2+7{\cdot}93\times4{\cdot}4+4{\cdot}73\times8{\cdot}7}{1{\cdot}88+7{\cdot}93+7{\cdot}93+4{\cdot}73}=\frac{101{\cdot}5}{22{\cdot}47}=4{\cdot}51\text{ m}$$

Reaction $R_2=\dfrac{22{\cdot}47\times5{\cdot}355}{10{\cdot}6}=11{\cdot}75$ t

Bending moment under load D

$=R_2\times5{\cdot}355-4{\cdot}73\times4{\cdot}3$

$=11{\cdot}75\times5{\cdot}355-4{\cdot}73\times4{\cdot}3$

$=63-20{\cdot}35=42{\cdot}65$ mt

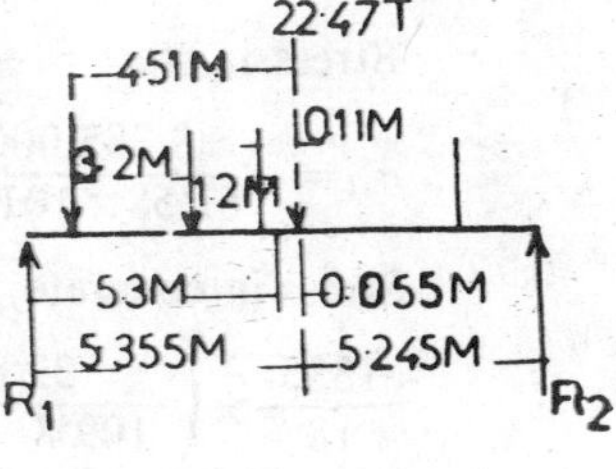

2·7·6. Bending moment due to Impact on Live Load

Impact factor

$$=\frac{4{\cdot}5}{6{\cdot}0+L}=\frac{4{\cdot}5}{6+10{\cdot}6}=\frac{4{\cdot}5}{16{\cdot}6}$$

$=0{\cdot}27$

Bending moment due to impact$=0{\cdot}27\times42{\cdot}65=11{\cdot}5$ mt

Total bending moment due to *DL*, *LL* and impact

$=29{\cdot}7+42{\cdot}65+11{\cdot}5=83{\cdot}85$ mt

2·8. T-Beam Design

Total depth of assumed trial section = 118 cm

Approximate effective depth = 106 cm

Approximate lever arm $= 106 - \frac{ds}{2} = 106 - 9 = 97$ cm

Approximate $A_t = \frac{8385000}{1400 \times 97} = 61{\cdot}6 \text{ cm}^2$

Use 10 bars of 28 mm ϕ in two layers (61·58 cm²)

Effective depth available

$= 118 - \left(4 + 2{\cdot}8 + \frac{2{\cdot}8}{2}\right) = 118 - 8{\cdot}2$

$= 109{\cdot}8$ cm

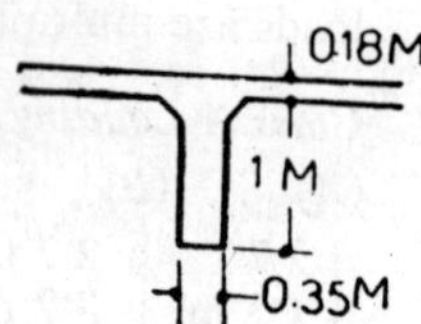

2·8·1. Effective Width of Flange

The effective width is least of:

(*i*) $\frac{1}{3}$ span $= \frac{10{\cdot}6}{3} = 3{\cdot}53$ m = 353 cm

(*ii*) Centre to centre of ribs = 2 m = 200 cm

(*iii*) $12\ ds + b_r = 12 \times 18 + 35 = 216 + 35 = 251$ cm

Therefore the effective width = 200 cm

Assuming *NA* to be below the flange

$B \times ds \left(n - \frac{ds}{2}\right) = m\ A_t\ (d - n)$

$200 \times 18\ (n - 9) = 18 \times 61{\cdot}58\ (109{\cdot}8 - n)$

$3600n - 32{,}400 = 121{,}900 - 1110\ n$

$4710\ n = 154{,}300$

$n = \frac{154300}{4710} = 32{\cdot}8$ cm

$\sigma_{cb}' = \frac{n - ds}{n} \times \sigma_{cb} = \frac{(32{\cdot}8 - 18)}{27{\cdot}35} \times \sigma_{cb}$

$\sigma_{cb}' = \frac{14{\cdot}8}{32{\cdot}8}\ \sigma_{cb} = {\cdot}452\ \sigma_{cb}$

$\bar{y} = \frac{\sigma_{cb} + 2\sigma_{cb}'}{\sigma_{cb} + \sigma_{cb}'} \times \frac{ds}{3}$

$= \frac{\sigma_{cb} + 2 \times 0{\cdot}452\ \sigma_{cb}}{\sigma_{cb} + 0{\cdot}452\ \sigma_{cb}} \times \frac{18}{3} = \frac{1{\cdot}904}{1{\cdot}452} \times 6 = 7{\cdot}85$ cm

Actual lever arm $= a = d - \bar{y} = 109{\cdot}8 - 7{\cdot}85 = 101{\cdot}95$ cm

2·8·2. Stresses

Stress in steel $= \frac{BM}{A_t \times \text{lever arm}}$

$\sigma_{st} = \frac{8{,}385{,}000}{61{\cdot}58 \times 101{\cdot}95} = 1335 \text{ kg/cm}^2 < 1400 \text{ kg/cm}^2$

Stress in concrete, $\sigma_{cb} = \frac{\sigma_{st}}{m}\left(\frac{n}{d - n}\right)$

$\frac{1335}{18} \times \left(\frac{32{\cdot}8}{109{\cdot}8 - 32{\cdot}8}\right) = \frac{1335}{18} \times \frac{32{\cdot}8}{77{\cdot}0}$

$= 31{\cdot}55 \text{ kg/cm}^2 < 50 \text{ kg/cm}^2$

2·8·3. Shear Force due to live load

Maximum shear force occurs when the first axle load of 7·93 t is pn the support

$$SF = 7{\cdot}93 + \frac{7{\cdot}93}{10{\cdot}6}(10{\cdot}6 - 1{\cdot}2) + \frac{4{\cdot}73}{10{\cdot}6} \times 5{\cdot}1 + \frac{4{\cdot}73}{10{\cdot}6} \times 2{\cdot}1$$

$= 7{\cdot}93 + 7{\cdot}025 + 2{\cdot}29 + 0{\cdot}935 = 18{\cdot}18$ t

2·8·4. Due to Impact

SF due to impact on live load

$= 18{\cdot}18 \times 0{\cdot}27 = 4{\cdot}9$ t

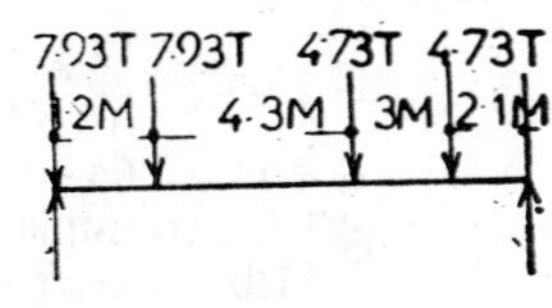

2·8·5. Due to Deadload

SF due to dead load

$= \frac{2{,}110 \times 10{\cdot}6}{2} = 11{\cdot}2$ t

Total shear force at support $= 18{\cdot}18 + 4{\cdot}90 + 11{\cdot}20 = 34{\cdot}28$ t

2·8·6. Shear Stress

Shear Stress $= \frac{Q}{b \times jd} = \frac{34280}{35 \times 101{\cdot}95}$

$= 9{\cdot}56$ kg/cm² > 5 kg/cm²

2·8·7. Shear that concrete can carry

Shear force that concrete section can carry $= 5 \times 101{\cdot}95 \times 35$

$= 17{,}900$ kg $= 17{\cdot}9$ t

2·8·8. Shear Force at 0·15 L (1·59 m)

Shear force:

Shear force due to *L.L.* at 0·15 *L* from support

$0{\cdot}15\, L = 0{\cdot}15 \times 10{\cdot}6 = 1{\cdot}59$ m

$$SF = 7{\cdot}93 \times \frac{9{\cdot}01}{10{\cdot}6} + 7{\cdot}93 \times \frac{7{\cdot}81}{10{\cdot}6} + 4{\cdot}73 \times \frac{3{\cdot}51}{10{\cdot}6} + 4{\cdot}73 \times \frac{0{\cdot}51}{10{\cdot}6}$$

$= 6{\cdot}75 + 5{\cdot}85 + 1{\cdot}57 + 0{\cdot}228 = 14{\cdot}398$ t

SF due to impact on live load

$= 14{\cdot}398 \times 0{\cdot}27 = 3{\cdot}88$ t

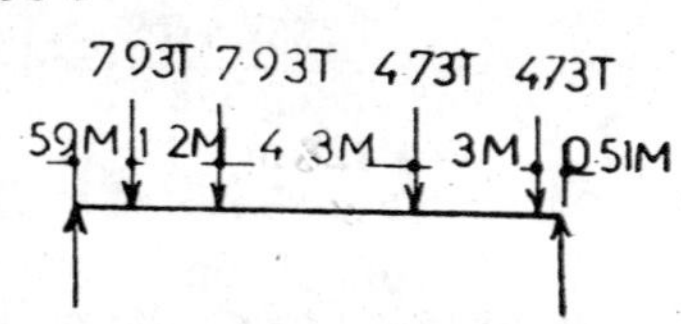

SF due to *DL* $= 11{\cdot}20 - 2{\cdot}11 \times 1{\cdot}59$

$= 11{\cdot}20 - 3{\cdot}335 = 7{\cdot}85$ t

Total *SF* at 1·59 m from support

$= 14{\cdot}398 + 3{\cdot}88 + 7{\cdot}85 = 26{\cdot}128$ t

Shear force at (0·15 + ¾ d) = 2·4 m

Distance $= 0{\cdot}15 \times 10{\cdot}6 + \frac{3}{4} \times 1{\cdot}098$

$= 1{\cdot}59 + {\cdot}822 = 2{\cdot}412$ m Say 2·4 m

SF due to *LL* at 2·4 m distance from support

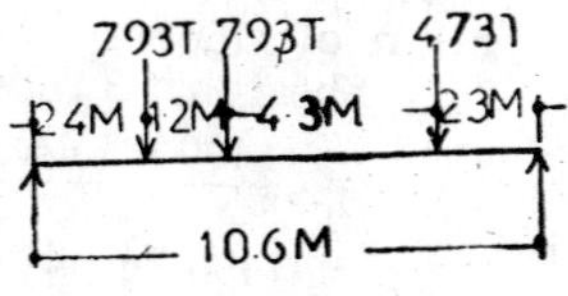

$$= 7{\cdot}93 \times \frac{8{\cdot}2}{10{\cdot}6} + 7{\cdot}93 \times \frac{7}{10{\cdot}6} + 4{\cdot}73 \times \frac{2{\cdot}7}{10{\cdot}6}$$

$= 6{\cdot}14 + 5{\cdot}24 + 1{\cdot}205 = 12{\cdot}585$ t

SF due to impact on *LL* at 2·4 m distance from support

$= 12{\cdot}585 \times 0{\cdot}27 = 3{\cdot}39$ t

SF due to dead load $= 11{\cdot}2 - 2{\cdot}11 \times 2{\cdot}4 = 11{\cdot}2 - 5{\cdot}06 = 6{\cdot}14$ t

Total $SF = 12{\cdot}585 + 3{\cdot}39 + 6{\cdot}14 = 22{\cdot}115$ t

Shear Force at ($0{\cdot}15\ L+2\times\frac{3}{4}\ d$) 3·2 m

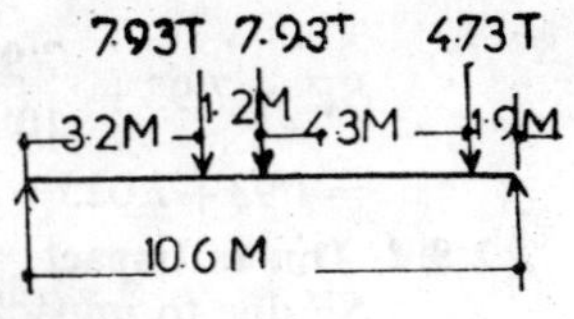

Distance $=0{\cdot}15\times10{\cdot}6+2\times\frac{3}{4}\times1{\cdot}098$
$=1{\cdot}59+2\times{\cdot}822=1{\cdot}59+1{\cdot}644$
$=3{\cdot}234$ m Say 3·2 m

SF due to *LL* at 3·2 m distance from support

$$=7{\cdot}93\times\frac{7{\cdot}4}{10{\cdot}6}+7{\cdot}93\times\frac{6{\cdot}2}{10{\cdot}6}+\frac{4{\cdot}73\times1{\cdot}9}{10{\cdot}6}$$

$=6{\cdot}54+4{\cdot}64-{\cdot}848=11{\cdot}028$ t

SF due to impact on live load at 3·2 m distance from support
$=11{\cdot}028\times0{\cdot}27=2{\cdot}98$ t

SF due to *DL* $=11{\cdot}2-2{\cdot}11\times3{\cdot}2=11{\cdot}2-6{\cdot}75=4{\cdot}45$ t

Total *SF* at 3·2 m distance $=11{\cdot}028+2{\cdot}98+4{\cdot}45=18{\cdot}458$ t

Shear Force at ($0{\cdot}75\ L+3\frac{3}{4}d$) 4 m

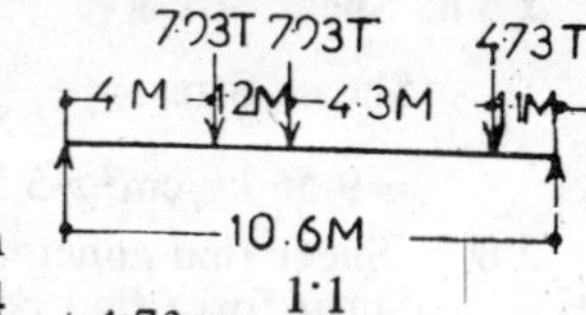

Distance $=0{\cdot}15\times10{\cdot}6+3\times\frac{3}{4}\times1{\cdot}098$
$=1{\cdot}59+3\times8{\cdot}22=1{\cdot}59+2{\cdot}466$
$=4{\cdot}056$ m Say 4 m

SF due to *LL* at 4 m distance from

$$\text{the support}=7{\cdot}93\times\frac{6{\cdot}6}{10{\cdot}6}+7{\cdot}93\times\frac{5{\cdot}4}{10{\cdot}6}+4{\cdot}73\times\frac{1{\cdot}1}{10{\cdot}6}$$

$=5{\cdot}08+4{\cdot}04+{\cdot}49=9{\cdot}61$ t

SF due to impact on *LL* $=9{\cdot}61\times0{\cdot}27=2{\cdot}6$ t

SF due to *DL* $=11{\cdot}2-2{\cdot}11\times4=11{\cdot}2-8{\cdot}44=2{\cdot}76$ t

Total SF at 4 m distance $=9{\cdot}61+2{\cdot}6+2{\cdot}76=15{\cdot}06$ t

Distance beyond which theoretically shear stirrups are not required from the end

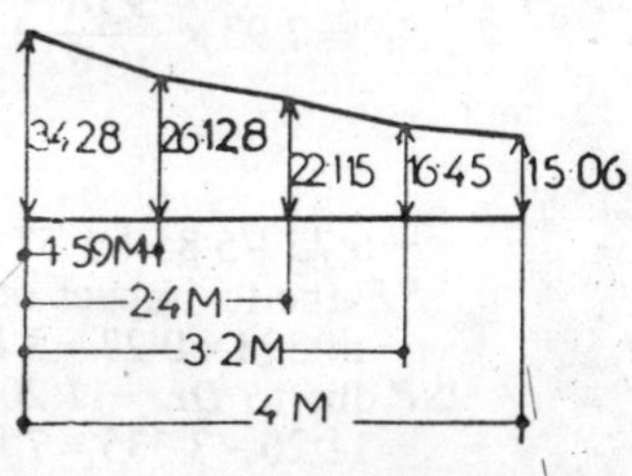

$$=4+\frac{18{\cdot}458-17{\cdot}9}{18{\cdot}458-15{\cdot}06}\times{\cdot}8$$

$$=4+\frac{{\cdot}558\times0{\cdot}8}{3{\cdot}398}=4+{\cdot}128$$

$=4{\cdot}128$ m

2·8·9. Shear Resisted by Bent up Bars

3 bars of top layer are bent at 0.15 *L* and the remaining 2 bars at ($0{\cdot}15\ L+\frac{3}{4}\ d$) 2·4 m from support. The spacing of the bent up bars along the length of beam is $(2{\cdot}4-1{\cdot}59)=0{\cdot}81$ m. The bent up rods will take shear in the region from 0·78 m to 2·4 m from the centre of support.

Shear that 3 rods of 28 mm ϕ can take for a span of 81 cm

$$=\frac{1400\times3\times6{\cdot}16\times102{\cdot}3\times1{\cdot}414}{81}$$

$=41400$ kg - 41·4 t

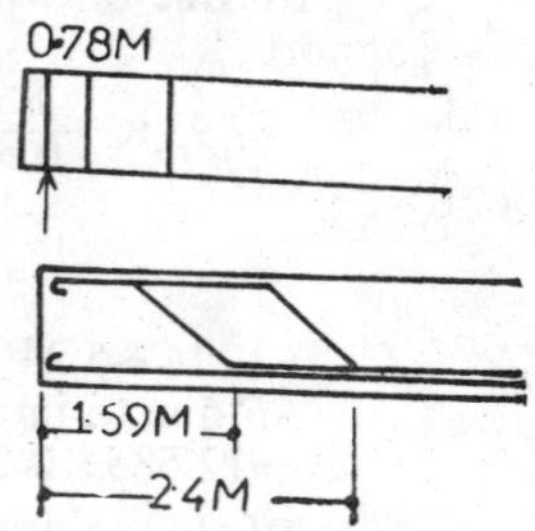

Shear that 2 rods of 28 mm ϕ can take for a spacing of 81 cm

$$=\frac{41{\cdot}4\times2}{3}=27{\cdot}6\text{ t}$$

The bend up bars can take all the shear in the region from 0·78 m to 2·4 m

2·8·10. Shear Stirrups

From the centre of support to 0·78 m

Using 4 legged —10 mm ϕ stirrups

$$\text{spacing} = \frac{A_w\, t_w\, jd}{Q} = \frac{4 \times 0{\cdot}79 \times 1400 \times 101{\cdot}95}{34{,}280} = 13.3 \text{ cm.}$$

Use 4—legged 10 mm ϕ stirrups at 13 cm c/c, 6 Nos from support upto ·78 m

For the length from ·78 m to 2·4 m bent up bars will take all the shear. However, stirrups at spacing required beyond 2·4 m shall be used for this region.

For the length from 2·4 m to 3·2 m

Spacing of 4 legged —10 mm ϕ stirrups

$$= \frac{4 \times 0{\cdot}79 \times 1400 \times 101{\cdot}95}{22115} = 20{\cdot}4 \text{ cm.}$$ Use 4 Nos at 20 cm c/c

Use 8 Nos at 20 cm c/c from ·78 m to 2·4 m

From 3·2 to 4·128 m spacing $= \dfrac{4 \times 0{\cdot}79 \times 1400 \times 102{\cdot}3}{18458} = 24{\cdot}5$ cm

Use 5 Nos at 24 cm c/c

2·8·11. Arrangement of Stirrups

2 stirrups behind the support, the first one at 5 cm from end and the second at 20 cm

6 Nos at 13 cm c/c from support up to ·78 m
8 ,, ,, 20 ,, ,, ,, ·78 m ,, 2·38 m
4 ,, ,, 20 ,, ,, ,, 2·38 m ,, 3·18 m
5 ,, ,, 24 ,, ,, ,, 3·18 m ,, 4·38 m
2 ,, ,, 46 ,, ,, ,, 4·38 m ,, 5·3 m.

3 T-Beam Bridge–Two Lane IRC Class-A or Single Lane Class-AA Loading

3·1. Data

Clear span = 10 m, Bearing for the beam = 0·60 cm

Number of lanes = two, Width of safety kerb = 60 cm

Loading IRC class—*AA* single lane, IRC class—*A* two lane

Wearing coat = 8 cm (average)

Materials available concrete *M* 150 grade, steel Grade – *I*

3·2. Relevant Codes

IRC standard specifications and code of practice for road bridges section (*i*) to (*iv*).

3·3. Width of Roadway

Width of road way = 3·8 + 3 + 1·2 = 8 m

3·4. Impact Factor

Upto 5 m span for tracked vehicles = 25%. For spans over 9 m tracked vehicle 10%

3·5. Characteristic Strengths

$\sigma_{cb} = 50$ kg/cm² $\quad \sigma_{st} = 1400$ kg/cm²

$m = 18$ $\quad jd = \cdot 87\, d$

$q = 5$ kg/cm² $\quad R = 8\cdot 7$ $\quad \sigma_{bt} = 10$ kg/cm²

3·6. Deck Slab Arrangements of Ribs

Use 4 ribs with 2 m c/c and a overhang of 1 m on either side.

Assume depth of beam $= (\frac{1}{8} \times 1060) = 130$ cm, Use $d = 125$ cm

depth of slab $= (\frac{1}{8} \times 165) - 20$ cm

depth of *T* beam rib = 105 cm

width of rib $= (\frac{1}{3} \times 105) = 35$ cm

0.35M 1.65M 1M
1M 2M

For the assumed arrangement clear span of deck slab = 2 – 0·35 = 1·65 m

Using 15 cm × 15 cm fillets clear span can be taken as effective span for slab.

Effective span for slab = 1·65 m

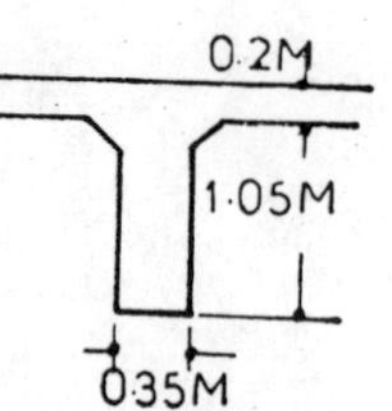

3·6·1. Dead Loads

Weight of slab (20 cm) $= \frac{20}{100} \times 1 \times 1 \times 2400$

= 480 kg/m²

Weight of wearing coat (8 cm)

$= \frac{8}{100} \times 1 \times 1 \times 2400 = 176$ kg/m²

Total dead load = 656 kg/m²

3·6·2. Dead Load Bending Moment

Bending moment due to dead load

$$= \frac{wL^2}{10} = \frac{656 \times 1\cdot 65^2}{10} = 178 \text{ m kg}$$

= 178,000 cm kg.

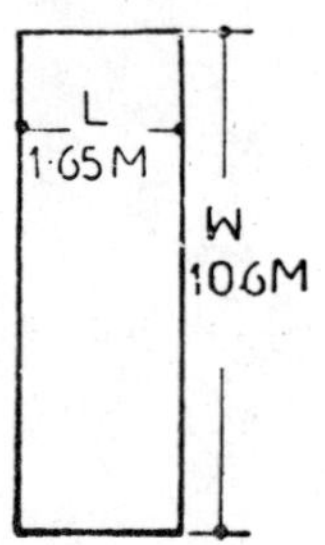

3·6·3. Live Loads

The slab is spanning in one direction only, but the width of slab is more than three times the span.

That is, $\frac{W}{L} = \frac{10\cdot 6}{1\cdot 65} > 3$

Therefore Pigeaud's method of distribution of live load can be used

The maximum BM due to live loads will occur when one of the tracks, of the tracked vehicle is at the centre of span.

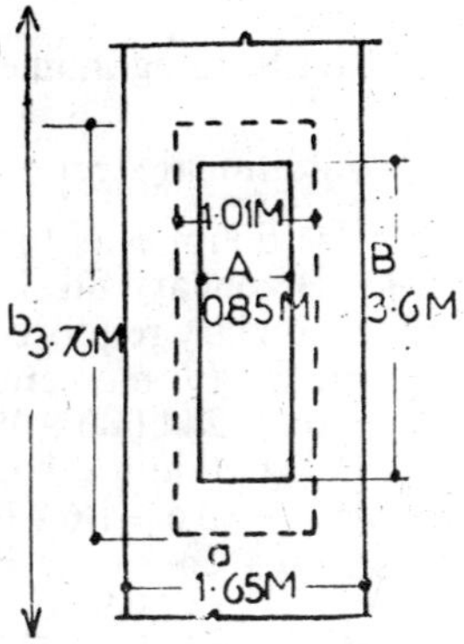

A=track contact in the direction of transverse span, L=0·85 m

B=track contact in the direction of longitudinal span, w = 360 cm

D=thickness of wearing coat=8 cm

$\frac{L}{w}=\frac{1·65}{\infty}=0$

u=width of load spread in the transverse direction=$A+2D$=85+2×8 =85+16=101 cm=1·01 m

v=width of load spread in longitudinal direction=$B+2D$ =360+2×8=376 cm=3·76 m

$\frac{u}{L}=\frac{101}{165}=0·612$, $\frac{v}{L}=\frac{376}{165}=2·28$

From ρ=0 curve corresponding to these ratios

$M_1\times10^2=6·9$, $M_1=·069$

$M_2\times10^2=·505$, $M_2=·00505$

3·6·4. Bending Moments

Magnitude of the track load=$\frac{70}{2}$=35 t

Bending moment in the transverse direction

$M_1=(M_1+0·15\,M_2)\times$intensity of the load

=(·069+0·15×·00505)×35,000=·06975×35,000=2440 m kg.

Bending moment in the longitudinal direction=m_2

=$(M_1\times0·15+M_2)\times35,000$=(·069×0·15+·00505)35,000

=·01540×35,000=539 m kg.

Making allowance for continuity, maximum bending moment due to live loads in the direction of transverse span=0·8×2440 =1952 m kg

Bending moment due to impact on Live Load =1952×0·25=488 m kg.

Bending moment due to DL=178 m kg.

Total maximum bending moment due to LL, impact and dead loads=1952+488+178=2638 m kg=263,800 cm kg

3·6·5. Thickness of Deck Slab

$8·7\,bd^2=263,800$

$$d=\sqrt{\frac{263,800}{8·7\times100}}=\sqrt{303}=17·4 \text{ cm}$$

Using 16 mm ϕ with a clear cover of 2·5 cm, Total depth required=17·4+2·5+0·8=20·7 cm>20 cm assumed

Use d=21 cm and d_e=21−3·3=17·7 cm

3·6·6. Main Steel

$$A_t=\frac{263,800}{1400\times0·87\times17·7}=12·21 \text{ cm}^2.$$ Use 16 mm ϕ at 16 cm c/c (12·57 cm²)

Negative bending moment at end ribs=$\frac{wl^2}{16}$

if half the bars are bent up, the bending moment resisted by these bars$=\dfrac{wl^2}{20}$

Net bending moment to be resisted$=\dfrac{wl^2}{80}$

Additional steel required$=\dfrac{A_t}{8}=\dfrac{12{\cdot}21}{8}=1{\cdot}53\ \text{cm}^2$

Use 6 mm ϕ at 16 cm c/c

3·6·7. Secondary Steel

Secondary reinforcement is provided to resist a moment equal to
$=0{\cdot}3\times LL$ moment$+0{\cdot}2\times$dead load moment
$=0{\cdot}3\times 244{,}000+0{\cdot}2\times 17800=73200+3560=76760$ cm kg
Using 10 mm ϕ bars effective depth available
$=21-(2{\cdot}5+1{\cdot}6+0{\cdot}5)=(21-4{\cdot}6)$
$=16{\cdot}4$ cm

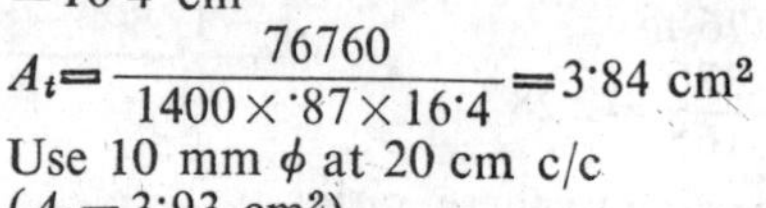

$$A_t=\frac{76760}{1400\times{\cdot}87\times 16{\cdot}4}=3{\cdot}84\ \text{cm}^2$$

Use 10 mm ϕ at 20 cm c/c
$(A_t=3{\cdot}93\ \text{cm}^2)$
This steel will be sufficient to take bending moment occuring in the direction of longitudial span.

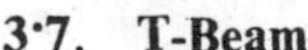

3·7. T-Beam

Clear span$=10$ m,
Effective span (centre to centre of bearing)$=10{\cdot}60$ m
Impact factor$=10\%$

3·7·1. Trial Section

Depth of beam$=125$ cm
Depth of slab $=21$ cm
Depth of rib$=104$ cm,
Width of rib$=35$ cm
Loads on intermediate *T*-beam
Centre to centre of rib$=2$ m

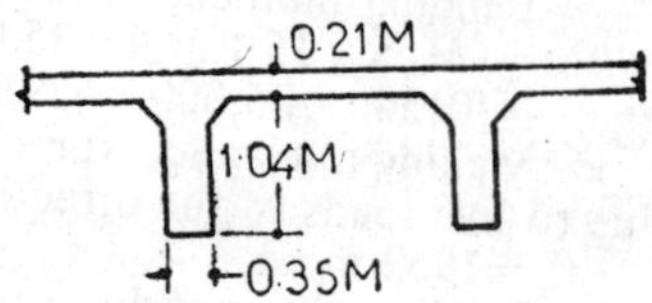

3·7·2. Dead Loads

Deck slab$=\dfrac{21}{100}\times 2\times 2400\times 1=1008$ kg/m

Wearing coat$=\dfrac{8}{100}\times 2\times 1\times 2200=352$ kg/m

T-beam rib$=\dfrac{35}{100}\times\dfrac{104}{100}\times 1\times 2400=875$ kg/m

Fillets 15 cm$\times$15 cm$=\frac{1}{2}\times\dfrac{15}{100}\times\dfrac{15}{100}\times 2\times 2400=54$ kg/m

Total dead loads$=1008+352+875+54=2289$ kg/m

3·7·3. Bending Moment Due to DL

Maximum bending moment due to dead loads

$$=\frac{2289\times 10{\cdot}6^2}{8}=32200 \text{ m kg}=32{\cdot}2 \text{ mt}$$

3·7·4. Live Loads

With one of the tracks directly on the intermediate *T*-beam is considered for maximum bending moment due to Live Load.

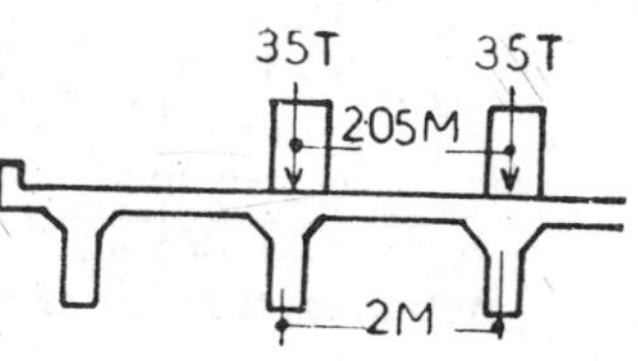

Maximum bending moment at the centre of the span

$=\frac{35}{2}\times 5{\cdot}3-\frac{35}{2}+\frac{3{\cdot}6}{4}$

$=\frac{35}{2}\ (5{\cdot}3-0{\cdot}9)$

$=\frac{35}{2}\times 4{\cdot}6=80{\cdot}5$ mt

Bending moment due to impact on Live load $=80{\cdot}5\times 0{\cdot}1=8{\cdot}05$ mt

3·7·5. Total LL Bending Moment

Total bending moment due to *DL*, *LL* and impact

$=32{\cdot}2+80{\cdot}5+8{\cdot}05=120{\cdot}75$ mt

$=12{,}075{,}000$ cm kg

3·8. T-Beam Design

Depth of assumed trial section = 125 cm

Approximate effective depth $=125-(4+6)$

$=115$ cm

Approximate lever arm

$=115-\frac{ds}{2}=115-\frac{21}{2}=104{\cdot}5$ cm

Approximate $A_t=\frac{12075{,}000}{1400\times 104{\cdot}5}=82{\cdot}6\ \text{cm}^2$

6 Nos – 40 mm $\phi=75{\cdot}40\ \text{cm}^2$ and

2 Nos – 28 mm $\phi=12{\cdot}31\ \text{cm}^2$

Total $A_t=87{\cdot}71\ \text{cm}^2$

C.G. of steel $=\frac{4\times 12{\cdot}57\times 6+2\times 12{\cdot}57\times 14+2\times 6{\cdot}16\times 13{\cdot}4}{87{\cdot}71}$

$=\frac{302{\cdot}0+352+165{\cdot}5}{87{\cdot}71}=\frac{819{\cdot}5}{87{\cdot}71}=9{\cdot}4$ cm

Effective depth available $=125-9{\cdot}4=115{\cdot}6$ cm

3·8·1. Effective Width of Flange

The effective width of flange is the least of

(*i*) $\frac{1}{4}$ span $=\frac{10{\cdot}6}{4}=2{\cdot}65$ m $=265$ cm

(*ii*) Centre to centre of rib $=200$ cm

(*iii*) $12\ ds+b_r=12\times 21+35=252+35=287$ cm.

Therefore the effective flange width = 200 cm

Assuming *NA* to be below the flange

$B\times d_s\times\left(n-\frac{ds}{2}\right)=m\ A_t\ (d-n)$

$200\times 21\times(n-10{\cdot}5)=18\times 87{\cdot}71\ (115{\cdot}6-n)$

$4200\ n-44100=182{,}000-1580\ n$

$5780\ n=226{,}100$

$n=\frac{226{,}100}{5780}=39{\cdot}2$ cm

$\sigma_{cb}'=\frac{n-ds}{n}\times\sigma_{cb}=\frac{39{\cdot}2-21}{39{\cdot}2}\times\sigma_{cb}=\frac{18{\cdot}2}{39{\cdot}2}\times\sigma_{cb}={\cdot}465\ \sigma_{cb}$

$$\bar{y}=\frac{\sigma_{cb}+2\sigma_{cb}'}{\sigma_{cb}+\sigma_{cb}'}\times\frac{ds}{3}=\frac{\sigma_{cb}+2\times{\cdot}465\,\sigma_{cb}}{\sigma_{cb}+{\cdot}465\,\sigma_{cb}}\times\frac{21}{3}$$

$$=\frac{\sigma_{cb}+{\cdot}930\,\sigma_{cb}}{\sigma_{cb}+{\cdot}465\,\sigma_{cb}}\times 7=\frac{1{\cdot}930}{1{\cdot}465}\times 7\quad 9{\cdot}2 \text{ cm}$$

Actual lever arm $=jd=d-\bar{y}=115{\cdot}6-9{\cdot}2=106{\cdot}4$ cm

Stress in steel $=\dfrac{BM}{A_t\times\text{lever arm}}=\dfrac{12075{,}000}{87{\cdot}71\times 106{\cdot}4}$

$=1290 \text{ kg/cm}^2 < 1400 \text{ kg/cm}^2$

Stress in concrete $\sigma_{cb}=\dfrac{\sigma_{st}}{m}\left(\dfrac{n}{d-n}\right)=\dfrac{1290}{18}\,\dfrac{(39{\cdot}2)}{(115{\cdot}6-39{\cdot}2)}$

$=\dfrac{1290}{18}\times\dfrac{39{\cdot}2}{76{\cdot}4}=28{\cdot}6 \text{ kg/cm}^2 < 50 \text{ kg/cm}^2$

3·8·2. Shear Force

Maximum shear force occurs when the load touches the support

Maximum shear force due to live load

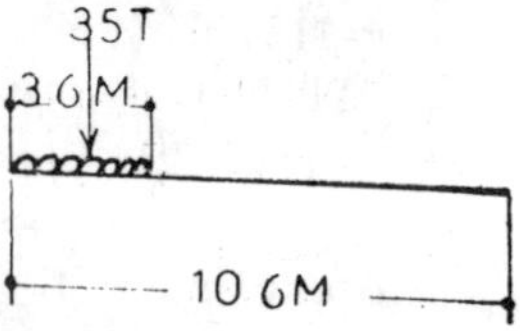

$=\dfrac{35\times(10{\cdot}6-3{\cdot}6)}{10{\cdot}6}=\dfrac{35}{10{\cdot}6}\times 8{\cdot}8$

$=29{\cdot}1$ t

Shear force due to impact on live load

$=29{\cdot}1\times\dfrac{10}{100}=2{\cdot}91$ t

Shear force due to $DL=\dfrac{2289\times 10{\cdot}6}{2}=12150$ kg $=12{\cdot}15$ t

Total shear force $29{\cdot}1+2{\cdot}91+12{\cdot}15=44{\cdot}16$ t

3·8·3. Shear Stress

Shear Stress $=\dfrac{Q}{b\times jd}=\dfrac{44160}{35\times 106{\cdot}4}=11{\cdot}85 \text{ kg/cm}^2>5 \text{ kg/cm}^2$

3·8·4. Shear Force that Concrete can carry

Shear force that concrete can carry

$=5\times 106{\cdot}4\times 35=18650$ kg $=18{\cdot}65$ t

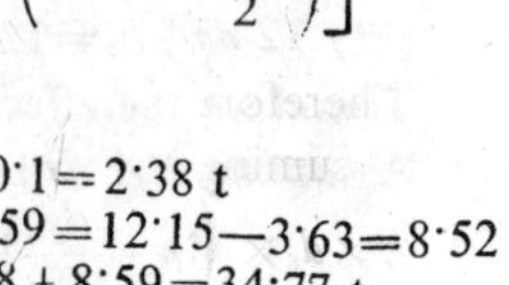

3·8·5. Shear Force at (0·15 L) = 1·59 m

Shear force at $(0{\cdot}15\times L$

$=0{\cdot}15\times 10{\cdot}6)=1{\cdot}59$ m

SF due to LL at 1·59 m $=\dfrac{35}{10{\cdot}6}\left[10{\cdot}6+\left(1{\cdot}59+\dfrac{3{\cdot}6}{2}\right)\right]$

$=\dfrac{35}{10{\cdot}6}\times 7{\cdot}21=23{\cdot}8$ t

SF due to impact on live load $=23{\cdot}8\times 0{\cdot}1=2{\cdot}38$ t

SF due to dead load $=12{\cdot}15-2{\cdot}289\times 1{\cdot}59=12{\cdot}15-3{\cdot}63=8{\cdot}52$ t

Total shear force at 1·59 m $=23{\cdot}8+2{\cdot}38+8{\cdot}59=34{\cdot}77$ t

3·8·6. Shear Force at (0·15 L + ¾d = 2·44 m)

Distance $=0{\cdot}15\times 10{\cdot}6+\frac{3}{4}\times 1{\cdot}156=1{\cdot}59+0{\cdot}857=2{\cdot}447$ m

Say 2·4 m

SF due to LL at 2·4 m from support $=\dfrac{35}{10{\cdot}6}\left[10{\cdot}6-\left(2{\cdot}4+\dfrac{3{\cdot}6}{2}\right)\right]$

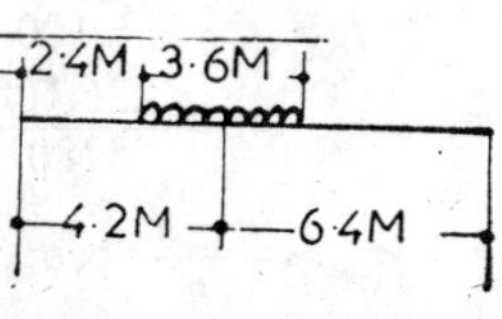

$=\dfrac{35}{10{\cdot}6}\times 6{\cdot}4=21{\cdot}1$ t

SF due to impact in live load $= 21{\cdot}1 \times 0{\cdot}1 = 2{\cdot}11$ t
SF due to dead load $= 12{\cdot}15 - 2{\cdot}289 \times 2{\cdot}4 = 12{\cdot}15 - 5{\cdot}5 = 6{\cdot}65$ t
Total SF at 2·4 m distance $= 21{\cdot}1 + 2{\cdot}11 + 6{\cdot}65 = 29{\cdot}86$ t

3·8·7. Shear Force at $(0{\cdot}15\ L + 2 \times \frac{3}{4}\ d = 3{\cdot}3\ m)$

Distance $= 0{\cdot}15 \times 10{\cdot}6 + 2 \times \frac{3}{4} \times 1{\cdot}156 = 1{\cdot}59 + 2 \times {\cdot}857$
$= 3{\cdot}304$ m Say 3·3 m

Shear force due to *LL* at 3·3 m distance $= \dfrac{35}{10{\cdot}6}\left[10{\cdot}6 - \left(3{\cdot}3 \times \dfrac{3{\cdot}6}{2}\right)\right]$

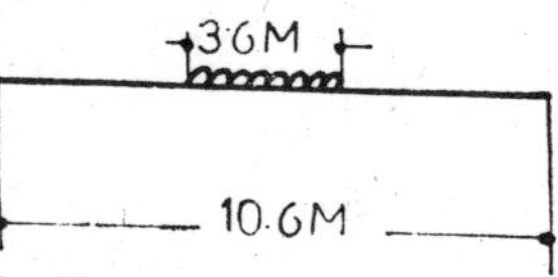

$= \dfrac{35}{10{\cdot}6} \times 5{\cdot}5 = 18{\cdot}20$ t

Shear force due to impact on *LL*
$= 18{\cdot}20 \times 0{\cdot}1 = 1{\cdot}82$ t
Shear force due to *DL* $= 12{\cdot}15 - 2{\cdot}289 \times 3{\cdot}3 = 12{\cdot}15 - 7{\cdot}55 = 4{\cdot}60$ t
Total SF $= 18{\cdot}2 + 1{\cdot}82 + 4{\cdot}6 = 24{\cdot}62$ t

3·8·8. Shear Force at $(0{\cdot}15\ L + 3\frac{3}{4}\ d = 4{\cdot}15\ m)$

Distance $= 0{\cdot}15 \times 10{\cdot}6 + 3 \times \frac{3}{4} \times 1{\cdot}156 = 1{\cdot}59 + 2{\cdot}6 = 4{\cdot}19$ m
Say 4·15 m

Shear force due to *LL* at 4·15 m distance

$= \dfrac{35}{10{\cdot}6}\left[10{\cdot}6 - \left(4{\cdot}15 + \dfrac{3{\cdot}6}{2}\right)\right]$

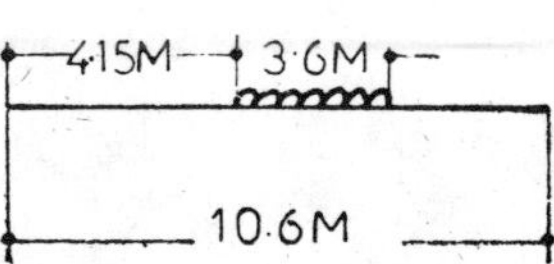

$= \dfrac{35}{10{\cdot}6}\,[10{\cdot}6 - 5{\cdot}957] = \dfrac{35}{10{\cdot}6} \times 4{\cdot}65$

$= 15{\cdot}36$ t

Shear force due to impact on live load $= 15{\cdot}36 \times 0{\cdot}1 = 1{\cdot}536$ t
Shear force due to *DL* $= 12{\cdot}15 - 2{\cdot}289 \times 4{\cdot}15 = 12{\cdot}15 - 9{\cdot}51$
$= 2{\cdot}64$ t
Total SF $= 15{\cdot}35 + 1{\cdot}535 + 2{\cdot}64 = 19{\cdot}525$ t

3·8·9. Shear Force at $(0{\cdot}15\ L + 4 \times \frac{3}{4}\ d)$

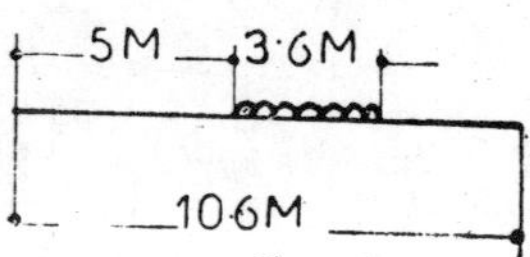

Shear force
Distance $= 0{\cdot}15\ L \times 4 \times \frac{3}{4}\ d$
$= 1{\cdot}59 + 3 \times 1{\cdot}156 = 1{\cdot}59 + 3{\cdot}468$
$= 5{\cdot}058$ m Say 5 m

Shear force due to *LL* at 5 m from support $= \dfrac{35}{10{\cdot}6}\left[10{\cdot}6 - \left(5 + \dfrac{3{\cdot}6}{2}\right)\right]$

$= \dfrac{35}{10{\cdot}6} \times 3{\cdot}8 = 12{\cdot}5$ t

Shear force due to impact on live load $= 12{\cdot}5 \times 0{\cdot}1 = 1{\cdot}25$ t
SF due to *DL* $= 12{\cdot}15 - 2{\cdot}289 \times 5$
$= 12{\cdot}15 - 11{\cdot}45 = 0{\cdot}70$ t
Total SF $= 12{\cdot}5 + 1{\cdot}25 + 0{\cdot}7 = 14{\cdot}45$ t

Adopt the same shear-inforcement as in the previous example 2.